MAKING SENSE OF ELEMENTARY ALGEBRA
Data, Equations, and Graphs
Instructor's Version

Elaine A. Kasimatis
California State University, Sacramento

Cindy L. Erickson
Cosumnes River College

CPM Educational Program
*College Preparatory Mathematics: Change from Within
directed by Elaine A. Kasimatis, Tom Sallee, Judith Kysh and Brian Hoey*

An imprint of Addison Wesley Longman, Inc.

Reading, Massachusetts • Menlo Park, California • New York • Harlow, England
Don Mills, Ontario • Sydney • Mexico City • Madrid • Amsterdam

Publisher: Jason A. Jordan
Project Editors: Kari Heen and Ruth Berry
Development Editor: Roberta Lewis
Managing Editor: Ron Hampton
Production Supervisors: Jane DePasquale and Kathleen Manley
Text Designers: Ellen Pettengell and Susan Carsten
Cover Designer: Susan Carsten
Composition: Scott Silva and WestWords
Technical Artist: Techsetters, Incorporated
Illustrator: Dan Clifford
Chapter Opener Illustrator: Leslie Haimes
Marketing Managers: Craig Bleyer and Laura Rogers
Manufacturing Supervisor: Evelyn Beaton

Cover: photo of Burnside Bridge, Antietam National Battlefield, Maryland. © Superstock
The bridge icon on the cover and title page is a registered trademark of CPM.

Copyright © 2000 Addison Wesley Longman. All rights reserved. No part of this publication may be reproduced, stored in a retrieval system, or transmitted, in any form or by any means, electronic, mechanical, photocopying, recording, or otherwise, with the prior written permission of the publisher. Printed in the United States of America.

ISBN 0-201-35196-X — Instructor's Version

ISBN 0-201-85900-9 — Student Version

1 2 3 4 5 6 7 8 9 10—CRW—0302010099

CONTENTS

Chapter Outlines	vii
Preface	ix
Note to the Student	xii
Note to the Instructor	xvii

CHAPTER ONE — OPENINGS (OP): DATA ORGANIZATION — xxvi

1.1	Building Corrals: Working Collaboratively *Diamond Problems* *Algebra Tool Kit*	OP-1 – OP-12	3
1.2	The Five-Digit Problem: Order of Operations *Calculator Tool Kit*	OP-13 – OP-28	11
1.3	Tiling with Squares: Area and Perimeter *"What's My Rule?" game*	OP-29 – OP-42	17
1.4	Using Guess-and-Check Tables to Solve Problems *"What's My Rule?"*	OP-43 – OP-59	24
1.5	Predicting Shoe Size: Making a Graph from Data	OP-60 – OP-75	34
1.6	The National Debt and Teraflops: Large Numbers and Scientific Notation *"What's My Rule?"*	OP-76 – OP-91	40
1.7	Driving to Redwood City: Data from a Graph *"What's My Rule?"*	OP-92 – OP-104	46
1.8	Summary and Review	OP-105 – OP-126	53

CHAPTER TWO — PROFESSOR SPEEDI'S COMMUTE (SP): PATTERNS AND GRAPHS — 60

2.1	The Algebra Walk: Coordinate Graphing	SP-1 – SP-14	63
2.2	Polygon Perimeter Pattern *Area as a Product = Area as a Sum*	SP-15 – SP-35	74
2.3	Getting to Work on Time: Situations and Graphs	SP-36 – SP-52	81
2.4	Hawaiian Punch Mix	SP-53 – SP-72	91
2.5	Slicing a Pizza *Algebra Tiles and the Distributive Property*	SP-73 – SP-89	98
2.6	The Egg Toss: A Quadratic Graph	SP-90 – SP-107	108
2.7	Folding Bills: More Nonlinear Graphs *"Head Problems"*	SP-108 – SP-122	115

2.8	The Burning Candle Investigation (Optional)	SP-123 – SP-124	122
2.9	Summary and Review
More Head Problems	SP-125 – SP-153	125

CHAPTER THREE — LIONS, TIGERS, AND EMUS (LT): WRITING AND SOLVING EQUATIONS — 136

3.1	Writing Equations from Guess-and-Check Tables
More Head Problems	LT-1 – LT-18	140
3.2	Composite Rectangles and Algebraic Expressions for Area
More Head Problems	LT-19 – LT-35	149
3.3	Solving Linear Equations at the Candy Factory	LT-36 – LT-56	157
3.4	Doing and Undoing
Solving Head Problems Puzzles	LT-57 – LT-78	172
3.5	Solving Literal Equations
The Zero Product Property	LT-79 – LT-101	182
3.6	Modeling Situations with Linear Equations
Slides: A Way to Represent Change	LT-102 – LT-123	191
3.7	Summary and Review	LT-124 – LT-153	197

CHAPTER FOUR — ESTIMATING FISH POPULATIONS (EF): NUMERIC, GEOMETRIC, AND ALGEBRAIC RATIOS — 206

4.1	Enlarging and Reducing Geometric Figures	EF-1 – EF-20	209
4.2	More Enlarging, Reducing, and Ratios	EF-21 – EF-40	218
4.3	Area and Subproblems: Geometric Factoring	EF-41 – EF-62	226
4.4	Similar Triangles: More Geometric Ratios	EF-63 – EF-83	236
4.5	More Similarity: Right Triangles and Missing Sides
Fraction Busters	EF-84 – EF-108	247
4.6	Fruit Punch: Using Equivalent Ratios to Graph	EF-109 – EF-129	256
4.7	Estimating Fish Populations: A Simulation	EF-130 – EF-142	263
4.8	Summary and Review	EF-143 – EF-166	272

CHAPTER FIVE — MANAGING WATER RESOURCES (WR): GRAPHS, SLOPE RATIOS, AND LINEAR EQUATIONS — 280

5.1	Ratios and the Slope of a Line	WR-1 – WR-25	283
5.2	Graphing Linear Equations using Two Points and a Slide	WR-26 – WR-55	293
5.3	Slopes of Parallel Lines and the Slope-Intercept Pattern in $y = mx + b$	WR-56 – WR-82	303
5.4	Using Graphs to Investigate Linear Relationships	WR-83 – WR-102	313

	5.5	Practice with $y = mx + b$	WR-103 – WR-121	**320**
	5.6	Using Graphs to Write Equations of Lines	WR-122 – WR-141	**329**
	5.7	Summary and Review	WR-142 – WR-164	**335**

CHAPTER SIX — FITNESS FINANCES (FF): SOLVING SYSTEMS OF EQUATIONS — 344

6.1	Olympic Records: Writing an Equation of a Line Given Two Points	FF-1 – FF-21	**346**
6.2	Using Graphs or Algebra to Solve Systems of Linear Equations	FF-22 – FF-37	**354**
6.3	Solving Systems of Linear Equations by the Substitution Method	FF-38 – FF-53	**360**
6.4	Using Graphs to Solve Systems of Nonlinear Equations	FF-54 – FF-73	**366**
6.5	Solving Quadratic Equations: Graphing and the Zero Product Property	FF-74 – FF-91	**371**
6.6	Solving Systems of Equations by the Addition Method	FF-92 – FF-112	**379**
6.7	Investigation: Using Algebra to Model and Analyze Linear Data (Optional)	FF-113 – FF-116	**385**
6.8	Summary and Review	FF-117 – FF-139	**388**

CHAPTER SEVEN — THE BUCKLED RAILROAD TRACK (RT): FROM WORDS TO DIAGRAMS TO EQUATIONS — 396

7.1	Right Triangles and the Pythagorean Relationship	RT-1 – RT-24	**399**
7.2	Picturing Square Root Lengths	RT-25 – RT-50	**409**
7.3	Models, Diagrams, and Right Triangles	RT-51 – RT-75	**421**
7.4	Distances, Times, and Rates: More Diagrams and Equations	RT-76 – RT-99	**432**
7.5	Calculating with Square Roots	RT-100 – RT-123	**439**
7.6	Summary and Review	RT-124 – RT-156	**449**

CHAPTER EIGHT — THE GRAZING GOAT (GG): DEALING WITH COMPLICATED SITUATIONS — 458

8.1	Solving Quadratic Equations Revisited: The Quadratic Formula	GG-1 – GG-26	**461**
8.2	Complicated Fractions: Using Subproblems to Calculate Average Speed	GG-27 – GG-52	**472**
8.3	The Election Poster: Using the Quadratic Formula	GG-53 – GG-72	**482**
8.4	Using Subproblems in Complicated Situations	GG-73 – GG-91	**488**
8.5	Quiz Scores: Data Points and Lines of Best Fit	GG-92 – GG-112	**495**

8.6	The Cookie Cutter and the Lunch Bunch: Two Investigations (Optional)	GG-113 – GG-122	**502**
8.7	Summary and Review	GG-123 – GG-155	**507**

APPENDIXES   515

A.	Debits and Credits: An Introduction to Integer Tiles	A-1 – A-7	**515**
B.	Using Suproblems to Derive the Quadratic Formula	B-1 – B-7	**522**
C.	Exploring Quadratic Equations and Their Graphs	C-1 – C-6	**525**

Resource Pages	**I-RP-1**
Some Answers and Some Ways To Get Started	**A-1**
Glossary	**G-1**
Mathematical Symbols	**MS-1**
Index	**I-1**

CHAPTER OUTLINES

CHAPTER ONE

OPENINGS (OP): Data Organization
OP-1 through OP-126

In this chapter, students work in groups or pairs as they collect, organize, graph, and interpret data. Guided problem sets help students strengthen their problem solving, pre-algebra, and calculator skills (order of operations, arithmetic skills with integers and fractions, using the meaning of exponents to rewrite expressions, and scientific notation). Guess-and-check tables provide a way to solve word problems and prepare students to write equations that model situations.

CHAPTER TWO

PROFESSOR SPEEDI'S COMMUTE (SP): Patterns and Graphs
SP-1 through SP-153

Using scientific calculators and patterning, students explore relationships among graphs, tables, and rules. They use the strategy of finding, describing, and using patterns to solve problems and make predictions. The Burning Candle Investigation provides an opportunity to collect data, represent it on a graph, and then use the graph to make a prediction.

CHAPTER THREE

LIONS, TIGERS, AND EMUS (LT): Writing and Solving Equations
LT-1 through LT-153

Building on patterning skills, students use guess-and-check tables to write equations for situations described in word problems. They then use a two-pan balance scale model to solve simple linear equations and use the ideas of doing and undoing to solve more complicated equations. Using algebra tiles and sketches of rectangles, students work with the distributive property of multiplication over addition. In preparation for solving quadratic equations, students encounter the zero product property. They also start to work on the concept of "slope of a line" by using the idea of a "slide" along grid lines to represent a constant rate of change on a graph.

CHAPTER FOUR

ESTIMATING FISH POPULATIONS (EF): Numeric, Geometric, and Algebraic Ratios
EF-1 through EF-166

Students explore the concept of ratio in several settings and learn to solve complicated problems by breaking them into simpler problems, or subproblems. They examine similarity by enlarging and reducing simple figures on dot paper and then compare ratios of corresponding sides, perimeters, and areas of similar figures. Students then use similar right triangles in connection with ratios, writing and solving equations, graphs of linear equations, and the concept of slope. The Estimating Fish Populations problem simulates how wildlife biologists use equivalent ratios to estimate the size of a given population.

CHAPTER FIVE

MANAGING WATER RESOURCES (WR): Graphs, Slope Ratios, and Linear Equations

WR-1 through WR-164

This chapter focuses on the algebraic and geometric meanings of the slope of a line. Students use their understanding of slides, similar right triangles, ratios, and writing equations to investigate linear relationships. They learn how to write equations of lines from data or graphs.

CHAPTER SIX

FITNESS FINANCES (FF): Solving Systems of Equations

FF-1 through FF-139

Students solve more complicated equations and investigate linear and nonlinear systems, emphasizing the relationship between the graphs of systems and their common solutions. In particular, they see that the zero-product property and solutions of quadratic equations are tightly linked to the graphs of the equations. The emphases are on understanding the physical interpretation of a graph and solving systems of linear equations graphically and algebraically. Students have the opportunity to collect and analyze sets of linear data in an extended investigation.

CHAPTER SEVEN

THE BUCKLED RAILROAD TRACK (RT): From Words to Diagrams to Equations

RT-1 through RT-156

In this chapter students concentrate on the problem solving strategy of drawing a picture or a diagram to represent information in a problem. Many times the diagram can be used to help write an equation that models the problem. Solving the original problem then becomes a matter of solving an equation and checking that the solution(s) makes sense in the problem. Quadratic equations and square roots arise naturally from problems written using the Pythagorean theorem.

CHAPTER EIGHT

THE GRAZING GOAT (GG): Dealing with Complicated Situations

GG-1 through GG-155

Students tie together many ideas and skills they have learned in the course as they solve quadratic equations, use subproblems to deal with complicated situations, and pull together their graphing and equation writing skills to write equations that model sets of linear data. The Grazing Goat Problem is a good example of a problem that involves several subproblems. The Cookie Cutter Investigation and the Lunch Bunch Investigation provide students opportunities to apply their problem solving and algebra skills to two complicated situations.

PREFACE

The mathematics course that follows is an adaptation of *College Preparatory Mathematics: Change from Within, Math 1 (Algebra 1)*. It reflects the experiences, vision, and philosophy of the teachers and university professors who originally developed and taught CPM Math 1 at the secondary level, and the guidance and needs of instructors at the college level. All who contributed to this text—from the initial vision to the final details—saw deep reasons for changing the way we teach and what we teach in beginning algebra courses.

We are not alone in wanting change: in the past decade, virtually every major document on improving mathematics instruction has called for a restructuring of the entire mathematics curriculum with more emphasis on understanding and less emphasis on routine drill. In the late 1980s the National Council of Teachers of Mathematics (NCTM) issued its *Curriculum and Evaluation Standards*, one of the early guides of the development of the original CPM Program. More recently, the American Mathematical Association of Two-Year Colleges (AMATYC) issued a parallel document, *Crossroads in Mathematics: Standards for Introductory College Mathematics before Calculus*.

Today, the mathematics that students need to know is changing dramatically. Relatively inexpensive calculators now on the market can do all of the routine manipulations that have historically made up the bulk of algebra instruction. So there is an urgent need to rethink what Elementary Algebra should be, and this course represents our efforts.

Most students have taken an algebra class in high school. However, they may not have been able to solve problems, or may not have understood what they were doing when they did solve a problem. We believe that by shifting the focus of learning to the students, this course will provide them with a better understanding of algebra through guided concept development, improved problem solving abilities, an increased sense of mathematical power, and strengthened oral and written communication skills.

ACKNOWLEDGMENTS

Many teachers helped develop the original course on which this text is based, and they generously shared their ideas, enthusiasm, and classes with us. For all their contributions, we thank them.

Our efforts in adapting the original course to one suitable for college students would not have been possible without the contributions of many colleagues. Most especially we are grateful for all that Roberta Gehrmann, Mathematics Coordinator for the Learning Skills Center at California State University, Sacramento, did to help make this text a reality. Robbie was one of the first college instructors to start adapting the CPM Math 1 text for use with college students. Over the years we had many far-ranging and deep discussions about the structure and content of the course. Robbie shared her experiences, insights, and ideas about meeting the needs of students enrolled in an elementary algebra course. Throughout this time our collaborative work with her has helped us greatly in understanding what it means to make sense of algebra. Robbie, in turn, would like to acknowledge the technical and research assistance of Eric Langdon and Priya Asnani.

We also benefited from the expertise of colleagues in other fields of work. Wil Iley, Professor of Physics at Cosumnes Community College, taught us about coefficients of linear thermal expansion and provided a great photo of buckled railroad tracks. We thoroughly enjoyed talking with Michael Kwasnicki and Dr. Fred Hanson, who generously shared some of their vast knowledge about salmon and the monitoring of animal populations.

We would also like to recognize and thank the instructors who reviewed and tested initial drafts of this text, sharing their successes and telling us what needed to be fixed based on their classroom experiences. Our discussions with them and their detailed comments played a large role in shaping this version of the text:

Bernardo Bernardi, *California State University, Sacramento; Sacramento, California*

Lindsey Bramlett-Smith, *Santa Barbara City College; Santa Barbara, California*

Derek Lance, *Chattanooga State Technical Community College; Chattanooga, Tennessee*

Lap Ly, *California State University, Sacramento; Sacramento, California*

Mike Mallen, *Santa Barbara City College; Santa Barbara, California*

Karla Martin, *Walters State Community College; Morristown, Tennessee*

Elizabeth Mefford, *Walters State Community College; Morristown, Tennessee*

Robin Solloway, *California State University, Chico; Chico, California*

We also wish to thank all those instructors and students who used the class test edition of this text in their classrooms. Their insightful feedback and recommendations for improvement have been invaluable in making the text more useful and student-friendly.

Class Test Edition Users

Anoka-Ramsey Community College
California State University, Chico
California State University, Sacramento
Calvin College
Chattanooga State Technical Community College
Cosumnes River College
Franklin College
Georgia State University
North Hennepin Community College
Northeast State Technical Community College
Ridgewater College
San Joaquin Delta College
Santa Barbara City College
Stockton State College
Walters State Community College

We hope our efforts to incorporate suggestions based on the experiences and discussions of the people we've acknowledged will prove to be successful.

We have also learned—and continue to learn—a lot from our students! We thank them for their candid comments, their patience, and their enthusiasm for learning.

We'd also like to thank the editorial, production and marketing staff at Addison Wesley Longman for all their help in producing this text.

And most of all, we'd like to acknowledge the support, love, and understanding of our families and friends throughout the development of this course.

EAK and CLE

NOTE TO THE STUDENT

You may be taking elementary algebra for a number of reasons—perhaps to prepare for a technical field or for a college-level course. Both cases require you to use the mathematics you have learned. The skills you learn in this course will be useful to you in many areas that require you to use mathematics, including chemistry, economics, psychology, and zoology courses, as well as advanced mathematics courses. For this reason,

YOUR AIM FOR THIS COURSE SHOULD BE UNDERSTANDING THE MATHEMATICS.

LEARNING FOR UNDERSTANDING

Only you will know if you have understood an idea. We all know how easy it is to memorize something and even do well on a test without having any real idea about what is going on. If you settle for just performing well, you are cheating yourself, and you will likely have difficulty completing the course or using what you have learned in another environment.

We have done our best to design a course that will make it comfortable for you to learn and to understand what you have learned. However, learning takes hard work. If you want to do well in this course and get a good start in college, there are four things you will have to do:

- Make understanding the mathematics your highest goal in this course.
- Attend class regularly.
- Discuss questions with your group.
- Keep up with the class work and the homework.

On your way to understanding the mathematics you are learning, this course will help you become better at three important learning skills—asking questions, solving problems, and explaining your reasoning. In order to ask questions, solve problems, and explain your reasoning, you need to have something about which to think and ask and talk. That's why this course is built around problems, or, more specifically, it's built around **students doing** problems. You also need to talk over your questions with someone. That's why the course is built around **students working together in groups**.

WORKING IN GROUPS

If you want to understand the mathematics you are doing, you need to be willing to spend time thinking and trying out alternative approaches. Often you won't be able to come up with a correct answer on the first try, so you need to be willing to persist. At this level there are generally several ways to think about a topic, and you should try to see more than one of them. Working in groups provides a natural opportunity to see and hear several approaches to one problem. That is one reason we want you to work in groups.

A second reason for working in groups is that most job situations today demand that employees work in groups discussing ideas, listening, and testing. Colleagues must be able to combine the good parts of one person's idea with the good parts of someone else's idea to get a solution. Many of the problems in this book will ask you to discuss your ideas with your group and listen to other people's ideas. This is an important skill to learn, so do not skip over that part of the assignment.

A third and probably most important reason for learning to work collaboratively is that we want the mathematics to make sense to *you*. Mathematical techniques should not seem random. That is why the problems in this book have been structured so that you and your group, with support from your instructor, can construct, and therefore understand, much more mathematics than you could from being given a rule and assigned a bunch of exercises which all look the same.

Working effectively in groups doesn't just "happen." It takes the conscious efforts of every student and the instructor. Here are some guidelines to help make your experiences in group work as rewarding as possible:

Guidelines for Working in Groups

1. You are responsible for your own behavior.
   (No student has the right to interfere with another student's right to learn.)

2. You must try to help anyone in your group who asks.
   (But don't give the answer or tell someone how to work a problem unless you are asked.)

3. You may ask the instructor for help only when all of the members of your group have the same question.

4. You must use a "group voice" that only members of your group can hear.
   (The volume of each student's voice should remain reasonable and within the hearing range of his/her group only.)

ASKING QUESTIONS AND DOING HOMEWORK

No one expects basketball players to become good (or even decent) if they just watch others play. All players have to practice diligently. But they also need to know what to practice. In mathematics, you might get stuck sometimes and not know what to do and have no one to ask for help. In such a situation, write down what you do know about the problem and write down what you tried. Then figure out what your question is and write it down, too. When you get a chance, share your notes and questions with your group members (and your instructor) so they can see how you tried to solve the problem.

Doing this kind of work is what we mean by doing homework. In fact, the homework exercises which you can easily answer do not help you learn anything new; they are just skill maintenance. Getting stuck on a problem is *a big opportunity* to learn something. Analyze the problem to find out just what the hard part is. And then when you do find out how to do the problem, ask yourself, "What was it about that problem that made it so difficult?" Answering that question will get you ready to handle the next difficult problem without getting stuck, and you will have learned something.

WHAT DO THE BAR GRAPHS AT THE BEGINNING OF EACH CHAPTER REPRESENT?

The six bars in the opening of each chapter represent the major mathematical areas in which you will be working throughout the course. One way to check your progress is to examine the bar graph whenever you start a chapter. For each major area, ask yourself, "How am I doing? What are some ideas I truly now know (or skills I truly now have) in this area? And what do I still need to do more work on to be really successful?"

Here is the bar graph from Chapter 3:

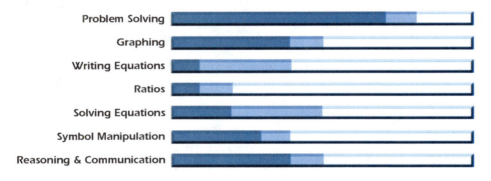

The darker blue part of a bar represents the portion of work in that area that you've already done. The lighter blue part shows how much emphasis you'll put on the area in the present chapter. The unshaded portion of a bar shows how much there is to come in future chapters.

RESOURCE PAGES

Some problems are marked with the icon to indicate that there is a Resource Page at the end of the text for use with the problem. When it is needed, simply remove a Resource Page from the text and use it as a convenient work space for completing the problem. Keep the completed page with the rest of the assignment.

At the beginning of the Resource Pages there are some Resource Page Templates to be used for making photocopies of pages that you will need to use more than once, such as Grid Sheets, Dot Paper, Chapter Summary Draft Pages, and Tool Kit pages. The number of copies you'll need are suggested in problem OP-11 Keeping a Notebook.

THE ALGEBRA TOOL KIT AND THE CALCULATOR TOOL KIT

Problems that are marked with the icon include instructions for you to write notes about important information in either your Algebra Tool Kit or your Calculator Tool Kit. Notes in your Algebra Tool Kit will be about concepts such as order of operations, evaluating algebraic expressions, variables, probability, area, the distributive property, and any other information you find helpful. Your Calculator Tool Kit will include your notes on how to use your calculator to change the sign of an entry, raise a number to a power, enter a number in scientific notation, and other information about using your calculator that is useful to you.

Creating your Algebra and Calculator Tool Kits will help you develop a handy reference for use as you work through this course. It will also help you develop effective study skills that will be useful in other courses you take.

THE APPENDIXES

There are three appendixes to the text; one provides background experience for integer skills used in the text while the other two contain extension exercises related to quadratic equations. Appendix A demonstrates a way to see why operations with signed (positive and negative) integers work the way they do. Appendix B uses subproblems to guide you through deriving a formula for solving quadratic equations. Appendix C extends the text's examination of quadratic equations and parabolas by investigating which part of a quadratic equation determines the steepness of the parabola and which part determines whether it opens upward or downward. It also explores moving parabolas horizontally and vertically on a coordinate grid.

SOME ANSWERS AND SOME WAYS TO GET STARTED

At the end of the text are answers to many of the Extension and Practice problems and hints for getting started on others. Some of the hints guide you through finding a solution; others refer you to a similar or related problem for guidance. Try to resist looking at the answers or hints until you've worked through a problem as far as you can on your own. Then, after doing as much as you can, check the back of the book.

If a problem is meant to be done in class—in collaboration with classmates and under the guidance of your instructor—then there is probably no answer or hint for it at the end of the text. You should verify your work on in-class problems by discussing your questions with classmates and getting clarification, if needed, from your instructor. This means that your regular attendance and participation in class are important for your successful completion of this course.

STRATEGIES FOR SOLVING PROBLEMS

Much of what you learn in this course will not be brand new, but will build on mathematics which you already know. A useful way to help yourself learn is to use such problem solving strategies as

Guess and Check,
Look for a Pattern,
Organize a Table,
Use Manipulatives,
Write an Equation,
Draw a Diagram or Graph,
Work Backwards,
Find a Sub-problem or an Easier Related Problem.

These strategies are not only useful for solving problems, but for *learning* as well. Throughout this course you will continue to build your skills at using a variety of problem solving strategies to progress from what you know toward the solution you seek.

Three major goals for this course are **for you to become better at asking questions, at solving problems, and at explaining your reasoning**, all of which aim at developing understanding of mathematics. On your way to achieving these goals, you will be learning how to take responsibility for your learning. One of your responsibilities will be to seek extra practice or assistance outside of the class to help you increase your confidence level. You might exchange phone numbers with other students in the class so you can discuss questions outside of class, or talk with your instructor about extra resources for help. With patience and persistence, your efforts will pay off!

Elaine A. Kasimatis

Cindy L. Erickson

NOTE TO THE INSTRUCTOR

As you look over this Elementary Algebra course, you will see that it differs from a traditional course in *what* algebra is taught and in *how* algebra is taught.

Calculators now on the market have "solve" keys that give numerical solutions to any equation with one variable. So there is little incentive for teaching the mechanics of doing algebra—but there is an urgent need to help students understand the key ideas of algebra better. And these ideas need to be reinforced over time.

Our course is designed to focus on these understandings and use them as natural places to practice the more traditional algebraic skills. Thus you will see these ideas used repeatedly throughout the course. In contrast to many courses, once a chapter is over a student cannot safely forget a topic which has been introduced—once an idea is introduced, it is used throughout the course.

Just as technical developments have motivated our change in topic emphasis, so has research on learning motivated our change in teaching emphasis. It is now clear that *lecturing has little to do with promoting learning*—students must construct their own understanding of ideas. In this course we often include problems which the students are expected to attack prior to any instruction on that particular type of problem. We do this deliberately, both to motivate students and to force them to think about what is challenging about the problem. Often students will invent a technique on their own and *understand* what they have done, rather than wait for you. For most situations, if you ask questions, rather than tell answers, your students will learn better and retain knowledge longer.

You will find this approach uncomfortable at first and so will your students. They have become accustomed to having a teacher tell them what to do so that they don't have to think. Now the burden falls on them. *Expect serious resistance* to this change. You will need to be quite firm during the first few weeks of class on insisting that students work with each other first, before seeking your help.

To facilitate these kinds of learning, this course

- develops important algebraic concepts early-on and builds on them all along;
- uses problem-solving strategies not only for understanding how to do mathematics, but also as a vehicle for learning mathematics;
- guides students to summarize regularly the ideas they've learned, assess their ability to apply the ideas to problems, and use algebraic skills.

GOALS OF THE COURSE

This course has six major goals that are content themes and six goals for developing students' learning and problem solving strategies. On completing the course, students should be able to do the following:

1. Use various problem solving strategies in order to analyze problems and formulate appropriate solution strategies.
2. Express, interpret, and graph linear and quadratic functions, and experience some other algebraic relations.
3. Use variables to represent relations from tables, graphs, verbally stated problems, and geometric diagrams, and understand that algebraic relations can be tested by substituting numbers.
4. Solve linear and quadratic equations and systems of linear equations, and understand their relationship to the graphs of equations.
5. Use ratio, proportion, and direct variation from numerical, geometric, and algebraic perspectives (percent, similarity, right triangles, slope, and probability).
6. Use the distributive property and order of operations to reorganize algebraic expressions into more useful forms.

The course also provides students the opportunity to do the following:

1. Move away from a rule-applying approach to a rule-generating approach.
2. Learn to use a scientific calculator effectively and efficiently.
3. Develop confidence in pursuing several mathematical approaches to a problem, as well as the ability to assess when an approach is not working and a new direction is needed.
4. Become more aware of their own thinking about problems and describe their efforts both orally and in writing.
5. Develop a positive attitude toward mathematics: "Math is important, useful, desirable, and very much connected. And I can learn it!"
6. Develop organizational skills in study habits as well as in doing mathematics.

For information on obtaining algebra titles, contact Cuisenaire Company of America, Inc. by calling 1-800-273-0338.

Information on calculator-related products is available from Texas Instruments at http://www.ti.com/calc/docs/dealers.htm or 1-800-TI-CARES.

STRUCTURE OF THE COURSE

In addition to teaching algebra in this course, we are also trying to de-mystify mathematics for the students and empower them. Students should feel that *mathematics makes sense* and that they can figure out most of the ideas without being told by the instructor *if they are willing to spend the time*. So the *focus of learning shifts to the student*. Since your students will spend much of their class time working in groups, you will be spending more of your time working with groups rather than individuals or the whole class. If you have never had your class working in groups, we've included some thoughts on groups and making them effective.

The course itself is built around problems ranging from routine review exercises, to problems designed to teach concepts, to short writing assignments which ask the student to integrate and explain. We have tried to design the course so that the time students work in class can be spent doing the more challenging problems and those that introduce new material, while the problems which are designed to reinforce previously-learned skills may be done for homework.

These review problems implement two other principles of the course: (1) certain skills need to be practiced frequently so that they are retained; and (2) for many major ideas (such as the notion of ratio) many weeks may pass before the concepts finally get internalized. Thus students need to see important ideas in many contexts and be asked to explain the connections many times before these ideas become second nature.

There are fewer problems in this course than in a traditional one, but most problems should be done in more depth, especially those that include a written reflection by the student. Students need to see why one particular approach may work better than another. They need time to consider different ways of looking at a problem so that they can truly understand its underlying structure. The answer is less important than the quality of thinking that is taking place. The quality of the students' thinking is improved in several ways, including by focused discussions within the groups, by students' presentations to the whole class, and by students and the instructor modeling good thinking by "thinking out loud" to the whole class.

SOME THOUGHTS ON USING COOPERATIVE LEARNING GROUPS

Activities in this course will require much more work in small groups and much less introduction and explanation of ideas by the instructor. Group interaction is an integral part of the learning process. As Tom Sallee and

Carol Meyer Waters explain in *Make It Simpler*, the group setting has many advantages for both the instructor and the student. From the teaching perspective, groups of four can reduce the number of students in the class from thirty-six to nine. Moreover, since members of the group respond to routine inquiries, more of the instructor's time can be spent dealing with questions of substance. Students benefit because they have the continuous presence of three others to whom they can explain their ideas. They get the opportunity to teach and refine their ideas, questions, and approaches in the security of a small group. They become more willing to try because there is less to risk in a group of four than in a whole-class setting.

When using groups, **the role traditionally assumed by instructors must change**. The instructor **must** give up the central role of "sage on the stage" and really become a "guide on the side," allowing students to explore problems and gradually construct their own understandings. To do this, **students need to explain to each other** and **listen to each other**. Sometimes their solutions will not work, but our goal is to have them find this out for themselves. On the other hand, instructor-led discussions are essential to summarize results of group activities and tie together big ideas after the students have had a chance to work on the pieces. These guided discussions should be conducted in an open, accepting way with the purpose of helping the students make connections to see where and how the pieces of information fit together.

To insure optimum interaction, we recommend pairs or groups of four students. In these situations, there is always an open line of communication. In the case of groups of four, if Student #1 and Student #2 are discussing a problem solving exercise, Students #3 and #4 still have the opportunity to discuss the activity with each other. They may choose not to, but the opportunity is available. In groups of three, if two people are talking the third person may be isolated or shut out.

Groups work best when group members have their desks in close proximity. For pairs, this means that the two desks are placed touching each other, either side-by-side or facing each other. For groups of four, all four desks should be touching (or nearly touching) and oriented so that all four group members can easily see each other's faces. Depending on the instructor and facilities, the desks (or tables) may remain arranged in pairs or fours, or the instructor may have students move their desks from rows into tight configurations for group work, then back into rows again. Note the operative word "tight." If the group members are too far apart, when they want to confer with group mates they either have to get out of their seats or shout. Both of these activities are distracting to the rest of the class. The alternative is perhaps more detrimental to the success of the class: it's so much bother to try to work together that group members give up and try to work alone, if they work at all. To avoid this scenario, instructors need to get their students to form tight-knit groups so they will have someone nearby to talk with about math.

It is equally important to have the group configurations well-spaced throughout the classroom. There needs to be enough space between groups for the instructor to circulate freely about the classroom. This space also provides something of a buffer zone between groups so that a conversation in one group doesn't interfere with a neighboring group's work.

The two most common configurations for groups of four are formed by either having pairs of students from adjacent rows (say Row 1 Students #1 and #2, and Row 2 Students #1 and #2) turn their desks 90° to face the aisle between their rows and lift-slide so the table tops are touching, or by having one pair (say Students #1 in Rows 1 and 2) turn-lift-slide and one pair simply lift-slide so they share a common work surface. ("Lifting and sliding" is less destructive to floors and nerves than "dragging.") For classrooms furnished with desk-chairs, some instructors prefer a spiral arrangement where the front of one desk touches the left side of the next.

While it is not particularly difficult to get students to physically move their desks into group configurations, it does take practice at first. Instructors who are fairly expert at using groups report that it is not enough to show their students how to configure their groups the first day of class—they must keep insisting that the desks be arranged as they have shown. This insistence may be in the form of verbal reminders, or the instructor may help group members push their desks together (perhaps while checking homework or taking care of other business), or help clustered groups separate from each other. The time and effort an instructor spends during the first few weeks of class pays off throughout the course.

Early in the course it is most efficient to randomly assign students to groups. During the first few weeks of class, try to make sure that all students have met and worked with each other. After that you could randomly assign students to groups, or ask them to write down their preferences and set up groups with their choices in mind.

One of your first tasks will be to share your expectations for behavior when the students are working in groups. We use the following guidelines:

GUIDELINES FOR WORKING IN GROUPS

1. You are responsible for your own behavior.
   (Disturbances are going to occur whether in rows or groups, but the seating should not be an excuse. Basically this is a translation of the rule, "No student has the right to interfere with another student's right to learn.")

2. You must try to help anyone in your group who asks.
   (This is one of the more difficult ideas for competitive students to accept.)

3. You may ask the instructor for help only when all of the members of your group have the same question.
   (You will probably have as much difficulty following this as the students will. You must break the habit of directly answering a question when asked, rather than checking if the group has discussed the question first.)

4   You must use a "group voice" that only members of your group can hear. *(The volume of the students' voices should remain reasonable within the hearing of their group only. When moving from group discussions to a whole-class discussion, be sure to alert groups to the transition and give them time to finish their thoughts before changing their focus.)*

While students are working in groups, the instructor's job is to *circulate*—moving from group to group—and *listen*. Try to listen to a group you are not looking at or currently working with. Two rules of thumb that are very difficult to follow, and are also very important in developing students' independence and responsibility for their own thinking are:

> Never carry or grab a writing implement.
> Try to respond with a leading question.

If you can't write or draw you will have to get the students to understand well enough to write or draw what is needed, and if you try not to tell, and try to ask questions, they will have to do the thinking. The hard part, and what makes these two rules almost impossible to follow, is time. But remember you can leave the group with a leading question, go to another, and come back later, so you don't have to spend the time telling anyone how to do a whole problem.

For the first few weeks after you start using groups, carry around a reminder to yourself: "Does everyone have the same question?" Even (especially) if the question is as simple as "What did you tell us to do?" refer it back to the group. You, and especially the students who like to get your attention, will find it HARD at the beginning. But being firm and consistent at the beginning will give you better group dynamics later in the year.

When joining a group, come in at their eye-level by bending your knees, sitting on a chair, or kneeling on the floor. Don't lean over. Leaning over makes you the boss, not the catalyst; besides, some of us have learned the hard way that it's not good for our backs.

It takes time and effort at the beginning of a term to get groups to work effectively, but the effort pays off later. As the term goes on, you will find by roving the classroom that students are asking better questions, doing some work even after getting stuck, helping each other with routine problems, and asking for help only on the more difficult parts of problems.

Following this note is a list of five practices that facilitate effective group work. We call it our "Five-Finger Check List" because as an easy memory device, we associate each item with a finger. We can then quickly run down the list before and after each class.

CHAPTER SUMMARIES

The last problem in most chapters is a Chapter Summary problem (in some cases, followed by a Course Summary problem). The purposes of a Chapter

Summary problem are to help students to pull together the main ideas of a chapter and to help them see how much they've learned.

Because at this level many students are not yet used to summarizing their learning and still have difficulty expressing themselves in writing they find the Chapter Summary problems challenging at first. To help them, we've broken the summarizing process into three parts: the rough draft of individual student's brainstorming, the group discussion of rough drafts, and the revision of the draft into a Chapter Summary.

Some instructors have students draft their chapter summaries two or three days before the end of the chapter, in preparation for a chapter review. Others use the Chapter Summary problem in lieu of a chapter review. Some start the summary by brainstorming with the class. Instructors are fairly directive at the beginning of the course; however, as time progresses and students become better able to identify and summarize the important ideas of a chapter, the instructor becomes more of a recorder and facilitator, rather than a director or guide.

Another idea for helping students understand what you expect in a chapter summary is to show them a model—ideally a copy of one written by a student from a previous term or from another instructor's class—but you may have to make up one yourself. Seeing a sample helps students get going in the right direction from the start.

ASSESSMENT

Assessment is a major issue. This course attempts to balance the traditional emphasis on facts and manipulative skills with an emphasis on problem solving and understanding concepts. If we continue to assess our students using traditional tests, we could face the following consequences:

- We'll never know whether students have succeeded in developing problem solving skills or conceptual understanding.

- We'll leave our students with the feeling that regardless of what we say, the important stuff is facts and algebraic manipulations.

- We'll not be able to convince our colleagues, let alone our critics, of the value of using a different approach when our students continue to score close to, but slightly below, their peers on traditional assessment instruments.

In addition to individual tests and group tests, we encourage the use of portfolios to encourage student self-assessment. The Chapter Summaries may serve as a way for you to have your students assemble portfolios of their work throughout the course. Each entry can be viewed as a "snapshot" of the student's understanding and skills at that time, so the whole collection is a record of the mathematics that is becoming a part of the student.

CHANGE TAKES TIME

This course requires students to make changes in the way they learn and requires instructors to make changes in the way they teach. Because change takes time, it is important to get off to a good start. Right from the beginning, the most effective instructors emphasize to their students that both the mathematical content and the pedagogy of the class will differ from those in a traditional class. They report that during the first few weeks it is crucial to focus on making groups function effectively. This requires as great a change in the instructor's behavior and thinking as it does in the students'.

One important factor which contributes to groups working effectively is that students must complete their assignments. This can also be a huge challenge to instructors. However, the time and effort you spend at the beginning of the term focusing on these two issues will make the rest of the term go much more smoothly.

We welcome you to the course and hope that you and your students find the success many others have.

Elaine A. Kasimatis
Cindy L. Erickson

FOR THE INSTRUCTOR

GETTING GROUPS TO WORK TOGETHER
A FIVE-FINGER CHECK LIST

Here are five things YOU can do to get your algebra class off to a good start. After each class, do a quick mental check to see how well each of these guidelines worked.

1. **START PROMPTLY.** Groups get organized quickly, quietly, and tightly so that students can begin working right away.

    Be firm and consistent all term in insisting that the desks be tightly configured so that all group members can make eye contact with, talk to, and hear each other, and so there is space for walking between adjacent groups. Expect students to arrange their desks and have materials out and ready for use at the beginning of class. It takes practice to "quickly and quietly" make the transition from a seating arrangement for individuals to one for groups, and vice versa, but the time spent at the start of a term reinforcing smooth transitions into optimal desk configurations is well worth it. After just a few minutes of socializing, groups should begin working on mathematics.

2. **PEER SUPPORT.** Group members consult each other before consulting you.

    Students are used to having teachers transmit information, not *facilitate thinking. It takes time for them to learn to work together, to trust and support each other, and to feel comfortable as generators of knowledge. Although you may empathize with their frustrations and find it difficult yourself to follow this group guideline, do it! It may take several weeks for students to accept this responsibility, but your early diligence will pay off for the rest of the term.*

3. **ASSIGNMENTS.** All students should attempt to do the assignment each day.

    Be especially vigilant at the start of the term that your students develop a sense of responsibility and make a serious attempt to do their assigned work every day. You will probably need to grade the work daily for some period of time. You will also need to guard against "covering" the previous assignment for students before they begin the current section; otherwise, you foster dependence on you.

4. **RESPOND TO THE GROUP.** You address your responses to the whole group, not just the individual who voices the question.

    One of the most effective ways to facilitate cooperative group work is to address responses or questions to the entire group. A way to do this is to stand opposite the student who has a hand raised, and try to make eye contact with each group member while you talk. If all group members are not listening because they do not all share the same question, get them to talk to each other about the problem.

5. **CIRCULATE.** You visit all groups regularly, not just those with raised hands.

    Your circulation pattern around the classroom should include pauses to make sure all groups are talking about mathematics. You should make contact, even if it is only a quick "Any problems?" at least three times every period. While it is important to respond to groups who have questions, waving hands should not determine your circulation pattern. Acknowledge raised hands by making eye contact with group members, or by saying, "I'll be right with you," and then continue your classroom "cruise." Get back to the group whose raised hands you acknowledged within one or two minutes. If your interactions with the groups is solely reactive—responding to signals for help—you will reinforce the students' dependence on you and undermine your goal of fostering student-centered learning.

CHAPTER ONE—FOR THE INSTRUCTOR

OPENINGS
Data Organization

MATERIALS

- Clue cards for Building Corrals cooperative logic problem
- Flat toothpicks
- Overhead transparencies and pens for groups
- Calculators, scientific or graphing
- Four large sheets of butcher paper with spaces numbered from 0 to 5, 6 to 10, 11 to 15, and 16 to 20
- Calculator Tool Kit Resource Page
- Small squares from commercial sets of algebra tiles, or cut from Centimeter Grid Resource Page (We use algebra tiles that are manufactured by Cuisenaire Company of America, Inc. They may be ordered by phoning 1-800-273-0338. Algebra tiles for the overhead projector are also available. Students will also use algebra tiles in later chapters to model solving equations, to show the distributive property, and to factor quadratic expressions.)
- Flip-chart grid paper for the class Predicting Shoe Size graph
- Sticky dots (20 each of two different colors)
- Graph paper for OP-62
- Resource Pages for OP-7, OP-8, and OP-29; also, Algebra Tool Kit Resource Page, Calculator Tool Kit Resource Page and Summary Draft Template

For Appendix A:
- Small square tiles for integer work
- Integer tiles for the overhead projector

This chapter includes activities to introduce students to organizing data in different ways, establishing routines for working in groups, and using calculators and manipulatives in the classroom. It also reviews several important skills that students should be prepared to use: order of operations, integer arithmetic, and interpreting graphs.

This chapter is especially important because it sets the style for the course in terms of group work, student investigations, problem solving, and the use of calculators and manipulatives. Emphasize to your students that this is a "get acquainted" chapter. We don't expect students to master every concept or process at this point. Students will continue to encounter problems that build on this initial exposure throughout the course.

Students may feel frustration when they do not reach mastery at the end of each chapter, since they have learned to expect short-term mastery from previous math classes. Their frustration with this new approach to learning will need to be addressed periodically throughout the course.

Assignments are listed with class work first, followed by work that can be continued at home. The first portion of an assignment generally requires group effort, whole-class discussion, or some instructor guidance. The remainder of an assignment consists of follow-up, practice problems, or problems that set the stage for future work based on skills students already have. Students will have opportunities to polish their skills in practice problems throughout the first few chapters of the course.

If most of the students in your class need help understanding integer arithmetic, we suggest you do Appendix A after completing Section 1.2. If only a few students need concrete help building an understanding of arithmetic with signed numbers, you could refer them to Appendix A, and have them work through it under your separate guidance. See the Note below.

The last problem, OP-126, asks students for an organized summary of the chapter. Each chapter has a summary problem designed to help students pull together the main ideas of the chapter and to help them see their progress. Since it will take time for students to develop summarizing skills, several problems are devoted to the chapter summary: a rough-draft problem (OP-104), a group discussion problem (OP-105), and additional review problems (OP-106 through OP-125) before they finally revise their summaries in OP-126.

NOTE This chapter includes practice work with integer arithmetic. For students whose skills are weak, a concrete model for developing the meaning of integer arithmetic is included as Appendix A. If most students in your class have not worked with integer arithmetic before, or if they need help understanding how to calculate with signed numbers, we suggest you include the introductory integer work by introducing the sections as follows: 1.1, 1.2, Appendix A, and then 1.3 through 1.8.

Students will need integer tiles for the work in Appendix A. If a class set of tiles is not available, students will need to bring their own set. As part of the homework prior to class, have students make integer tiles using the Integer Tiles Resource Page (in the appendix), or use the small squares from a purchased algebra tiles set.

CHAPTER ONE

OPENINGS
Data Organization

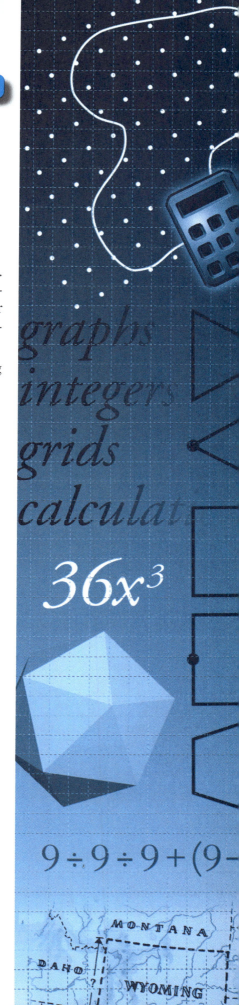

In this chapter, you will investigate different ways to organize data. Working collaboratively in groups, you will explore several problem-solving strategies as you develop and review your graphing and calculator skills. Many of the problems, especially those that involve very large numbers, require the use of a scientific or graphing calculator.

Throughout the chapter, you will get better at solving problems, asking questions, and explaining your reasoning.

IN CHAPTER 1 YOU WILL HAVE THE OPPORTUNITY TO:

- establish routines for working collaboratively to solve problems;
- collect and organize data or information;
- use a guess-and-check table to solve word problems;
- read and interpret graphs;
- practice computing with integers;
- rewrite exponential expressions by using the meaning of positive integer exponents;
- work with numbers expressed in scientific notation;
- use your calculator more efficiently;
- calculate simple probabilities.

Chapter 1 Openings: Data Organization

MATERIALS	CHAPTER CONTENTS	PROBLEM SETS
For OP-1: Clue Cards for Building Corrals, flat toothpicks, overhead transparencies and pens for groups; For OP-7 Dot Paper Corrals Resource Page; For OP-8: Perimeter Resource Page; Algebra Tool Kit Resource Page	**1.1** Building Corrals: Working Collaboratively Diamond Problems Algebra Tool Kit	OP-1–OP-12
Calculators, scientific or graphing; For OP-16: Four large sheets of butcher paper with spaces numbered from 0 to 5, 6 to 10, 11 to 15, and 16 to 20 for recording solutions to the Five-Digit Problem (OP-15); Calculator Tool Kit Resource Page;	**1.2** The Five-Digit Problem: Order of Operations Calculator Tool Kit	OP-13–OP-28
For OP-29: Small squares from sets of algebra tiles or cut from Centimeter Grid Resource Page, OP-29 Area and Perimeter Resource Page	**1.3** Tiling with Squares: Area and Perimeter "What's My Rule?"	OP-29–OP-42
	1.4 Using Guess-and-Check Tables to Solve Problems "What's My Rule?"	OP-43–OP-59
For OP-60: Flip-chart grid paper for class graph, sticky dots (20 each of two different colors); For OP-62: Graph paper	**1.5** Predicting Shoe Size: Making a Graph from Data	OP-60–OP-75
	1.6 The National Debt and Teraflops: Large Numbers and Scientific Notation "What's My Rule?"	OP-76–OP-91
Summary Draft Template Resource Page	**1.7** Driving to Redwood City: Data from a Graph "What's My Rule?"	OP-92–OP-104
	1.8 Summary and Review "What's My Rule?"	OP-105–OP-126

Problem Solving
Graphing
Writing Equations
Ratios
Solving Equations
Symbol Manipulation
Reasoning & Communication

1.1 BUILDING CORRALS: WORKING COLLABORATIVELY

SECTION MATERIALS

FOR OP-1: CLUE CARDS FOR BUILDING CORRALS COOPERATIVE LOGIC PROBLEM MADE FROM THE CLUE CARDS FOR OP-1 MASTER IN THE RESOURCE PAGES

FLAT TOOTHPICKS TO REPRESENT FENCE SECTIONS

OVERHEAD TRANSPARENCIES AND PENS FOR GROUPS

FOR OP-7: DOT PAPER RESOURCE PAGE

FOR OP-8: PERIMETER RESOURCE PAGE

ALGEBRA TOOL KIT RESOURCE PAGE

Building Corrals (OP-1) is a cooperative logic problem that focuses on the process of students working together in groups in which every student contributes. The activity provides an opportunity to acquaint your class with your expectations about moving into groups and group behaviors. (Additional cooperative logic problems may be found in Get It Together *by EQUALS, and* United We Solve *by Tim Erickson, both available from Dale Seymour Publications.)*

We recommend doing OP-1 at the beginning of the class before administrative details consume too much time. A master for the clue cards for OP-1 is in the Instructor's Resource Pages. You'll need to make one set of cards for each group of four students. Note that there are four clue cards and one "check" card on the master. The problem has multiple solutions. If you anticipate using a second cooperative logic problem, you will also need to make up the sets of cards for OP-4, Cars and Bikes, from its Resource Page.

To each group, distribute at least 20 flat toothpicks (or matchsticks with the heads removed) with the set of four clues. (Don't give out the check card until the group indicates it has solved the problem.) Tell the students that they are to solve the problem cooperatively. Each member of a group receives a different clue. Students must read their clues to each other, but they may not show each other the written clues, so careful listening is important. Tour the room while you carefully watch what the groups do, and listen to their conversations. Some students may not know what a corral is. If no one knows, tell the whole class; otherwise let a student tell. If groups get stuck on the problem, ask some guiding questions to help them focus: "Have you read all the clues?" "How can you organize the information on the cards?" "Is there a clue (or clues) that helps you get started?" "What are you trying to find?" "Is it possible to build another rectangle with the same number of toothpicks?" "Does the arrangement fit all the information?" and "Can you find another solution?"

When a group wants to know "Are we right?" hand them the check card with a response along the lines of "Oh, are you ready to check your solution?" or "Did you find an arrangement that fits all the clues?" This will convey to them your belief that they have the ability and information to decide correctness for themselves.

Since there is more than one answer to the Building Corrals problem, this activity provides a nice opportunity for groups to present their solutions to the class. You could provide each group with a transparency and an overhead projector pen, or you could wander around the room, choose groups that have different solutions, and ask them to send a representative to put their solution on the board.

OP-1. Building Corrals Many people find it helpful to work with others when trying to solve problems in mathematics. Working collaboratively can also be a challenge. Building Corrals is a cooperative logic problem that requires your group to work together. Each of you will be given only part of the information needed to solve the following problem:

Students who attend the University of California at Davis can board their horses at the Campus Equestrian Center. This year the Equestrian Center is home to three student horses, which until now have been boarded together in a large rectangular corral. However, the three horses have been fighting recently, and need to be separated from each other.

Cole, who works at the Equestrian Center, has decided to subdivide the large rectangular corral to form three new corrals. Use the information on the clue cards to figure out how he can arrange the new corrals.

Each member of your group will receive a different clue card. Share your clue by reading it aloud, but do not show your clue card to other group members. After your group has used all of its clue cards to solve the problem, answer each of the following questions:

a. How could Cole arrange the corrals? Sketch your solution, and then write your response in complete sentences. *Cole should add six fence sections to the original corral. He should use two sections to build a one-by-one square in one corner of the rectangle and the other four sections to enclose a two-by-two square in another corner, so that the two new squares do not touch.*

Building Corrals solutions:

b. Are there other solutions to the problem? If so, find them. If not, explain why not.

As the class gets started on OP-2, be sure to circulate among the groups. Listen in and, if necessary, remind students to reflect silently on the questions before starting their group discussions. The group discussion is likely to proceed with more focus and not bog down if individuals spend a few moments putting their thoughts together before sharing them.

OP-2. We begin the course by focusing on problem solving and cooperative learning, skills that apply in many college courses and technical fields.

Read the Guidelines for Working in Groups in the Note to the Student at the front of the text.

a. Think about the following questions on your own:

- Which guideline will be easiest for you personally to follow?
- Which guideline will be the most difficult?
- What previous experiences, if any, have you had working in groups?
- What types of things could you do to contribute to positive and productive cooperative learning experiences throughout the course? Another way to think of this is: What are the attributes of someone you would like to work with in a group?

b. Now discuss your responses in part a with your group. Take turns sharing your personal experiences and commitments.

OP-3. Now that your group has solved the Building Corrals problem, think about how you went about finding an answer. Discuss the process your group used: How did you decide which clue to use first? Then what did you do? Write a brief outline of your group's approach to solving the problem. Be ready to explain to the rest of the class the way your group solved the problem.

If a group finishes early, you might assign a second cooperative logic problem, OP-4 Cars and Bikes, as an additional opportunity for a fast group to work together while other groups catch up.

OP-4. Cars and Bikes The Ferrari family and the Schwinn family share a garage. There are four questions your group should answer according to the information given on the clue cards. There are different ways to organize this information, so share your ideas with your group.

a. Use the clue cards to help your group solve the problem. *Ferrari: two cars and two bikes; Schwinn: three cars and one bike*

b. How did your group solve the problem? Describe the process you used.

c. Do you think there is more than one solution to the problem? Explain why you think so. *There is only one solution.*

The Diamond Problems format is introduced in this chapter to prepare students for finding product/sum combinations, a skill they will use later in factoring quadratic polynomials. It is important that students see the answers to the Diamond Problems of OP-5 in context, which is why they are told to copy the diamonds. If they don't see the shortcut of drawing a large "X", you could share it with them so they don't get caught up in the busywork of drawing the whole diamond.

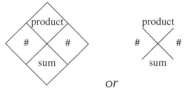

or

*Students could start these problems without an introduction. However, we like to use the chalkboard to introduce them to the whole class to get things rolling. As you circulate, use the terms **sum** and **product** so students can start getting used to them in this context.*

Parts e through l are included for diagnostic purposes. As groups work, you'll have an opportunity to get a sense of each student's arithmetic skills with negative integers, fractions, and variables.

OP-5. Diamond Problems

If you know the numbers in the west and east diamonds, can you find the numbers that go in the north and south ones?

Look at these completed diamonds and see if you can discover the pattern. How are the north and south diamonds related to the east and west diamonds? Explain the pattern you see to a partner in your group.

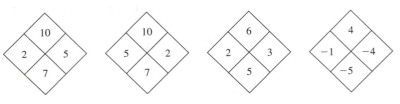

Copy these Diamond Problems, and use the pattern you found to complete each of them:

Hint: You do not need to sketch the outside edge of the diamonds—you'll save time if you sketch only a large "X":

instead of

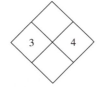

a.
$N = 12, S = 7$

b. (6, 11)
$N = 66, S = 17$

c. (12 top, 7 bottom)
3, 4

d. (3 right, 4 bottom)
$N = 3, W = 1$

e.  (−2, −3)
$N = 6, S = -5$

f. (−1, 4)
$N = -4, S = 3$

g. 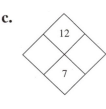 (6 top, −2 right)
$W = -3, S = -5$

h. 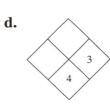 (−8 top, 7 bottom)
−1, 8

i. ($\frac{1}{2}$, $\frac{1}{2}$)
$N = \frac{1}{4}, S = 1$

j. ($\frac{1}{3}$, $\frac{3}{4}$)
$N = \frac{1}{4}, S = \frac{13}{12}$

k. (x, x)
$N = x^2, S = 2x$

l. (a, b)
$N = ab, S = a + b$

EXTENSION AND PRACTICE

OP-6. Do you know what **multiplication** means? You probably can easily recite multiplication facts, but you may not be as comfortable explaining to someone what it means to multiply two numbers.

To demonstrate how well you understand something, you must do more than simply use symbols or get correct numerical answers. You must also be able to explain your thoughts in words and in pictures. The following rectangle illustrates these four ways to make sense of a mathematical concept:

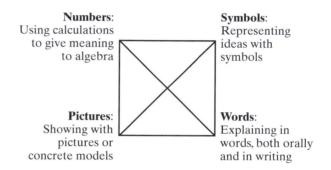

Numbers: Using calculations to give meaning to algebra

Symbols: Representing ideas with symbols

Pictures: Showing with pictures or concrete models

Words: Explaining in words, both orally and in writing

Suppose you're a volunteer in an elementary school classroom. Max, a third grader who is learning about multiplication, asks you to explain what "3 times 5" means. He says that he knows the answer is 15, but he doesn't understand why.

a. Pictures Draw a picture that you could use to show Max what "3 times 5" means.

b. Words In a complete sentence (or sentences), describe how you could explain verbally to Max what "3 times 5" means.

c. Symbols There are several ways to use symbols to represent multiplication. Each of the following expressions represents "3 times 5":

$$(3)(5) \quad \text{or} \quad 3(5) \quad \text{or} \quad (3)5 \quad \text{or} \quad 3 \cdot 5$$

In algebra, we rarely write "3 × 5" for "3 times 5." The reason we don't use × for "times" is that it easily could be confused with the letter x, which we reserve for another use. (In other math books, you may come across the × symbol, especially in certain expressions like 3×10^4. But in this book we'll stick to one of the forms shown in the above display.)

On your paper, use appropriate symbols to represent the expression

"450 times 0.5 times 72."

Possible answer: 450(0.5)(72)

8 Chapter 1 Openings: Data Organization

NOTE Whenever you see a problem marked with the symbol , you'll find a **Resource Page** at the end of the text for you to use to complete the problem. The Resource Pages are designed to make your work easier and more efficient.

OP-7. More Corrals This problem builds on the work you did in OP-1.

a. Suppose Cole wanted to build a new rectangular corral using exactly 12 fence sections. How many different rectangles could he build? Experiment using toothpicks (or something similar) and then record what you find on a copy of Dot Paper. Instructions are included in the Resource section. (The dots represent fence posts.) *1 by 5, 2 by 4, and 3 by 3*

b. Repeat part a with exactly 15 fence sections. *None is possible.*

c. Now, sketch as many *non*rectangular corrals as you can using exactly 10 fence sections. Explain why you are sure there are no other possibilities. Note: A fence section will not reach diagonally between two dots. *There are four possible figures:*

OP-8. Perimeter In determining the number of fence sections to form a corral, you have been dealing with the mathematical concept of *perimeter*.

To find the **perimeter** of a figure such as a triangle, a rectangle, or a hexagon, add the lengths of all its sides.

Example 1: The perimeter of a rectangle that is 2 units wide and 4 units long can found by adding

$$4 \text{ cm} + 2 \text{ cm} + 4 \text{ cm} + 2 \text{ cm}$$

to get 12 centimeters.

Example 2: The perimeter of this irregular hexagon can be found by adding

$$5 + 1 + 2 + 1 + 3 + 2$$

to get 14 units. ▲

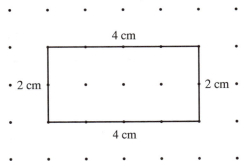

For each of the following figures, label each side with its length and then find the perimeter of the figure.

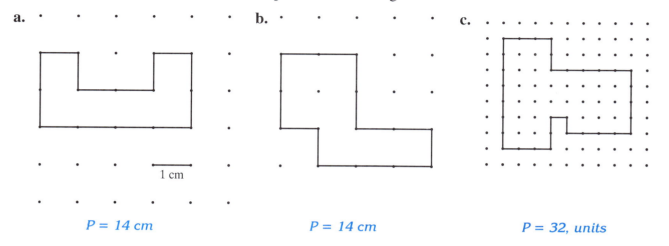

a. $P = 14$ cm

b. $P = 14$ cm

c. $P = 32$, units

NOTE Problems that require **Tool Kit** entries are marked with a . You'll find Algebra Tool Kit sheets with the Resource Pages at the end of the text.

OP-9. Algebra Tool Kit An **Algebra Tool Kit** is a useful way to keep track of methods and topics you would like to be able to find easily. The idea is to have at your fingertips a resource for figuring out what to do when you don't remember what to do. Your Algebra Tool Kit is for your own use, so make it something *you* understand. Directly copying a definition or explanation from the book will not be helpful unless it makes sense to you.

You'll find some Algebra Tool Kit sheets in the Resource Pages at the end of the text. Several topics, including *perimeter*, are listed on the first few sheets.

To start to make your Algebra Tool Kit a personal source of information, write on the Resource Page any information or examples that will be useful to you in understanding and remembering the topic *perimeter*. You will be filling in the other boxes as we reach the related topics in subsequent sections of the text.

You may want to tell students to write their responses to OP-10 on a separate sheet of paper so that you can collect and read them.

OP-10. Your Mathematical Autobiography Reread the Note to the Student in the beginning of the text, and think about the following questions, and then describe your mathematics background and goals in several paragraphs.

What is your mathematical background? Of all the math classes you've taken, which did you enjoy the most? In which math class (or classes) did you learn the most? What are your mathematical strengths? What are your mathematical weaknesses? Include both formal and informal experiences and training.

What are *your* goals as you work through this course? What goals do you want to reach when you've completed this course?

OP-11. Keeping a Notebook You'll need to keep an organized notebook for this course following a system that works for you. Here's one method of keeping a notebook that has worked for many students:

- The notebook should be a sturdy three-ring loose-leaf binder with a hard cover.

- The binder should have dividers to separate it into sections:

    class work and homework

    class notes and Tool Kits

    chapter summaries

    tests and quizzes, with corrections

    supplies (notebook paper, graph paper, blank Resource Pages, and so on)

- For each class meeting, you will need to bring:

    a calculator, either scientific or graphing, as required by your instructor

    lined, hole-punched notebook paper

    graph paper

    a ruler marked with both centimeters and inches

    pencils (Colored pencils will be helpful, especially with some of the graphing problems.)

    an eraser

    a small stapler

- You'll need more than one copy of some of the resource pages because you'll be using them for more than one problem. Find the Resource Page Templates at the beginning of the Resource Pages and make photocopies as indicated:

    Algebra Tool Kits—about 10 copies

    Calculator Tool Kits—about 5 copies

    Grid Sheets—at least 8 copies

    Dot Paper—at least 8 copies

    Summary Draft Sheets—about 26 copies

A well-organized, complete notebook is a big asset for learning in this course. It can be especially helpful when studying for tests.

To get off to a good start in this course, get a suitable binder and start organizing your notebook *now*!

OP-12. Copy and complete each of these Diamond Problems:

a.
$N = -21,$
$S = 4$

b.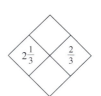
$N = \frac{14}{9},$
$S = 3$

c.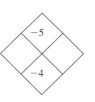
$-5, 1$

d.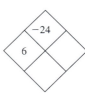
$E = -4,$
$S = 2$

e.
$N = -81,$
$S = 0$

f.
$W = 0,$
$S = \frac{4}{7}$

g.
$E = 0.5,$
$S = -0.25$

h.
$E = -8,$
$S = -23$

1.2 THE FIVE-DIGIT PROBLEM: ORDER OF OPERATIONS

SECTION MATERIALS

FOR OP-16: POSTERS FOR RECORDING FIVE-DIGIT PROBLEM ANSWERS. TO AVOID CROWDING, USE FIVE LARGE PIECES OF BUTCHER PAPER, NUMBERED FROM 0 TO 5, 6 TO 10, 11 TO 15, AND 16 TO 20. POST THESE CHARTS IN DIFFERENT PARTS OF THE ROOM. LEAVE PLENTY OF ROOM ON THE CHARTS FOR STUDENTS TO RECORD MULTIPLE SOLUTIONS.

CALCULATOR TOOL KIT RESOURCE PAGE

*Some students may not have a clear idea about what a **digit** is. One way of letting them know without just telling is to ask the class how many digits our base 10 number system uses. (One instructor suggests "talking with your hands and wiggling your fingers" while you ask.) Once someone responds "Ten," ask for an example of one of the digits, and write it on the board. Then ask for another example, and write the response so that the two digits appear in increasing order from left to right on the board with blank spaces left for missing digits. For example,*

2,   5,   7,

Try to get as many students as possible involved by taking only one response from each volunteer. After a few examples (when it seems the entire class understands what a base 10 digit is), have them tell you the rest of the digits in chorus as you point to the blanks, and then write the responses. When all 10 digits are written, have the class choose any digit except 0 (between 1 and 9, inclusive), to be used later in OP-15. Then let the groups get going on OP-13, which provides an opportunity for students to review (or encounter) order of operations by asking them to simplify a few problems in the five-digit format before they start OP-15.

As students work on OP-13, you can informally assess their understanding of order of operations if you resist the temptation to jump in too quickly to explain or clarify. Students can use calculators to check their answers and discuss differences. (Both scientific and graphing calculators perform the standard order of operations.) You'll need to have enough calculators on hand—at least one calculator per student pair. Take this opportunity to emphasize the importance of calculators in this course.

OP-13. Five 4s When you see a mathematical expression such as $4 + 7$, it is clear what is to be done—simply add the two numbers. When several operations are used in an expression, however, the situation is more complicated and the results could vary depending on the order in which the operations are performed.

a. Each of the following expressions uses five 4s and involves more than one operation. Use *mental math* (solving the problem in your head) or pencil and paper to calculate a value for each expression:

$$4 + [4(4) - 4] \div 4 \quad 4 + 4(4 + 4) - 4 \quad (4 \cdot 4 + 4 \cdot 4) \div 4$$

$$4(4 + 4) \div (4 + 4) \quad 4 + 4 - (4 + 4 \div 4)$$

7, 32, 8, 4, 3

b. Compare your results in part a with those of other group members. For those expressions in which your results are not the same, compare the order in which the operations were performed.

Students should continue on to OP-14 as you circulate among them, making suggestions and assessing what they know and don't know. Save discussion of the order of operations with the whole class until after OP-14 or until the end of the class session.

OP-14. Order of Operations Imagine two people sitting at the movies waiting for the feature to start when on the screen flashes the expression

$$3 \cdot 12 + 2 \cdot 10$$

They both calculate the result but do it in different ways: One does both the multiplications first and then adds the results, while the other first multiplies 12 by 3, then adds 2, and finally multiplies the result by 10. Compare the results. Are they the same? The second movie goer argues that his way is correct because when there is more than one operation to do, the rule is to work from left to right. The first insists that her result is correct because the rule is to do all the multiplications first.

a. Follow the two ways in which the expression $3 \cdot 12 + 2 \cdot 10$ was evaluated by writing down each step and the result. Which answer is correct? Explain how you know. **$36 + 20 = 56$, $(36 + 2) \cdot 10 = 38 \cdot 10 = 380$**

b. Working with expressions that contain more than one operation such as $3 \cdot 12 + 2 \cdot 10$ or $7 + 2(3 + 15) - 26$ can be confusing. Just as there are rules of spelling, grammar, and punctuation to make communication clear in English, there are conventions (or rules) in mathematics to make it clear what value a numerical expression should have (or represent). Some of these rules for the order in which arithmetic operations are to be done are probably familiar to you.

With your group, write out the rules for the order of operations. To do this, you can start by filling in the blanks in the following paragraph using words from the list at the left.

addition

division

grouping symbols:
( ) or { } or [ ]

left

multiplication

powers

right

subtraction

Order of Operations Rules

First, simplify each expression within the _____.

Second, compute the _____, if there are any.

Third, do the _____ and _____ in

order from _____ to _____.

Finally, do the _____ and _____

in order from _____ to _____.

grouping symbols, powers, multiplication, division, left, right, addition, subtraction, left, right

If students ask whether they can use exponents in the Five-Digit Problem, OP-15, they may so long as the exponent is the class digit.

$$1+1-(1\div 1)(1) \qquad \frac{22}{2+2-2}$$

$$(3-3)(3\cdot 3\cdot 3)$$

$$6+6-6+6\div 6$$

$$4^4+(4+4)(4)$$

$$(5)5-5(5+5)$$

$$7(7+7)-7+7$$

$$(8+8)(8+8-8)$$

$$9\div 9-9+(9-9)$$

 OP-15. The Five-Digit Problem In this problem, you will practice working with order of operations in a collaborative setting. You'll find a record sheet for the Five-Digit Problem in the Resource Pages at the end of your text. Before you start, the class will need to select a single digit between 1 and 9.

Here's the Five-Digit Problem: Write expressions for the integers from 0 to 20 by using *five* copies of the class digit (but no other numbers) along with any mathematical symbols you like, such as $+$, $-$, $\cdot$, and $\div$.

Example: Suppose the class chose to use the digit 4 as in OP-13. Then you could use the 4 five times with division and subtraction to obtain $4 - \frac{4}{4} - \frac{4}{4}$.

This expression would be recorded on the "2" line of your record sheet because $4 - \frac{4}{4} - \frac{4}{4} = 2$.

Example: If Jorge wrote the expression $4 + [4 + (4 \cdot 4)] \div 4$, he would record it on the "9" line of your record sheet. (Check to see why this is the correct line.) ▲

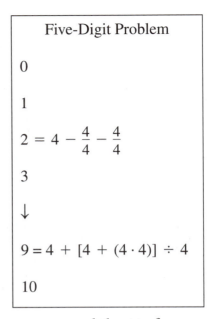

a. Write five-digit expressions that are equal to at least 10 of the integers from 0 to 20. Record each expression next to the corresponding number on your record sheet to form an equation. There may be more than one expression for each number. Use a scientific or graphing calculator to check your answers.

b. Compare your solutions with those of other group members. Record their results on your record sheet also.

For OP-16, have groups record solutions on large Five-Digit Problem charts (described in the materials list at the beginning of the section) posted on walls around the room.

OP-16. Your instructor will post several class record charts for the Five-Digit Problem.

a. Choose a member of your group to record your solutions on the class record charts.

b. Now compare your expressions with those from the rest of the class. Which line has the most different expressions? Which has the fewest? Is there any number for which it is impossible to create a five-digit expression? Write several sentences about your observations.

In OP-17, the form of students' answers—decimal or scientific notation—will depend on the type of calculators each student uses. The results should stimulate some questions and class discussion when students check their homework answers ("Are the answers equal?" "What does the exponent mean?") and provide an opportunity to discuss scientific notation—Why, when, where, and by whom was it developed? Who uses it now, and why?—and perhaps stimulate some library research. In part b we want to pique students' curiosity about large numbers. Later we will tie the large numbers in with follow-up on exponent work. Some students may not know how to write 1 billion in standard decimal form (1,000,000,000). Use this as a chance to quickly list the powers of 10 in standard form.

Important: When students are using a calculator in scientific mode, emphasize that the displayed exponent is a power of 10, not an exponent of the base 1. For example, although the display "1^{12}" looks like "1 to the power 12," it really represents $1 \cdot 10^{12}$. Be prepared to remind students of the calculator's hidden base 10 throughout the course.

OP-17. Dollar Bill Facts According to information from the U.S. Mint, 454 U.S. $1 bills weighs about 1 pound. Use this information and your calculator to answer each of the following questions.

a. About how much does a single $1 bill weigh? *About 0.0022 lb in standard decimal form, or 2.2×10^{-3} lb in scientific notation. If some students' calculators display scientific notation, this gives you an opportunity to talk about what the negative exponent represents.*

b. About how much does 1 billion $1 bills weigh? *about 2,200,000 lb, or about 1100 tons*

EXTENSION AND PRACTICE

OP-18. In Max's class, 6 was chosen as the digit to be used in the Five-Digit Problem.

a. Max wrote the expression "$(6 \div 6 \cdot 6) \cdot 6 + 6$" on the "7" line on the class chart. Do you agree that it should go there? Explain why or why not. *disagree; expression = 42*

b. Joyla wrote the expression "$6 + 6 - 6 + 6 \div 6$" on the "10" line on her paper. However, after a discussion with her group, she erased it and wrote it on another line. On which line does $6 + 6 - 6 + 6 \div 6$ belong? Show how you know. *belongs on the "7" line*

c. By inserting parentheses, ( ), change the order of operations in the expression

$$6 + 6 - 6 + 6 \div 6$$

to obtain a result of 10. *$6 + 6 - (6 + 6) \div 6$*

OP-19. On the Algebra Tool Kit sheet, find the topic *order of operations*. Complete that entry with any information or examples that will help you remember the order in which you should make calculations when combining several operations in an expression.

NOTE The 🖩 symbol indicates that you should do the problem without using a calculator.

OP-20. Complete each of the following computations without using a calculator. Write each answer in a complete statement; for example, $(-12)(-3) = 36$. (To refresh your understanding of the arithmetic of positive and negative integers, refer to Appendix A.)

a. $-15 + 7$ *−8*   b. $8 - (-21)$ *29*   c. $\left(\frac{-3}{8}\right)\left(\frac{16}{15}\right)$ $\frac{-2}{5}$

d. $-9 + (-13)$ *−22*   e. $\frac{-1}{4} - \frac{3}{4}$ *−1*   f. $\frac{5.2}{-0.4}$ *−13*

g. $12 + 15 \div (-3)$ *7*   h. $\frac{-3 - 4}{7}$ *−1*   i. $3 - 2(-5)$ *13*

NOTE Problems marked with a symbol require you to use a calculator.

 OP-21. Use a calculator to check your work in OP-20. To enter a negative number, use either the change-sign ⊞ key, or the negation ⊟ key, but *not* the subtraction ⊟ key. For example:

To enter −9 using the change-sign key, first enter "9" and then press ⊞.

To enter −9 using the negation key, first press ⊟ and then enter "9".

OP-22. Calculator Tool Kits Like an Algebra Tool Kit, a **Calculator Tool Kit** can be a useful way to keep at your fingertips a set of instructions for using your calculator. Even the manual that came with your calculator can be difficult to understand—so, write the entries in your Tool Kit carefully, so that they *make sense* to *you*!

On the Calculator Tool Kit sheet from the Resource Pages at the end of the text, you will see several calculator keys. Add helpful notes about the *change-sign* ⊞ key or the *negation* ⊟ key to your Calculator Tool Kit.

As you need to use other keys in the next few chapters, you will fill in the other boxes.

OP-23. Which of the calculations in OP-20 were the most challenging for *you* to complete without a calculator? *Answers will vary.*

OP-24. Imagine you are a volunteer in a fifth-grade class. Molly is starting to learn about exponents and asks you, "Why does $2^3 = 8$?" How would you respond? Write your answer in a complete sentence. *Possible response: "2^3 is the same as $2 \cdot 2 \cdot 2$. How did I know how many 2s to write? What does $2 \cdot 2 \cdot 2$ equal?"*

OP-25. Big Spender

a. If you spent $1 million at the rate of $1000 per day, how many days would it take you to spend it all? About how many years would that be? Show how you know. *1000 days, or about 2 years and 9 months*

b. How many days would it take to spend $1 billion at the rate of $1000 per day? About how many years would that be? Show how you know. *1,000,000 days, or about 2740 years*

OP-26. Copy and solve these Diamond Problems:

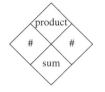 a. 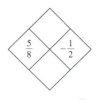 b. c. d.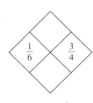

a. $N = -\frac{5}{16}$, $S = \frac{1}{8}$

b. $-2, 3$

c. $E = -4$, $S = -7$

d. $N = \frac{1}{8}$, $S = \frac{11}{12}$

OP-27. Describe in a complete sentence how the north and south diamonds in a Diamond Problem are related to the east and west diamonds. *The east and west numbers are multiplied to get the north number, and added to get the south one.*

OP-28. Reread the Note to the Student, focusing on the section titled "Asking Questions and Doing Homework." You have probably noticed by now that the homework in this book is different from homework in other math books. In this course, "homework" consists of class work, working (and perhaps struggling) with old and new ideas at home, and asking questions to further your understanding.

Write two or three questions that come to mind about the exercises you've done so far. Be prepared to discuss your questions with your group or with the whole class at the next class meeting.

1.3 TILING WITH SQUARES: AREA AND PERIMETER

SECTION MATERIALS

SMALL SQUARES FROM SETS OF ALGEBRA TILES OR CUT FROM CENTIMETER GRID PAPER RESOURCE PAGE (NINE SQUARES FOR EACH PAIR OF STUDENTS) FOR OP-29: DOT PAPER RESOURCE PAGE

"What's My Rule?" (A 10-Minute Chalkboard Game)

We use "What's My Rule?" games to introduce the idea of making a table, finding a pattern, and writing a rule in symbols. This game, which helps develop the concept of variable as a generalized number, is a useful way to start a day's class.

Before the start of class, draw a game board on the chalkboard that looks like this:

Input	3	1	5	2	7	-1	$\frac{1}{2}$		-2		
Output	6	2			14		1		$\frac{1}{2}$	3	

Notice that the entries in the top row of the game board are not in the usual order (that is, increasing from left to right). This arrangement is deliberate, because we want students to pattern vertically, from the input values in the top row to the output values in the bottom row. If the input values were in numerical order, it would encourage students to pattern horizontally and thus miss the whole point of the game—using patterns to discover and write a functional rule. Also, we suggest leaving some columns blank at first, because you may need to add more clues during the game.

The first time the class plays "What's My Rule?" have the students work quietly with a partner. Ask them to raise their hands if they can fill in a part of the game board. Hand the chalk to a volunteer who will then go to the board, write his or her answer in the chart, and then wait at the board until he or she hands off the chalk to the next volunteer. Try to get as many different students involved as possible by limiting each to one entry. If the game moves slowly because only a few students see the pattern, then add some more input-output pairs to the chart to provide more hints.

Tell students you'll quietly erase answers that don't fit the rule you have in mind. Be sure to do so as unobtrusively as possible so that the "error" is not associated with the person who wrote it. Students may also discreetly erase and replace incorrect answers.

After completing the numerical table, ask students to pair off and share with each other their answers to the question, "How did you compute the output?" ["multiply x by 2," or "double x," and so on]. After pairs discuss, solicit all the computation methods the class used. Then ask, "What's my rule?" and get volunteers to state it in a number of ways. Finally ask, "What if the input were x?" as you write x for the input in the column farthest to the right. Write the response (in this case, 2x or x + x) below it.

Play another round of the game with students working in pairs. Try $y = 5x$ or $y = 3x + 1$, depending on the class's skills in seeing patterns. Finish the game with a round of questioning as you did for the previous game.

Wrap up the game by telling students that you'll play more games in the next few weeks to develop skills in finding patterns and describing rules using variables.

Problem OP-29 provides an opportunity to informally assess student understanding of the concepts of perimeter and area, two notions used throughout the course. Problem OP-30 formally introduces the concept of area. Students will use small square tiles from commercial algebra tiles set (or squares cut from the Centimeter Grid Paper Resource Page), nine tiles for each pair of students. Before students start, it would be a good idea to show correct and incorrect configurations for five or six tiles and discuss the area and perimeter of the figures formed. You may also want to show their dot-paper representations. Allow students time to explore with the tiles, although some students may prefer to create their configurations directly on paper rather than using tiles first. The dot paper will help them draw their configurations more quickly and accurately.

 OP-29. Area and Perimeter You used the concept of perimeter when you were solving the Building Corrals problem. In this problem you'll examine how the area and perimeter vary among configurations that you form using nine square tiles.

a. Working with a partner, arrange nine congruent square tiles so that they touch along entire sides but don't overlap. Then draw the arrangement on dot paper.

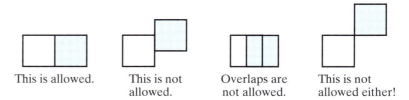

This is allowed.   This is not allowed.   Overlaps are not allowed.   This is not allowed either!

b. Make at least five more different configurations (so you have at least six configurations in all). Each configuration should use all nine squares. Draw each arrangement on dot paper or graph paper, and indicate the area and perimeter of each configuration next to its drawing.

c. What are the largest and smallest perimeters you found? What are the largest and smallest areas you found? Did any of the figures have an odd number as its perimeter? *12 cm ≤ P ≤ 20 cm; A = 9 cm²; no*

d. Compare your configurations with others in your group. Are they the same? What conclusions can you make about the perimeters and areas of your group's configurations? Write your observations using complete sentences. *Although the areas are all the same, the perimeters may differ.*

OP-30. Area In the previous problem, you dealt with the mathematical concept of **area**.

To find the area of a figure such as a triangle, a rectangle, or a circle, determine the number of **square units** within the figure.

Example 1: The area of this irregular hexagon (a six-sided polygon) can be found by counting the number of square units to get

8 square units. ▲

Example 2: The area of this triangle can be found by adding the number of square centimeters to get

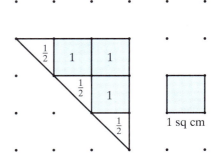

$\frac{1}{2} + 1 + 1 + \frac{1}{2} + 1 + \frac{1}{2} = 4.5$ square centimeters. ▲

Use the dot grid to find, or estimate, the area of each of the following figures.

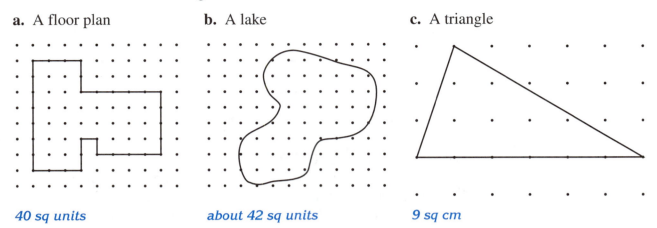

a. A floor plan

40 sq units

b. A lake

about 42 sq units

c. A triangle

9 sq cm

OP-31. The King's Reward In an old tale, a king wishes to reward his humble advisor with a special gift in appreciation of his years of able assistance. The advisor argues that he needs no compensation other than the satisfaction of knowing his work has pleased his master. However, at the insistence of the king, the advisor finally agrees to accept one penny on the first day of the month, two pennies on the second day, four pennies on the third day, and so on, for 30 days. Each subsequent day the advisor is to receive twice as many pennies as he did the previous day. How many pennies will the advisor receive on the 10th day? the 20th day? the 30th day? Show how you know. *On the 10th day he will receive $2 \cdot 2 \cdot 2 \cdot 2 \cdot 2 \cdot 2 \cdot 2 \cdot 2 \cdot 2 = 512$ pennies; on the 20th day, 524,288 pennies; and on the 30th day, 536,870,912 pennies.*

OP-32. Notation: Base, Exponent, and Exponential Form The basic operation in arithmetic and algebra is addition. Moreover, using addition repeatedly with the same positive integer leads to the operation we call multiplication. In OP-6, you might have explained to Max, the inquisitive third grader, that we can think of multiplication of positive integers, such as $3 \cdot 5$, as repeated addition of the same integer: $3 \cdot 5 = 5 + 5 + 5$.

In a similar way, we can generate a third operation, *exponentiation,* out of repeated multiplication by the same number. For example, in OP-31 to find the number of pennies the advisor received on the 10th day, you could have written

$$2 \cdot 2 \cdot 2 \cdot 2 \cdot 2 \cdot 2 \cdot 2 \cdot 2 \cdot 2,$$

which takes a lot of time and space. To be more efficient, you could instead write this repeated product in **exponential form** as

$$2^9,$$

which is read

"2 raised to the power 9"   or   "2 to the ninth power."

Section 1.3 Tiling with Squares: Area and Perimeter **21**

We call 2 the **base** and 9 the **exponent**. The process of finding a repeated product such as $2 \cdot 2 \cdot 2 \cdot 2 \cdot 2 \cdot 2 \cdot 2 \cdot 2 \cdot 2$ is called **exponentiation**. The numeric value of 2^9 is 512.

a. Identify the base and the exponent for the repeated product $7 \cdot 7 \cdot 7 \cdot 7 \cdot 7$, and then write it in exponential form. *base is 7; exponent is 5; 7^5*

b. Write 64 as a repeated product and then in a corresponding exponential form. Identify the base and the exponent. *There are several possibilities. Encourage those who show $8 \cdot 8$ to write it another way, such as $4 \cdot 4 \cdot 4 = 4^3$ with base 4, exponent 3; or $2 \cdot 2 \cdot 2 \cdot 2 \cdot 2 \cdot 2 = 2^6$ with base 2, exponent 6.*

EXTENSION AND PRACTICE

OP-33. While trying to figure out how much carpet and baseboard to buy for the new family room addition to her house, Jana drew this floor plan.

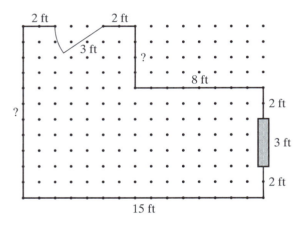

a. Copy the plan on dot paper and label the lengths of the sides. Compute the length of the missing sides. Then use the plan to calculate the area of the floor and the amount of baseboard Jana will need to buy for the room. (Baseboards are not needed in the doorway or on the fireplace!) *11 ft, 4 ft; baseboard = 46 ft, area = 133 sq ft*

b. Add information and examples about *area* to your Tool Kit.

OP-34. Rewrite each of the following powers as a product, and compute the numeric value.

a. 8^3 **b.** 1^8 **c.** 7^1 **d.** 3^5 **e.** 10^4
 512 1 7 243 10,000

f. Which is larger: 2^6 or 6^2? Explain how you know. *2^6 is larger: $2^6 = 64$, but $6^2 = 36$.*

OP-35. Add explanations and examples for *base*, *exponent*, and *exponential form* to your Tool Kit.

The focus of OP-36 is for students to practice using the meaning of integer exponents. We don't expect them to use memorized exponent rules, which are so often misused, until later.

OP-36. Use the meaning of integer exponents to rewrite each of the following expressions with a single exponent. For example,
$$4(4^2) = 4 \cdot (4 \cdot 4) = 4^3.$$

a. $(10^3)(10^4)$ *10^7* b. $(5^2)(5^2)(5^2)$ *5^6*

c. $(7^3)(7^2)(7)$ *7^6* d. $(5^2)^3$ *5^6*

OP-37. Rowan and Martin were interrupted in the middle of playing "What's My Rule?" and left the following game board on the table:

Input	10	6	13	−1	6	−5	0	9	$\frac{1}{2}$
Output	29	17	38	−4	17	*−16*	*−1*	26	$\frac{1}{2}$

Look for a rule that fits the data, and then copy the table on your paper and use the rule to complete it. Finally, using the words "input" and "output," write the rule you found in a complete sentence. *To find the output, multiply the input by 3 and then subtract 1.*

OP-38. Variables To generate the clues for this "What's My Rule?" game board, Giovanna used the following rule: To find the output, double the input and then add 6.

Input	2	13	0	−3	138	−5	10	16	−21
Output	10	32	6	0	282	*−4*	*26*	*38*	*−36*

a. Copy the table and then use Giovanna's rule to complete it. Use any integer you like in place of "?".

b. It is often more convenient to write mathematical rules in symbols than in words. To do this for "What's My Rule?" pick a letter, say x, to represent the input value. Now, apply the rule to x just as you did the other input values:

Input	Apply the Rule		Output
2	2(2) + 6	=	10
13	2(13) + 6	=	32
138	2(138) + 6	=	282
x	2(x) + 6	=	$2x + 6$

So, Giovanna's rule could then be written in symbols as $2x + 6$. Because the x in a symbolic rule can be replaced with a variety of numbers, it is called a **variable**. Although x is probably the letter most often used as a variable, there is nothing special about it—any letter could be used. It is the use of variables that extends the power of algebra beyond arithmetic and makes it such a valuable mathematical tool.

Add information about *variables* to your Algebra Tool Kit.

c. Look at OP-37 again. Choose a variable to represent the input, and then write the rule in symbols. *3x − 1*

OP-39. Representing Multiplication You have already seen that there are several ways to represent multiplication when using numbers: To represent "3 times 5," you could write (3)(5), 3(5), (3)5, or 3 · 5. The parentheses or the dot are necessary to avoid confusion—if you omit the symbols for multiplication, you get a different meaning: "35" means "thirty-five" rather than "3 times 5."

With variables, it is not necessary to use parentheses or a dot to represent multiplication. For example, to represent "3 times x," you could write $(3)(x)$, $3(x)$, $(3)x$, or $3 \cdot x$ as you did with "3 times 5," or you could simply write $3x$.

Write in words what each of the following expressions means:

a. $10x$
10 times x

b. $\frac{1}{2}bh$
half of b times h

c. $-2 + 7N$
−2 plus 7 times N

d. Add information about how to *represent multiplication* with variables to your Algebra Tool Kit.

OP-40. Write an expression to represent each of the following quantities:

a. The product of 3 and some number. *3x*

b. Twice Emily's age, if Emily's age is a. *2a*

c. The total cost of purchasing n folding chairs that cost $22 each. *22n*

24 Chapter 1 Openings: Data Organization

OP-41. Evaluating Algebraic Expressions In expressions such as $2x + 6$, the variable x represents any number you choose, just as it did in Giovanna's "What's My Rule?" game in OP-38. You could think of the variable as a "generalized" or "generic" number. If you pick a specific number for the variable, say $x = 7$, you can find a numeric value for the expression by replacing the x with 7:

$$\text{Evaluating } 2x + 6 \text{ for } x = 7,$$
$$\text{you get } 2(7) + 6 = 14 + 6 = 20.$$

Or if $x = -5$, the value of $2x + 6$ is

$$2x + 6 = 2(-5) + 6 = -10 + 6 = -4.$$

Evaluate each of the following expressions for the indicated value of the variable:

a. $4x + 7$ for $x = 9$ **43**

b. $2R + 3R^2$ for $R = 8$ **208**

c. $-x + 2$ for $x = 5$ **−3**

d. $-4N$ for $N = -1$ **4**

e. The process of replacing the variable(s) in an expression with a specific number is often called **substituting for x** or **evaluating an expression**. Add information about *evaluating an expression* to your Tool Kit.

OP-42. Copy and complete each of these Diamond Problems:

a.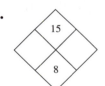

3, 5

b.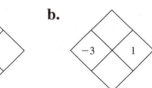

$N = -3$,
$S = -2$

c.

$E = 2$
$S = -3$

d.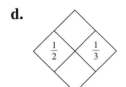

$N = \frac{1}{6}$

$S = \frac{5}{6}$

e.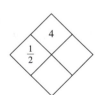

$E = 8$

$S = 8\frac{1}{2}$

f.

$N = 2 \cdot a$ or $2a$

$S = a + 2$

1.4 USING GUESS-AND-CHECK TABLES TO SOLVE PROBLEMS

Start off with another game of "What's My Rule?" only this time explain that it will be done in total silence. You might try $y = x^2$, in which case before the start of class, draw this game board:

Input	3	1	5	2	7	−1	4		−2		
Output	9	1			49		16			0	

Instead of asking questions after the numerical table is complete, simply insert x in the table and silently offer the chalk to a volunteer.

Silent board games such as this can be used throughout the semester to make constructive use of the last few minutes of a class period.

Demonstrate the use of guess-and-check tables for solving problems by modeling the example (OP-43) at the overhead projector or chalkboard as you ask guiding questions and record students' responses. We find it helpful to use two colors, one for the number guessed (whenever it appears in the table) and the other for the rest of the table numbers.

Many students are reluctant to risk guessing. Emphasize the notion that no guess is wrong, because all guesses give us information that leads to a solution. You can even avoid looking for a good first guess, by requiring students to always use the same number (10, for example) for the first guess for each problem.

*We suggest that students make a minimum of three guesses for each problem, even if they find the solution with fewer guesses. (Some instructors require four guesses.) We use this approach because we want students eventually to be able to use the patterns in the tables to write equations. With fewer than three guesses, students do insufficient patterning to develop the skill of generalizing and representing their patterns. If you require students to make at least three guesses right from the start, the transition they make in Chapter 3 to using a guess-and-check table to write an equation will be much easier and make more sense. It is also essential that students carefully identify and label **every** column in the table, including the initial guess that will later be represented by a variable, and the check of whether their result is too large, too small, or just right. Let groups set up the tables in their own ways, provided each table is complete and clearly labeled.*

The word problems that follow give students practice in setting up guess-and-check tables and using them to solve problems. This work, if done thoroughly, will set the stage for writing equations.

Important: *OP-59 is an essential homework problem, since the data from it will be used in Section 1.5.*

You've probably done it before—guessed what an answer is, and then used clues to figure out if you were right or wrong. In algebra, the idea of systematic guessing and checking is extremely important because it provides a first step to seeing patterns and mastering the power of algebra and solving problems.

Organizing guesses in a *systematic* way is the key to making the transition from trial-and-error methods to algebraic methods of solving problems. An effective way to organize guesses is to make a table. To be effective, the table must be

- well-organized,
- as detailed as possible,
- neat and easy to understand,
- complete.

The organization of the table is important. Be meticulous about setting up clear and complete tables. Using the table to detect patterns in the columns is the key to making the transition to the more abstract skill of writing an equation for the problem. To have enough information to detect patterns, you need to have the data from *lots* of guesses.

Keep this in mind as you are learning to set up guess-and-check tables: your goal is not only to solve the problem at hand, but also to develop the skill in recognizing patterns that will make it easier for you to write equations later.

NOTE Problems that are especially important are marked with a ⚷ symbol. They are designed either to help you develop understanding of mathematical ideas or processes, or to help you consolidate understanding. Pay careful attention to these problems, and be sure to revise your work when necessary. Many of these problems will give you ideas for Tool Kit entries.

OP-43. Using a Guess-and-Check Table to Solve a Problem Copy the following example in your notebook:

Example: The length of a rectangle is 3 centimeters more than twice the width. The perimeter is 45 centimeters. Use a guess-and-check table to find how long and how wide the rectangle is.

Before starting a table it would be helpful to draw and label a picture:

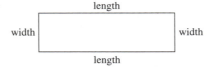

STEP 1 **Decide which quantity to use for a guess:** Start a table. Label the first (far left) column as your "Guess" along with a description of what you are guessing, as in the table below. Why is the width a reasonable thing to guess? *The length is described in terms of the width.*

STEP 2 **Guess a number:** Make a guess—any guess—for the width. It does not need to be a good guess; any number will do. The idea is to get started. Our goal is to look for patterns, not just to get the answer. (It is often useful to guess "10" because calculations with 10 are easy to do.)

Guess: Width			
10			

STEP 3 **Develop descriptions for the remaining columns:** To test your first guess, write down all of the steps you take to work through the problem. These steps will help you write column descriptions in words. *It is essential that you carefully identify and label each column, including the Guess and the Check columns.*

Column 2 Label the next column "Length." Using the width 10, how could you calculate the length of the rectangle?

Guess: Width	Length		
10	$2 \cdot 10 + 3 = 23$		

Column 3 Label the third column "Perimeter." Find the perimeter, and write out your calculations as shown. (Remember, rectangles have two "width" sides and two "length" sides.)

Guess: Width	Length	Perimeter	
10	$2 \cdot 10 + 3 = 23$	$2 \cdot 10 + 2 \cdot 23 =$	

The last column Label the far right column "Check," and include a description of what you are checking. Check the perimeter against 45 centimeters, and identify it as "correct," "too high," or "too low."

Guess: Width	Length	Perimeter	*Check:* Perimeter = 45?
10	$2 \cdot 10 + 3 = 23$	$2 \cdot 10 + 2 \cdot 23 =$	66 (too high)

STEP 4 **Make another guess:** Use clues from the Check column to try to make a better guess.

Guess: Width	Length	Perimeter	*Check:* Perimeter = 45?
10	$2 \cdot 10 + 3 = 23$	$2 \cdot 10 + 2 \cdot 23 =$	66 (too high)
5			

Step 5 **Test your new guess:** Calculate the quantities in each column to see if your guess is correct.

Guess: Width	Length	Perimeter	Check: Perimeter = 45?
10	$2 \cdot 10 + 3 = 23$	$2 \cdot 10 + 2 \cdot 23 =$	66 (too high)
5	$2 \cdot 5 + 3 = 13$	$2 \cdot 5 + 2 \cdot 13 =$	36 (too low)

Using the clues from the Check column to make better and better guesses, repeat the process until you find the correct answer.

Step 6 **Complete the solution:** Be sure to check your answer in the original word problem. Then write your answer to the problem in a complete sentence. For example,

Check: $2 \cdot 6.5 + 2 \cdot 16 = 45$

Answer: The width is 6.5 centimeters and the length is 16 centimeters.

OP-44. Here's an example to help you set up a guess-and-check table to solve another problem. Read through it carefully.

Duke cleared the cash register of nickels and dimes. There were 22 coins in all, and the total value of the coins was $1.45. How many of each type of coin were in the cash register?

Step 1 **Decide which quantity to use for a guess:** Before starting a table, you need to identify the quantities in the problem that might be useful as a guess; that is, decide which quantities could be used to build toward a solution to the problem. (In this case you could guess the number of nickels, or you could guess the number of dimes.) From all the possible candidates, decide which one to use—each table can only have one Guess column. Label the first column "Guess," and underneath it write a description of what you are guessing:

Guess: Number of Nickels	

Step 2 **Guess a number:** Make an easy guess for the number of nickels. It is often useful to guess "10" because calculations with 10 are easy to do.

Guess: Number of Nickels	
10	

STEP 3 **Develop descriptions for the remaining columns:** Continue to develop the columns of the table by writing down each step you take to work through the problem. Each step identifies a new column. Be sure to label each column with a description of what it represents.

Column 2   Ask yourself, "How could I calculate how much of the total value comes from 10 nickels?" Decide which unit of measure to use, cents or dollars. (If you choose cents, how should the total value of the coins be represented?) Don't forget to label the column with a description in words.

Guess: Number of Nickels	Value of Nickels
10	10(5¢) = 50¢

Columns 3 and 4   Add two more columns by asking yourself, "How many dimes are there?" (Use the fact that there are 22 coins in all, and you've guessed that 10 of them are nickels.) Then ask, "How much are the dimes worth?" Label each of these columns with words.

Guess: Number of Nickels	Value of Nickels	Number of Dimes	Value of Dimes
10	10(5¢) = 50¢	22 − 10 = 12	12(10¢) = 120¢

Column 5   Assuming there were 10 nickels, calculate the total value of the coins, and label that column.

Guess: Number of Nickels	Value of Nickels	Number of Dimes	Value of Dimes	Total Value of Coins
10	10(5¢) = 50¢	22 − 10 = 12	12(10¢) = 120¢	10(5¢) + 12(10¢) =

The last column   Now you're ready to check whether the guess was correct: Does the total value of the coins equal 145¢? Be sure to label the Check column and include a description. Identify the result as "correct," "too high," or "too low."

Guess: Number of Nickels	Value of Nickels	Number of Dimes	Value of Dimes	Total Value of Coins	Check: Total = 145¢?
10	10(5¢) = 50¢	22 − 10 = 12	12(10¢) = 120¢	10(5¢) + 12(10¢) =	170 (too high)

a. Copy the last version of the table from Step 3 onto your paper.

b. Make another guess. To check your new guess, calculate the quantities in each column. Using the clues from the Check column to make better and better guesses, repeat the process until you find the answer. Check your answer in the original word problem. As the final step, write your answer in a complete sentence. *7 dimes and 15 nickels*

Problems OP-45 through OP-48 are group problems. Provide each group of students with an overhead transparency and a marking pen to write up how they solved one of the problems. Then have volunteer groups share their solutions with the whole class.

Guide to Setting Up Guess-and-Check Tables

Before you start to make a guess-and-check table for a problem, read the problem and answer these questions for yourself:

1. What is the problem asking me to find?

2. What part of the problem could serve as a guess and be used to generate more information?

When setting up your guess-and-check table, remember the following guidelines:

- Only one column (usually the far-left column) should be labeled "Guess."

- Label each column, including the Guess column, with a description of what it represents.

- The first number you choose as a guess should be one that will make the calculations easy, such as 10 or 100.

- Make at least three (3) guesses, even if you solve the problem in fewer guesses.

- As you fill in the columns to check a guess, write out explicitly how you make the computations. This will be essential later when you look for a pattern in order to write an equation.

- Only one column (usually the far-right column) should be labeled "Check." It holds a place for the desired result of all the computations.

- Check your answer in the original problem.

- To complete the solution, record these two important pieces of information:

    the final *check equation*

    your *answer* written in a complete sentence

Solve each of the problems OP-45 through OP-48 by using a guess-and-check table. If you need help, look again at OP-43 and OP-44, and review the "Guide to Setting Up Guess-and-Check Tables."

OP-45. The cover of Dina's algebra text is a large rectangle. The width of the cover is $\frac{2}{3}$ its length and the perimeter of the cover is 90 centimeters. What are the dimensions of the cover? *18 cm by 27 cm*

OP-46. Drew keeps only quarters, dimes, and nickels in her coin bank. The bank contains the same number of dimes and nickels, and it has twice as many dimes as quarters. If there is a total of $3.85 in the bank, how many nickels are there? *14 nickels*

*Problem OP-47 uses the term **consecutive even numbers**. For students who may not be familiar with the concepts **even**, **odd**, and **consecutive**, both the student answer section and the glossary contain definitions and examples.*

OP-47. Find two consecutive even numbers whose sum is 142. *70 and 72*

OP-48. Admission to the WaterWorld Water Slide and Wave Pool Park is $3.95 for children aged 6 to 11 years, and $8.95 for people 12 years and older. On an unusually cold day in July, the park was nearly deserted, and there were only 66 more paid admissions at the higher rate than at the children's rate. When the receipts were counted at the end of the day, they totaled a meager $1326. How many visitors paid for each type of admission on that cold day? *57 admissions at $3.95, and 123 at $8.95*

EXTENSION AND PRACTICE

OP-49. Use a guess-and-check table to solve the following problem. Refer to the "Guide to Setting Up Guess-and-Check Tables" that precedes OP-45.

Jorge says, "I'm thinking of two positive integers. One of the numbers is five more than the second number. The product of the numbers is 3300. What two integers do I have in mind?" Solve Jorge's problem. *55 and 60*

OP-50. Use a guess-and-check table to solve the following problem:

The perimeter of a triangle is 76 centimeters. The second side is twice as long as the first side. The third side is 4 centimeters shorter than the second side. How long is each side? *16, 32, and 28 cm*

OP-51. Use a guess-and-check table to solve the following problem:

At the start of the Saturday morning farmers' market, Dianne has 27 more bouquets of freesias than bouquets of daffodils to sell. In all, there are 301 bouquets. How many bouquets of freesias does Dianne have to sell? *137 bouquets of daffodils, 164 bouquets of freesias*

OP-52. Raising a Number to a Power on a Calculator

a. A simple four-function calculator has keys for the basic four math operations:

$$\boxed{+} \quad \boxed{-} \quad \boxed{\times} \quad \boxed{\div}$$

Describe a way to calculate 2^3 using only the multiplication $\boxed{\times}$ and equals $\boxed{=}$ keys.

b. To compute 2^3 quickly with a scientific or graphing calculator, use the exponentiation $\boxed{y^x}$ key by pressing this sequence of keys:

$$\boxed{2} \quad \boxed{y^x} \quad \boxed{3} \quad \boxed{=}$$

(On some calculators the exponentiation key is $\boxed{x^y}$ or $\boxed{\wedge}$, but all three keys work the same way.)

The exponentiation key can be used to calculate other exponential values too. Show how you could use it to calculate 3^6, and then write out the keystrokes for your calculator as shown above. *729*

c. In your Calculator Tool Kit, write the keystrokes or examples that will be useful to you in understanding and remembering how to use the *exponentiation key* ($\boxed{y^x}$, $\boxed{x^y}$, or $\boxed{\wedge}$) on *your* calculator.

OP-53. How Far Away Is the Storm? You can calculate how far you are from a thunderstorm if you know the number of seconds it takes the sound of thunder to reach you once the lightning appears. A formula that describes that relationship is

$$D \approx \tfrac{1}{5} S,$$

where the variable S represents the number of seconds between the appearance of the lightning and the sound of the thunder, and D represents the distance in miles that you are from the lightning. Notice from the formula that the distance in miles is approximately one-fifth of the number of seconds. How far away is the storm if you hear the thunder 18 seconds after you see the lightning flash? *3.6 mi*

OP-54. Write each of the following numbers in exponential form with an exponent greater than 1:

a. 8 *2^3*   b. 121 *11^2*   c. 81 *9^2 or 3^4*

d. 243 *3^5*   e. 49 *7^2*   f. 125 *5^3*

OP-55. Felix and Oscar are trying to figure out what $(7^2)^3$ means.

Oscar says, "We've got seven squared to the third power. That's seven to the fifth, right?" Felix replies, "No, you've got to use the meaning of the exponent. It means to use 7^2 as a factor three times." So Oscar, thinking he has got it, writes

$$3 \cdot 7^2$$

"No!" Felix shouts, "Seven squared as a factor three times means repeating the 7^2, like this:

$$(7^2)^3 = 7^2 \cdot 7^2 \cdot 7^2."$$

"Oh," says Oscar, and as Felix breathes a sigh of relief, Oscar finally writes

$$(7^2)^3 = (7 \cdot 7)(7 \cdot 7)(7 \cdot 7) = 7^6.$$

Use what Oscar learned, along with the meaning of integer exponents, to do the following exercises:

a. $(3^4)^2$ *3^8* **b.** $(3^4)(3^2)$ *3^6* **c.** $(2^3)^3$ *2^9*

d. $(2xy)^3$ *$8x^3y^3$* **e.** $2(xy)^3$ *$2x^3y^3$* **f.** $(10^2)(10^3)$ *10^5*

OP-56. Skills Practice Without using a calculator, decide whether each equation is true or false. If false, rewrite the equation with the correct result.

a. $4 + 5 \cdot 6 = 54$ *F, 34* **b.** $2 - (-3) + -17 = -12$ *T*

c. $6 \cdot 6 - 6 \cdot 6 = 0$ *T* **d.** $4 + 2 \div (-1) = -6$ *F, 2*

e. $4 + -2(3) \div -3 = 6$ *T* **f.** $2^3 - 15 = 1$ *F, -7*

g. $\frac{2}{3}(3) = 6$ *F, 2* **h.** $-4(\frac{1}{4}) = -1$ *T*

OP-57. Use the digit your class chose in OP-15 to write five more five-digit expressions that will result in the numbers 21 through 25.

OP-58. You've been working in groups for several class meetings now. As you get to know each other, you will probably develop your own style of working together. What works for one group may not for another. However, it may help you develop as a group to consider methods that have worked for others.

Some groups collaborate effectively when different people in the group assume different roles and no person occupies a particular role for more than a few minutes. To help you identify the four roles, here are examples of the kinds of questions asked and statements made by people in each role. As you read them, think about the ways your group has been working together to solve problems.

Questioner: "What are we supposed to do?" "I don't understand." "Where do I find the paper clips we need?" "What does ____ mean?"

Organizer: "I'll get the paper and colored pencils; you get the rulers." "You do the first two parts of the problem, and I'll do the next two parts." "All right, let's get this chart made."

Prober: "Hey, wait a minute! Maybe there's a pattern in this table." "What if we put it in this order?" "Maybe this works for all of them." "Can we write an equation?"

Summarizer: "Let's look at what we've done." "Remember to label the axes." "Do you have all three pages together, Jennifer?" "Did we answer all the questions?" "Don't forget—we need to write our solution in a sentence."

Choose a problem that your group has worked on in class, and recall how your group worked together. Identify the problem you've selected, and then use complete sentences in parts a and b below to describe what happened in your group while you were solving the problem:

a. Which of the roles described above did you take on? Which was the most comfortable for you, and why? How about the other members of your group?

b. In your group, what aspects of solving the problem worked well? What could you do to help the group work more effectively?

Important: The data from OP-59 will be used in Section 1.5, so it is an essential homework problem.

OP-59. Measure your height to the nearest inch, and record your shoe size. You will need this information for the next class meeting.

1.5 PREDICTING SHOE SIZE: MAKING A GRAPH FROM DATA

SECTION MATERIALS

FOR OP-60:
FLIP-CHART GRID PAPER FOR THE CLASS PREDICTING SHOE SIZE GRAPH STICKY DOTS (20 EACH OF TWO DIFFERENT COLORS; ORANGE AND GREEN SUGGESTED, BUT NOT REQUIRED)

FOR OP-62: GRAPH PAPER

About the Materials:

Prior to the class doing the Predicting Shoe Size problem, OP-60, you'll need to make a large set of coordinate axes on flip-chart grid paper for a class graph. To easily accommodate the sticky dots, scale the axes in 1-inch increments. Label the horizontal axis "Height (inches)" and the vertical axis "Shoe Size." Don't foreshorten the axes (by starting the Height axis at 49 inches, for example). Such shortcuts at this time will contribute to students' misunderstanding. Be meticulous in labeling the scales, as shown below.

Write these instructions above the graph:

"Place a dot above your height to indicate your shoe size."

Below the instructions place a colored dot (e.g., orange) labeled "women" and a dot of another color (e.g., green) labeled "men" to make a legend for the graph.

For OP-60, give each student a sticky dot, one color for women and another for men as indicated on the graph. Let students place their dots on the graph on their own with no further instructions. After this is done, ask the class to identify features of the graph such as origin, horizontal axis, *and* vertical axis, *adding these labels to the graph as they do so. Discuss what it means to have a consistent scale on each axis. Point out the title of the graph. We suggest leaving the class graph posted on the wall throughout the term to remind students of the features that should be included on every graph.*

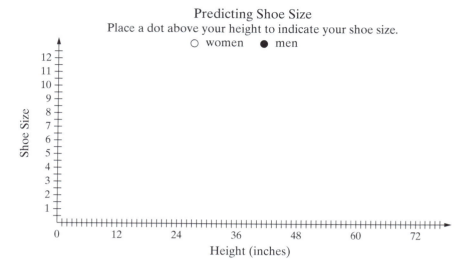

OP-60. On the coordinate grid posted on the wall, place a sticky dot above your height to indicate your shoe size.

For OP-61, working in groups, each student will record the class data from OP-60 onto his or her own table. One relatively efficient way to do this is to ask one student to come up to the class graph and read off the coordinates of the points aloud to the class. After the first few readings, pause and let the students discuss the best way to organize their papers before continuing. Having a student in the role of reader may help to uncover faulty placements of points in the class graph.

Another, perhaps more expedient, way to do this is to have students put their entries (height and shoe size) in a table you set up on the chalkboard, or one that they propose as a way to best organize the class data. Students will then use this table to construct their own tables.

OP-61. Predicting Shoe Size Although he has very little retail experience, a recently retired football player has decided to open an athletic shoe store, Lou's Shoes. Lou has no idea how many pairs of shoes of each type to order, but he believes that taller people wear larger shoe sizes and shorter people wear smaller sizes. Lou wonders if he could figure out a person's shoe size by estimating his or her height. If so, he could base his orders of new lines of shoes on the heights of his customers.

To investigate his theory, Lou decides to survey customers by asking each one to plot his or her shoe size and height on a graph he has taped to the wall; the result is the class graph generated in OP-60. After looking at the graph, Lou thinks it would also be helpful to have a table representing the same data.

a. Discuss how the table should be set up and how your group could make effective use of its members so that each one can produce an accurate table of the data presented on the graph.

b. Carry out your group's plan from part a so that each group member makes a table that represents the data from the Predicting Shoe Size graph. Verify the entries in your table with other group members.

OP-62. Use the table you made in OP-61 to make your own graph of the class data. On graph paper, scale the horizontal (Height) axis so that one tick mark represents 1 inch of height. On the vertical (Shoe Size) axis, let each tick mark represent half a shoe size. Each scale must be consistent; that is, tick marks on an axis must be equally spaced and the distance between marks should represent the same amount.

Parts b and c of OP-63 may stimulate a brief class discussion of the concept of average. Later in the course, students will draw lines of "best fit" for graphs to allow more precision in their predictions for paired data.

OP-63. Use your Predicting Shoe Size graph to answer the following questions:

a. What is the range in shoe sizes—that is, what are the smallest and largest shoe sizes represented? What are the shortest and tallest heights represented?

b. What is the average height of students in the class? Explain and show how you found the average.

c. What is the average shoe size of students in the class? Explain and show how you found the average.

d. What differences, if any, do you notice between the data for males and the data for females?

e. Which height is represented most frequently on the graph? Calculate the average shoe size for classmates of this height. Could there be a student whose shoe size is equal to the average? Why or why not? *Probably not, since the average shoe size is not likely to be a multiple of $\frac{1}{2}$.*

OP-64. In OP-61, Lou's idea was to use a Predicting Shoe Size graph to predict a customer's shoe size based on his or her height.

a. The 1997 inaugural season of the Women's National Basketball Association brought many professional women basketball players back to the United States, including Lisa Leslie, a 6-foot 5-inch forward for the L.A. Sparks. Explain how you could use the Predicting Shoe Size graph to predict Ms. Leslie's shoe size. *Possible response: Find 6 ft 5 in. on the Height axis, then go vertically to where the data are clumped, and finally go horizontally to the Shoe Size axis to read the size.*

b. In over 120 years of thoroughbred horse racing history, only 11 horses have won the Triple Crown, which consists of the Kentucky Derby, the Preakness, and the Belmont Stakes. The most recent Triple Crown winner, Affirmed, was ridden in 1978 by 5-foot 2-inch jockey Steve Cauthen. Explain how you could use the Predicting Shoe Size graph to predict Mr. Cauthen's shoe size. *Possible response: Find 5 ft 2 in. on the Height axis, then go vertically to where the data are clumped, and finally go horizontally to the Shoe Size axis to read the size.*

OP-65. Recall Lou's theory in OP-61 about shoe size and height. Based on your Predicting Shoe Size graph, is his theory correct? Can he predict a person's shoe size if he knows the person's height? Explain your response in complete sentences. *Most students will say "no," because for any given height, shoe sizes vary. Some may say Lou can make general predictions.*

EXTENSION AND PRACTICE

OP-66. The 1997 (Fortieth Anniversary) edition of *The Guinness Book of World Records* recognizes Matthew McGrory, a West Chester University student from Pennsylvania, as the living person with the biggest feet in the world. At age 19, McGrory was 7 feet 4 inches tall.

a. Use your Predicting Shoe Size graph from OP-62 to predict McGrory's record-winning shoe size. (You may need to extend the axes on your graph.) *Answers will vary.*

b. *Guinness* reports McGrory's shoe size to be 26. How close was your estimate in part a? (McGrory's shoe size beats the previous record, which was held by Haji Mohammad Alam Chamma of Pakistan. The 7-foot $7\frac{1}{4}$-inch Chamma would wear a size-24 American shoe and is the coholder of the world's record height.) *Answers will vary.*

OP-67. Frank Zippa and his wife Celeste have three children: Moon, Komet, and Star. Use the information from the following graph to describe each member of the Zippa family according to height and age:

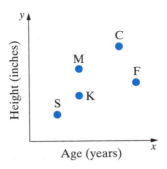

Frank is oldest, Celeste is tallest. Moon and Komet are twins, but Moon is taller than Komet. Star is the youngest and the shortest.

OP-68. Use a guess-and-check table to solve the following problem. State your answer in a complete sentence.

A precinct reports that 327 people voted in an election. The number of Republican voters is 30 more than one-half the number of Democrats. How many Republicans and how many Democrats voted in that precinct? *198 Democrats, 129 Republicans*

OP-69. Use a guess-and-check table to solve the following problem. State your answer in a complete sentence.

Find four consecutive integers such that the product of the two largest is 74 more than the product of the two smallest. *17, 18, 19, 20*

OP-70. Fifteen-Percent-Off Sale! Using the formula $S = P - P(0.01r)$, you can calculate the sales price S if you know the regular price P and the discount percentage r. How much will a mountain bike cost if it is regularly priced at $350 but it is on sale at 15 percent off? *$297.50.*

OP-71. Skills Check Copy each of the following problems and use mental math to compute the result:

a. $(-4)(7)(0)$ *0*
b. $\frac{4}{3} \cdot \frac{9}{20} - (-3)$ *$3\frac{3}{5}$*
c. $(5 \cdot 5 - 5) \cdot 5$ *100*
d. $2^1 \cdot 2^3$ *16*
e. $14 + (-2)(-50) \div (-50)$ *12*

OP-72. Joe has never worked with integers before. Explain to him your rules for doing each of the following problems. Answer in complete sentences.

a. $-4 + (-2) = -6$
b. $3 + (-7) = -4$
c. $2 - 5 = -3$
d. $3 - (-2) = 5$

Possible solution for part a: When adding two negative numbers, add the absolute values of the numbers and make the answer negative.

OP-73. Use the meaning of positive-integer exponents to match each expression in column A with equivalent expressions in columns B and C. Write the three forms on your paper as in the example.

Example: $(6x^3y)^2 = (6 \cdot x \cdot x \cdot x \cdot y)(6 \cdot x \cdot x \cdot x \cdot y) = 36x^6y^2$ ▲

A Expression	B Expanded Form	C Simplified Form without Parentheses
a. $(6xy)^2$	$(6 \cdot x \cdot x)(6 \cdot x \cdot x)(6 \cdot x \cdot x)(6 \cdot x \cdot x)$	$6x^8$
b. $(6x^2)^4$	$(x \cdot x)(6 \cdot x)(6 \cdot x)$	$36x^3$
c. $6(x^2)^4$	$(6 \cdot x \cdot y)(6 \cdot x \cdot y)$	$36x^4$
d. $(6x)(6x^2)$	$(6 \cdot y)(6y \cdot y)$	$36xy^2$
e. $(x^2)(6x)^2$	$(6 \cdot x)(6 \cdot x \cdot x)$	$36x^2y^2$
f. $(6x)(6y^2)$	$6(x \cdot x)(x \cdot x)(x \cdot x)(x \cdot x)$	$1296x^8$

a. $(6xy)^2 = (6 \cdot x \cdot y)(6 \cdot x \cdot y) = 36x^2y^2$
b. $(6x^2)^4 = (6 \cdot x \cdot x)(6 \cdot x \cdot x)(6 \cdot x \cdot x)(6 \cdot x \cdot x) = 1296x^8$
c. $6(x^2)^4 = 6(x \cdot x)(x \cdot x)(x \cdot x)(x \cdot x) = 6x^8$
d. $(6x)(6x^2) = (6 \cdot x)(6 \cdot x \cdot x) = 36x^3$
e. $(x^2)(6x)^2 = (x \cdot x)(6 \cdot x)(6 \cdot x) = 36x^4$
f. $(6x)(6y^2) = (6 \cdot x)(6 \cdot y \cdot y) = 36xy^2$

OP-74. Oscar and Felix were working on the previous problem together and got into a heated discussion. Felix claims that Oscar is sloppy in his work, and that he should have included an extra step in each equation.

Oscar's Work: $(6x^3y)^2 = (6 \cdot x \cdot x \cdot x \cdot y)(6 \cdot x \cdot x \cdot x \cdot y) = 36x^6y^2$

Felix's Work: $(6x^3y)^2 = (6 \cdot x \cdot x \cdot x \cdot y)(6 \cdot x \cdot x \cdot x \cdot y)$
$= (6 \cdot 6) \cdot (x \cdot x \cdot x \cdot x \cdot x \cdot x) \cdot (y \cdot y) = 36x^6y^2$

Explain why Felix thinks the extra step is needed, and also explain why it's true. *It's true because multiplication is both commutative and associative, though we wouldn't expect students to name the properties. An acceptable answer might be "Perhaps Felix likes to show all steps. It's true because the order in which numbers are multiplied doesn't affect the result."*

OP-75. Express three numbers between 30 and 40 using the class digit in the manner of the Five-Digit Problem (OP-15).

1.6 THE NATIONAL DEBT AND TERAFLOPS: LARGE NUMBERS AND SCIENTIFIC NOTATION

As a preliminary to OP-76, you could start off with another silent version of "What's My Rule?" Try one of the following rules: $y = 3x + 1$, $y = x^3$, or $y = 10x - 1$.

OP-76. The National Debt In 1995, the U.S. gross national debt was about $4.92 trillion. That's a fairly large number—in **standard decimal form,** with all the zeros included, it looks like this:

$$\$4,920,000,000,000.$$

Not only is this tedious to write out; because it's 13 digits long, if you tried to enter it into a calculator, it wouldn't fit in a standard 10-digit display. Long before calculators were common tools, scientists and mathematicians decided they needed a convenient way to express very large (and very small) numbers. The method they devised, called *scientific notation*, is based on the fact that our place-value system of writing numbers uses powers of 10. Here is what 10^n turns out to be for several values of the exponent n:

$10^1 = 10$ ten
$10^2 = (10)(10) = 100$ hundred
$10^3 = (10)(10)(10) = 1000$ thousand
$10^4 = (10)(10)(10)(10) = 10{,}000$ ten thousand
$10^5 = (10)(10)(10)(10)(10) = 100{,}000$ hundred thousand
$10^6 = (10)(10)(10)(10)(10)(10) = 1{,}000{,}000$ million
$10^9 = (10)(10)(10)(10)(10)(10)(10)(10)(10) = 1{,}000{,}000{,}000$ billion
$10^{12} = (10)(10)(10)(10)(10)(10)(10)(10)(10)(10)(10)(10) = 1{,}000{,}000{,}000{,}000$ trillion

a. How does the number of zeros in 10^n seem to relate to the exponent n?
 10^n has n zeros for $n \geq 0$.

b. Using powers of 10, 4,920,000,000,000 could be written as

$$4{,}920 \cdot 1{,}000{,}000{,}000 \text{ or } 4{,}920 \cdot 10^9.$$

It could also be written as

$$4.92 \cdot 1{,}000{,}000{,}000{,}000 \quad \text{or} \quad 4.92 \cdot 10^{12}.$$

Indeed, there are many ways to express 4,920,000,000,000 as a number times a power of 10. Find two more ways to express the national debt as a number times a power of 10. *Some possibilities: $49.2 \cdot 10^{11}$, $492 \cdot 10^{10}$, $4{,}920{,}000 \cdot 10^6$*

c. Only one of the numbers written in part b is in *scientific notation,* namely $4.92 \cdot 10^{12}$. A number is in **scientific notation** if it is written in the form

$$A \cdot 10^n$$

where A is a number that is at least 1 but less than 10, and n is an integer. Here are three more numbers written in both standard decimal form and in scientific notation:

Standard Decimal Form		Scientific Notation
4,920,000,000,000	=	$4.92 \cdot 10^{12}$
1995	=	$1.995 \cdot 10^{3}$
263,000,000	=	$2.63 \cdot 10^{8}$

Add this information about *standard decimal form* and *scientific notation* to your Algebra Tool Kit.

OP-77. Displaying Scientific Notation on a Calculator

a. Follow your calculator's instruction manual (or ask another group member) to learn how to use scientific notation on your calculator. Experiment with 4,920,000,000,000, 1995, and 263,000,000. What does the calculator display? Copy the following table, and fill it in with what your own calculator displays. *Answers will vary with calculators; possible answers shown.*

Standard Decimal Form		Scientific Notation	Calculator Display
4,920,000,000,000	=	$4.92 \cdot 10^{12}$	*4.92^{12}*
1995	=	$1.995 \cdot 10^{3}$	*1.995^{3}*
263,000,000	=	$2.63 \cdot 10^{8}$	*2.63^{8}*

b. In your Calculator Tool Kit, explain the *shorthand notation* your calculator uses *to represent scientific notation*, and describe how to enter a number in scientific notation in your calculator.

c. Debra's calculator showed $\boxed{5.34\ ^{3}}$ when she was figuring the cost of her vacation in Kenya. What number is represented by the display $\boxed{5.34\ ^{3}}$? Write the number in standard decimal form and in scientific notation. *5,340, $5.34 \cdot 10^{3}$*

OP-78. In 1995 the U. S. population was about 263 million. If everyone in the United States had given the government $19,950, how close would their total contribution have been to paying off the national debt? Write the problem in scientific notation, and then use your calculator to solve it. (Reread OP-76 to find the rest of the data you need for this problem.) *The contributions would total about $5.25 trillion, more than enough to pay off the debt, leaving almost $330 billion in overpayments.*

OP-79. Heavy Debt In 1995 the U. S. population was about 263 million and the gross national debt was about $4.92 trillion.

a. If all the people had been required to contribute equal amounts to pay off the national debt, about how much would each person's share have been? Use your calculator to find out. (To make sure your answer is correct, use what you know about order of operations and the ( ) keys on your calculator. Also be sure your answer makes sense—it should not be larger than the national debt!) Show how you solved the problem by writing down the keys you used. *about $18,707*

b. While she was thinking about the 1995 gross national debt, Sara recalled that 454 U.S. $1 bills weighs about 1 pound. Suppose you paid your share of the 1995 national debt in $1 bills. Could you actually carry all the bills? About how much would one person's share weigh? *One share would weigh about 41.2 lb.*

OP-80. Teraflops When the world's most powerful computer, Teraflops, came fully online in June 1997, it shattered the previous computing speed record by performing at a rate of $1.34 \cdot 10^{12}$ calculations per second. One way to put this speed into perspective is to imagine that all 5.8 billion of the world's people have split a job into parts and are working on a single problem at the same time. If that were possible, how many calculations per second would each person have to do to keep up with the current Teraflops record? Use scientific notation and your knowledge of exponents to make your calculation easier. (By the way, this computer gets its name from the prediction that when it reaches its top speed, it is expected to run at 1.8 trillion calculations per second: *tera* is a scientific prefix that means "trillion," and *flops* is computer jargon for "calculations.") *about 231 calculations per second*

After completing OP-81, if students don't have a sense of how long a kilometer is, they could use calculators to convert the kilometer distances into miles; 1 km ≈ 0.62 mi. The idea is for them to get a concrete sense of 1 billion.

OP-81. More Dollar-Bill Facts According to data from the U.S. Mint, 454 U.S. $1 bills weighs about 1 pound, a stack of 454 $1 bills is about 4.96 centimeters thick, and a newly made $1 bill is 15.6 centimeters long and 6.63 centimeters wide. (There are 100 centimeters in a meter, and 1000 meters in a kilometer.)

Use these data and your calculator to answer the following questions. Show your work and express your answers in both standard decimal form and scientific notation.

a. About how much would 1 billion $1 bills weigh in pounds? *≈ 2,200,000, or $2.2 \cdot 10^6$ lb*

b. If placed end to end lengthwise, about how far (in kilometers) would 1 billion $1 bills extend? *156,000 or $1.56 \cdot 10^5$ km; about 97,000 mi*

c. If placed side by side widthwise, about how far (in kilometers) would 1 billion $1 bills extend? *66,300 or $6.63 \cdot 10^4$ km; about 41,100 mi*

EXTENSION AND PRACTICE

OP-82. Big Bucks Imagine that your rich great-uncle has just died and has left you $1 billion! However, there's a catch: If you accept the money, you must count it for eight hours a day at the rate of $1 per second. Only when you are finished counting is the $1,000,000,000 yours, and then you may start to spend it.

a. What's your gut feeling: Would you accept your great-uncle's offer? Why or why not?

b. At the rate of $1 per second for eight hours a day, about how long would it take to count the billion dollars? What's your reaction to this fact? *about 95 years*

OP-83. Grade Reports Jack, Nancy, Donna, and Lee are all students in Professor Speedi's math class. Professor Speedi, who chairs the math department, has a reputation for being stern yet thorough. Along with informing students of their progress halfway through each semester, she always makes the following graph to help her students see how they might improve. On the graph, each student's midsemester grade is represented in relation to the amount of time spent studying.

- Jack never studies and has a poor test percentage.
- Nancy is very able, but because of her active social life, she seldom studies outside of class. Her test percentage is okay, but could be better.
- Donna has worked a lot both in and outside of class. Her test grades are very good.
- Lee often studies outside of class and has done reasonably well on the tests.

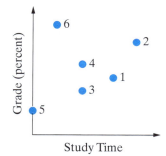

a. Which description corresponds to which point? *Jack = 5, Nancy = 3 or 4, Donna = 2, and Lee = 1. Nancy's position is purposely ambiguous so that students can discuss their interpretations of the relative values "okay" and "reasonably well."*

b. Make up descriptions for the remaining two points on the graph. *Answers will vary. Some possibilities: 6 represents a student who seldom studies but does well on tests; points 3 and 4 correspond to students who study occasionally and get average grades.*

c. Think about your own study habits, and predict your math grade at midterm. Copy the graph, and place a point on it that reflects your prediction and your study habits. Explain your choice of location. *Answers will vary.*

OP-84. Use a guess-and-check table to solve the following problem. State your answer in a sentence.

The state of Wyoming is notable for several "firsts": Wyoming women were the first in the country to obtain the right to vote, and Wyoming voters elected the first woman governor in the United States. Wyoming has the world's largest deposits of sodium carbonate (natrona).

Wyoming's borders form a rectangle, and its area is about 97,109 square miles. If the length of Wyoming is 92 miles more than its width, about how long are the borders? *about 269 by 361 mi*

OP-85. Use the meaning of positive-integer exponents to rewrite each of the following in expressions in exponential form with no parentheses and a single exponent.

Example: $(5x^3)^2 = (5x^3)(5x^3)$
$= 5 \cdot x \cdot x \cdot x \cdot 5 \cdot x \cdot x \cdot x$
$= 5 \cdot 5 \cdot x \cdot x \cdot x \cdot x \cdot x \cdot x = 25x^6$ ▲

a. $(4x^2)^3$ $64x^6$ **b.** $4(x^2)^3$ $4x^6$

c. $(x)(4^2)(4^3)$ 4^5x, or 1024 **d.** $(4x^2)(4x^3)$ $16x^5$

e. $(2x)(3x^2)$ $6x^3$ **f.** $8(xy)^2$ $8x^2y^2$

OP-86. Make each of the following calculations using mental math only (no paper, pencil, or calculator). Then write down the results of each calculation as a complete statement.

Example: $17 - (-12) = 29$ ▲

a. $32 + (-7)$ 25 **b.** $-5 + (-10)$ -15

c. $[-13 + (-12)](-4)$ 100 **d.** $(-8)(-8) - (-8)$ 72

e. $-1\frac{1}{2} + \frac{1}{2}$ -1 **f.** $\frac{1}{2}(-68)$ -34

g. $(1\frac{1}{2})(6)(0)$ 0 **h.** $\frac{3}{4}(100)$ 75

OP-87. Which of the calculations in OP-86 were the most challenging for *you* to do without a calculator? Explain why. *Answers will vary.*

OP-88. Evaluating Algebraic Expressions Evaluate each of the following expressions for the given value of the variable. If you need help getting started, look back at OP-41.

a. $3.14x^2$ for $x = 8$ *200.96*   **b.** $(3.14x)^2$ for $x = 8$ *631.0144*

c. $3T + 2$ for $T = 4.5$ *15.5*   **d.** $\frac{x^4}{8}$ for $x = -10$ *1250*

OP-89. Katja and Lukas enjoy solving Diamond Problems.

a. Lukas claims he's found the world's easiest Diamond Problem: Just put a zero in each of the four diamonds. Explain why this solution fits the Diamond Problem pattern. *It fits because the product of 0 and 0 is 0, and the sum of 0 and 0 is also 0.*

b. Katja thinks a Diamond Problem with all zeros is boring. She wonders if she can create three more interesting Diamond Problems as follows:

- a Diamond Problem that has just one zero in one of the four diamonds *$E = n$, $W = -n$, $N = -n^2$, $S = 0$*

- another Diamond Problem with zeros in just two spaces *$E = 0$, $W = n$, $N = 0$, $S = n$*

- another with zeros in exactly three spaces *None, because zeros in any three diamonds force a zero in the fourth.*

Do you think that any of these are possible? If so, show how. If not, explain why not.

OP-90. Reflections on Group Work Throughout this chapter you have been working in your group daily. Write two paragraphs that address the following questions:

- How have you contributed to the group?

    Have you asked questions?

    Have you helped organize group activity?

    Have you helped to look for patterns?

    Have you kept your group on task?

    Have you been an active contributor, or did you just listen and copy down answers?

    In what ways are you an effective group member?

    How could you have done better?

- How well has your group worked? Explain.

46 Chapter 1 Openings: Data Organization

Problem OP-91 helps students develop patterning and explanation skills. The problem requires learning to predict results without drawing all the figures in the pattern that come between the last one given and the one requested. Allow a fair amount of latitude in how the students describe and explain their solutions. Note that OP-91 is not essential to the development of other mathematical content in the text.

OP-91. Here are the first five figures in a geometric pattern:

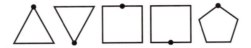

a. Copy these five figures onto your paper.

b. Draw the next four figures that would follow the first five figures.

c. If the pattern continued, how many sides would the 16th figure have? *10*

d. If the pattern continued, where would the dot in the 27th figure appear? *top*

e. Explain, so that a new student in the class would understand, how you determined your answers for parts c and d. *Possible student response: The pattern consists of pairs of figures with the same number of sides. Each figure in a new pair has one more side than a figure in the previous pair. The first figure in the pair has a dot at the top; the second is upside down from the first and has a dot at the bottom.*

1.7 DRIVING TO REDWOOD CITY: DATA FROM A GRAPH

Start off with another silent version of "What's My Rule?" Then set groups to work on OP-92.

When groups have completed OP-92, you can use this opportunity to have a class discussion of the concept **slope of a line** *in terms of a constant rate of change. Students will continue to work toward a formal definition of slope of a line in Chapter 5.*

OP-92. Professor Speedi drove from Ukiah to Redwood City by the route shown on the following map. The graph beside it depicts her trip.

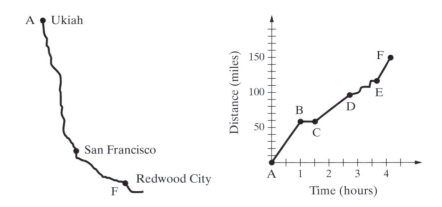

a. How far did Professor Speedi travel on her trip? *150 mi*

b. During which intervals did Professor Speedi drive at a steady rate of speed? *from A to B, from C to D, and from E to F*

c. Without making any calculations, use the graph to determine the answer to the following question: In going from C to D, did Professor Speedi travel faster or slower than she did from A to B? Explain how you know. *She traveled fewer miles in more time—that is, she went slower.*

d. During which interval did Professor Speedi drive at the fastest rate? Explain how you know. *E to F; steepest slope*

e. In the interval from A to B, how fast did Professor Speedi drive? How fast did she drive in the interval from C to D? Explain how you know. *(60 mi)/(1 hr) = 60 mph; (100 − 60 mi)/(2.75 − 1.5 hr) = 32 mph*

f. In the interval from B to C, how fast was Professor Speedi moving? What might explain this? *0 mph; a rest stop*

g. Something different happened as Professor Speedi traveled from D to E. Explain what you think might have happened. *She hit stop-and-go traffic in San Francisco. She is not just going up a bumpy hill.*

Problem OP-93 is a preparatory problem for Professor Speedi's Commute, SP-38, in Chapter 2, in which students will graph the "distance = rate · time" relationship. For now, let students work out the relationship numerically so that it makes sense to them. Don't simply tell them that "d = r · t" to make parts a though d easier. Working through these problems should provide students with the ability to express the relationship in part e.

OP-93. Imagine you are driving south on Interstate 5.

a. If you drive at an average speed of 60 miles per hour for 2 hours, how far will you travel? Show how you know. *120 mi*

b. If you drive at 65 mph for 3 hours, how far will you travel? Show how you know. *195 mi*

c. If you drive at 48 mph for 30 minutes, you will be on the road for only half an hour. How far will you travel? Show how you know. *24 mi*

d. If you drive at 57 mph for 40 minutes, for what fraction of an hour will you travel? How many miles will you travel? Show how you know. *$\frac{2}{3}$ hr, 38 mi*

e. Explain in words how to find the number of miles you've traveled if you know your car's speed and the amount of time you've traveled. Add this explanation to your Algebra Tool Kit. *distance = speed · time*

OP-94. While driving on Interstate 5, Luther Leadfoot usually sets the cruise control at 80 miles per hour. However, as soon as he notices a California Highway Patrol (CHP) car, he drives 64 mph. Recently, Leadfoot drove 456 miles on I-5 to Los Angeles. During the trip he drove at 80 mph for $1\frac{1}{2}$ hours less than the time he drove at 64 mph.

a. Did Luther spend more time driving at 64 mph, or at 80 mph? *at 64 mph*

b. If Luther drove at 64 mph for 5.5 hours and at 80 mph for 4 hours, how far did he drive in all? *(64 mph)(5.5 hr) + (80 mph)(4 hr) = 672 mi*

c. Use a guess-and-check table to find how many hours Luther drove at 80 mph during his recent trip to L.A. State your answer in a sentence. *2.5 h*

For OP-95, students should work in groups. Recommend that they take out a penny, a nickel, and a dime to use while working through the problem. After they do parts a and b in their groups, bring the class back together and have one student from each group, in turn, write a possible outcome on the board or overhead projector. Use this list to stimulate a discussion of part b. Then have students return to their groups to move on to parts c through e.

OP-95. Suppose you toss three coins: a penny, a nickel, and a dime. One possible outcome is for the coins to come up "heads, heads, heads," which we abbreviate as HHH. Another possibility is all tails TTT.

a. What other possible outcomes might occur? Make a list of *all* the different possibilities, including HHH and TTT. *HHH, HHT, HTH, THH, HTT, THT, TTH, TTT*

b. How can you be *sure* that you listed all the possibilities in part a? Describe a strategy for checking your list. *Be systematic—list all possibilities with three heads, then with just two heads; etc.*

Definition: When all outcomes are equally likely to occur, the **probability** (or likelihood) that a particular event will occur, which we denote $P(\text{event})$, is the fraction

$$P(\text{event}) = \frac{\text{number of ways that the event occurs}}{\text{total number of possible outcomes}}.$$

For example, when rolling a fair standard die, the probability of getting an odd number is written $P(\text{odd}) = \frac{3}{6}$.

c. If you randomly toss three coins, what is the probability that exactly one coin will come up heads? $\frac{3}{8}$

d. If you randomly toss three coins, what is the probability that at least one coin will come up heads? $\frac{7}{8}$

EXTENSION AND PRACTICE

OP-96. Here's some more practice with probability.

a. For a standard six-sided die, what is the probability that you will roll a number that is a factor of 30? $\frac{5}{6}$

b. For a standard six-sided die, what is the probability that you will roll a number that is at least 4? $\frac{3}{6}$ or $\frac{1}{2}$

c. In a well-shuffled standard deck of playing cards, what is the probability that you will draw a red card (either a heart or a diamond)? $\frac{26}{52}$ or $\frac{1}{2}$

d. In a well-shuffled standard deck of playing cards, what is the probability that you will draw an ace? $\frac{4}{52}$ or $\frac{1}{13}$

e. Add the definition of *probability* and an example to your Algebra Tool Kit.

OP-97. Car Comparison The graph in the margin describes three cars: car 1, car 2, and car 3.

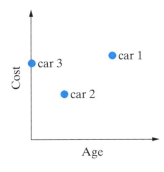

a. Which car is the most expensive? Explain how you know. *1, because its point is highest on the graph.*

b. Which car is cheapest? Explain how you know. *2, because its point is lowest on the graph.*

c. Which car is oldest? Explain how you know. *1, because its point is farthest to the right.*

d. How old is car 3? Explain how you know. *It's brand new because its age is 0.*

In OP-98, the range of a car depends on its fuel capacity and fuel efficiency. The speed refers to maximum highway speed. Expect some discussion about these interpretations.

OP-98. More Car Comparison The following three graphs describe two cars, car A and car B. ("Range" in the third graph means the distance a car can travel on one tank of gas.)

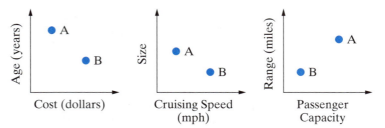

Decide whether each of the following statements is true or false. Explain your reasoning.

a. The newer car is more expensive. *True, because the point that is lower is farther to the right.*

b. The slower car is larger. *True, because the point that is higher is farther to the left.*

c. The larger car is newer. *False, because car A is larger but car B is newer.*

d. The less expensive car carries more passengers. *True, because car A is less expensive and also carries more passengers.*

OP-99. Use the data from OP-81 and your calculator to find approximately how much 1 billion $1 bills would weigh in ounces. Show your work and express your answers both in standard decimal form and in scientific notation. *about 35,000,000 or $3.5 \cdot 10^7$ oz*

OP-100. Use a guess-and-check table to solve the following problem. State your answer in a sentence.

The sum of three numbers is 61. The second number is five times the first, and the third is two less than the first. Find the three numbers. *9, 45, 7*

OP-101. Creating Diamond Problems

a. Create two more Diamond Problems of your own, and solve them on your homework paper. For example:

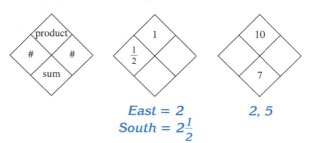

Answers: *East = 2*   *2, 5*
            South = $2\frac{1}{2}$

b. Can you create a Diamond Problem for which all four numbers would be negative? If so, do it. If not, explain why not. *No. When the E and W numbers are negative, the N number must be positive.*

OP-102. Kristina thinks that $(6x^5)^2$ and $6(x^5)^2$ are equal, but you know they are not the same. Clear up her confusion by explaining to her what is being squared, and then showing her, step by step, how to rewrite each expression without parentheses. *$36x^{10}$, $6x^{10}$*

 or

OP-103. Evaluate these expressions. Use your calculator only on those parts for which you really need it.

a. $-50 - 30$ *-80*
b. $4(-9 - 6)$ *-60*
c. $(-13)(-2)$ *26*
d. $-178 - (-3)$ *-175*
e. $\frac{-1}{2} - 2\frac{2}{4}$ *$-3\frac{1}{4}$*
f. $(-2)^2 + \frac{3}{2}[\frac{-1}{2} + (\frac{-1}{6})]$ *3*
g. $2(0 - 6)^2(\frac{1}{2})$ *36*
h. $(0.005) \div (-0.021) + 8.31$ *≈ 8.07*
i. $(2.35)(-4.01)^2$ *≈ 37.79*
j. $(137)(-229)(0)(3)$ *0*

*The summary for Chapter 1 is developed in three problems: OP-104 (Summary Rough Draft), OP-105 (Summary Group Discussion), and OP-126 (Summary Revision). All three are all **very** important and so are marked with the key problem symbol.*

We have broken the summary process into three parts to emphasize its importance and to make it a bit easier for students to handle. For the rough draft (OP-104), students should spend time at home thinking and writing down ideas for what to include in their own summaries. For the group discussion (OP-105), they should come to class prepared to discuss and share with their group what they have written for OP-104. The final problem of the chapter (OP-126) provides a chance for students to revise their written summaries into a form appropriate for inclusion in a portfolio or for use on an exam.

This work will reinforce for students (or, in some cases, begin for them) the process of summarizing information. Students will have this assignment near the end of each chapter. The goal is for students by the end of the course to find this process not only essential but also satisfying. Some instructors allow students to use these summaries as a resource when preparing for quizzes and tests, a practice we recommend. Make sure student summaries have explanations and descriptions along with examples or sample problems.

At this level, students are not yet used to summarizing their learning and still have difficulties expressing themselves in writing. Therefore, most students will probably find the chapter summary problems difficult at first. One way to help students meet this challenge is to have them read the summary problem (which essentially repeats the introduction to the chapter) on the first day of work on that chapter. There are two reasons for doing this: First, it tells the students what they're going to be doing in the chapter, and second, it serves as a guide. Students will know that the problems will be summarized at the end of the chapter, and some may learn to make notes for their summaries as they work through the chapter.

Another idea for helping students understand what you expect in a chapter summary is to show them an example. Ideally this would be one written by a student from a previous year or from another instructor's class, but you may have to make one up yourself. Seeing a sample helps students go in the right direction from the start.

At the beginning of the course, you may have to be fairly directive about how to put together a summary. As students progress through the course and become better able to identify and explain the important ideas from a chapter, you can move into the role of recorder and facilitator, rather than director or guide.

Because students have been doing a lot of problems, at this point they see the chapter as a lot of pieces without necessarily seeing the whole. It's time to pause and reflect: What are the big pieces, and how do they fit together?

OP-104. Chapter 1 Summary Rough Draft In this course, as in most courses you study, you will continue to work on many ideas, concepts, and skills. There is a lot of information in this course. It may be difficult to process this information and organize it for yourself, so that you can easily remember *all* of it. It often helps to summarize it in the context of big ideas.

Writing a summary is an important part of the learning process, and you will be expected to write a summary for every chapter in this course. A well-organized, complete summary will also help you study for quizzes and tests. Therefore, creating a good summary is very important. To help you do this well, we have divided the process into three parts: writing a rough draft (this problem) discussing the main ideas in your group (OP-105) and revising your rough draft (OP-126). (You will encounter this three-part process in every chapter.) Your final presentation in OP-126 should be thorough and neat. For now, though, you need to brainstorm ideas, write them down, and be ready to discuss them with your group at the next class meeting.

These are the main ideas of Chapter 1:

- working collaboratively to solve problems
- collecting and organizing data or information
- using a guess-and-check table to solve word problems
- reading and interpreting graphs

Some ideas should have been review while others may be new to you. On copies of the Summary Draft Template, which is in the Resource Pages (or using the same format on your own paper), do the following for each of the four main ideas of the chapter:

- State the main idea.
- Find and describe a problem from the chapter that you feel illustrates the main idea well.
- Include a completely worked-out solution.
- Write one or two sentences describing each idea.

1.8 SUMMARY AND REVIEW

Play another Silent Board Game of "What's My Rule?" Then get groups started right away on OP-105. Circulate and listen to group discussions of their drafts from OP-104. After groups have finished OP-105, lead a whole class discussion to pull the main ideas together.

OP-105. Chapter 1 Summary Group Discussion Take out your draft of the Chapter 1 Summary from OP-104. For each of the four main ideas of the chapter—

- working collaboratively to solve problems,
- collecting and organizing data or information,
- using a guess-and-check table to solve word problems, and
- reading and interpreting graphs—

choose one member of the group to lead a short discussion. When it is your turn to be the discussion leader, you should

- explain the problem you chose to illustrate your main idea, and
- tell why you chose that particular problem.

This is also a good opportunity for you to make sure your chapter summary is complete—later you will revise it for homework.

OP-106. Use a guess-and-check table to solve the following problem:

Woody is now 28 years old, and Susie is eight years old. In how many years will Woody be twice as old as Susie? *12 yr*

OP-107. Professor Speedi carries only bills in her book bag, no coins. As she raced off to work this morning, she tossed in some ones, fives, and tens. There are eight bills in all, and their total value is $38.

a. How many of each kind of bill are in Professor Speedi's bag? *3 ones, 3 fives, and 2 tens*

b. If Professor Speedi randomly grabs one bill out of her book bag, what is the probability that it is a $5 bill? Explain how you know. *$\frac{3}{8}$*

OP-108. Proxima Centauri After our Sun, the star nearest to Earth is Proxima Centauri, about 4.22 light-years away. (A light-year is the distance light travels in one year.)

a. If light travels at the rate of about 186,250 miles per second, how many miles does light travel in one year? *about $5.87 \cdot 10^{12}$ mi*

b. How many miles from Earth is Proxima Centauri? Show your work, and express your answer in both standard decimal form and scientific notation. *about $2.48 \cdot 10^{13}$ mi*

c. The 1997 NASA Pathfinder mission (which sent a rover named Sojourner to explore the surface of Mars) cruised on its 212-day journey at an average speed of 12,000 miles per hour (relative to the Sun). If a spaceship could travel as fast as 25,000 mph, about how long would it take to reach Proxima Centauri? *more than 113,200 yr*

OP-109. Tool Kit Check Throughout this chapter you've been told to make entries in your Algebra and Calculator Tool Kits. Your Tool Kit should now include entries for at least the following topics:

Algebra

- perimeter *OP-8, 9*
- area *OP-30, 33*
- order of operations *OP-14, 19*
- variables *OP-38*
- base, exponent, and exponential form *OP-32, 35*
- scientific notation *OP-76*
- representing multiplication *OP-39*
- standard decimal form *OP-76*
- how to calculate distance traveled *OP-93*
- evaluating algebraic expressions *OP-41*
- probability *OP-95, 96*

Calculator

- exponentiation key *OP-52*
- changing the sign of a number *OP-22*
- entering number in scientific notation *OP-77*
- calculator representation of scientific notation *OP-77*

a. As part of the process of making sense of the mathematics you've been doing, it's time to check that your Tool Kit entries are clear and accurate. Exchange Tool Kits with members of your group, and read each other's entries to check for clarity and accuracy.

b. At home, make sure that your Tool Kit includes examples and explanations that help you remember and understand the ideas. Make any revisions, clarifications, or updates that will reflect your developing understanding of mathematics.

EXTENSION AND PRACTICE

Problem OP-110 is meant to be used for informal diagnosis of a student's familiarity with solving simple linear equations, a skill that is usually introduced in pre-algebra courses.

OP-110. Without the aid of a calculator, or of paper and pencil, use your knowledge of arithmetic to solve each of the following equations. Then write the equation and its solution on your paper.

a. $x + 3 = 7$ *4*

b. $x + 3 = -1$ *−4*

c. $3x = -12$ *−4*

d. $-24 = -3x$ *8*

e. $x - 3 = -7$ *−4*

f. $10 = x + 14$ *−4*

g. $8x = 4$ *1/2*

h. $0 = 5x$ *0*

In OP-111, the exact locations of A and B cannot be determined. It is their relative positions that matter.

OP-111. Another Car Comparison

a. Copy the pairs of axes shown here. On each graph, label two points that would represent cars A and B as described in the More Car Comparison problem (OP-98).

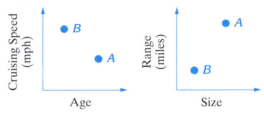

b. Which car would you buy? Why? Explain your answer in a complete sentence. *Car A is older, larger, and slower, and has a greater range relative to car B.*

OP-112. If the class number had been 6, on what lines would each expression go?

a. $6 \cdot 6 + \frac{6}{6} - 6$ *31*

b. $\frac{6 + (6 - 6) \cdot 6}{6}$ *1*

c. $6 + 6^{[(6+6) \div 6]}$ *42*

OP-113. Using the variable N as the input, express each of the following rules in symbols:

a. Add three to the input, and then multiply the result by seven. *7(N + 3)*

b. Square the input, and then divide the result by 4. *(N²)/4*

Solution:

OP-114. Copy the axes that were used for the graph in OP-83, but leave off dots 1 through 6. Now put a dot on the graph to indicate each of the following:

Student A, who studies hard but gets only average grades

Student B, who studies little but gets good grades

Student C, who *never* studied, not even for one minute, and got poor grades

OP-115. In the novel *Casino*, author Nicholas Pileggi describes the count room where money by the ton is bundled, boxed, and stacked. Pileggi claims that, rather than count the daily fortune, the casino sorts the bills and coins by denomination and weighs them.

  a. A quarter is about 0.18 centimeter thick. If the counting room contained $1 million in quarters, and if all those quarters could be stacked in a single stack, about how high (in centimeters) would it reach? (By the way, $1 million in quarters weighs about 21 tons.) *about 720,000, or $7.2 \cdot 10^5$ cm*

  b. Use the data from OP-81 and your calculator to decide approximately how tall (in kilometers) a stack of 1 billion $1 bills would be if the bills were stacked one on top of each other. Show your work, and express your answer in both standard decimal form and scientific notation. *109, or $1.09 \cdot 10^2$ km; students may be interested to know that this is about 68 mi*

OP-116. Use the meaning of positive-integer exponents to rewrite each of the following expressions without parentheses:

  a. $(4x \cdot x \cdot x)(9x \cdot x \cdot x)$ *$36x^6$*
  b. $(2xy)(2xy)(2xy)(2xy)$ *$16x^4y^4$*
  c. $(5^2)(5^2)^3$ *5^8, or 390,625*
  d. $7(xy^2)^3$ *$7x^3y^6$*
  e. $(x^3)(x^5)(x)$ *x^9*
  f. $(-3x^2y)^2$ *$9x^4y^2$*

OP-117. Without using a calculator, compute each of the following:

  a. $(4982)(-556)(0)$ *0*
  b. $(-2)(-2)(-2)(-2) - 2$ *14*
  c. $-12 + (-18) - 15$ *-45*
  d. $\frac{3}{8}[-50 - 9(\frac{-2}{9})]$ *-18*

OP-118. Which of the calculations in OP-117 was the most challenging for *you* to do mentally? *Answers will vary.*

OP-119. Evaluate each expression for the given value of the variable:

  a. $3x^2$ for $x = -4$ *48*
  b. $(5x)^3$ for $x = -2$ *-1000*
  c. $(4x^2)^3$ for $x = -5$ *1,000,000*
  d. $-8 - x$ for $x = 5$ *-13*
  e. $(c - 2)^2$ for $c = 9$ *49*
  f. $7m^3 - 7$ for $m = -1$ *-14*

OP-120. Use a guess-and-check table to solve the following problem. State your solution in a sentence.

A cable 84 meters long is cut into two pieces so that one piece is 18 meters longer than the other. Find the length of each piece of cable. *33 m, 51 m*

OP-121. Use a guess-and-check table to solve the following problem. State your solution in a sentence.

The total cost for a chair, a desk, and a lamp is $562. The desk costs four times as much as the lamp, and the chair costs $23 less than the desk. Find the cost of the chair and the desk. *The chair costs $237; the desk $260.*

OP-122. Copy and solve these Diamond Problems:

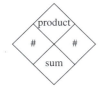

a.

N = −24, S = 2

b.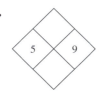

N = 45, S = 14

c.

N = −2a, S = a − 2

d.

−6, −2

e.

$N = \frac{-3}{16}$, $S = \frac{1}{2}$

f.

$\frac{1}{2}$, $\frac{1}{2}$

g.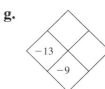

N = −52, E = 4

OP-123. Alice and Leroy were figuring out how much linoleum they would need to cover their kitchen floor. They also needed to know how much baseboard stripping to buy. They sketched a floor plan and wrote their measurements on it. Based on their floor plan, calculate the area and perimeter of their floor. *perimeter = 44 ft; area = 79 sq ft*

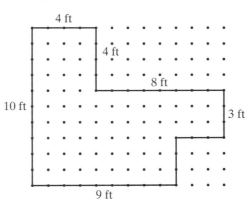

OP-124. Self-Evaluation In this chapter you've worked on the following calculation skills:

- computing with integers
- using the order of operations
- rewriting expressions using the meaning of positive-integer exponents
- working with large numbers and with numbers in scientific notation
- evaluating expressions for particular values of the variables
- using your calculator more efficiently
- figuring out areas and perimeters
- calculating simple probabilities

a. Of these topics, choose one that you understand better now than before you worked through the chapter, and describe what you learned. *Answers will vary.*

b. Write a realistic self-evaluation: Which skills in the list above do you feel confident in performing? Which ones do you still need to work on?

Problem OP-125 is included for practice in patterning skills but is not crucial to developing any of the major math content areas in the text. Finding the position of the dot in each box can be used to raise the issue of examining remainders, in this case from dividing by 4. For example, in the first box, $1 = 4 \cdot 0 + 1$ tells us that the dot will be in the upper left-hand corner of the first box; $2 = 4 \cdot 0 + 2$ indicates a dot in the upper right-hand corner of the second box; and, continuing with this pattern, $15 = 4 \cdot 3 + 3$ predicts a dot in the lower right-hand corner of the 15th box. Thus $n = 4q + r$ tells us that a dot will appear in the corner of the nth box, corresponding to $r = 1, 2, 3,$ or 0.

OP-125. The three figures in the margin are the start of a sequence of squares with a dot in one corner.

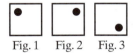

Fig. 1 Fig. 2 Fig. 3

a. Copy the three figures. Then draw five figures to the right of them that could reasonably follow in the sequence. State in a sentence or two the rule for your pattern. *The dot in the corner rotates clockwise 90° in each successive figure.*

b. Without drawing any more figures, determine how Figure 27 should be drawn. Explain how you made your decision so a student new to the class could understand your reasoning. *The position of the dot is related to the figure number and moves clockwise from corner to corner. If a figure number (such as 24) is a multiple of 4, then the dot will be in the lower left-hand corner. Add 3 to 24 to determine that the 27th figure will have a dot in the lower right-hand corner.*

OP-126. Chapter 1 Summary Revision This is the final summary problem for Chapter 1. Using your rough draft from OP-104 and the ideas you discussed in your groups from OP-105, spend time revising and refining your Chapter 1 Summary. Make your presentation thorough and organized, and write it on pages that are separate from the rest of your homework.

CHAPTER TWO—FOR THE INSTRUCTOR

PROFESSOR SPEEDI'S COMMUTE
Patterns and Graphs

MATERIALS

- Graph paper—needed daily
- Chalk, or tape, or rope (about 20 m) to set up outdoor coordinate axes
- Input-output cards and sticky dots in five colors: blue, green, yellow, orange, and red (nine cards and nine dots in each color)
- Poster-size coordinate axes drawn on 1-in.-grid flip-chart paper and colored sticky dots for class graphs
- Punch bowl
- Algebra tiles
- Plastic or hard-boiled egg
- Burning Candle video;
  or birthday candles, a timer, and electronic scales with LED readout;
  or data from the Instructor Notes
- Resource Pages for SP-1, SP-20, SP-21, SP-22, SP-73, and SP-76; also, Grid Sheets, Birthday Problem (optional), and Summary Draft Template

This chapter focuses on the problem-solving strategy of detecting, describing, and using patterns in various situations, including numeric and geometric contexts. Students apply their patterning skills to graphing different types of equations, including linear, quadratic, and exponential. The aim is to strengthen students' patterning skills and reinforce the notion that patterning is an important problem-solving strategy. Indeed, they will use patterning throughout the course to develop concepts and reinforce ideas. Many problems in the chapter also provide opportunities for practicing calculator skills.

In working with a variety of equations and their representative curves, students see that graphs have many different shapes. They also become aware that different types of equations have differently shaped graphs.

The in-class graphing problems can be approached in two ways: For the easier equations, individuals make their own graphs. For the more complicated graphs, each group produces a large, neatly done, appropriately labeled group graph that can be posted on the wall. These can serve to remind students about types of graphs or be used for class discussions. All in-class graphing (whether students produce individual or group graphs) should be done in groups, with the members sharing the work of generating the tables of points to plot. Sharing the work this way really decreases some of the tedium of graphing because each student needs to calculate only two or three y-values rather than eight to twelve. Students working in groups can also compare their individual graphs to make sure they've plotted all points correctly.

Students will start to connect types of equations with general shapes; for example, linear equations produce lines as graphs, and quadratic equations produce parabolas. However, we don't expect them to make any generalizations about families of curves at this point. That realization will come later.

Continue to reinforce the importance of group work and discussion. As often as time allows, have groups present their results and explain their thinking to the class.

Note on Materials: Commercial sets of algebra tiles may be ordered from the Cuisenaire Company of America, Inc., 1-800-273-0338. Algebra tiles for the overhead projector are also available. Alternatively, students can make algebra tiles at home by gluing Algebra Tiles Resource Page to construction paper or lightweight cardboard (cereal boxes work well) and carefully cutting. Store tiles in a resealable plastic bag. (The algebra tiles are, for all practical purposes, "non-commensurable" because no "small" multiple of the length of the side of the big square aligns with any "small" multiple of the length of the side of the small square).

A nine-minute video presentation of a burning candle experiment entitled "Patterns and Functions—An Application of Graphing" produced by San Juan High School, San Juan Unified School District (©1990 by Studio E-3 Productions, SECC/SJUSD) may be ordered by writing CPM Educational Program, 1233 Noonan Drive, Sacramento, CA 95822. The cost is $11, including postage and handling.

CHAPTER TWO

PROFESSOR SPEEDI'S COMMUTE
Patterns and Graphs

In the previous chapter, you encountered a number of problems that asked you to figure out the pattern in a sequence of numbers or shapes. Patterning is an important problem-solving strategy that you will use throughout this course. This chapter introduces you more formally to the idea. You will look for and describe patterns, and use them to solve a variety of problems. As part of the problem-solving process, you will use patterns to make conjectures and see the importance of checking conjectures. You will also apply your patterning skills to making graphs of rules, and will see that some rules have graphs that are lines, while others result in nonlinear curves.

IN THIS CHAPTER YOU WILL HAVE THE OPPORTUNITY TO:

- use the important problem-solving strategy of looking for, describing and using patterns;
- make and test conjectures;
- see how graphs, tables, and rules are related to the pattern or situation they describe;
- graph different kinds of rules to produce lines, parabolas, and other curves.

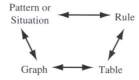

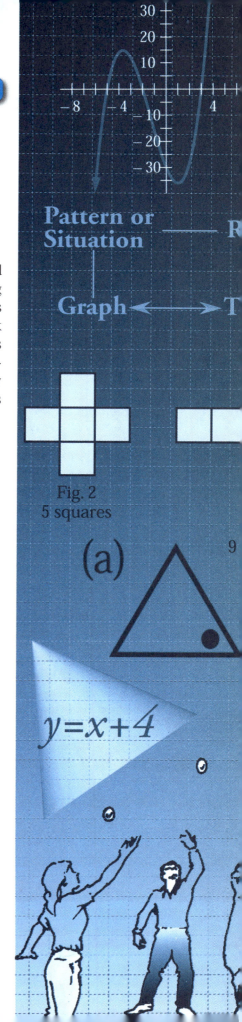

Chapter 2 Professor Speedi's Commute: Patterns and Graphs

MATERIALS	CHAPTER CONTENTS		PROBLEM SETS
Outdoor coordinate axes; colored input-output cards; SP-1 Algebra Walk Observations Resource Page; five sets of poster-size coordinate axes (for SP-2); colored sticky dots (nine each of blue, green, yellow, orange, and red); (SP-2) Grid Sheets Resource Page	2.1	The Algebra Walk: Coordinate Graphing	SP-1–SP-14
SP-20 Area as a Product and Area as a Sum Resource Page; SP-21 and SP-22 Resource Pages; Grid Sheets	2.2	Polygon Perimeter Pattern Area as Product = Area as a Sum	SP-15–SP-35
Poster-size coordinate axes (for SP-38); 10 to 12 sticky dots in one color	2.3	Getting to Work on Time: Situations and Graphs	SP-36–SP-52
Punch bowl, poster-size coordinate axes (for SP-56); sticky dots in one color—at least one per student; Algebra Tiles Resource Page (for SP-72)	2.4	Hawaiian Punch Mix	SP-53–SP-72
SP-73 Resource Page and transparency; algebra tiles	2.5	Slicing a Pizza Algebra Tiles and the Distributive Property	SP-73–SP-89
Plastic or hard-boiled egg; poster-size coordinate axes (for SP-92); sticky dots in one color	2.6	The Egg Toss: A Quadratic Graph "Head Problems"	SP-90–SP-107
Summary Draft Template Resource Page (for SP-122)	2.7	Folding Bills: More Nonlinear Graphs	SP-108–SP-122
Birthday Problem Resource Page; burning candle video (or birthday candle, timer, electronic scale with LED readout, or data from instructors' notes)	2.8	The Burning Candle Investigation (Optional)	SP-123–SP-124
	2.9	Summary and Review More Head Problems	SP-125–SP-153

- Problem Solving
- Graphing
- Writing Equations
- Ratios
- Solving Equations
- Symbol Manipulation
- Reasoning & Communication

2.1 THE ALGEBRA WALK: COORDINATE GRAPHING

SECTION MATERIALS
- OUTDOOR *XY*-COORDINATE AXES
- COLORED INPUT-OUTPUT CARDS
- ALGEBRA WALK OBSERVATIONS SHEETS
- POSTER-SIZE COORDINATE AXES
- COLORED STICKY DOTS
- GRID SHEETS

About the Materials: *The materials listed here are for SP-1 and SP-2.*

The coordinate axes (for the Algebra Walk, SP-1) should be marked off before class. For the two axes, some instructors have used two pieces of rope (about 8 meters for the x-axis and about 12 meters for the y-axis), the units marked by wrapping pieces of color tape in equally spaced intervals. Alternatively, you could use chalk to mark and label the axes on the pavement (or use paint to draw a permanent set of axes on the campus grounds). However you produce large-scale axes, the units should be 24 to 30 in. apart to give students enough room to stand comfortably. Setting up the axes near a hill or a set of stairs makes it easier for student observers to see the graph. In case of poor weather, you could set up the axes inside if you have the space.

For the Algebra Walk, you'll need to make colored input-output cards, with the colors keyed to the equations in SP-1. One way to do this is to copy the instructor resource page—Instructor Resource: Colored Cards for use in Algebra Walk (SP-1)—onto colored paper, and then cut out the individual cards. Alternatively, you could use the Instructor Resource as a guide and write out your own cards. In either case, you'll need nine cards each of blue, green, yellow, orange, and red. Each card should have a different input value—use integers from -4 to 4, inclusive—on the left-hand side, and space for an output value on the right.

Copy the Algebra Walk Observations sheet from the Resource Pages, or make them from your own design.

The poster-size coordinate axes are for class graphs (SP-2) after the Algebra Walk. Make one set of coordinate axes on 1-in.-grid flip-chart paper for each of the five Algebra Walk rules, color-coding the axes and the scale to the rule. Delay writing the rule on the poster until the class discussion following the graphing.

The sticky dots are also for the class graphs (you'll need nine dots each of blue, green, yellow, orange, and red).

Remind students that they may use the Grid Sheet Resource Page for SP-2, but should make copies of the Grid Sheet Template for use in future graphs.

The Algebra Walk (SP-1): Most students will have some experience with the xy-coordinate system from previous classes. Regardless of a student's experience, however, our intention is to have the students learn how the system works by seeing patterns in graphs.

The Algebra Walk gives students an unconventional introduction to the xy-coordinate system. It is a vivid experience that you can refer to throughout the chapter ("Remember when we were outside ... ?"). It is important for students to have an opportunity to graph in the large-scale setting (and for you to have an opportunity to informally assess their graphing skills) before any formal class discussions occur. After the outdoor portion of the activity, students record the results on large coordinate grids that you've posted in the classroom. Use the "human graphing" and the posters as catalysts for class discussion and clarification of any definitions the class needs.

It is important for you to use proper terminology, such as slope *and* intercepts, *but don't expect students to use it right away. Relatively concrete descriptions such as "tilt," "slant," or "uphill/downhill," and "where the line crosses the axis" are acceptable from students. As you continue to use mathematical vocabulary in class discussions, students will become more comfortable with it and gradually start using it themselves.*

Before going outside:

At the start of class you might play one more round of "What's My Rule?" (from Chapter 1) to help students see the connection between inputs, outputs, rules, and their graphs.

Then give each student an Algebra Walk Observations sheet (copied from the Resource Pages, or make your own in a less-structured form), and at least one colored input-output card. Each student will also need a pencil or pen, and a book, binder, or clipboard on which to write while outdoors.

The outdoor graphing and indoor follow-up will be more effective if you go over logistics briefly with the class before going outside. While students follow along, you or a student can read aloud the Instructions for Graphers and the Instructions for Observers in SP-1 and the instructions for plotting sticky dots on the class graphs in SP-2.

Once outside (SP-1):

Position observers near the coordinate grid (on a hill or stairs, if possible) so that they are facing the x-axis and looking toward the positive-y direction. This orientation is important because it corresponds to the standard one we use when graphing on paper.

Call for students with the blue cards to find their places along the x-axis. The graphers should stand with both feet on the x-axis facing the positive-y direction and with their backs to the rest of the class. Start with $y = 2x + 1$, and give the following directions:

Be sure you are standing on the mark that corresponds to the number on your card. This is your **input** *number. Look at the input number. Multiply it by 2, and then add 1. Got it? Write the resulting number on the right-hand side of your card. This is your* **output** *number. When I say "Go," walk that number of spaces forward or backward, depending on the sign of the result. A "space" is the distance between two tick marks on the y-axis. Ready?* **Go!***"*

Mistakes will probably be made. In most cases, students will handle corrections themselves. Resist the urge to manage the activity too much— **once you've taken care of logistical matters, let the students help each other take care of the math.**

While the human graph is forming, you might want to take the opportunity to informally introduce two concepts—slope and intercepts—that will be introduced in a formal way later (intercepts in this chapter, and slope in Chapter 5). To introduce slope informally, you could have a student perform a "slide" (formally introduced in Section 3.6) from their *spot to the next person on the line by walking along imaginary grid lines parallel to the x- and y-axes. For example, for the graph of $y = 2x + 1$, to move from the position (2, 5) to the position (3, 7), a person could move "up" (forward) two steps and then to the right one step. Each person on the graph could use the same slide—namely "up 2, and to the right 1"—to move to the next person's position, a casual exposure to the slope ratio examined later in the course. To introduce intercepts, you might simply have the students at the x- and y-intercepts take turns raising their hands.*

Observers also have an important learning task—to complete the appropriate sections of their Algebra Walk Observations sheets. Be sure the observers are writing down the information they need while the graph is in place.

Follow the above process for each of these rules:

Blue rule:	$y = 2x + 1$	*Double your input number, then add 1.*
Green rule:	$y = -2x$	*Multiply your x-value by −2.*
Yellow rule:	$y = x + 4$	*Add 4 to your input number.*
Orange rule:	$y = -x + 4$	*Multiply your number by −1, then add 4.*
Red rule:	$y = x^2 - 4$	*Subtract 4 from the square of your number.*

Before the group of graphers for $y = x + 4$ (the Yellow rule) leave, ask them to remember their final position because you will have them return to it shortly. After graphing $y = -x + 4$ (the Orange rule), ask the students on the Yellow graph to return, and ask what the observers notice—the point of intersection of the two graphs can't be missed, as two students try to occupy the same spot.

Some instructors use this as an opportunity to prepare students for the concepts of independent and dependent variables by mentioning that once an input value is chosen, the rule determines the output value, so the output "depends" on the input (and the rule).

(If you want to extend the activity further, have a set of students take two steps to their right after they have created a graph of a rule. Ask them what features of the graph change and what features stay the same. This is an intuitive introduction to translations.)

Back in class (SP-2):

As they reenter the classroom, have the students use sticky dots to plot their positions in the human graphs on the large coordinate axes you have posted around the room.

(If poor weather and lack of indoor space prevent your class from doing human graphing, an alternative is to do the activity, as described, entirely on poster-size coordinate grids. However, this replacement activity does not have the same powerful learning impact as the kinesthetic exercise.)

It will be beneficial to spend some time at the beginning of the next class session discussing with the students what they learned from the Algebra Walk and from the follow-up activities. As students draw their own graphs in SP-2, some will probably want to connect the points. Don't discourage them from doing this, but do call their attention to the fact that so far we can only be sure of the points represented in the tables. Throughout the chapter, we are building up to the notion of discrete and continuous graphs, which are described in Section 2.5 (SP-82).

SP-1. The Algebra Walk Many algebra students have had some experience with the xy-coordinate system in previous classes. Regardless of your background, however, the Algebra Walk will add to your understanding of graphing by focusing on graphs from a new and unconventional perspective.

The Algebra Walk is an exercise in "human graphing," in which people represent points on a graph. Your instructor will have marked off a large set of axes on the ground outside your classroom. Before going outside, you will receive at least one colored card marked with an integer (the input number). You will also need an Algebra Walk Observations sheet from your Resource Pages, a pencil, and a book or other hard surface on which to write while outdoors. During this exercise, you will assume two different roles—sometimes a grapher, and sometimes an observer. Graphers follow oral instructions on how to form their graph. Observers have three tasks: to write the graph's rule in words, to sketch what they see, and to describe in words what they see.

Instructions for Graphers: Look at the integer on your card. It is an **input** number. At a signal given by your instructor, find your place on the horizontal axis (also called the x-axis). Stand with both feet on your integer so that you are facing the positive-y direction and your back is to the rest of the class. Your instructor will call out instructions similar to the ones below. Follow the instructions to form a human graph.

"Look at the input number. Multiply it by 2, and then add 1. Got it? Write the resulting number on the right-hand side of your card. This is your **output** number. When I say 'Go,' walk that number of spaces forward or backward, depending on whether your result is positive or negative. (A *space* is the distance between two tick marks on the y-axis.) Ready? *Go!*"

Keep your cards with the input and output numbers on them; you will need them when you return to the classroom.

Instructions for Observers: Find a place where you can see the entire outdoor grid while facing the positive-y direction. Listen to the instructions given to the graphers. Then on your Algebra Walk Observations sheet, do these three things: Write the graph's rule in words, make a *rough* sketch of the graph, and, as clearly as you can, describe in words the shape of the human graph.

Here are the Algebra Walk rules so that you can use them as a check.

Blue rule: Double your input number, then add 1.

Green rule: Multiply your number by -2.

Yellow rule: Add 4 to your number.

Orange rule: Multiply your number by -1, then add 4.

Red rule: Subtract 4 from the square of your number.

After students work in groups to complete SP-2d, ask them to share their observations in a whole-class discussion. If you don't have time to finish the discussion today, continue at the start of the next class session. Students will need their graphs from problem SP-2 to do SP-16 (in Section 2.2) in class.

  SP-2. Algebra Walk Follow-Up When you return from the Algebra Walk, you will see several large coordinate grids posted around the classroom. Find the grid that matches the color of your integer card, and put a matching sticky dot to represent your "point" on the human graph.

Work in groups to complete the following graphs, using Grid Sheets from your Resource Pages. Make each graph on a separate grid. (Save the Grid Sheet Template to make copies to use for future graphs.)

NOTE Graphing can quickly become tedious when there are many points to plot. It is easy if you are working alone to get so bogged down in drudgery that you miss making important observations and connections. Here's a case where "group power" can increase your learning by minimizing tedium and redundancy. Share the work!

---

a. On a Grid Sheet, complete a table for the rule $y = 2x + 1$, then neatly graph each point.

Input (x)	-4	-3	-2	-1	0	1	2	3	4
Output ($y = 2x + 1$)	-7	-5	-3	-1	1	3	5	7	9

b. Now, repeat part a for each of the following rules. Use a new grid for each graph.

		Input							
Rule	−4	−3	−2	−1	0	1	2	3	4
$y = -2x$	8	6	4	2	0	−2	−4	−6	−8
$y = x + 4$	0	1	2	3	4	5	6	7	8
$y = -x + 4$	8	7	6	5	4	3	2	1	0
$y = x^2 - 4$	12	5	0	−3	−4	−3	0	5	12

c. Match each of the five symbolic rules in parts a and b with the words that describe them from SP-1.

d. Compare the five graphs in parts a and b. How are they similar? How are they different? Write several sentences to describe what you notice. *Possible observations: Only one of the graphs (red: $y = x^2 - 4$) is curved, the rest are straight lines. Of the lines, two slant upward (blue: $y = 2x + 1$ and yellow: $y = x + 4$) and two slant downward (green: $y = -2x$ and orange: $y = -x + 4$). Two of the lines (yellow: $y = x + 4$ and orange: $y = -x + 4$) cross the y-axis at the same point. Two of the lines (blue: $y = 2x + 1$ and green: $y = -2x$) have a steeper slant than the other two lines (yellow: $y = x + 4$ and orange: $y = -x + 4$). Only one of the graphs (red: $y = x^2 - 4$) crosses the x-axis in two different points.*

EXTENSION AND PRACTICE

SP-3. The graph shown here provides some information about a soda machine located in the lunch room of a large office building. For two different times during an average day, it shows the number of cans of soda in the soda machine.

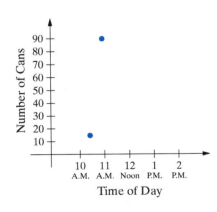

a. Each dot represents particular information regarding the soda machine. Write two specific pieces of information given by each of the two dots. *There are about 15 cans at 10:30 A.M. and about 90 cans at 11:00 A.M.*

b. What might have happened to the soda machine between the times indicated by the two dots? *It was refilled.*

c. Most of the office workers have lunch between noon and 12:45 P.M. How many cans of soda do you think would be in the machine after lunch? Copy the graph onto your paper, and then place a dot corresponding to the number of cans you think would be in the machine at 1:00 P.M. Explain why you placed the dot where you did. *Between 15 and 90 sodas*

SP-4. Coordinate Graphs The Predicting Shoe Size graph you made in Chapter 1, the Algebra Walk graphs (from SP-2), and the soda machine graph (from SP-3) are all examples of **coordinate graphs**. On each of these graphs, for a specific input value—such as height, a given integer, or time of day—there is an associated output value (or values) that depends on the input. The coordinate graph is a useful mathematical tool because it gives us another way—in addition to descriptions, rules, and tables of data—to picture relationships between two numbers.

a. Take out the SP-4 Grid Sheet from your Resource Pages. (You'll use it again in SP-6 and SP-7.) On each grid are two perpendicular number lines, which form the **coordinate axes**. What is a number line? What does it mean for two lines to be perpendicular? (The Glossary at the end of the text may be helpful.)

b. **Labeling the Axes** In graphs that represent data (with or without a rule), such as the soda machine graph, each axis is labeled with the name of the quantity represented on the axis, and the units of measure when necessary. For example, the vertical axis in the Soda Machine Graph is labeled "Number of Cans," and the horizontal axis for the Predicting Shoe Size graph is labeled "Height (inches)."

In graphs that represent mathematical rules, the horizontal axis is the line where we locate input values and is usually called the *x*-**axis**. This is because x is usually chosen to represent the input (or independent variable) in a mathematical rule. The vertical axis is the axis that indicates output values and is usually called the *y*-**axis** because y usually represents the output (or dependent variable). For each pair of coordinate axes on your Grid Sheet, label the horizontal axis with an "*x*" near the far right arrow, and label the vertical axis with a "*y*" near the top arrow.

Whenever you graph, make a habit of labeling the axes—either with the appropriate input and output variables or with the name of the quantity (and units of measure) represented on each axis. In some cases it's helpful to label them with both.

c. **Scaling the Axes** The point where the two coordinate axes cross is called the **origin**. Beginning with zero at the origin, each coordinate axis is **scaled** (numbered) like a ruler so that the numbers increase in size to the right along the horizontal axis and go up along the vertical axis. Scale the axes on your Grid Sheet by labeling each tick mark.

 d. **Consistent Scales** Look at the ruler shown in the accompanying figure. The scale on this ruler is *inconsistent*—each space on the ruler represents a different distance. Why would measuring with this ruler be difficult?

The scale on a coordinate axis, like a ruler, should always be *consistent*—every space should represent the same quantity. Explain what is wrong with the scaling on each of the following coordinate axes. Then describe how you could fix each inconsistent scale.

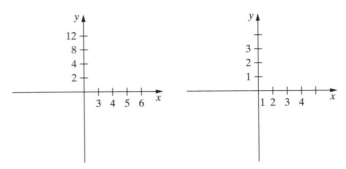

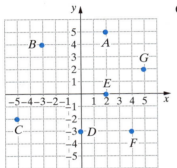

e. **Plotting Points** To locate a point on a coordinate graph, first find the input (or x-value) on the x-axis. Then move vertically up or down, counting units along the way, until you reach the output (or y-value)—just as you did in the Algebra Walk. The two numbers used to locate a point are called the **coordinates** of the point, and are written together as an **ordered pair**, in parentheses. For example, the coordinates of point A in the graph are written (2, 5)—the 2 is the x-value and the 5 is the y-value.

To avoid confusion, the coordinates in an ordered pair are *always* given with the input (x-value) first: (x, y). So (5, 2) describes the point whose x-coordinate is 5 and y-coordinate is 2. Which point in the accompanying graph is represented by (5, 2)? Name the coordinates of all the other labeled points on the grid. G; A(2, 5), B(−3, 4), C(−5, −2), D(0, −3), E(2, 0), F(4, −3)

(Coordinate graphs were introduced in 1637 by René Descartes (pronounced "day-CART"), a French mathematician and philosopher who tried to use mathematics to explain all thought.)

f. Add information about *coordinate graphing* to your Algebra Tool Kit.

SP-5. Making a Graph from Data The label on a bottle of concentrated windshield washer solvent includes the following directions for diluting the concentrate for use as an antifreeze in the washer pumps and lines:

For each quart of water in the windshield washer reservoir, add the following amounts of concentrate to protect your car to the given temperature:

Amount of Concentrate (ounces)	For Protection To
4	22°F
8	10°F
12	0°F
14	−4°F

a. Which of the following four pairs of horizontal and vertical axes would you use to graph the data from the table? Explain your choice in complete sentences. For each of the pairs of axes you did not select, state your reasons for rejecting it. *Best is graph iv because the scales are consistent and of appropriate sizes. Acceptable is graph iii but the vertical-scale increments are too large for accurate representation. The horizontal scale in graph i does not start at 0, and neither scale in graph ii is consistent, so both pairs are unacceptable.*

i.
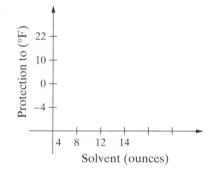

ii.
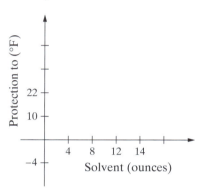

iii.
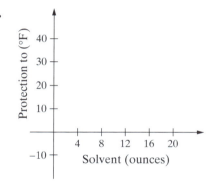

iv.
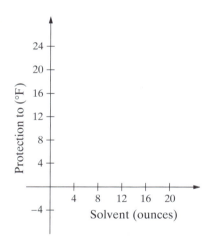

b. Copy the most appropriate pair of axes from part a onto graph paper. Then use it to graph the data points from the table.

SP-6. Follow these steps to make a graph of the rule $y = -x + 2$.

a. On a Grid Sheet copied from your Grid Sheet Template Resource Page (or on your own graph paper), complete the following table using the rule $y = -x + 2$:

Input (x)	-4	-3	-2	-1	0	1	2	3	4
Output ($y = -x + 2$)	*6*	*5*	*4*	*3*	*2*	*1*	*0*	*-1*	*-2*

b. Label the axes on your Grid Sheet and indicate the scale on each axis. Then plot the points from the table in part a on the coordinate axes. Finally, label the graph with its rule.

SP-7. Follow the same steps you did in SP-6, but this time use this rule:

To find y, triple x and add 1.
Output values: $-11, -8, -5, -2, 1, 4, 7, 10, 13$

SP-8. Write a paragraph describing how you think the Algebra Walk will help your understanding of graphing. Be as specific as you can.

SP-9. Mystery Rule The output values in the following table are the result of applying a "mystery rule" to the input values. Copy the table and look for a pattern. Use the pattern to complete the table, and then write the mystery rule in words and in symbols. *Add 4 to the input number.*

Input (x)	Output ($y = $ _____)
3	7
1	5
5	9
2	6
6	10
-10	-6
17	*21*
1000	*1004*
x	*$x + 4$*

SP-10. Reciprocals When Max was in fifth grade, he learned that to find the reciprocal of a number, you just "flip it over." He recently learned the following definition, which is more precise:

Two numbers are **reciprocals** of each other if their product equals 1.

For example, 2 and $\frac{1}{2}$ are reciprocals of each other, since
$$2 \cdot \frac{1}{2} = 1.$$

a. Which of the following pairs of numbers are reciprocals of each other? Check each pair by using the definition of *reciprocal*.

i. 3 and 0.3 *no*   ii. $\frac{-1}{4}$ and -4 *yes*

iii. $\frac{7}{5}$ and $\frac{5}{7}$ *yes*   iv. $1\frac{2}{5}$ and $\frac{5}{12}$ *no*

b. Use the precise definition of reciprocal to explain why $1\frac{1}{3}$ is the reciprocal of $\frac{3}{4}$. *They are reciprocals because* $(1\frac{1}{3})(\frac{3}{4}) = 1$.

c. Use the precise definition of reciprocal to explain why, for any *nonzero* number x, the reciprocal of x is $\frac{1}{x}$. *Because for $x \neq 0$, $(x) \cdot (\frac{1}{x}) = 1$*

d. According to Max's method, what is the reciprocal of $\frac{3}{4}$? Use part c to help explain why Max's method works. *The reciprocal of $\frac{3}{4}$ is $\frac{4}{3}$, which is true, because $\left(\frac{3}{4}\right) \cdot \left(\frac{4}{3}\right) = 1$.*

e. Add the definition of *reciprocal* to your Algebra Tool Kit.

SP-11. Does 0 have a reciprocal? Try $\frac{1}{0}$ on your calculator. What happens? Use the definition of reciprocal to explain why. *Calculator produces an error message—0 times its reciprocal must equal 1, but 0 times any number is equal to 0.*

SP-12. Plot the points $(5, 4)$, $(-3, 4)$, $(-3, -2)$, and $(5, -2)$ on regular graph paper. Use a ruler to connect the points to form a rectangle.

a. What are the length and the width of the rectangle? *8 units by 6 units*

b. What is the area of the rectangle in square graph-paper units? *48 sq units*

SP-13. Use the meaning of exponents to write each of the following expressions using a single exponent:

a. $(4^2)(4^5)$ *4^7* **b.** $(4^3)^2$ *4^6*

c. $4^3 \cdot 4^5$ *4^8* **d.** $4^2 \cdot 4^3$ *4^5*

e. One of the expressions in parts a through d is not like the others. Which one is it? Describe how it is different. *$(4^3)^2$ is different because it represents "something times itself" while the others do not.*

SP-14. Aneka was given this problem about money:

When Alfred emptied the slot machine, there were only nickels and dimes. He noticed that there were twice as many dimes as nickels, and that the total value of the coins was $3.25. How many of each type of coin were in the slot machine?

To solve this problem, Aneka set up the following guess-and-check table, but forgot to label the columns:

Guess: # Nickels	Value Nickels	# Dimes	Value Dimes	Total Value of Coins	Check: Total = $3.55?
10	$0.50	20	$2.00	$0.50 + 2.00 =	$2.50 (too low)
15	$0.75	30	$3.00	$0.75 + 3.00 =	$3.75 (too high)
12	*$0.60*	*24*	*$2.40*	$0.60 + 2.40 =	*$3.00 (too low)*

a. Study Aneka's table for clues to the quantity that each column represents. Copy the table onto your paper. Then identify each column by labeling it with a heading to describe what it represents.

b. Aneka's next guess is 12. Test this guess by filling in the rest of the columns in your table to check it.

c. Finish solving the problem, and write your answer to the problem in a complete sentence. *There were 13 nickels and 26 dimes in the slot machine.*

2.2 POLYGON PERIMETER PATTERN

SECTION MATERIALS

SP-20 AREA AS A PRODUCT AND AREA AS A SUM RESOURCE PAGE

SP-21 AND SP-22 RESOURCE PAGES

GRID SHEET RESOURCE PAGE (FOR SP-23)

Have groups get right to work on completing their Algebra Walk graphs from SP-2. They should compare graphs with each other to check for mistakes, and then complete SP-15 and SP-16. After everyone has completed both problems, it would be worthwhile to lead a brief class discussion to consolidate ideas and address questions generated by the Algebra Walk and its follow-up work.

SP-15. Inputs and Outputs Use your graphs from the Algebra Walk Follow-Up (SP-2) to answer the following questions:

a. How can the graph for the rule $y = 2x + 1$ be used to predict the output value for an input (*x*-value) of 5? How can the graph be used to predict the output value associated with an input of $3\frac{1}{2}$? *Locate 5 on the x-axis, move up vertically to the graph, and then turn toward the y-axis and move horizontally to the number 11. Follow a similar procedure for $x = 3\frac{1}{2}$ to get to $y = 8$.*

b. If you wanted an output of 7 for the rule $y = -x + 4$, what would you need as an input? Use your graph to decide. *−3*

SP-16. Point of intersection We say that two (or more) graphs **intersect** when they meet or cross each other. A point where graphs intersect is called a **point of intersection**.

a. For each of the rules in parts a and b of SP-2, identify the point where the graph intersects the *y*-axis. Write the coordinates near each of these points on your Grid Sheets. *(0, 1), (0, 0), (0, 4), (0, 4), (0, −4)*

b. For each of the rules in parts a and b of SP-2, identify the point (or points) where the graph intersects the *x*-axis. Write the coordinates near each of these points on your Grid Sheets. *$(\frac{-1}{2}, 0)$, (0, 0); (−4, 0); (4, 0); (2, 0) and (−2, 0)*

c. Add information about *point of intersection* of two graphs to your Tool Kit.

SP-17. Polygon Perimeters Pattern Recall from Chapter 1 that the perimeter measures the distance around a two-dimensional figure. The perimeter of the polygon in Figure 1 below is 8 units, and the perimeter of the polygon in Figure 2 is 14 units:

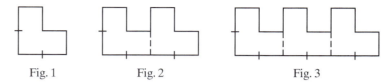

Fig. 1   Fig. 2   Fig. 3

a. Find the perimeter of the polygon in Figure 3. Find the perimeter of the polygon that would be drawn for Figure 4. *20 units, 26 units*

b. If the pattern of forming polygons continued as shown here, what would be the perimeter of the polygon in Figure 7? What would be the perimeter of the polygon in Figure 9? Make a table of data (see SP-18 for a hint about how to set up the table), and look for a pattern. *44 units, 56 units*

c. Explain in a sentence or two how to determine the perimeter of any polygon in this pattern. *Multiply the figure number by 6, and then add 2. Or subtract 1 from the figure number, multiply this result by 6, and then add 8.*

d. For some generic figure number N, state a rule (formula) for finding the perimeter P of the polygon in Figure N. *$P = 6N + 2$, or $P = 6(N - 1) + 8$*

SP-18. Graphing the Polygon Perimeters Pattern Complete the data table you created in SP-17, by including data for all of the first 10 figures. Copy the axes shown here onto your graph paper. Then use the data from the table to make a graph. You will need to plan ahead to decide how to scale the perimeter axis. (Think about the largest perimeter you will need to include when you scale the vertical axis.) *The largest perimeter is 62 units.*

Figure Number, N	Perimeter, P
1	8
2	14
3	20
4	26
5	32
6	38
7	44
8	50
9	56
10	62

SP-19. Use your graph from SP-18 to answer the following questions.

a. Is there a figure in the Polygon Perimeters Pattern that would have a perimeter of 17 units? Show how you know by drawing on the graph you made for SP-18. *no*

b. If the points of a graph lie on a line, we say that the graph is **linear**. Otherwise, we say the graph is **nonlinear**. Does the Polygon Perimeters Pattern graph seem to be linear or nonlinear? Explain why you think so.
linear because all 10 graphed points lie on a line

c. Add information about *linear* and *nonlinear graphs* to your Algebra Tool Kit.

 SP-20. Area as a Product, and Area as a Sum Derek and Deanne want to tile the bottom surface of their rectangular lap-pool with 1-foot-square tiles, and they are trying to find the simplest way to find the area of the bottom surface (shown here). How will knowing the area help them to know how many tiles to buy?

Using an old 10-foot-long measuring tape that Derek found in the garage, they measured to find the lengths of the sides. They found that the width was 3 feet. To find the length, they first measured 10 feet, and then another 2 feet:

To compute the area of the rectangular bottom, Derek first added the 10 feet and 2 feet to find the total length. He then found the *product* of the width and length to get the area:

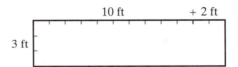

Area as a Product = (3')·(10' + 2') = 3' · 12' = 36' sq ft
Derek's expression

Deanne solved the problem by first dividing the rectangle into two parts and finding the area of each part. She then *added* to find the total area, as shown below:

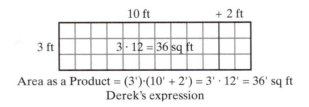

Area as a Sum = 3'(10') + 3'(2') = 30' sq ft + 6 sq ft = 36 sq ft
Deanne's expression

Both methods give the same result because the area itself did not change; only the ways in which the area was computed were different. So the area of the bottom of the rectangular pool could be computed by multiplying the width and length, (3')(10' + 2'), or by adding the areas of parts, (3 ft)(10 ft) + (3 ft)(2 ft), since both expressions are equal to 36 square feet.

a. On the Resource Page (Area as a Product and Area as a Sum), label the sides of the following rectangle with their lengths:

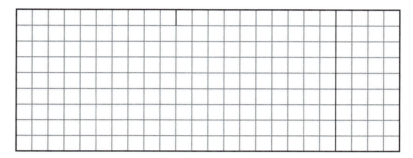

b. The dark line splits the large rectangle into two sections. Label each of the two sections with its area. *9 · 20 = 180 sq ft, 9 · 4 = 36 sq ft*

c. To express the area of the rectangle as a *product*, multiply the length by the width:

$$9 \cdot (20 + 4)$$

To express the area of the rectangle as a *sum*, add the areas of the parts:

$$9(20) + 9(4)$$

Check to see that these two expressions (given in square feet) are equal by evaluating each of them using the proper order of operations:

Area as a Product	=	**Area as a Sum**
9 · (20 + 4)	=	9(20) + 9(4)
9(24)	=	*180 + 36*
216	=	*216*

EXTENSION AND PRACTICE

SP-21. For each of the following rectangles, write an equation showing that the area (in square units) can be computed two ways—as the *product* of its dimensions (as Derek did in SP-20), and as a *sum* of its parts (as Deanne did). On the SP-21 Resource Page, label the side lengths and the area of each part. Then write an area equation below each rectangle.

a.

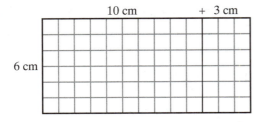

In sq cm:
   area as a product = area as a sum
          6(10 + 3) = 60 + 18
             6(13) = 78

In sq cm:
   area as a product = area as a sum
          8(20 + 1) = 160 + 8
             8(21) = 168

b.

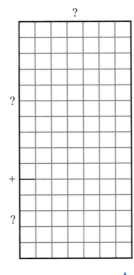

In sq units:
   area as a product = area as a sum
         (10 + 5) · 7 = 70 + 35
            (15) · 7 = 105

c.

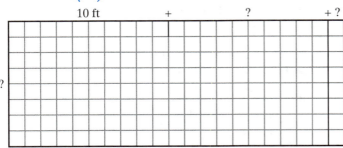

SP-22. In each of the following equations, the area of a rectangle is represented both as a product and as a sum. Using the SP-22 Resource Page or a piece of graph paper, make a sketch of the rectangle represented by each equation, and label each section with its area:

a. 4(10 + 5) = 4(10) + 4(5) = 60

b. 7(20 + 1) = 7(20) + 7(1) = 147

c. (10 + 2)(9) = 10(9) + 2(9) = 108

Solutions:

(a)

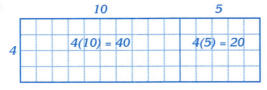

(b)

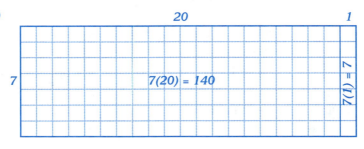

(c)

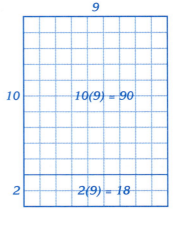

SP-23. Follow these steps to make a graph of the rule $y = -x + 1$.

a. On a Grid Sheet copied from your Grid Sheet Template Resource Page, complete a table using the rule $y = -x + 1$. You could also use your own graph paper and make a table like this:

Input (x)	−4	−3	−2	−1	0	1	2	3	4
Output ($y = -x + 1$)	5	4	3	2	1	0	−1	−2	−3

b. Label the axes on your Grid Sheet, and indicate the scale on each axis. Plot the points from the table on the coordinate axes. Then label the graph with its rule.

SP-24. Follow the same steps as in SP-23, but this time use the rule $y = x^2 - 2$. *Outputs: 14, 7, 2, −1, −2, −1, 2, 7, 14*

SP-25. Plot the points (0, 0), (0, 6), (18, 6), and (18, 0) on regular graph paper. Use a ruler to connect the points to form a rectangle.

a. What are the length and the width of the rectangle? *6 by 18 units*

b. To compute the area of the rectangle easily, divide it into two sections by thinking of the dimensions as 6 by (10 + 8). (In other words, draw a vertical line through the point (10, 0).) Then find the area of each section. *6(10) = 60 sq units, and 6(8) = 48 sq units*

c. Use part b to find the area of the rectangle in square graph-paper units *60 + 48 = 108 sq units*

SP-26. Write 2^6 as a product of powers of 2 in three different ways. For example, $2^6 = 2^2 \cdot 2^4$. *$2^6 = 2^1 \cdot 2^5$, etc.*

SP-27. Use the meaning of exponents to write each of the following expressions using a single exponent:

a. $(10^2)(10)$ *10^3*

b. $(10^5)(10^2)$ *10^7*

c. $(10^2)^3$ *10^6*

d. $10^2 \cdot 10 \cdot 10^3$ *10^6*

SP-28. One of the expressions in SP-27 is not like the others. Which one is it? Describe how it is different. *$(10^2)^3$, because it represents "something used as a factor three times" while the others do not*

For students who do not know how to start SP-29, the note in the Student Answer section reminds them of the list they made for the three coins and gives some further direction.

SP-29. Lance is taking a geography quiz with only three questions. For each question, he must choose "true" or "false." Unfortunately, he missed the last two weeks of his geography class, and did not study for the quiz.

a. Make an organized list of all the possible answer sheets Lance could turn in. *Have students look back at their work for OP-95 if they want help getting started. TTT, TFF, TFT, TTF, FTT, FTF, FFT, and FFF.*

b. If Lance randomly guesses answers, what is the probability that he will score 100% on the quiz? Explain how you know. $\frac{1}{8}$

SP-30. Professor Speedi has another rule in mind to use for next semester's Algebra Walk. Write the rule

$$y = \frac{1}{2}x + 3$$

in everyday words she could use to give instructions to the student graphers. *Compute half of the input, then add 3.*

SP-31. Copy and solve these Diamond Problems:

a.

$N = 28,$
$S = -11$

b.

$6, -2$

c.

$E = -16,$
$S = -15\frac{1}{2}$

d.

$N = \frac{1}{10},$
$S = \frac{7}{10}$

SP-32. Copy each of the following expressions, and carry out the calculations using mental math:

a. $-415 + (-85.3)$ -500.3

b. $-12 - (-492)$ 480

c. $\left(\frac{-2}{3}\right)\left(\frac{3}{4}\right)$ $\frac{-1}{2}$

d. $\left(4\frac{1}{5}\right)(-5)$ -21

e. $18 + (-9) \div (-3) + 72$ 93

f. $-2^4 + 2^3$ -8

SP-33. Use a guess-and-check table to solve the following problem. Write your answer in a complete sentence.

Find four consecutive odd integers such that the sum of the second integer and twice the fourth integer is 65. *17, 19, 21, 23*

SP-34. Mystery Rule Copy the table and look for a pattern. Use the pattern to complete the table, and then write the mystery rule in words and in symbols.

Input (x)	Output (y = ____)
3	−6
7	−14
6	−12
4	−8
10	−20
−1	2
100	−200
x	−2x

SP-35 could be assigned to enrich students' patterning skills, but it is not crucial to the development of the other ideas in the course. You might assign it to all students, or as extra credit, or only to students who are ready for more problem-solving exercises.

SP-35. Examine this pattern:

a. Draw the eighth figure in the pattern.

b. Draw the 30th figure in the pattern.

c. Explain how to decide what the 366th figure would look like.

Solutions: (a) (b) (c)

Use the formula figure # = 3q + r (where r = 0 indicates a dot on the left, r = 1 indicates a dot at the top, and r = 2 indicates a dot on the right).

2.3 GETTING TO WORK ON TIME: SITUATIONS AND GRAPHS

SECTION MATERIALS
POSTER-SIZE COORDINATE AXES
COLORED STICKY DOTS

About the Materials: Both of these materials are for SP-39. Draw coordinate axes on flip-chart paper and put them on the wall. For plotting data on the axes, you'll need at least 10 sticky dots (any color will do).

Problems SP-36 and SP-37 are intended to help to prepare students for SP-38, where they encounter the inverse relationship of traveling a fixed distance at various rates. Problem SP-36 reinforces OP-86, where students worked with the relationship d = r · t in numeric contexts:

"If we drive 60 mph for 2 hours, we've traveled 120 miles."

"If we drive 60 mph for 30 minutes, we've driven for one-half hour, so we've traveled...."

Problem SP-37 gives students practice with the calculations before they graph the time = distance ÷ rate relationship in SP-38. Right now, let students work out the relationship numerically so that it makes sense to them. In particular, don't tell them "t = d ÷ r" to make parts a through c easier. By working through the first three parts, students should be able to express the relationship in part d.

SP-36. Consider the following problem:

a. Discuss and solve Professor Speedi's problem with your group:

On her way to a mathematics conference, Professor Speedi was driving at an average speed of 58 miles per hour. She figured that at that rate it would take her a total of three and a half hours to get to the conference. How far did she drive? *203 mi*

b. Describe what you did to solve part a. Using that knowledge, work together in your groups to write an explanation for how to find the distance you've traveled if you know your speed and the amount of time you traveled. *To get the distance, multiply the speed and the time.*

 SP-37. Imagine you are driving north on Interstate 5.

a. If you drive at an average speed of 50 miles per hour, how long will it take you to travel 200 miles? Show how you know. *4 hr*

b. If you drive at 20 mph, how long will it take you to travel 70 miles? Show how you know. *3.5 hr*

c. If you drive at 24 mph, how long will it take you to travel 30 miles? Give your answer in hours and minutes, and show how you know. *1.25 hr, or 1 hr and 15 min*

d. Explain in words *how to determine how long it takes to travel a certain distance if you know the average speed.* Write your explanation in your Algebra Tool Kit. *hours = distance ÷ speed*

In problem SP-38, many students look only for a pattern in the numbers and ignore the units of measure. Because the situation "45 miles per hour for 40 minutes" mixes units of measure, it is troublesome and so forces students to consider units. It may be helpful to direct groups to make the table with only the "mph" and "hours" columns, and to add the "minutes" column at the end. The class should discuss the data table before making their graphs.

It is common for students to create linear graphs in this early stage of graphing, so watch for inconsistent scaling. We want students to be aware from the start that not all graphs are linear, and that graphing skills are an important part of algebra and other subjects.

As groups complete SP-38, distribute several sticky dots to each group for plotting on a coordinate grid you've prepared on poster-size graph paper. After the easier points are plotted, encourage students to fill in points they may not have chosen—for example (25, 72)—to get as many points as possible. Many instructors find that the class graph is helpful for focusing a class discussion after groups complete SP-41. The class graph will also be useful later when discussing discrete and continuous graphs in SP-82.

SP-38. Professor Speedi's Commute Professor Speedi, the math instructor, has figured out that if she drives to work at 60 miles per hour, it will take her half an hour to get there. Unfortunately that speed exceeds the speed limit, so Professor Speedi must drive at a slower rate. Some days she might drive at 30 mph. Other days she could drive at 45 mph. Since Professor Speedi doesn't want to be late for class, she needs to know the speed and time combinations that might allow her to get to campus on time.

a. How many miles away from work does Professor Speedi live? *Hint: The problem states that it takes Speedi half an hour when she drives 60 mph.*

b. Calculate how long it would take Professor Speedi to get to work if she drove at 30 miles per hour. Calculate how long it would take if she drove at 45 mph. (Review your work from OP-93, SP-36, and SP-37 if you need help.) *At 30 mph, it would take 1 hr; at 45 mph it would take $\frac{2}{3}$ hr, or 40 min.*

c. Record your data from part b in a table. Add at least five more speeds to the data table, and calculate the time associated with each speed. Then convert each time in hours to time in minutes.

84 Chapter 2 Professor Speedi's Commute: Patterns and Graphs

Speed (mph)	Time (hours)	Time (minutes)
60	$\frac{1}{2}$ hr	30
30	1	60
45	$\frac{2}{3}$	40
40	$\frac{3}{4}$	45
25	$1\frac{1}{5}$	72
20	$1\frac{1}{2}$	90
15	2	120
10	3	180
65	$\frac{6}{13}$	27.7
70	$\frac{3}{7}$	25.7
80	$\frac{3}{8}$	22.5

d. On a sheet of graph paper, copy the axes shown here, making them large enough that they use up most of the page. Make a prediction: Do you think the data will lie on a straight line or on a curve? *Students will check their predictions in SP-38e and in SP-39d.*

e. Graph all of your data. How does the shape of the graph fit your prediction?

Solution:

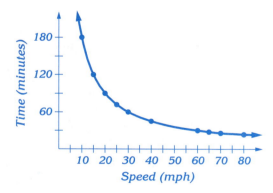

SP-39. Use your graph of Professor Speedi's data from SP-38 to answer the following questions. Write your descriptions in complete sentences.

a. When reading the graph from left to right, does the graph slope upward always, or downward always, or change direction? *The graph slopes downward. We are using "slope" in a generic way.*

b. At what speed would Professor Speedi have to travel to reach the college in 0 minutes? Do you think the graph should cross the horizontal axis? Explain. *It's impossible! The graph will not meet the horizontal (speed) axis. To help students see this, discuss what it would mean to travel a fixed distance in 0 minutes.*

c. How long would it take Professor Speedi to get to the campus if she drove at 0 mph? Do you think the graph should cross the vertical axis? Explain. *The graph will not meet the vertical (time) axis. To see this, discuss what it would mean to travel a fixed distance at 0 mph.*

d. Susan claims that the points she plotted for Professor Speedi's graph in SP-38 lie on a straight line. Could a straight line represent the Speedi data? Explain your answer to Susan. (*Hint:* Is it possible to draw a straight line that does not cross either axis?) *The points do not lie on a line. (The curve is part of a hyperbola.)*

SP-40. Suppose Professor Speedi tells you that she has only 15 minutes to get to the college. Describe how you could use your graph to quickly predict the speed at which she must drive. How could you find the speed more accurately? Write your responses in complete sentences. *Use the graph by starting at the desired time on the vertical axis, moving horizontally across to the graph, and then down to read the speed off the horizontal axis. For a more accurate answer, divide the distance to the college by the amount of time given, to get the (average) rate at which she must drive: $r = \frac{d}{t} = (30 \text{ mi})/(\frac{1}{4} \text{ hr}) = 120$ mph*

SP-41. If Professor Speedi's gas pedal is stuck so that she can only drive at 8 mph, how long will it take her to get to work? Use your graph of Professor Speedi's data to make an estimate, and then calculate the time more accurately. *$3\frac{3}{4}$ hr*

EXTENSION AND PRACTICE

If there is time in class, have the groups start on SP-42 and/or SP-43. You could have them work on the "Mystery Rules" in class and then assign the graphing parts (SP-42a and SP-43c) as homework. Be sure that students write the rule both in words and in symbols.

SP-42. Belinda made the following table, but then she lost the rule she used to calculate the y-values:

Input (x)	−4	−3	−2	−1	0	1	2	3	4
Output ($y = $ ___)	17	10	5	2	1	2	5	10	17

a. Plot the points Belinda found on a graph. Be sure to label the axes and to include the scale. *The points lie on the parabola $y = x^2 + 1$.*

b. Explain in words what Belinda did to the input value, x, to produce the output value, y. *multiplied the input by itself and added 1; or squared the input and added 1*

c. Use algebraic symbols to write the process you described in part b. *$y = x^2 + 1$*

SP-43. Here's another Mystery rule. Copy and complete this table, and then answer parts a, b, and c.

Input (x)	0	2	6	−8	4	−10	−28		x
Output ($y = $ ___)	0	1	3	−4	2	−5	−14		$\frac{x}{2}$

a. Explain in words what is done to the input value, x, to produce the output value, y. *divide by 2*

b. Write the process you described in part b in algebraic symbols. *$y = \frac{x}{2}$*

c. Graph the data. Use the same scale on both the x-axis and the y-axis. (You may not be able to fit all the points from the table on your graph because the x- or y-values are too large or too small.)

SP-44 and SP-45 follow up SP-38 (Professor Speedi's Commute) and give students more experience using calculators and graphing nonlinear relations.

SP-44. Calculating Reciprocals with $\boxed{\tfrac{1}{x}}$ **or** $\boxed{x^{-1}}$ In SP-10, you saw that as long as $x \neq 0$, the reciprocal of any number x is $\frac{1}{x}$, so their product equals 1:

$$x \cdot \text{reciprocal of } x = 1, \quad \text{or} \quad x \cdot \frac{1}{x} = 1.$$

Using this idea and your calculator, you have two ways to find the reciprocal of a number.

Method 1: Use division.
Example: If $x = 2$, the reciprocal of x is $\frac{1}{2}$. To calculate, press $\boxed{1}$, $\boxed{\div}$, $\boxed{2}$, $\boxed{=}$. The calculator display should be 0.5, which is the correct reciprocal according to the definition in SP-10, since $(2)(0.5) = 1$. ▲

Method 2: Use the $\boxed{\tfrac{1}{x}}$ or $\boxed{x^{-1}}$ key.

Example: If $x = 8$, the reciprocal of x is $\frac{1}{8}$. To calculate, press $\boxed{8}$, then the $\boxed{\frac{1}{x}}$ or $\boxed{x^{-1}}$ key. The calculator display should be 0.125, which is the correct reciprocal since $(8)(0.125) = 1$. ▲

a. Update your Calculator Tool Kit on *how to calculate reciprocals*.

b. Copy and complete the following table. For each value of x, use your calculator to compute the reciprocal of x by using either of the methods described above. You will use this table for the graph in SP-45.

x	−5	−3	−2	−1.5	−1	−0.33	−0.2	0	0.2	0.33	1	1.5	2	3	5
$y = \frac{1}{x}$	−0.2	−0.$\overline{3}$	−0.5	−0.6	−1	−3.03	−5	not defined	5	3.03	1	0.6	0.5	0.$\overline{3}$	0.2

SP-45. The Graph of $y = \frac{1}{x}$

a. Use the table of reciprocals in SP-44 to sketch a graph of $y = \frac{1}{x}$. Scale the axes so that two marks on the graph paper represent one unit.

b. Consider now just positive values of x. How is the y-value affected as x gets larger? What happens to the y-value as x gets smaller? *y gets smaller; y gets larger.*

c. What happens to the y-value as x decreases from 1 to 0? What happens to the y-value as x increases from −1 to 0? *y gets larger; y gets smaller.*

d. What would happen if $x = 0$? *There is no y-value for x = 0.*

e. Use the graph to *estimate* a value for x that will make $\frac{1}{x}$ equal to each of these numbers:

  i. 2 *x = 0.5*
  ii. $\frac{1}{2}$ *x = 2*
  iii. −2 *x = −0.5*
  iv. 0.3 *x ≈ 3.3*
  v. $\frac{-1}{4}$ *x = −4*
  vi. 2.5 *x = 0.4*

f. Compare the graph of $y = \frac{1}{x}$ that you made for part a to your graph for Professor Speedi's Commute (SP-38). Describe at least two similarities and at least two differences between the graphs. *Answers will vary. Some similarities: neither is linear, neither intersects either axis, they have similar shapes. Some differences: SP-38 has only positive values (one branch of hyperbola), the points are different, the scales are different.*

SP-46. Write each of the following expressions using only one base and one exponent, if possible. Otherwise, write the result in standard form.

a. $2^3 \cdot 2^4$ 2^7
b. $(2^4)^3$ 2^{12}
c. $2^3 + 2^4$ $8 + 16 = 24$
d. $10^5 \cdot 10^4 \cdot 10^2$ 10^{11}
e. $(x^3)(x^5)$ x^8
f. $(-x^3)^2$ x^6

SP-47. For each of the following expressions, sketch a rectangular figure whose area is described by the expression. Then use your figure and the given expression to write an equation representing the idea that "area as a product = area as a sum," as you did in SP-20.

a. $10(10 + 3)$
b. $2(10 + 3)$

Solutions:

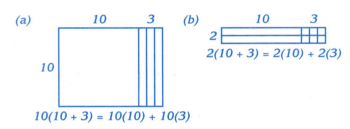

(a) $10(10 + 3) = 10(10) + 10(3)$

(b) $2(10 + 3) = 2(10) + 2(3)$

SP-48. Imagine you have two fair standard dice—one is orange and the other is yellow.

a. Suppose you roll a 1 on the orange die, and then roll the yellow die. What numbers could you get on the yellow die? Make a list of all the possibilities. *1, 2, 3, 4, 5, or 6*

b. Suppose you leave the 1 showing on the orange die and roll the yellow die again. What combinations of numbers might come up for the two dice? Make an organized list of possible pairs of numbers. *(1, 1), (1, 2), (1, 3), (1, 4), (1, 5), or (1, 6)*

c. If you rolled a 1 or a 2 on the orange die, what combinations of numbers would be possible for the two dice? Make an organized list of possible pairs of numbers. *(1, 1), (1, 2), (1, 3), (1, 4), (1, 5), (1, 6), (2, 1), (2, 2), (2, 3), (2, 4), (2, 5), or (2, 6)*

d. If you rolled both dice at the same time, how many different outcomes would be possible? (The answer is *not* 12.) Make an organized list of all the possible combinations. *There are 36.*

e. Suppose you roll both dice one more time, and then add the two numbers that come up. What is the probability that the sum of the two numbers is 5? $\frac{4}{36} = \frac{1}{9}$

SP-49. Use a guess-and-check table to solve the following problem. Write your answer in a complete sentence.

Melissa cut a 150-centimeter board into two pieces. One piece is 24 centimeters longer than the other piece. How long is each piece of board? *63 cm, 87 cm*

Section 2.3 Getting to Work on Time: Situations and Graphs **89**

SP-50. Patterns with Exponents In this problem, you will look at expressions that involve two or more powers of 2 to see which of the expressions can be rewritten more "simply" as a single power of 2.

a. First, copy and complete this table of powers of 2:

Power of 2	Standard Decimal Form
$2^{10} =$	*1024*
$2^9 =$	*512*
$2^8 =$	*256*
$2^7 =$	*128*
$2^6 =$	*64*
$2^5 =$	*32*
$2^4 =$	*16*
$2^3 =$	*8*
$2^2 =$	*4*
$2^1 =$	*2*

b. Now, compute the value of each of the following expressions using the order of operations and your calculator.

  i. $2^5 + 2^2$
  36

  ii. $2^5 \cdot 2^2$
  128

  iii. $2^5 - 2^2$
  28

  iv. $2^5 \div 2^2$
  8

  v. $2^7 - 2^3$
  120

  vi. $2^7 + 2^3$
  136

  vii. $2^7 \div 2^3$
  16

  viii. $2^7 \cdot 2^3$
  1024

c. Compare the value of each of the expressions in part b with the values of the powers of 2 in the table for part a. Which of the part b expressions are equal to (and therefore can be rewritten as) a power of 2? Write the expression and its equivalent power of 2 in an equation. For example, if $2^3 \cdot 2^2$ were one of the expressions in part b, then

$$2^3 \cdot 2^2 = \underbrace{8 \cdot 4 = 32}_{\text{by calculations}} = \underbrace{2^5}_{\text{from the table}}$$

so $2^3 \cdot 2^2 = 2^5$.

(ii) $2^5 \cdot 2^2 = 2^7$; (iv) $2^5 \div 2^2 = 2^3$; (vii) $2^7 \div 2^3 = 2^4$; (viii) $2^7 \cdot 2^3 = 2^{10}$

d. List the expressions in part b that can*not* be rewritten as a single power of 2. *(i) $2^5 + 2^2$; (iii) $2^5 - 2^2$; (v) $2^7 - 2^3$; (vi) $2^7 + 2^3$*

For SP-51 we want students to notice from parts c and d of SP-50 that products and quotients of powers of 2 can be rewritten with a single exponent, while sums and differences cannot. After more patterning work and practice, students will summarize their understanding of integer exponents by describing the pattern for multiplying powers of like bases (at the end of this chapter) and for raising a power to a power (in Chapter 3).

SP-51. Look again at your work in SP-50 and look for patterns among the expressions that can be rewritten as powers of 2, and among those that cannot.

a. Write one or two sentences to summarize what you observed in SP-50.
Answers will vary: "If powers of the same base are multiplied or divided, the result can be rewritten with a single exponent; if powers of the same base are added or subtracted, the result cannot be rewritten with a single exponent."

b. Make up several of your own examples to illustrate your observations from part a.

SP-52 could be assigned as reinforcement of exponent work and enrichment of patterning skills, but it is not crucial to the development of other main ideas of the course. You might assign it to all students, or as extra credit, or only to students who are ready for more problem-solving exercises.

SP-52. In this problem you will write powers of 4 in standard decimal form, and then look for a pattern.

a. Copy and complete this table for powers of 4:

Exponential Form		Standard Form	Ones Digit
4^1	=	4	4
4^2	=	16	6
4^3	=	*64*	
4^4	=	*256*	
4^5	=	*1024*	
4^6	=	*4096*	

b. Look for patterns in the completed table: Which powers of 4 have a 6 for the ones digit? Use what you observe to list the exponential form of the next three powers of 4 that have a 6 for the ones digit. *4^8, 4^{10}, 4^{12}*

c. What digit is in the ones place of the standard form for 4^{128}? (Your calculator won't be of any help; use a pattern in the table.) *6*

d. Explain how you figured out the answer to part c so that a high school student could identify the ones digit for 4^{997}. *If the exponent is odd, the ones digit is 4. For even exponents, the ones digit is 6.*

2.4   HAWAIIAN PUNCH MIX

SECTION MATERIALS

PUNCH BOWL (SP-53); OR OTHER MATERIALS, AS DESCRIBED HERE

POSTER-SIZE COORDINATE AXES POSTED ON THE WALL (FOR SP-56)

STICKY DOTS IN ONE COLOR, AT LEAST ONE PER STUDENT (FOR SP-56)

ALGEBRA TILES RESOURCE PAGE (FOR SP-72)

Problems SP-53 through 55 are a series of problems that reinforce the connections between patterns, graphs, and rules. The notion of a rate of change is informally introduced in SP-55. After students complete the series of problems, help them summarize what they've learned in a whole-class discussion.

To get groups going on SP-53, it is helpful to model the start of the problem. One fun and effective way we've tried is to use a real punch bowl and have students imagine that there are 5 quarts of guava juice already in it. We then repeatedly "add" 2 quarts each of other types of (imaginary) juice, as suggested by the class. We've also illustrated the situation on the board with colored chalk and given each group slips of colored paper to represent the different kinds of punch (one slip each of at least nine different colors for each group). To extend the problem and to generate more questions, you might use different values for the total amount of punch and each type of juice.

Students will need algebra tiles for the next class (Section 2.5). If you do not intend to provide them with commercially made algebra tiles, then be sure to assign SP-72, which directs them in making their own.

SP-53. How About a Nice Hawaiian Punch? Matthew was mixing fruit punch on a hot Maui summer day. Because he liked guava juice so much, he first poured 5 quarts of guava juice into the punch bowl. Matthew then added 2 quarts each of several different types of fruit juice. Eventually, he had a total of 21 quarts of mixed-fruit punch.

# types	# quarts
1	5
2	7
3	9
4	11

a. Make a table showing the relationship between the number of types of fruit juice Matt used (the input values, x) and the total amount of punch in the bowl after he added each type (the output values, y). The 5 quarts of guava juice represent a "punch mix" with just one type of juice. Use your data table to make a graph of the relationship between the number of types of fruit juice and the total amount of punch.

b. How many types of fruit juice did Matt use to make 21 quarts of fruit punch? *9 types*

c. If Matt wanted to make 27 quarts of punch, how many types of fruit juice would he need to use? Explain how you know. *12 types; Answers will vary: follow the pattern; $5 + 12 \cdot 2 = 27$ qt.*

d. If Matt used 15 types of juice, how much punch would he make? Explain how you know. *33 quarts; Answers will vary: guava + 14 other types make $5 + 14 \cdot 2$ qt.*

e. Use the variables x and y to write a rule (equation) that describes the relationship shown in the table and on the graph. *$y = 2x + 3$ for $x \geq 1$*

f. If Matt had 0 types of juice, he wouldn't be able to make any punch. What point on the graph would represent this idea? Does it follow the pattern of the other points on your graph? Explain. *(0, 0); No, it does not lie on the line.*

SP-54. Reread the Hawaiian Punch problem (SP-53). Then answer the following questions about your graph:

a. Should your graph show negative values for *x*? for *y*? Explain your answers. *No, a negative number of quarts of punch would not make sense.*

b. If Matt continues using the same recipe, would he be able to make exactly 10 quarts of punch? Explain your answer. *No, since that would correspond to adding 3.5 types of juice.*

c. Is it possible for Matt to end up with an output (*y*-value) that is a fraction? Could the input (*x*-value) be a fraction? Describe as specifically as possible what type of number the inputs could be. Also describe as specifically as possible what type of numbers the outputs could be. *Neither the inputs nor the outputs can be fractions. The inputs must be nonnegative integers, and the outputs must be odd numbers greater than or equal to 5.*

After groups complete SP-55, conduct a class discussion on what they've learned in SP-53 through SP-55 before moving on to SP-56.

SP-55. Using the recipe in SP-53, the total amount of punch that Matt makes increases at the rate of 2 quarts for each type of juice added. We can express this rate more briefly as "2 qt/type."

a. If Matt decided to add 3 quarts of each new type of juice, he would add punch at the rate of 3 quarts per type. If you were to graph this new recipe, would the graph be steeper than the graph in SP-53, or less steep? Explain how you know. *Steeper, because the graph would go up 3 units for each type of juice added.*

b. Matt found a cookbook with 3 new recipes for punch. Study the graphs for each recipe, and determine the rate at which each punch mix grows.

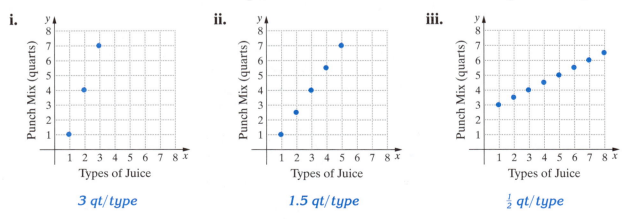

i. *3 qt/type*

ii. *1.5 qt/type*

iii. *½ qt/type*

In SP-56 students see that the graph of $y = x^2$ is not a line, but is a curve symmetric about the y-axis. Some students will recall the term "parabola"; you should reinforce the use of this term or introduce it in part c. Students will develop understanding of the graphs of quadratic functions throughout the course. Some students may need help with the operation of their calculator's $\boxed{x^2}$ key if it involves using a second function mode.

Post coordinate axes on flip-chart grid paper on the wall, and distribute sticky dots, at least one per student, so that students can plot the data from their tables to create one large class graph. Assign each group at least four rational numbers (for example, $-2.9, -1\frac{1}{3}, -0.1, 0.27, 1\frac{1}{3}$), different ones for each group, and have them add their points to the large graph. The class graph is a handy reference as students graph parabolas throughout the course.

SP-56. In this problem you'll examine the graph of the squaring rule, $y = x^2$.

a. Copy and complete the following data table for $y = x^2$. Leave room for several additional inputs that your instructor will assign.

x	-4	-3	-2	-1	-0.5	0	0.5	1	2	3	4			
$y = x^2$	16	9	4	1	0.25	0	0.25	1	4	9	16			

b. Using all the values from your table, graph the equation $y = x^2$. Check your graph with other members of your group. Have one person plot your group's additional points on the class graph.

Discuss the following questions with your group and then write your answers in complete sentences:

c. Describe the shape of the graph of $y = x^2$. (Is it a straight line? Is it a "V"?) *The graph is U-shaped, a parabola.*

d. Suppose everyone in your group used three more x-values to find three more points on the graph. If all the new points were plotted on the class graph, what could you say about where these points would be plotted? *They'd fit into the pattern of points already on the graph.*

e. Can any value for x be used? Explain. *Yes, since any number can be squared.*

f. Can any y-value result from this rule? Explain. *no; $y \geq 0$*

g. Connect the points on your graph with a smooth curve, and describe the result. How does the shape of this graph compare with your graph of $y = x^2 - 4$ from the Algebra Walk (SP-2b)? *Both are parabolas.*

h. Show how you can use your graph to approximate the x-value(s) that correspond to $y = 5$. *≈ ±2.2*

i. Use your calculator to find more precisely (to the nearest hundredth) the x-value(s) that correspond to $y = 5$. *≈ ±2.24*

SP-57. Write an expression for the length of each composite line segment shown here.

Example: The segment has a length of x units, while the composite segment has a length of $a + a + 5 + 3$, or $2a + 8$ units. ▲

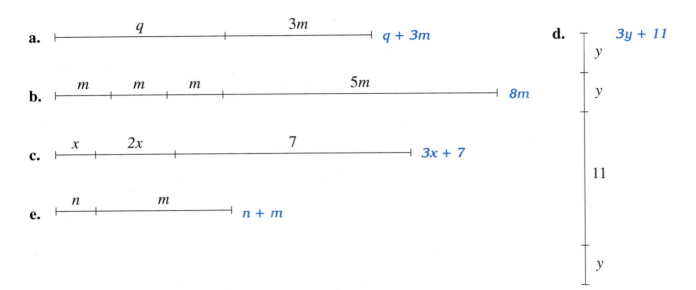

EXTENSION AND PRACTICE

SP-58. Follow these steps to make a graph of the mystery rule.

a. Copy and complete the following table:

x	1	8	3	0	−1	9	$2\frac{1}{3}$	$\frac{1}{2}$	x
$y =$ ___	4	−3	2	5	6	−4	$2\frac{2}{3}$	$4\frac{1}{2}$	$5 - x$

b. Use the variable x to write a rule for these data as an equation. $y = 5 - x$

c. Set up and label coordinate axes, and then graph the rule. Label the graph with its rule.

SP-59. Follow the same steps as in SP-58, but use the following table:

x	8	6	18	−10	$\frac{1}{2}$	10	−4	x
$y =$ ___	5	4	10	−4	$\frac{5}{4}$	6	−1	$\frac{1}{2}x + 1$

SP-60. Follow the same steps as in SP-58, but use the following table:

x	2	10	6	7	−3	−5	−10	100	x
$y =$ ___	8	32	20	23	−7	−13	−28	302	$3x + 2$

SP-61. Compare the graphs you made in SP-58, SP-59, and SP-60. Describe how the graphs are alike and how they are different. You might consider features such as shape, direction, and steepness.

SP-62. Look again at the graphs you made in SP-58, SP-59, and SP-60.

a. Estimate the coordinates of the point where each graph crosses the *x*-axis. *(5, 0), (−2, 0), $\left(\frac{-2}{3}, 0\right)$*

b. What do the three points in part a have in common? Write what you observe in a complete sentence. *The y-coordinate is zero.*

c. In part b, you made an observation (or *conjecture*) about the point where a graph crosses the *x*-axis. To see whether your conjecture applies to other graphs, you need to check more data. Look back at the Algebra Walk graphs you made in SP-2. Check the points where these graphs cross the *x*-axis. Is your observation true for these graphs as well?

SP-63. Definition: An **x-intercept** of a graph is a point where the graph crosses the *x*-axis.

a. What is the *y*-coordinate of *any* *x*-intercept?

b. Add the definition for the *x-intercept* of a graph to your Tool Kit.

SP-64. Write the coordinates of the *x*-intercepts, if any, of the following four graphs:

a.   b.

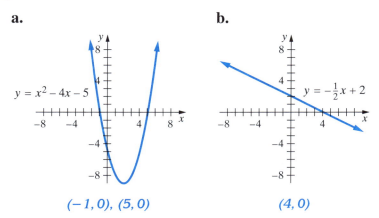

(−1, 0), (5, 0)   *(4, 0)*

c. $y = (x + 2)(x - 3)(x + 5)$

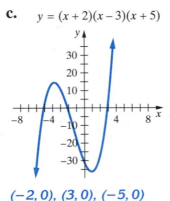

$(-2, 0), (3, 0), (-5, 0)$

d.

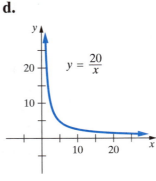

There are no x-intercepts.

SP-65. Suppose you randomly toss four coins—a penny, a nickel, a dime, and a quarter. They might land with "heads, heads, heads, heads" (HHHH) showing.

a. What other possible outcomes might occur? Describe a strategy you could use to make sure you find *all* the possibilities. Using your strategy, list all the ways the four coins could land, including HHHH. *HHHH, HHHT, HHTH, HHTT, HTHH, HTHT, HTTH, HTTT, THHH, THHT, THTH, THTT, TTHH, TTHT, TTTH, TTTT (Have students look back at their work for OP-95 if they want help getting started.)*

b. Find the probability that exactly one coin lands "heads" up. $\frac{4}{16}$ or $\frac{1}{4}$

c. Find the probability that at most two coins land "heads" up. $\frac{11}{16}$

SP-66. Marybeth wanted to multiply 35 by 9 without using a calculator, so she drew the following rectangle:

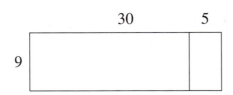

a. Explain why the area of the large composite rectangle is equal to $35 \cdot 9$. *To find the area of a rectangle, we multiply length times width.*

b. Find a value for $35 \cdot 9$ by finding the areas of the two smaller rectangles and summing. *270 + 45 = 315*

SP-67. Professor Speedi wants to include two additional graphs when her students do the Algebra Walk activity.

a. Write the rule $y = 3x^2$ in words that Professor Speedi could use when she gives Algebra Walk instructions. *Square your number, and then multiply the result by 3.*

b. Write the rule $y = (3x)^2$ in words that she could use for Algebra Walk instructions. *Multiply your number by 3, and then square the result.*

SP-68. The following two expressions are different. Explain how the meaning of exponents can be used to rewrite one of the expressions as a power of 2, and then find the value of each expression:

a. $2^5 \cdot 2^3$ *$2 \cdot 2 \cdot 2 \cdot 2 \cdot 2 \cdot 2 \cdot 2 \cdot 2 = 2^8 = 256$* **b.** $2^5 + 2^3$ *$32 + 8 = 40$*

SP-69. Legend has it that on July 4, 1947, several thousand motorcyclists took over the small California farm town of Hollister, inspiring the movie *The Wild Ones* with Marlon Brando. On July 4, 1997, Hollister invited bikers back for a reunion, and it was estimated that 110,000 bikers rode in from all over the world for the weekend event.

a. If the average motorcycle costs $22,000, what was the total worth of the motorcycles in town that weekend? Write your answer in both standard form and scientific notation. *$2,420,000,000 or $2.42 \cdot 10^9$*

b. Officials in Hollister estimated that $150 million might be spent at the weekend event. What would the average biker have spent? *about $1,363.64*

c. An impressive 16,000 bikers from Los Angeles rode together up Interstate 5 on Thursday night. If each motorcycle is about 5.5 feet long, and all the bikers had stopped single-file, tire to tire, just outside of town, how long would the line of bikers have been? (There are 5280 feet in a mile.) What if the bikers had stopped side by side in pairs? *≈16.7 mi single-file; ≈8.3 mi in pairs*

SP-70. Exponent Pattern

a. You know that x^3 means $x \cdot x \cdot x$. Explain what $y^3 \cdot y^2$ means. *(yyy)(yy)*

b. Rewrite the product $y^3 \cdot y^2$ using a single exponent. *y^5*

c. Explain what the product $y^4 \cdot y$ means. *Then* rewrite $y^4 \cdot y$ more simply, using a single exponent. *(yyyy)(y) = y^5*

d. Look for a pattern in parts a through c. Then, without explaining the meaning, use the pattern to rewrite $y^{35} \cdot y^{29}$ using a single exponent. *y^{64}*

SP-71. Use a guess-and-check table to solve the following problem. Write your answer in a complete sentence.

Find two consecutive whole numbers such that the sum of their squares is 265. *11 and 12, or −12 and −11*

For the next class session (Section 2.5), each student will need a set of algebra tiles. If you do not plan to provide the manufactured tiles for use during class, students will need to prepare their own sets to bring. The following problem gives directions for using the Resource Pages to make the tiles.

SP-72. At your next class session you will be using algebra tiles to investigate different ways to represent the areas of rectangles. To be ready for class, prepare a set of algebra tiles according to these directions:

Make a copy of your Algebra Tiles Resource Page. Glue or paste the copy to a piece of construction paper or lightweight cardboard (old cereal or tissue boxes will do). Carefully cut out the tiles. Use an envelope or a resealable plastic bag to store and carry the tiles.

2.5 SLICING A PIZZA

SECTION MATERIALS
- TRANSPARENCY OF SP-73 SLICING A PIZZA RESOURCE PAGE
- ALGEBRA TILES
- SP-76 COMPOSITE RECTANGLES AND AREA RESOURCE PAGE

As a brief opener, give everyone an opportunity to see the different ways classmates generated and organized the data for SP-65 by asking several students to put their solutions on the board at the beginning of class.

SP-73 is an important problem in terms of patterning because the obvious pattern that evolves in the first five terms, # regions = $2^{(\text{\# points} - 1)}$, falls apart for the sixth term. (The relation between the number of points and number of regions in this problem can indeed be expressed mathematically, but it is not as obvious as the false pattern involving exponents that we first see. The relation involves a combinatorial argument: for positive integer n, the number of regions formed when n points on a circle are connected pairwise by chords is $\binom{n}{4} + \binom{n}{2} + 1$, or $\frac{1}{24}(n^4 - 6n^3 + 23n^2 - 18n + 24)$.) The problem illustrates the need to test patterns carefully.

Be sure to have groups share their descriptions from SP-74.

About the Materials: *For SP-75, manufactured algebra tiles may be ordered from the Cuisenaire Company of America, Inc., 1-800-273-0338. For clear and easy class demonstrations, you can also purchase a set of algebra tiles for the overhead projector. Students can also make their own sets of tiles using the Algebra Tiles Resource Page.*

SP-73. Slicing a Pizza In this problem you'll explore the following question about slicing a round pizza into pieces by making straight cuts through points on the edge of the crust:

Suppose you have a perfectly round (circular) pizza. Around the edge of the crust are some olives. (There are no other olives on the pizza.) If you slice the pizza by making cuts that connect the olives, so that every pair of olives is connected by a cut, how many pieces of pizza will you make?

To explore this problem, you'll consider a succession of circles with points on them to represent the pizzas with olives. Each successive circle will have one more point on it than the previous circle. To cut each circle into pieces, you will connect each possible pair of points with a line segment.

a. On your copy of the SP-73 Slicing a Pizza Resource Page, mark one point on a circle as in Figure 1, shown here. On another circle, add a second point and connect the points as in Figure 2. On a third circle, mark three points and connect pairs of points as in Figure 3. Repeat the process with a fourth circle as in Figure 4.

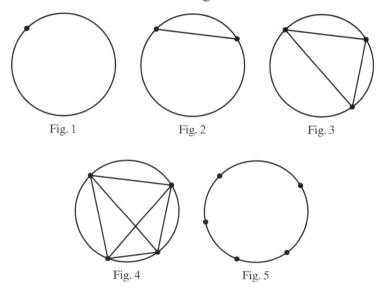

Fig. 1 Fig. 2 Fig. 3

Fig. 4 Fig. 5

b. Now count the number of pieces of pizza (regions) in each figure. For example, Figure 3 has four regions.

c. Complete the fifth circle in the pattern, and count the regions. *16*

d. Make a table to organize the data you've collected so far.

No. of Points	1	2	3	4	5
No. of Regions	$2^{1-1}=1$	$2^{2-1}=2$	$2^{3-1}=4$	$2^{4-1}=8$	$2^{5-1}=16$

e. Suppose a circle had six points marked on it and you connected each pair of points with a line segment. Use your data from part d to *predict* the number of regions that would be formed in the circle. Don't fill in Fig. 6a at this time. *Students usually predict 32.*

f. Test your prediction from part e by completing Fig. 6a and Fig. 6b on your Resource Page and then counting the regions. *There are usually 31, but students may find only 30 if three of the lines meet at a point inside the circle (because their figures are symmetric).*

SP-74. Write a short paragraph to describe what you did and your results in SP-73. What did the problem help you understand about looking for and finding patterns?

Problems SP-75 through SP-77 introduce students to the distributive property by modeling it geometrically using algebra tiles. The idea of writing the area of a rectangle as a product and as a sum will help them link the distributive property to factoring quadratic expressions in later chapters.

*We find it helpful to point out to students that they have used the distributive property since elementary school. The vertical method of multiplying $3 \cdot 52$ uses the distributive property because we compute the product by computing the sum of $3 \cdot 2$ and $3 \cdot 50$. Give an example on the board where you compare the **vertical** method of multiplying a two-digit number by a single-digit number and the less familiar **horizontal** format used in the distributive property; for example, $3 \cdot 52 = 3(50 + 2) = 3 \cdot 50 + 3 \cdot 2$.*

SP-75. Algebra Tiles For this problem, you will need algebra tiles in two shapes: rectangles and small squares. Suppose the small square has a side of length 1, and the rectangle has a side of unknown length, say x.

a. Use this information to find the area of each of the figures:

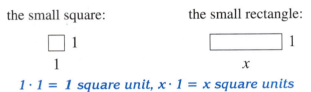

b. Trace each of the tiles in your Algebra Tool Kit. Mark the dimensions along each side, and clearly label each tile with its area. Algebra tiles are often referred to by their areas.

Problem SP-76 develops the idea of using algebra tiles to form composite rectangles and builds on the idea of representing area as a product and as a sum. Students may decide that the two composite rectangles shown here are different:

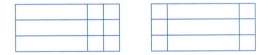

The purpose is to develop familiarity with the tiles and the formation of composite rectangles. Answers will vary tremendously. However, students should be able to explain their answers.

SP-76. Composite Rectangles and Area You can use algebra tiles to build **composite** rectangles. A composite rectangle is a rectangle made with at least two tiles. For example, you can build a composite rectangle using three rectangles and six small squares like this:

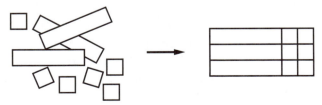

a. Determine whether you could build a composite rectangle with each set of tiles indicated in the following chart. If so, on your SP-76 Composite Rectangles and Area Resource Page, sketch a composite rectangle you could build; if not, answer "no."

b. For each composite rectangle, label each tile with its area as shown, and then label the length and width of the rectangle.

c. Use what you have done to write the area of each set of tiles as a sum of the areas of the pieces, and then as a product of the length and the width, if possible.

Number of $1x$ Tiles	Number of 1 Tiles	Is a Composite Rectangle Possible?	Sketch	Algebraic Expression for the Area as a Sum	Algebraic Expression for the Area as a Product
2	10	yes	$2\begin{array}{c} x \;+\; 5 \\ \begin{array}{\|c\|c\|c\|c\|c\|c\|} \hline 1x & 1 & 1 & 1 & 1 & 1 \\ \hline 1x & 1 & 1 & 1 & 1 & 1 \\ \hline \end{array} \end{array}$	$2x + 10$	$= (2)(x + 5)$
3	12	yes	$3\begin{array}{c} x \;+\; 4 \\ \begin{array}{\|c\|c\|c\|c\|c\|} \hline 1x & 1 & 1 & 1 & 1 \\ \hline 1x & 1 & 1 & 1 & 1 \\ \hline 1x & 1 & 1 & 1 & 1 \\ \hline \end{array} \end{array}$	$3x + 12$	$= 3(x + 4)$
3	5	no		$3x + 5$	
1	3	yes	$1\begin{array}{c} x \;+\; 3 \\ \begin{array}{\|c\|c\|c\|c\|} \hline 1x & 1 & 1 & 1 \\ \hline \end{array} \end{array}$	$1x + 3$	$= 1(x + 3)$
4	3	no		$4x + 3$	
4	4	yes	$4\begin{array}{c} x \;\;\;\;+1 \\ \begin{array}{\|c\|c\|} \hline 1x & 1 \\ \hline 1x & 1 \\ \hline 1x & 1 \\ \hline 1x & 1 \\ \hline \end{array} \end{array}$ or $2\begin{array}{c} 2x \;\;\;\;+ 2 \\ \begin{array}{\|c\|c\|c\|c\|} \hline 1x & 1x & 1 & 1 \\ \hline 1x & 1x & 1 & 1 \\ \hline \end{array} \end{array}$	$4x + 4$ $4x + 4$	$= 4(x + 1)$ $= 2(2x + 2)$

SP-77. On your Resource Page for SP-77, fill in the missing parts of the table with algebraic expressions and composite rectangles. The first row is done for you. Read the column headings carefully before you begin.

Sketch of Composite Rectangle	Area as a Product	=	Area as a Sum
Example:	$3(x + 5)$	=	$3(x) + 3(5)$, or $3x + 15$
	$4(2x + 3)$	=	$4(2x) + 4(3)$, or $8x + 12$
	$2(x + 3)$	=	$2(x) + 2(3)$, or $2x + 6$
	$5(x + 4)$	=	$5(x) + 5(4)$, or $5x + 20$
	$(x + 2)(6)$	=	$x(6) + 2(6)$, or $6x + 12$

SP-78. Distributive Property In SP-76 and SP-77 you saw that

$$2(x) + 2(5) = 2(x + 5)$$

because both sides of the equation represent the same area, and

$$3(x + 5) = 3(x) + 3(5)$$

because both sides of the equation also represent the same area. These are examples of the **distributive property of multiplication over addition.**

a. Add the *distributive property* to your Tool Kit with examples and explanations that are useful to you.

b. Use the distributive property to rewrite each of the following expressions, and draw a sketch of a composite rectangle represented by each equation.

i. 7(2) + 7(x) = ?   7(2 + x)

ii. 5(x + 3) = ?   5(x) + 5(3)

EXTENSION AND PRACTICE

SP-79. For each of the following figures, use the distributive property to write an equation that shows:

The area as a product is equal to the area as a sum.

Each figure is composed of algebra tiles, and it may help you to build the figures with tiles first.

a.

3(x + 2) = 3(x) + 3(2)

b.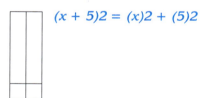

(x + 5)2 = (x)2 + (5)2

c.

5(x + 1) = 5(x) + 5(1)

d.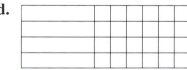

4(x + 6) = 4(x) + 4(6)

SP-80. Use the distributive property to match algebraic expressions in the left-hand column with equivalent expressions in the right-hand column. Not all expressions on the left can be matched with expressions on the right.

a. $4(x + 5)$ **D**   A. $x(x + 3)$

b. $4(2x) + 4(7)$ **F**   B. $3x + 3$

c. $3x + x$ **none**   C. $4x + 5$

d. $(x + 1)(3)$ **B**   D. $4x + 20$

e. $5(x + 4)$ **none**   E. $5(x + 4)$

f. $x^2 + 3x$ **A**   F. $4(2x + 7)$

Students will need their graphs for SP-81 to complete SP-112 (in Section 2.7).

SP-81. Complete this problem on separate sheets of paper. You will need to use them again in SP-112.

a. Copy and complete this table:

x	4	-1	5	0	-16	$\frac{1}{2}$	-3	0.3	2	$-\frac{1}{3}$	2.5	-11	-2	x
y	11	1	13	3	-29	4	-3	3.6	7	$\frac{7}{3}$	8	-19	-1	$2x + 3$

b. Describe in words the rule you used in part a. *Double the number and add 3.*

c. Write the rule using a variable such as x. *$y = 2x + 3$*

d. Graph the ordered pairs (x, y).

e. Find three more pairs of points that satisfy the rule without using any more whole numbers. *A few possibilities: $\left(\frac{-1}{2}, 2\right)$, $(-4.5, -6)$, $(3.14, 9.28)$*

f. If everyone on campus today used three different x-values, found the corresponding y-values, and put these points on the same graph, would the "line" formed by the points be complete? *No, since only finitely many points would be plotted.*

g. Are there any x-values that it would not make sense to choose? *no*

In SP-82, the descriptions of continuous and discrete graphs are not meant to be precise. (We implicitly assume the domain is an interval.) Our purpose in these problems is only for students to get an intuitive idea of differences between discrete and continuous graphs and situations they might represent. You could ask if an input (or output) value of 1.3 would make sense if the problem were about the number of cars in a parking lot, for example. From now on, students should draw continuous graphs whenever it makes sense to do so.

SP-82. Connected or Isolated? In this problem, you will compare two graphs (from SP-53 and SP-81) that look different even though they use the same rule, $y = 2x + 3$. You noticed in SP-53 that the Hawaiian Punch Mix graph has gaps or holes, since some numbers don't make sense in the context of the problem. The graph of the rule $y = 2x + 3$ in SP-81, on the other hand, can be filled in since any value of x can be chosen. So even though the two graphs look similar, one has isolated points and one does not. It is useful to have terminology to distinguish between these two types of graphs.

a. Read the following descriptions, and copy the information in your Tool Kit.

> **Continuous Graph** A graph is **continuous** if it has no breaks, jumps, or holes. The points you get from a table can be connected to form a line or curve without lifting the pencil from the paper. Points may only be meaningfully connected if every point on the line or curve you draw makes sense in the context of the problem. Mathematicians often assume a graph is continuous unless a specific context implies that it is not.
>
> **Discrete Graph** A graph is **discrete** if each point on the graph is isolated from the other points on the graph. In the context of a problem, it won't make sense to connect the points (x, y) you get from a table for the rule.

So, the Hawaiian Punch Mix graph is *discrete* (its points are isolated), because connecting its points would include some points that do not make sense for Matt's punch recipe. On the other hand, the graph of the rule $y = 2x + 3$ from SP-81 is *continuous* (its points are connected to form a line). For every value of x, you can find a point on the graph.

b. Should the points on your Polygon Perimeters graph (SP-18) be connected? Or are there points (N, P) that don't make sense in the context of the problem? Use your answers to these questions to explain whether the graph is discrete or continuous. *Discrete—both the figure numbers and the perimeters must be whole numbers.*

c. Should the points on your graph of the data for Professor Speedi's commute (SP-38) be connected? Or, are there points (x, y) that don't make sense in the context of the problem? Use your answers to these questions to explain whether the graph is discrete or continuous. *Continuous— because you can calculate a time for any speed > 0.*

 SP-83. Symbol: "..." The symbol "..." is called an **ellipsis**. It indicates that certain values in an established pattern, or sequence, have not been shown, although they are part of the sequence. For example, in the sequence

$$3, 3.5, 4, 4.5, \ldots$$

the ellipsis (...) means that the pattern continues with 5, 5.5, 6, 6.5, 7, 7.5, and so on, even though the rest of the sequence is not written.

a. Find a pattern for each of the following sequences, and list the next three numbers using the pattern you found:

7, 10, 13, 16, 19, ...   *22, 25, 28*

1, 1, 2, 3, 5, 8, ...   *13, 21, 34*

b. Find two different patterns that could describe the following sequence, and list the next three numbers using each pattern you found. Describe each pattern in words.

1, 2, 4, ...

Many possibilities: For example, 8, 16, 32, multiply the previous term by 2; 7, 11, 16, add the previous term and its term number.

c. Copy the information about *ellipsis* in your Tool Kit.

SP-84. Joyce and John each have a number. The sum of these numbers is 15.

a. Make a table showing several pairs of possible numbers, and graph your results. *(1, 14), (3, 12), (5, 10), (7, 8), (9, 6), (11, 4), (13, 2), (15, 0)*

b. If John's number is $9\frac{1}{3}$, what is Joyce's number? $5\frac{2}{3}$

c. If Joyce's number is 17, what is John's number? -2

d. Should the points on your graph be connected to form a line? Explain. *Yes, since any two numbers may be chosen as long as their sum is 15.*

SP-85. Use a guess-and-check table to solve the following problem. Write your answer in a complete sentence.

Dorothy and Frank left Wichita going in opposite directions. Dorothy drove 5 miles per hour faster than Frank, and in 4 hours they were 476 miles apart. How fast was each person traveling? *Dorothy, 62 mph; Frank, 57 mph*

SP-86. For parts a, b, and c, write an expression that represents each amount. Then explain the process in part d.

a. The college president cuts a 30-foot piece of ribbon at the dedication ceremony for the new Rose Garden. If one piece is x feet long, how long is the other piece? *$30 - x$ ft*

b. At the first WNBA game held in Sacramento, 4000 adult tickets were sold. If w women bought tickets to the game, how many men bought tickets? *$4000 - w$ men*

c. A 10-quart bottle of windshield cleaning solution is made of ammonia and water. If the bottle contains A quarts of ammonia, how many quarts of water are there? *$10 - A$ qt*

d. Samantha is working on a problem in which there is a quantity that is split into two parts, but all she knows is the total quantity. Using the words "one part" to represent one part of a whole amount, show Samantha how to write an expression for the other part.
whole amount − one part = other part

Take a few minutes at the start of the next class to have students compare answers to SP-87 and to make sure they are correctly stating the exponent pattern in their Tool Kit.

SP-87. Consider the following expressions:

i. $3^5 \cdot 3^2$ *3^7*   ii. $x^4 \cdot x^2 \cdot x^5$ *x^{11}*   iii. $2^{32} \cdot 2^{57}$ *2^{89}*

iv. $(3^5)^2$ *3^{10}*   v. $3^5 + 3^2$ *252*   vi. $2^5 \cdot 3^2$ *288*

a. Rewrite each of the expressions using a single exponent, if possible. Otherwise, write the number in standard form.

b. In part a what pattern do you see for *multiplying two or more exponential expressions with the same base*? Using the correct vocabulary ("base" and "exponent"), write your observations in your Tool Kit.
When you multiply two exponential expressions with the same base, the result is the base raised to the sum of the exponents.

SP-88. Symbols for Inequality There are several ways to express the relationship between two unequal numbers or expressions. You could represent the fact that Hakeem is more than 6 feet tall by writing

$$H > 6,$$

where H stands for Hakeem's height in feet. To make the statement that Spud is less than 6 feet tall, you could let S represent Spud's height in feet and write

$$S < 6.$$

Similarly, you could express the idea that …

x is less than or equal to 3	by writing	$x \leq 3$,
x is greater than or equal to 3	by writing	$x \geq 3$,
x is at most 3	by writing	$x \leq 3$,
x is at least 3	by writing	$x \geq 3$,
x is more than 3	by writing	$x > 3$,
x is less than 3	by writing	$x < 3$.

a. Copy the information about *how to use symbols to express inequality* in your Tool Kit.

b. Suppose x must fit the condition $x > 8$. List three integers that meet this condition. List three other numbers that are not integers, but that meet the condition. *Answers vary; e.g., 10, 230, 65; 8.01, $9\frac{1}{2}$, 27.99*

c. Suppose x must fit the condition $x \leq -2$. List four different numbers that meet this condition. *Answers vary; e.g., −2, −2.9, −10, $\frac{-5}{3}$*

In SP-89b, have students check that the pattern is valid for all lines up to and including the conjectured "last" line but fails for subsequent lines. (Note that an 8-digit display calculator cannot directly deal with the eighth and ninth lines.)

SP-89. Consider this sequence:

$1 \cdot 9 + 2 = $ _____ $1 \cdot 9 + 2 = 11$
$12 \cdot 9 + 3 = $ _____ $12 \cdot 9 + 3 = 111$
$123 \cdot 9 + 4 = $ _____ $123 \cdot 9 + 4 = 1111$
$\vdots$ $\vdots$
_____ $\cdot 9 + $ __ $ = $ _____ $12345678 \cdot 9 + 9 = 111111111$

a. Copy the sequence, fill in the blanks, and write the eighth line of the pattern without writing lines four through seven. A calculator will be helpful.

b. How far will this pattern continue? Experiment to check whether your conjecture is true. *Up to the ninth line: $123456789 \cdot 9 + 10 = 1111111111$*

2.6 THE EGG TOSS: A QUADRATIC GRAPH

SECTION MATERIALS

PLASTIC OR HARD-BOILED EGG (FOR SP-92)

POSTER-SIZE COORDINATE AXES DRAWN ON FLIP-CHART PAPER AND POSTED ON WALL (FOR SP-92)

STICKY DOTS (AT LEAST 10) FOR PLOTTING DATA IN SP-92

Take a few minutes at the start of class to have students compare their answers to SP-87 and to make sure they are correctly stating the exponent pattern in their Tool Kits.

The focus problem for the day is SP-92 (the Egg Toss). Its purpose is for students to look at parabolas in a physical context. We will revisit the ideas introduced here throughout the course. You may have to briefly discuss how to correctly input values on a calculator to evaluate the expression $80x - 16x^2$. Students might be interested in knowing that examples of parabolic shapes appear all around them: the reflector in a headlight of an older car, a satellite dish, the stream of water from a drinking fountain, the path of pouring milk, the path of a model rocket—each has the shape of a parabola.

SP-90. Rewrite each of the following so that the variable is on the left-hand side of the statement:

a. $6 \leq x$

$x \geq 6$

b. $100 > y$

$y < 100$

c. $-2 < x$

$x > -2$

SP-91. Betweenness Sometimes you may want to express in symbols the idea that a number is between two others, as in when the low temperature reading is between -3 degrees and 7 degrees. To express the idea that a number x is **between** -3 and 7, notice that there are two conditions on x:

"x is greater than or equal to -3 *and* x is less than or equal to 7."

We translate this as

$x \geq -3$ *and* $x \leq 7$,

or we can rewrite the left-hand statement to get

$-3 \leq x$ *and* $x \leq 7$.

We can express the idea in a single statement called a *compound inequality* by writing

$$-3 \leq x \leq 7.$$

This means that x is *both* greater than or equal to -3 *and* less than or equal to 7. In other words, x can be any number between -3 and 7, including -3 and 7. For example, x could be -2 because $-3 \leq -2 \leq 7$; or x could be 1.05 because $-3 \leq 1.05 \leq 7$.

a. Copy the information about how to express *betweenness* in your Tool Kit.

b. Suppose x must fit the restriction $5 < x < 8$. What are the only two integers that meet this condition? List four other numbers that are not integers, but that meet the condition. *Integers are 6, 7; other answers vary, e.g., 5.2, 5$\frac{1}{2}$, 7.99, 6.4.*

c. Suppose x must fit the restriction $-1 \leq x \leq 2$. List 10 different numbers that meet this condition. *Answers vary, e.g., -0.9, $\frac{-1}{3}$ 0, 1, 1$\frac{3}{4}$, 1.99.*

For SP-92, we like to bring a plastic or hard-boiled egg to class and demonstrate tossing it straight up into the air. This helps clarify confusion some students might have between the graph of the height of the egg versus time and the actual path of the tossed egg. It's useful to make a poster-size class graph using sticky dots for the students' data.

Here, and for the rest of the text, when the problem involves parabolas, rather than include entire tables and graphs in the instructor's answers, we've indicated the vertex and two other points for each graph. These are for your reference, to enable a quick visualization or sketch of the graph.

SP-92. The Egg Toss If an egg is tossed straight up into the air at a rate of 80 feet per second, its height y (measured in feet) above the ground after x seconds is given by $y = 80x - 16x^2$. (This is a formula used in physics.)

a. Make a table by choosing at least 10 x-values so that $0 \leq x \leq 5$. Be sure to include some noninteger values such as 0.25, 0.5, and 4.8. *(0, 0), (0.25, 19), (0.5, 36), (1, 64), (1.5, 84), (2, 96), (2.5, 100), (3, 96), (3.5, 84), (4, 64), (4.5, 36), (4.8, 15.36), (5, 0)*

b. Graph your information from part a. Scale the x-axis in increments of $\frac{1}{2}$ second and the y-axis in increments of 5 feet. *vertex: (2.5, 100); x-intercepts: (0, 0), (5, 0)*

c. How high is the egg after 1 second? *64 ft*

d. When is the egg exactly 96 feet high? When is the egg 200 feet high? *at 2 sec and at 3 sec after being tossed; never*

e. Describe the shape of the graph and what it represents in terms of tossing an egg into the air. *a parabola that opens downward; the height of the egg as it goes up and comes back down*

110 Chapter 2 Professor Speedi's Commute: Patterns and Graphs

    f. When does the egg reach its maximum height? What is the maximum height the egg reaches? (The highest point or lowest point of a parabola is called the **vertex** of the parabola.) *2.5 sec, 100 ft*

    g. What are the x-intercepts of the graph? What do these represent about the egg? *(0, 0) and (5, 0); the elapsed times when the egg is at ground level*

    h. If $x = 6$, what is the value of y? What might this number represent?
$y = -96$; e.g., 96 ft below a cliff

Have groups report on SP-93 to help avoid errors in subsequent problems.

SP-93. Scott says that $-x^2$ means the same thing as $(-x)^2$, but Ximena insists that they are different. In this problem you will help them settle the argument by graphing $y = -x^2$ and $y = (-x)^2$ on the same set of axes. Be sure to label each graph with its equation.

    a. The first expression, $-x^2$, can be translated as

        "first square x and *then* find the opposite (negative) of the square."

    Use these directions to make a table for $-3 \leq x \leq 3$ using at least 10 x-values. Then graph the data.

x	-3	-2	-1	-0.5	0	0.5	1	2	3	1.5
$y = -x^2$	-9	-4	-1	-0.25	0	-0.25	-1	-4	-9	-2.25

concave down parabola with vertex (0, 0)

    b. The second expression, $(-x)^2$, can be translated as

        "first find the opposite of x, *then* square it."

    Use these directions to make a table for $-3 \leq x \leq 3$ using at least 10 x-values. Then graph the data on the same set of axes you used in part a.

x	-3	-2	-1	-0.5	0	0.5	1	2	3	1.5
$y = (-x^2)$	9	4	1	0.25	0	0.25	1	4	9	2.25

concave upward parabola with vertex (0, 0)

    c. How could these two graphs be of help in settling Scott and Ximena's disagreement? Which claim is correct? Convince Scott and Ximena by explaining the difference in as many ways as you can.
Possible answers: The two expressions are different because the order of operations used for each expression is different. This makes the y-values different; specifically, except for $x = 0$, the y-values for $-x^2$ are always negative but the y-values for $(-x)^2$ are always positive. The graphs are different—one is a parabola opening downward and the other is a parabola opening upward.

You might begin SP-94 orally as a class exercise by asking students whether or not each graph is discrete.

SP-94. Sketch a graph to represent each of the situations or relationships described below.

a. As the number of gallons of gasoline you buy increases, the total cost of your gasoline purchase increases at a steady rate.

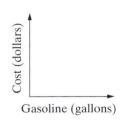

b. As the number of hours you have driven your car since you filled the tank increases, the amount of gasoline in your tank decreases at a steady rate. (Set up your axes with the number of hours on the horizontal axis.)

Solutions:

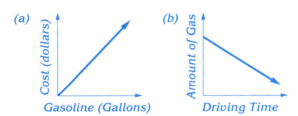

EXTENSION AND PRACTICE

SP-95. While working at his office, Ronnie eats jelly beans at a constant rate. He refills the jelly bean jar every morning. The number of jelly beans in the jar at any time is given by the equation

$$J = 80 - 16t,$$

where t represents the number of hours since Ronnie refilled the jar.

a. On a piece of graph paper, make a table for the rule $J = 80 - 16t$, with input values t and output values J. Complete the table using at least four different values for t. *(0, 80), (1, 64), (2, 48), (3, 32), (4, 16), (5, 0)*

b. Draw and label axes on your graph paper, and indicate the scale on each axis. Plot the points from the table, and then label the graph with its rule.

c. Find the x-intercept of your graph. (You may need to add more points.) What does the x-intercept represent in this situation? *(5, 0); jar is empty after 5 hr*

SP-96. Evaluate each of the following expressions for $x = 3$ and for $x = -1$:

a. $-3x^2$ *−27, −3*
b. $-x^2 + 2x$ *−3, −3*
c. $(-4x)^2$ *144, 16*
d. $(-x)^2 - 1$ *8, 0*

SP-97. So far in this chapter, many of the rules or equations you have graphed have been **linear** (in other words, their graphs are straight lines). However, not all of the equations you graph will produce lines. Look back at the graphing problems you've done. Find three equations whose graphs are straight lines. Copy each equation, and sketch its graph.

SP-98. Here are two composite line segments of equal lengths:

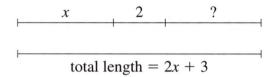

Since we know the total length is $2x + 3$, we can find the missing length:

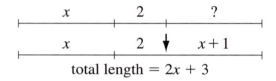

Find the missing length in each composite line segment.

a.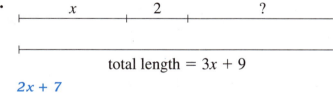

2x + 7

c.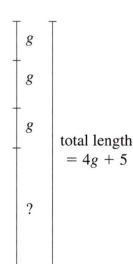

b.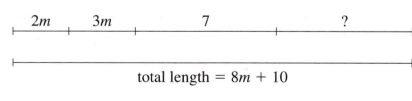

3m + 3

g + 5

SP-99. Copy the following table onto your paper and fill in the missing parts with algebraic expressions or composite rectangles.

Sketch of Composite Rectangle	Area as a Product	=	Area as a Sum
(rectangle 4 by (x+9), labeled x and 1)	4(x + 9)	=	4(x) + 4(9), or 4x + 36
(rectangle 2 by (x+1), labeled x and 1)	3(2x + 1)	=	3(2x) + 3(1), or 6x + 3
(rectangle labeled x, width divided into 3)	(x + x)(3)	=	3(x) + 3(x), or 3x + 3x, or 6x
(rectangle 6 by (x+1), labeled x and 1)	6(x + 1)	=	6(x) + 6(1)

SP-100. Use the distributive property to match each algebraic expression in the left-hand column with an equivalent expression from the two right-hand columns. (The expressions in the right-hand columns have been simplified.)

a. $4(x + 7)$ **D**   A. $x^2 + 1$ F. $6x + 5$

b. $x(x + 1)$ **B**   B. $x^2 + x$ G. $28 \cdot x$

c. $10(x + y)$ **H**   C. $y^2 - 3$ H. $10x + 10y$

d. $y(y - 3)$ **J**   D. $4x + 28$ I. $6x + 15$

e. $3(2x + 5)$ **I**   E. $10xy$ J. $y^2 - 3y$

SP-101. Deanne says she should be able to use the distributive property to write $3(2 \cdot 5) = 3(2) \cdot 3(5)$, but Derek insists she is wrong. Is Deanne's statement correct? Explain why or why not. *No. $3(2 \cdot 5) = 3(10) = 30$, but $3(2) \cdot 3(5) = 6 \cdot 3(5) = 18(5) = 90$.*

SP-102. Definition: A **y-intercept** of a graph is a point where the graph crosses the y-axis.

a. What is the *x-coordinate* of *any* y-intercept? *x-coordinate is 0.*

b. Add the definition for *y-intercept* to your Tool Kit.

c. Write the coordinates of the y-intercept, if any, of the following graphs:

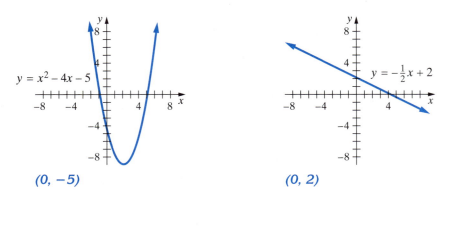

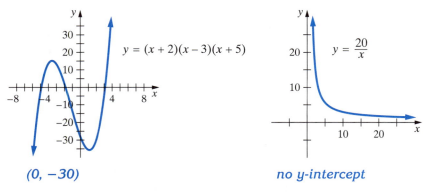

SP-103. Pair-a-Dice Lost? Suppose you have two standard fair six-sided dice.

a. If both dice are rolled, how many outcomes are possible? Make an organized list of the outcomes. (Refer to SP-48 if you need help.) *36*

b. How many ways are there of rolling a total of 6 on these two dice? *5*

c. What is the probability of rolling a total of 6? *$\frac{5}{36}$*

SP-104. Copy and solve these Diamond Problems:

a.

b.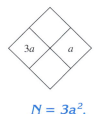

$N = 3a^2$,
$S = 4a$

c.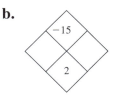

$5, -3$

d.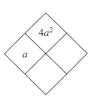

$E = 4a$,
$S = 5a$

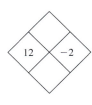

$N = -24$
$S = 10$

SP-105. Use a guess-and-check table to solve the following problem. Write your answer in a complete sentence.

Find two consecutive integers such that the sum of the first integer and five times the second integer is 107. *17, 18*

SP-106. Yu-Ping is keeping track of her checking account balance at the beginning of each month. If she wants to graph this data with months on the horizontal axis and dollars on the vertical axis, should she connect the points, or should she graph discrete, isolated points? Explain your answer. *The graph should be discrete. There is a skip, or gap, between the beginnings of two consecutive months.*

SP-107. Using information from OP-81 (Chapter 1), show how to calculate the approximate thickness of a single \$1 bill. $\frac{4.96}{454} \approx 0.011$ cm

2.7 FOLDING BILLS: MORE NONLINEAR GRAPHS

SECTION MATERIALS

SUMMARY DRAFT
TEMPLATE RESOURCE
PAGE (FOR SP-122)

Do these first two "head problems" with no discussion. The purpose of these problems is to continue to develop the idea of a variable as an unknown number. Discussion will be more appropriate later, after students have had several experiences with the problems.

Before you start, establish the following Rules for Head Problems:

1. *Listen carefully throughout the problem. The problem will be stated only once.*

2. *Only mental computation may be used, not pens, pencils, paper, or calculators.*

3. *Wait to be called on to give an answer.*

Head Problem 1: *Read the following lines to the whole class, pausing after each line to allow students time to perform calculations mentally.*

"Pick a number." [Pause to allow students a little time to choose a number.]

"Add 20." [Pause. Be sure to allow some wait time for students to do mental calculations.]

"*Multiply the sum by 2.*" *[Pause]*

"*Now subtract 30.*" *[Pause]*

"*Divide by 2, …* " *[Pause]*

"*… then subtract your original number.*" *[Pause]*

"*What result did you get?*" 5

Head Problem 2: As before, pause after each sentence to give students time to do the mental math:

"*Think of a number. Multiply it by 4. Add 8 to the product. Subtract 4. Divide by 2. Subtract 2, then divide by 2. What is the result?*"
the original number

About SP-108: *The Dollar Bill-Folds problem leads students to examine an exponential curve. Some students may need help with the conversion to kilometers in part e.*

SP-108. Dollar Bill-Folds You found in SP-107 that a $1 bill is about 0.011 centimeter thick. Suppose you folded it in half once—how thick would the two layers be? If you folded the dollar bill in half again, you'd have four layers. Imagine that you keep folding the dollar bill in half to make more and more layers.

a. Make a prediction. How thick do you think the folded dollar bill would be after 50 folds?

b. Make a table to record the results of your bill folding with "Number of Folds" as the input and "Thickness" of layers as the outputs. Do this for up to 10 folds.

Input	Number of Folds, F	0	1	2	3	4	…	10
	Number of Layers, L	1	2	$2^2 = 4$	$2^3 = 8$	$2^4 = 16$	…	$2^{10} = 1024$
Output	Thickness, T (in cm)	0.011	$(0.011)(2) = 0.022$	0.044	0.088	0.176	…	11.264

c. Plot the information from your table on a grid, with "Number of Folds" on the horizontal axis and "Thickness (in cm)" on the vertical axis. Does it make sense to connect the points on your graph? Explain. *No, number of folds must be a whole number.*

d. Describe in a few sentences a pattern that will allow you to find the thickness of the dollar bill after a given number of folds. *thickness = 0.011(2^F)*

e. *Really stretch* your imagination and suppose you could fold a dollar bill in half 50 times! Without filling in the rest of the table, determine how thick the dollar bill would be after 50 folds. Give this thickness in centimeters and in kilometers. (There are 100 centimeters in 1 meter, and 1000 meters in 1 kilometer.) *≈ 1.24 · 10^{13} cm, ≈ 1.24 · 10^8 km*

SP-109. Rowan and Martin made a careful graph of $y = \sqrt{x}$ in their algebra class last week. It looks like this:

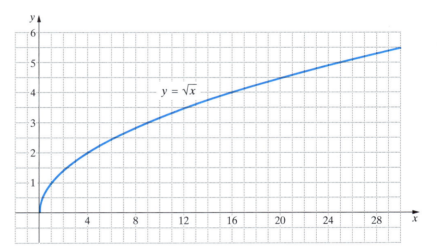

a. When Rowan made a table of values, the first value he chose for x was 9. Find the corresponding value of y, and label that point on the graph with its coordinates. *(9, 3)*

b. Without using a calculator, use the graph to estimate the value of ? that will make this equation true: $\sqrt{3} = ?$ *≈ 1.7*

c. Without using a calculator, use the graph to estimate the value of ? that will make this equation true: $\sqrt{?} = 5$ *25*

d. Without using a calculator, use the graph to estimate the value of ? that will make this equation true: $\sqrt{?} = 4.5$ *≈ 20*

e. Describe how to use the graph to estimate the value of ? in each of the following types of statements:

$$\sqrt{\text{a number}} = ? \qquad \sqrt{?} = \text{a number}$$

Left-hand equation: find given number on x-axis, go vertically up to curve, then move horizontally to y-axis to find the answer. Right-hand equation: start with given number on y-axis, go horizontally to curve, then down to x-axis.

SP-110. Sometimes the ellipsis symbol "..." is used even though a sequence does not go on indefinitely. Instead, we could use the ellipsis to represent the missing middle numbers of an established pattern. In this case, the first few numbers show how the pattern begins, and the last couple of numbers show where to stop.

Example: To indicate counting by 2s from 2 to 40, you could write out every number, like this:

2, 4, 6, 8, 10, 12, 14, 16, 18, 20, 22, 24, 26, 28, 30, 32, 34, 36, 38, 40.

Or, you could write only the first few and last few numbers, and let the ellipsis stand for the rest:

2, 4, 6, 8, ..., 38, 40. ▲

a. Add the new information about *ellipsis* to update your Tool Kit.

b. List all of the numbers in the sequence given by

8, 8.2, 8.4, 8.6, ..., 9.8, 10.

8, 8.2, 8.4, 8.6, 8.8, 9.0, 9.2, 9.4, 9.6, 9.8, 10

c. List all of the numbers in the sequence given by

−4, −3.5, −3, ..., 2.5, 3.

−4, −3.5, −3, −2.5, −2, −1.5, −1, −0.5, 0, 0.5, 1, 1.5, 2, 2.5, 3

SP-111. In this problem you will graph the equation $y = -x^2 + 2x - 3$ for $x = -2, -1, 0, ..., 4$.

a. Make a table using the input values $x = -2, -1, 0, ..., 4$. (You will have to fill in the missing x-values represented here by the ellipsis.) Then graph the equation and label your graph. *vertex: (1, −2); two points on parabola: (0, −3), (2, −3)*

x	−2	−1	0	1	2	3	4
$y = -x^2 + 2x - 3$	−11	−6	−3	−2	−3	−6	−11

b. Identify the vertex (the highest point or lowest point) of the parabola, and write its coordinates. *vertex: (1, −2)*

c. Identify the y-intercept and the x-intercepts, if any, of the parabola, and write the coordinates of each. *x-intercepts: none; y-intercept: (0, −3)*

EXTENSION AND PRACTICE

Problem SP-112 starts to make the connection between the geometric and algebraic meanings of an x-intercept.

SP-112. Look back at your sketch of the graph of $y = 2x + 3$ from SP-81.

a. Estimate the x-intercept of the graph. *(−1.5, 0)*

b. Find a value for x that makes the equation $2x + 3 = 0$ true. If you are unsure how to solve the equation algebraically, use the guess-and-check strategy shown in the following table. *$x = -1.5$*

Guess: Value of x	Value of $2x + 3$		Check: Does $2x + 3 = 0$?
0	$2(0) + 3$	=	3 (too high)

c. How do your answers to parts a and b compare? *The values for x are the same.*

SP-113. Robbie set up this guess-and-check table for a problem about a rectangle:

Guess: Width	Length	Perimeter		Check: Perimeter = 60?
5	11	$2(5) + 2(11)$	=	32 (too low)
7	15	*$2(7) + 2(15)$*	*=*	*44 (too low)*
10	21	*$2(10) + 2(21)$*	*=*	*62 (too high)*

a. Copy and complete the table, as it is shown, but do not solve the problem.

b. If the next guess for the width is 9, describe in words how you would calculate the length. *Double the 9, and then add 1.*

c. Use your description from part b to write the length in symbols if the next guess for the width is x. *$2x + 1$*

 SP-114. Ginger wants to multiply 47 by 4 without using a calculator.

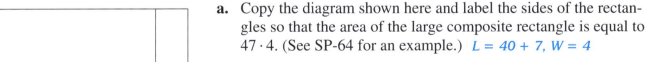

a. Copy the diagram shown here and label the sides of the rectangles so that the area of the large composite rectangle is equal to $47 \cdot 4$. (See SP-64 for an example.) *$L = 40 + 7, W = 4$*

b. Explain how Ginger can use the rectangle to mentally compute a value for $47 \cdot 4$. *$160 + 28 = 188$*

SP-115. Use the distributive property to fill in the following table with equivalent expressions for the areas of rectangles. Make a sketch or use algebra tiles if it will help.

	Area of Rectangle as a Product	=	Area of Rectangle as a Sum
a.	$3(x + 4)$	=	$3x + 12$
b.	$x(x - 3)$	=	$x^2 - 3x$
c.	$-7(c + 8)$	=	$-7c - 56$

(Reverse the process, or "undistribute")

d.	$4(\underline{x} + \underline{2})$	=	$4x + 8$
e.	$5\underline{x}(\underline{x} + \underline{2})$	=	$5x^2 + 10x$
f.	$\underline{3}(\underline{x} + \underline{3})$	=	$3x + 9$

SP-116. Patty's Hamburgers advertises that the company has sold an average of 405,693 hamburgers per day for the past 15 years.

  a. About how many hamburgers in total is this? Round your answer to the nearest 100,000. *≈ 2,222,400,000 or ≈ 2,222,800,000, depending on whether there are three or four leap years in the 15-year period*

  b. Express your answer in part a in scientific notation. *≈ 2.22 · 10⁹*

SP-117. Write the meaning of each expression below and *then* simplify using exponents. Write each answer in exponential form if possible; otherwise write it in standard form.

  a. $3^5 \cdot 3^2$ *3⁷*   b. $3^5 + 3^2$ *252*   c. $3^5 \div 3^2$ *3³*   d. $3^5 - 3^2$ *234*

SP-118. Here are two more rate problems:

  a. Ms. Escargot drove 50 miles per hour for 45 minutes. How far did she drive? Show how you got your answer. *$50\left(\frac{3}{4}\right) = 37.5$ mi*

  b. Mr. Snail drove 25 mph for 80 miles. How long did it take him? Show how you got your answer. *$\frac{80}{25} = 3.2$ hr, or 3 hr and 12 min*

SP-119. Sally says $10^3 \cdot 10^4 = 100^7$, but Elaine is sure that $10^3 \cdot 10^4 = 10^7$. Which calculation is correct? Use examples or numbers to explain your reasons. *(10 · 10 · 10)(10 · 10 · 10 · 10) = 10⁷*

SP-120. Find the perimeter of Mr. Rivera's garden. *46 in.*

$3\frac{5}{8}$ in.

$2\frac{3}{4}$ in.

$4\frac{3}{8}$ in.

Garden

$12\frac{1}{4}$ in.

Problem SP-121 could be assigned to enrich students' patterning skills, but it is not crucial to the development of other ideas of the course. You might assign it to all students, or as extra credit, or only to students who are ready for more problem-solving exercises.

SP-121. Lawson shaded some numbered squares to fit a certain pattern. The first five figures in his pattern are shown here.

a. Study the figures, and try to find Lawson's pattern.

b. Copy and shade the next three figures to continue the pattern you found.

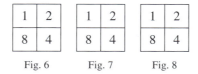

c. Describe your pattern in a sentence. *In each figure, the sum of the digits in shaded boxes equals the figure number.*

d. If you were to extend the pattern as far as you could, what would be the last figure in the sequence? *the 15th figure, since 1 + 2 + 4 + 8 = 15 is the maximum number that can be shaded.*

SP-122. Chapter 2 Summary Rough Draft You've been doing lots of problems. Now it's time to pause and to reflect on the main ideas of this chapter. By taking the time to carefully think and write about the focal ideas of the chapter, you are helping to organize your ideas about what you've learned in a way that should be meaningful and useful to you.

In Chapter 2 you've learned an important problem-solving strategy: to look for, describe, and use patterns. You've used patterns to make conjectures, and have seen that it is important to test conjectures. You have also learned to apply your patterning skills to make graphs of rules.

These are the main ideas of Chapter 2:

- An important problem-solving strategy is to look for, describe, and use patterns. We can use patterns to make conjectures. It is important to test conjectures.

- Graphs, tables, and rules are related to the pattern or situation they describe.

- The graph of an equation follows a pattern. Some graphs are *discrete* (made of isolated points), and some graphs that can be meaningfully connected with a smooth curve are *continuous*.

- A point on a graph that doesn't follow the pattern of the other points is probably miscalculated.

- The graphs of some equations are *lines*. Other rules produce *parabolas* as their graphs. Still other equations produce *curves* that are not necessarily parabolas.

a. Make photocopies of the Summary Draft Template Resource Page (or follow the same format on your own paper). Use a page to complete the following steps for each of the five main ideas of the chapter:

- State the main idea.

- Select and recopy a problem that is a good example of the idea and in which you did well.

- Include a completely worked-out solution to the selected problem.

- Write one or two sentences that describe what you learned about the idea.

You will have the opportunity to revise your work, so at this point you should focus on the content of your summary rather than the appearance. Be ready to discuss your responses with your group at the next class meeting.

b. What problem in this chapter did you like best, and what did you like about it?

c. What were the most difficult parts of this chapter? List sample problems, and discuss the hard parts.

2.8 THE BURNING CANDLE INVESTIGATION (OPTIONAL)

SECTION MATERIALS
BIRTHDAY PROBLEM
  RESOURCE PAGE
  (OPTIONAL)
BURNING CANDLE VIDEO,
  OR OTHER MATERIALS
  DESCRIBED HERE

About the Materials: *You'll need the following materials for SP-123 and SP-124: either (1) the video presentation of a burning candle experiment entitled* Patterns and Functions—An Application of Graphing *produced by San Juan High School, San Juan Unified School District (©1990 by Studio E-3 Productions, SECC/SJUSD); or (2) your own video; or (3) birthday candles, a timer, and electronic scales with an LED readout if you do the experiment yourself rather than use a video presentation. The nine-minute videotape produced at San Juan High School may be purchased for $11 (including postage and handling) by writing CPM Educational Program 1233 Noonan Drive, Sacramento, CA 95822.*

A Birthday (Head) Problem: *Rather than starting this section with a Head Problem, it's fun to introduce the Burning Candle Investigation with the Birthday Problem, which some students may be able to do in their heads, but most will need a calculator to do. If you come back to it later when students begin to write expressions using variables in Chapter 3, they'll see how the "trick" works. If students would like to have a copy of the instructions, copy the Birthday Problem Resource Page for them.*

Read the following instructions to the whole class. Pause after each line to allow students time to calculate.

"Start with the number of the month in which you were born. Add 4. Multiply the sum by 10. Subtract 15. Multiply by 5. Add the day of the month on which you were born. Subtract the number of days in April. Double the result. Subtract the day of the month in which you were born. Add 15. Tell me the result."

To learn the month and date of the student's birthday, subtract 205 from the number the student gives. For example, if the student's number is 634, then 634 − 205 = 429, so the student was born on 4/29.

About the Burning Candle Investigation (SP-123 and SP-124): *If your syllabus allows time for the Burning Candle Investigation, this will help students consolidate and extend their understanding. As with the other optional exercises in the book, it will give students an opportunity to deal with some fairly messy data collection and to use their analysis skills to make a prediction in a "realistic" setting. Students will collect, tabulate, and graph time and mass data, and then interpret their graphs.*

You might introduce the Burning Candle Investigation by asking students to discuss instances in which it is possible to predict whether, or when, an event will occur. Mention that constant or measurable rates of change make reasonable predictions possible. **Prior** *to showing the video, ask whether the students think it's possible to predict when a lit candle will go out. Most will say yes. Solicit reasons for this assertion, and ask how it could be verified.*

Show part of the video, and then have groups discuss how they want to organize the gathering of data. As a class, discuss the units for scaling the mass and time axes. Allow the students to choose their own way of scaling the axes, and perhaps make the mistake of having their graphs go off the page. Allow them to struggle and to see a need to revise their work. We would like students to learn that sometimes work must be revised, which is part of learning and **not** *wasted time. Show the video again (and maybe even a third time) to allow students to check their predictions.*

There may be confusion about the meaning of "elapsed time" in parts c through e in SP-123. Be sure to distinguish between elapsed time and the real time.

On the video of the burning candle, students will see the candle, a digital scale readout, and a digital watch.

Have each group write, in a list on the board, the time they think the candle will go out. Ask the class to devise a way to use these guesses to make a fairly accurate prediction. Students will usually come up with the idea of eliminating the extreme values and calculating the mean of the remaining values.

The rate at which the candle in the video burns is about 0.01 grams for every 9 seconds, or 0.067 g/min. The candle's mass, y, can be estimated from the linear relation

$$y = 0.82 - 0.067x \text{ where } 0 \leq x \leq 12.24 \text{ min.}$$

Thus, the candle should go out around 11:55:29, which is 12.24 min (or 0:12:14) after the candle is first shown with a mass of 0.82 g.

Because the scale readouts are not continuous and are given in increments of 0.01 g, the data are more accurately represented by the graph of a step function: The graph remains constant over intervals—of 8 to 10 seconds in length—then "steps down" one-hundredth of a gram at a time. For the purposes of this assignment, students should just draw a line approximately fitting their data points without necessarily hitting any of them. The candle does burn continuously, even though the scale measures in increments.

If you decide to run the experiment yourself rather than use the video presentation, the investigation will take about two class periods. Don't rush the students! The goal of this activity is for students to apply their data organization and analysis skills, and achieving that goal may take extra time.

Practice the experiment before trying it in class with an identical candle. Make sure you choose a candle that changes mass significantly in a short period of time. Measure the mass of the candle, light it, note the time, and observe the changes in mass over 10 minutes. If the candle loses mass too slowly, the experiment will not be effective.

While we don't recommend doing the Burning Candle Investigation without some sort of model (a video or your own experiment), it is possible to have students complete their investigation by simply using the table of data from the video given here in the instructor answer for SP-123c.

SP-123. The Burning Candle Investigation Suppose it's your friend's birthday and you want to surprise her by walking into the room carrying a piece of cake with a lighted candle. Could you predict how long it will be before the candle goes out?

To answer this question, you'll use a video presentation of a burning candle to collect data and then make a graph and look for a pattern.

a. Gather data from the burning candle video presentation. Note the mass of the candle at various times during the presentation, and write down your observations. You should make *at least* five observations, with at least 40 seconds between observations. Be sure to write down both the time and the associated candle mass.

b. Make a table from your data. *See part c for answer.*

c. Make a new column in your table to show elapsed time. Calculate the elapsed time for each of your observations, and add this information to your table.

Possible answer:

Real Time	Mass (in grams)	Elapsed Time (in seconds)
11:43:15	0.82	0
11:44:03	0.78	48
11:44:06	0.77	51
11:44:15	0.76	60
11:44:24	0.75	69
11:44:33	0.74	78
11:45:39	0.67	144
11:45:47	0.66	152
11:47:05	0.57	230

d. Set up a graph (using equal intervals) with elapsed time on the horizontal axis. *See answer in part e.*

e. Graph your data, comparing mass, m, to elapsed time, t.

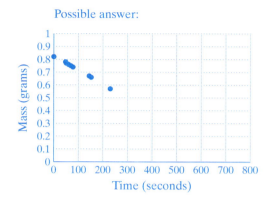

Possible answer:

f. Sketch a line or curve connecting your data points. In your group, compare graphs and decide which is the most accurate. Copy your group's choice. *The points on the graphs will not line up precisely, but the graph is essentially linear.*

g. Use your group's graph to predict the mass of the candle at the elapsed time of 1:20. Check the accuracy of your group's prediction by reviewing the burning candle video. *about 0.73 g*

h. Use your group's graph to predict the mass of the candle at the elapsed time of 2:47. Check the accuracy of your group's prediction by reviewing the burning candle video. *about 0.63 g*

SP-124. Use your graph from the Burning Candle Investigation to answer this question:

"If the candle continued to burn, when do you think it would go out?"

After discussing the question with your group, carefully explain your answer in complete sentences. *Once the total elapsed time is found, we can use it to calculate the actual time the candle will extinguish. A question for class discussion: "Is the candle's mass 0 grams when the candle goes out?"*

If you wish to assign practice problems for individual work, see the Extension and Practice exercises in Section 2.9.

2.9 SUMMARY AND REVIEW

The purposes of this section are to give students more experience with calculators and graphing, and to make them aware that not all graphs are linear or parabolic.

Start with Head Problems 3 and 4. Then go back and use "x" for the starting number and record the algebra as shown below. Spend no more than 15 minutes doing this.

Head Problem 3: *As with other Head Problems, pause after giving each direction aloud to the class. This time, though, as you read the instructions, write them in a column on the chalkboard. Then, when you go back to use **x** as a starting number, record the algebraic representation of each expression in the right-hand column one step at a time.*

Instructor's Directions	Recording of Discussion
Pick a number.	x
Multiply it by 2.	$2x$
Add 6.	$2x + 6$
Double the sum.	$2(2x + 6)$ or $4x + 12$
Divide by 4, then . . .	$(4x + 12) \div 4$
. . . subtract 3.	$x + 3 - 3$
What is your result?	
Same number as chosen	x

Questions that might help the discussion:

- *Did you all start with the same number?*
- *Did you all end with the same number?*
- *Describe the results you all got.*
- *How can we write our **guesses** when we "think of a number" for each of you? for all of you?*
- *How can we write the **process** down for each of you? for all of you?*

Head Problem 4:

"Think of a number. Double it. Add 3. Double the sum. Add your original number. Subtract 1. Divide by 5, then subtract your original number. What is the result?" 1

SP-125. Chapter 2 Summary Group Discussion Take out the rough draft summary you completed in SP-122. Take this time to discuss your work; use homework time to revise your summaries as needed.

For each of the five main ideas of the chapter, listed in SP-122, choose one member of the group to lead a short discussion. When it is your turn to be the discussion leader, you should do the following:

- Explain the problem you chose to illustrate your main idea.
- Explain why you chose that particular problem.

- Tell which problem you liked best, and what you liked about it.
- Tell what you thought were the most difficult parts of this chapter.

This is also your chance to make sure your summary is complete, to update your Tool Kit, and to work together on problems you may not be able to solve yet.

SP-126. Tool Kit Check In addition to your Tool Kit entries from Chapter 1, your Tool Kit should include entries for at least the following topics:

Algebra

- coordinate graphing *SP-4*
- reciprocal of a number *SP-10*
- point of intersection of two graphs *SP-16*
- linear and nonlinear graphs *SP-19, 97*
- calculating the time if you know the distance and average speed *SP-37*
- x-intercept of a graph *SP-62*
- algebra tiles: 1 and x *SP-75*

- distributive property *SP-78*
- continuous graph *SP-82*
- discrete graph *SP-82*
- ... (ellipsis) *SP-83, 109*
- multiplying expressions with like bases *SP-87*
- using symbols for inequality *SP-88*
- betweenness *SP-91*
- y-intercept of a graph *SP-102*

Calculator
- calculating the reciprocal of a number *SP-44*

You are not limited to topics on this list. If there are other ideas you'd like help remembering, add them to your Tool Kit as well.

a. To be sure that your Tool Kit entries are clear and accurate, exchange Tool Kits with members of your group, and read each other's entries.

b. At home, make any necessary revisions, clarifications, or updates to your Tool Kit.

SP-127. In the following pattern, each figure is composed of squares:

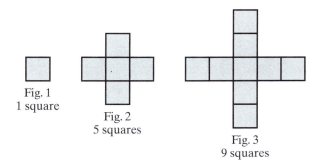

Fig. 1
1 square

Fig. 2
5 squares

Fig. 3
9 squares

a. Determine how many squares Figure 4 will have. *13*

b. Record all of your information in a table. Use a pattern you see in your table to predict how many squares there will be in Figure 10. How many squares do you think there will be in Figure 100? Write one or two sentences to explain how to find the number of squares for any figure in this pattern. *37; 397; possible response: multiply the figure number by 4 and subtract 3.*

If the figure number is F, state a formula (rule) for finding the number of squares, S, of the shape in Figure F. $S = 4F - 3$ or $4(F - 1) + 1$

c. Suppose you graphed the points in your table with the figure number on the horizontal axis, and the number of squares on the vertical axis. Make a conjecture: Do you think the graph would be linear, or not? Explain your answer. *linear, since the number of squares increases by the same amount for each successive figure.*

d. If you graphed the points in your table as described in part c, should your graph be discrete or continuous? Explain. *discrete; It doesn't make sense to connect points because figure numbers and number of squares must be whole numbers.*

e. To check your conjecture from part c, graph the points in your table with the figure number on the horizontal axis, and the number of squares on the vertical axis.

SP-128. Look at the equations and graphs from the Egg Toss problem (SP-92) and SP-97 and SP-112. Find a relationship between the shape of the graph and its equation. Explain how to tell if the graph of a formula (or rule) will be a straight line or not. *The graph of a formula with an x^1 term will be a line; the graph of one with an x^2 term will not.*

SP-129. *Without* drawing a graph, describe the shape of the graph of each of the following equations:

a. $y = x^2 - 3$ *parabola, or curve*

b. $y = 2x - 3$ *line*

There are enough exercises here to be used as homework assignments for Section 2.8 as well as for this section.

EXTENSION AND PRACTICE

SP-130. Make a table of values and graph the equation
$$y = x^3 \quad \text{for } x = -2, -1.5, -1, \ldots, 1.5, 2.$$

The graph passes through (−2, −8), (0, 0), and (2, 8).

SP-131. Make a table of values and graph the equation

$$y = x \quad \text{for } x = -4, -3, \ldots, 3, 4.$$

SP-132. Follow these steps to make a graph of the rule described by the table in part a.

a. Copy and complete the table:

Input (x)	2	1	-3.5	8.1	5.8	0.3	-10	x
Output ($y = $____)	0.2	0.1	-0.35	*0.81*	*0.58*	*0.03*	-1	*0.1x*

b. Explain in words what is done to the input value, x, to produce the output value, y. *Multiply the input value by $\frac{1}{10}$.*

c. Write the process you described in part b in algebraic symbols. *$y = 0.1x$*

d. Graph the data. Use a full sheet of graph paper, and scale your axes so that every square represents 0.1 unit. Use the same scale on both the x-axis and the y-axis. (You may not be able to fit all the points from the table onto your graph if some of their x- or y-values are too large or too small.)

SP-133. Do the same steps as in SP-132, but this time use the following table:

Input (x)	5	10	$6\frac{1}{2}$	$15\frac{1}{2}$	-9	2	$9\frac{1}{2}$		x
Output ($y = $____)	$7\frac{1}{2}$	$12\frac{1}{2}$	*9*	*18*	$-6\frac{1}{2}$	$4\frac{1}{2}$	12		$x + 2\frac{1}{2}$

SP-134. Write the meaning of each exponential expression, and *then* simplify using exponents. Show all the steps, as in the example.

Example: $(x^3)^2 = (x^3)(x^3) = (x \cdot x \cdot x) \cdot (x \cdot x \cdot x) = x^6$ ▲

a. $(x^4)^2$ *$x^4 \cdot x^4 = x^8$* b. $(x^2)^3$ *$x^2 \cdot x^2 \cdot x^2 = x^6$*

c. $(2x^2)^4$ *$16x^8$* d. $(x \cdot y)^2$ *$x^2 \cdot y^2$*

e. $(x^2 \cdot y^3)^3$ *$x^6 \cdot y^9$* f. $2(x^2)^4$ *$2x^8$*

SP-135. Walter says that $x^7 \cdot x^3 = x^{10}$. James says that $x^7 \cdot x^3 = x^{21}$. Which claim is correct? To settle the disagreement, write an explanation of how Walter and James could figure out the right answer. *One possible answer: $x^7 \cdot x^3 = x^{10}$ is correct, because $x^7 \cdot x^3 = x \cdot x \cdot x \cdot x \cdot x \cdot x \cdot x \cdot x \cdot x \cdot x = x^{10}$.*

SP-136. Copy the following table and fill in the missing parts with algebraic expressions and composite rectangles.

Sketch of Composite Rectangle	Area as a Product	=	Area as a Sum
(rectangle with x and 1 labeled, 2 rows tall)	$2(3x + 2)$	=	$2(3x) + 2(2)$, or $6x + 4$
(vertical rectangle with x and 1 labeled, 2 columns wide)	$(x + 4)2$	=	$x(2) + 4(2)$, or $2x + 8$
(rectangle with x and 1 labeled, 3 rows tall)	$3(x + 1)$	=	$3(x) + (3)1$

SP-137. Use the distributive property to match algebraic expressions in the left-hand column with equivalent expressions in the right-hand column. Not all expressions on the left can be matched with expressions on the right. (The expressions on the right have been simplified.)

a. $6(5x - 3)$ **F**   A. $28xy$

b. $(a + b)2$ **D**   B. $-3x(x + 3)$

c. $21x + 7y$ **none**   C. $3x - 5$

d. $5 + 5y$ **E**   D. $2a + 2b$

e. $-3x^2 - 9x$ **B**   E. $5(1 + y)$

f. $\frac{1}{2}(6x - 10)$ **C**   F. $30x - 18$

SP-138. Use the distributive property to rewrite each of the following expressions:

a. $4(x + 5)$ *4x + 20*

b. $5x + 15$ *5(x + 3)*

c. $24 + 8y$ *8(3 + y)*

d. $-3(2y - 5)$ *−6y + 15*

SP-139. Hakeem set up this guess-and-check table for a problem about money:

Guess: # Nickels	Value Nickels	# Dimes	Value Dimes	# Quarters	Value Quarters	Total Value of Coins	Check: total value = $3.55?
5	$0.25	7	$0.70	4	$1.00	$0.25 + 0.70 + 1.00 =	$1.95 (too low)
7	$0.35	9	$0.90	6	$1.50	$2.75	too low
12	$0.60	14	$1.40	11	$2.75	$4.75	too high

a. Complete the table as it is shown, but do not solve it.

b. If the next guess for the number of nickels is 10, describe how you would calculate the number of dimes. *Add 2 to 10.*

c. If the next guess for the number of nickels is 10, describe how you would calculate the value of the quarters. *$0.25(10 − 1)*

d. If the next guess for the number of nickels is x, write your descriptions from parts b and c in symbols. *# dimes = x + 2; value of quarters = $0.25(x − 1)*

SP-140. Use a guess-and-check table to solve the following problem. Write your answer in a complete sentence.

Eduardo used raspberry juice and lemon-lime soda to make $11\frac{1}{2}$ quarts of punch for his party. If there were four more quarts of soda than raspberry juice, how much of each beverage was used in the punch? *$3\frac{3}{4}$ qt raspberry, $7\frac{3}{4}$ qt lemon-lime soda*

SP-141. Compute the numerical value of each of the following five-digit expressions:

a. $3 + 3 \div (3 + 3) + 3$ *6.5*

b. $3 \cdot 3 + 3^3 \div 3$ *18*

SP-142. Copy the following table. Then rewrite each expression using exactly one set of parentheses in order first to maximize and then to minimize its value.

Expression	Greatest Value	Least Value
$2^2 \cdot 3^2 - 1^2$	$(2^2 \cdot 3^2 - 1)^2 = 1225$	$2^2 \cdot (3^2 - 1^2)^2 = 32$
$5 + 7 \cdot 3^2 - 4$	$(5 + 7 \cdot 3)^2 - 4 = 672$	$5 + 7 \cdot (3^2 - 4) = 40$
$4^2 \cdot 3 - 2^2 \cdot 5$	$(4^2 \cdot 3 - 2)^2 \cdot 5 = 10{,}580$	$4^2 \cdot (3 - 2^2 \cdot 5) = -272$

SP-143. Algebra Tiles and Perimeter The four figures in parts a through d are drawings of rectangles built with these two algebra tiles:

You have been practicing with finding the areas of figures. This time, find the *perimeter* of each figure. It may help you to build the figures with tiles first, and to label the side lengths of each figure.

a.
10

b.
2x + 6

c.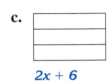
2x + 6

d.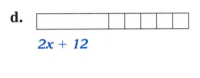
2x + 12

SP-144. Stan is sure that $2(x + 4) = 2x + 4$. Ollie is equally sure that $2(x + 4) = 2x + 8$.

a. Which claim is correct? *$2(x + 4) = 2x + 8$*

b. Use words and/or pictures to give an explanation that demonstrates the correct equation. *Sample response: Try it with a number for x. What if x were 5? $2(5 + 4) = 2(9) = 18$, which is what you get with $2(5) + 2(4)$ and is not what you get with $2 \cdot 5 + 4 = 14$.*

SP-145. Use the graph shown here.

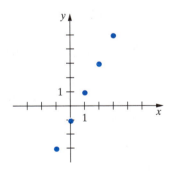

a. Copy and complete this table of values for the points in the graph:

x	−1	0	1	2	3
y	−3	−1	1	3	5

b. Find the rule $y = $ ____. *$2x - 1$*

SP-146. Copy and complete the given table for the rule $y = x^2 + 2x + 1$. Then use the table to graph the rule.

x	−3	−2	−1	0	1
y	4	1	0	1	4

vertex(−1, 0)

SP-147. For each of the following inequalities, list at least five values for x that meet the condition:

a. $-3 \leq x < 2$

Some: −3, −2, −1, 0, 1

b. $x < -10$

Some: −597, −25, −12, −11, −10.1

c. $0 < x \leq 1$

Some: 0.1, 0.2, 0.3, 0.5, 1.0

d. $0 < x$

Some: 0.5, 1, π, 7, 913

SP-148. Match each of these equations with its graph below:

a. $y = x^2 - 5$ *i*

b. $y = -2x + 5$ *iv*

c. $y = -x^2 + 5$ *ii*

d. $y = 2x + 5$ *iii*

i.
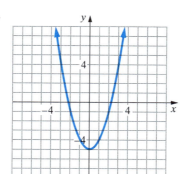

ii.
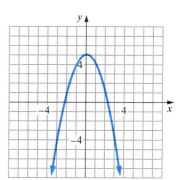

iii.
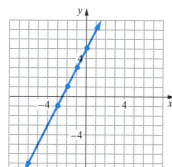

iv.
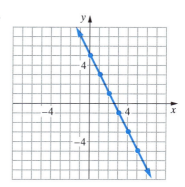

SP-149. In this chapter you have looked at the graphs of many equations. Write a paragraph to describe relationships you've noticed between equations and graphs. You may wish to include an example to illustrate your ideas.

SP-150. Sketch a graph to represent each of the situations or relationships described below. Label the axes. The quantity for the vertical axis is given first.

a. the height of a burning candle compared with elapsed time

b. the height of an unlit candle compared with elapsed time

Solutions: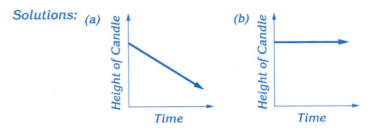

SP-151. Mel's Deli sells four kinds of sandwiches and three kinds of drinks. How many different combinations of one sandwich and one drink could a customer choose for lunch? (To be sure you've counted all possibilities, make a list.) *12*

Problem SP-152 (or SP-15) could be used as a bridge to Section 3.1, as students learn to write equations from guess-and-check tables. To do this, lead a whole-class discussion by using the overhead transparency master in the Resource Pages to help students review their work. Then help them see how to extend the table to write an equation for the problem.

SP-152. Brandon was given this guess-and-check table for the following problem about concert tickets:

Keisha sold 300 tickets to the Rolling Stones Reunion concert. General admission tickets sold for $35 each, and student tickets were $25 each. Total sales amounted to $8,280. How many students bought tickets to the concert?

Brandon set up the following guess-and-check table but forgot to label the columns:

Guess: # GA ticket	Value of GA ticket	# Student Tickets	Value of Student Ticket	Total Sales of Tickets	Check: Total = $8,280?
10	$350	290	$7,250	$350 + 7,250 =	$7,600 (too low)
45	$1,575	255	$6,375	$1,575 + 6,375 =	$7,950 (too low)
105	$3,675	195	$4,875	$3,675 + 4,875 =	$8,550 (too high)

a. Study the table and look for clues to the quantity that each column represents. Copy the table onto your paper, and then identify each column by labeling it with a heading to describe what it represents.

b. Brandon's next guess is "105." Test this guess by filling in the rest of the columns in your table to check it.

c. Finish solving the problem, and write your answer to the problem in a complete sentence. *78 general admission and 222 student tickets were sold.*

SP-153. Chapter 2 Summary Revision This is the final summary problem for Chapter 2. Using your rough draft from SP-122 and the ideas you discussed in your groups from SP-125, spend time revising and refining your Chapter 2 Summary. Your presentation should be thorough and organized, and done on a separate piece of paper.

CHAPTER THREE—FOR THE INSTRUCTOR

LIONS, TIGERS, AND EMUS
Writing and Solving Equations

MATERIALS
- Transparency of SP-152 Instructor Resource Page
- Pens and blank transparencies for the overhead projector, 1 or 2 per group (optional)
- Sets of algebra tiles
- Small resealable ("snack size") bags, jelly beans, and a two-pan balance scale
- Transparency of Two-Pan Balance Scale Resource Page, film canisters, and small square algebra tiles for the overhead projector
- About 20 jelly beans and 6 small resealable bags (or 20 small square algebra tiles and 6 film canisters) for each pair of students
- A wrapped package
- Resource Pages for LT-20, LT-26, LT-27, LT-36, LT-47, Two-Pan Balance Scale Resource Page, and Summary Draft Template

This chapter is one of the most powerful chapters in the course. In it students extend the organizational and patterning skills they developed using guess-and-check tables to include the process of writing algebraic equations to represent word problems. Students reinforce their understanding of the distributive property by using concrete models and diagrams. They also use hands-on models (which emphasize the notion of "balance" in an equation) and geometric representations (segments of equal lengths) to understand the process of solving linear equations.

By the end of the chapter, students will have only started feeling comfortable with writing equations, solving equations, and solving word problems, but they will have a solid foundation on which to build. Indeed, we don't expect students to master these skills within this chapter. Throughout the rest of the course they will work on strengthening and extending their abilities to write equations that fit given situations and to solve equations. Thus, there is no need to dwell on these skills to the point of mastery before moving on.

As students become more confident in their abilities to write and solve equations, they will become less dependent on using guess-and-check tables to solve problems. They may still set up a table to help write an equation, but will then solve the equation rather than filling in the table. Allow each student to proceed at a comfortable pace. As the course progresses, students will recognize that facility with solving equations is a powerful skill; they will see the efficiency of solving equations compared to the time-consuming trial-and-error approach of using a guess-and-check table. We have sequenced the problems to allow you to capitalize on student experiences to reinforce the strengths of each approach.

This chapter includes many problems for practice and consolidation, including ones that refresh and strengthen skills introduced in Chapters 1 and 2. There are also some problems which anticipate concepts developed in subsequent chapters.

CHAPTER THREE

LIONS, TIGERS, AND EMUS
Writing and Solving Equations

This chapter is a powerful one. In it you will tie together the organizational and patterning skills you developed in Chapters 1 and 2 using guess-and-check tables with the process of writing algebraic equations to represent word problems.

The power of the algebraic approach lies in the relative ease and efficiency with which you can solve equations. However, in the case of word problems, it may be difficult to figure out exactly how to write an equation that you should solve.

That's where a guess-and-check table can help. Although you can use guess-and-check tables to solve a wide variety of problems, many problems are extremely tedious, if not impossible, to solve by guessing and checking. The real power of a guess-and-check table is that it helps you—through organization and patterning—to figure out how to write an equation.

Many situations in daily life—such as scheduling deliveries, ordering supplies, and calculating costs of health insurance or phone service plans—can be represented by linear equations. The ability to write and solve equations associated with everyday problems is a useful, timesaving skill. You may have experienced a situation similar to the following one:

> You want to rent a kayak for an upcoming outing, but you can only afford to spend at most $50 on the cost of renting a kayak. Outdoor Adventures rents kayaks on a daily basis, and charges an $11.50 maintenance fee plus $6.50 per day. How long a trip would your budget limitation allow you to take?

Uncommon situations may also be modeled and solved using linear equations:

> Sydney, a ranger at Wild Animal Park who is known for his odd sense of humor and love of puzzles, noticed that there were twice as many tigers as lions in the reserve. He also noted that the lions, tigers, and emus had combined totals of 152 heads and 562 legs. How many of each type of animal were in the park?

You already know you can solve problems by organizing guesses and checks in a table. In this chapter you will learn to use a more efficient way. The skills you've developed using a guess-and-check table to solve a problem will help you write equations for situations. The ability to model a problem with an equation means that the amount of guessing you must do is reduced substantially—rather than guessing and checking to solve the problem, you'll solve an equation. Later in this chapter you'll learn how to solve equations in one variable that are linear*; that is, equations like

$$8x - 3 = 3x + 19 \quad \text{or} \quad 5(2x + 7) + 47 = 2(11 - 3x).$$

*In Chapter 2 you studied linear equations in two variables, like $y = 2x + 3$.

IN THIS CHAPTER YOU WILL HAVE AN OPPORTUNITY TO:

- use guess-and-check tables to write equations that model situations;
- use the distributive property to rewrite products of polynomials;
- solve linear equations;
- use the zero product property to solve equations.

Section 3.1 Writing Equations from Guess-and-Check Tables

MATERIALS	CHAPTER CONTENTS		PROBLEM SETS
Transparency of Instructor Resource Page for SP-152 (optional); overhead transparencies and pens (for LT-3 and LT-4)	3.1	Writing Equations from Guess-and-Check Tables *More Head Problems*	LT-1–LT-18
Sets of algebra tiles; LT-20 Composite Rectangles and Area Resource Page; LT-26 Resource Page; LT-27 Resource Page	3.2	Composite Rectangles and Algebraic Expressions for Area *More Head Problems*	LT-19–LT-35
For class demonstration: jelly beans, small resealable plastic bags, a two-pan balance scale, small square algebra tiles, film canisters, a transparency of the Two-Pan Balance Scale Resource Page; for each pair of students: about 20 jelly beans (or small square algebra tiles), 6 resealable plastic bags (or film canisters), Two-Pan Balance Scale Resource Page, LT-36 Using Packets and Jelly Beans to Solve Linear Equations Resource Page; LT-47 Representing Tiles with Polynomials Resource Page	3.3	Solving Linear Equations at the Candy Factory	LT-36–LT-56
Wrapped package to illustrate "doing and undoing" (for LT-57);	3.4	Doing and Undoing	LT-57–LT-78
algebra tiles (for LT-61 and LT-64)	3.5	Solving Literal Equations *The Zero Product Property*	LT-79–LT-101
Chapter Summary Draft Resource Page	3.6	Modeling Situations with Linear Equations: *Slides: A Way to Represent Change*	LT-102–LT-123
	3.7	Summary and Review	LT-124–LT-153

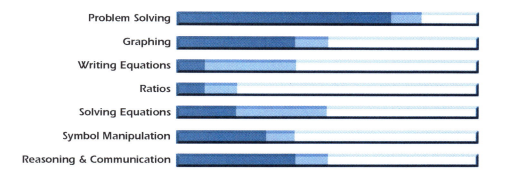

140 Chapter 3 Lions, Tigers, and Emus: Writing and Solving Equations

3.1 WRITING EQUATIONS FROM GUESS-AND-CHECK TABLES

SECTION MATERIALS
TRANSPARENCY OF RESOURCE PAGE FOR SP-152 (OPTIONAL)

ADDITIONAL TRANSPARENCIES, AND PENS FOR THE OVERHEAD PROJECTOR (OPTIONAL) FOR LT-3 AND LT-4

*This idea was presented by teacher Frank Gold at the first Summer Institute of the Northern California Mathematics Project in 1982.

In this section we use patterning in guess-and-check tables to develop and write equations.*

Head Problems: Start off with a couple of head problems for the opening 10 minutes of class. After students share their results, go back over the problem and record the algebraic steps as shown below.

Head Problem 5. "Pick a number. Add 5 to it. Multiply the sum by 10. Subtract 50, and then divide by 10. What is your result?" *the originally chosen number*

Instructor's Directions	Recording of Discussion
Pick a number.	x
Add 5 to it.	$x + 5$
Multiply the sum by 10.	$10(x + 5)$
Subtract 50.	$10(x + 5) - 50$
Divide by 10.	$\dfrac{10(x + 5) - 50}{10}$
What is your result?	*x, the original number*

Head Problem 6. "Pick a number. Multiply it by 3. Add 6 to the product. Divide by 3, and then subtract the original number. What is your result?" *2*

Instructor's Directions	Recording of Discussion
Pick a number.	x
Multiply it by 3.	$3x$
Add 6 to the product.	$3x + 6$
Divide by 3.	$\dfrac{3x + 6}{3}$
Subtract the original number.	$\dfrac{3x + 6}{3} - x$
What is your result?	*2*

About LT-1: You could use Problem SP-152 (or SP-15) as a bridge to Section 3.1, as students learn to write equations from guess-and-check tables. To do this, use the overhead transparency master for SP-152 in the Resource Pages to review students' work, and then extend it to writing an equation for the problem by using a variable for the guess and completing the table.

Carefully model the example, LT-1, for the whole class on an overhead transparency or on the board. Follow each step to show how to use patterns in a guess-and-check table to write an equation for a word problem.

You know how to use guess-and-check tables to solve word problems. Being able to organize your guesses is useful because an organized table helps you find a pattern that leads to an equation. Once you can model a problem with an equation, you're able to move beyond the tedious process of guessing and checking to utilizing the power of algebra in solving equations. In this section, you'll focus on using your organizational and patterning skills to write equations to model word problems.

LT-1. Using a Guess-and-Check Table to Write an Equation. In this problem you'll see how to extend the use of a guess-and-check table to include writing an algebraic equation to represent a situation. As you read each step of the following example, copy the table as it's developed into your notebook:

Example: The length of a rectangle is 3 centimeters more than twice the width. The perimeter is 45 centimeters. Use a guess-and-check table to find how long and how wide the rectangle is, and write an equation from the pattern developed in the table.

STEP 1 *Set up a Guess-and-Check Table.* In Chapter 1 we developed the following guess-and-check table. Notice that the computations in each column are written out explicitly to facilitate looking for a pattern.

Guess: Width	Length	Perimeter	*Guess:* Perimeter = 45?
10	$2 \cdot 10 + 3 = 23$	$2 \cdot 10 + 2 \cdot (23) =$	66 (too high)
5	$2 \cdot 5 + 3 = 13$	$2 \cdot 5 + 2 \cdot (13) =$	36 (too low)
7	$2 \cdot 7 + 3 = 17$	$2 \cdot 7 + 2 \cdot (17) =$	48 (too high)
6.5	$2 \cdot 6.5 + 3 = 16$	$2 \cdot 6.5 + 2 \cdot (16) =$	45 (correct)

STEP 2 *Use a variable for the next guess.* If we make W our next guess, what would the corresponding length be? Write an expression for the length in your table, and record what the variable represents.

Guess: Width	Length	Perimeter	*Guess:* Perimeter = 45?
10	$2 \cdot 10 + 3 = 23$	$2 \cdot 10 + 2 \cdot (23) =$	66 (too high)
5	$2 \cdot 5 + 3 = 13$	$2 \cdot 5 + 2 \cdot (13) =$	36 (too low)
7	$2 \cdot 7 + 3 = 17$	$2 \cdot 7 + 2 \cdot (17) =$	48 (too high)
6.5	$2 \cdot 6.5 + 3 = 16$	$2 \cdot 6.5 + 2 \cdot (16) =$	45 (correct)
W	$2 \cdot W + 3$		

Variable: Let $W =$ the width of the rectangle.

Note that it is important to record *in words* what your variable represents because you will need the information when you solve your equation.

STEP 3 *Look for a pattern.* Look at the pattern you developed in the Perimeter column. Use this numerical pattern to write an expression for perimeter using the variable W:

Guess: Width	Length	Perimeter	*Guess:* Perimeter = 45?
10	$2 \cdot 10 + 3 = 23$	$2 \cdot 10 + 2 \cdot (23) =$	66 (too high)
5	$2 \cdot 5 + 3 = 13$	$2 \cdot 5 + 2 \cdot (13) =$	36 (too low)
7	$2 \cdot 7 + 3 = 17$	$2 \cdot 7 + 2 \cdot (17) =$	48 (too high)
6.5	$2 \cdot 6.5 + 3 = 16$	$2 \cdot 6.5 + 2 \cdot (16) =$	45 (correct)
W	$2 \cdot W + 3$	$2 \cdot W + 2 \cdot (2 \cdot W + 3)$	

Variable: Let $W =$ the width of the rectangle.

STEP 4 *Write an equation.* Use your expression for the perimeter and information from the problem that the perimeter is 45 centimeters to write an equation:

Guess: Width	Length	Perimeter	*Guess:* Perimeter = 45?
10	$2 \cdot 10 + 3 = 23$	$2 \cdot 10 + 2 \cdot (23) =$	66 (too high)
5	$2 \cdot 5 + 3 = 13$	$2 \cdot 5 + 2 \cdot (13) =$	36 (too low)
7	$2 \cdot 7 + 3 = 17$	$2 \cdot 7 + 2 \cdot (17) =$	48 (too high)
6.5	$2 \cdot 6.5 + 3 = 16$	$2 \cdot 6.5 + 2 \cdot (16) =$	45 (correct)
W	$2 \cdot W + 3$	$2 \cdot W + 2 \cdot (2 \cdot W + 3) = 45$	

Variable: Let $W =$ the width of the rectangle.

Equation: $2 \cdot W + 2 \cdot (2W + 3) = 45$

STEP 5 *Complete the solution.* Finally, after checking your answer in the original problem, write your answer in a complete sentence. To reinforce algebraic thinking, you should always record these three key pieces of information—the variable, the equation you developed, and the answer you found—below your table, like this:

Variable: Let $W =$ the width of the rectangle.

Equation: $2 \cdot W + 2 \cdot (2 \cdot W + 3) = 45$

Answer: The width of the rectangle is 6.5 centimeters, and the length is 16 centimeters. ▲

Section 3.1 Writing Equations from Guess-and-Check Tables **143**

Later in this chapter you'll work on techniques for solving equations such as $2 \cdot W + 2 \cdot (2 \cdot W + 3) = 45$. The answer you find algebraically will correspond to the correct guess from a guess-and-check table that could be made to model the problem. Until you learn how to solve the equation, you will continue to use a guess-and-check table to find the solution.

Have groups do LT-2 while you tour the room to monitor progress and facilitate group interactions.

LT-2. Here's another example of using a guess-and-check table to write an equation. Read through it carefully.

Admission to the fair is $2 for children and $3 for adults. On Monday, 80 more children's tickets were sold than adults' tickets. Total ticket sales for Monday came to $1570. How many of each type of ticket were sold?

a. To make a guess-and-check table, you first need to decide which quantity would be useful as a guess. (In the table started here, we chose the number of adults' tickets as the quantity to guess, although we could just as well have chosen the number of children's tickets.) The first number you choose as a guess should be one that will make the calculations easy. (In this case, we'll guess that 100 adults' tickets were sold.)

Copy the table onto your paper, and leave room for at least three more rows. Fill in the appropriate description for the Check column. Then follow the column descriptions to complete the first row of the table for the guess "100," and check to see whether it's correct.

Guess: Number of Adults' Tickets	Value of Adults' Tickets Sold	Number of Children's Tickets	Value of Children's Tickets Sold	Total Value of Ticket Sales	Check: Total = $1570?
100	$3(100) = $300	180	$2(180) = $360	$300 + $360 =	$660 (low)
A	3(A)	A + 80	2(A + 80)	3(A) + 2(A + 80) =	$1570

b. If we had been lucky and guessed correctly on the first guess, we'd be only halfway finished since we still need to write an equation. To do this, first repeat the process by making as many more guesses as needed (at least two more) to solve the problem and filling in the columns for each guess. *The key is to write the expressions in a way that will help you see the patterns in your table, as shown in LT-1.*

c. Now use the variable A to represent the number of adults' tickets. Below your table, write the meaning of the variable A in words.   *Variable: Let A = the number of adults' tickets.*

d. Follow the numeric pattern in the table to write an expression for the total value of ticket sales. Use your expression for total value of ticket sales and information in the Check column description to write an equation. Complete the solution as you did in Step 5 of LT-1, writing your answer in a complete sentence.   *Variable: Let A = the number of adults' tickets. Equation: $3A + 2(A + 80) = 1570$. Answer: 282 adults' tickets and 362 children's tickets were sold.*

Guide to Setting Up Guess-and-Check Tables to Write Equations

Before you start to make a guess-and-check table for a problem, read the problem and ask yourself these questions:

1. What is the problem asking me to find?

2. What part of the problem could serve as a guess and be used to generate more information?

When setting up your guess-and-check table, remember the following guidelines:

- Only one column (usually the far-left column) should be labeled "Guess."

- Label each column, including the Guess column, with a description of what it represents.

- The first number you choose as a guess should be one that will make the calculations easy, such as 10 or 100.

- Make at least three (3) guesses, even if you solve the problem in fewer guesses.

- As you fill in the columns to check a guess, write out explicitly how you make the computations. These details will help you find a pattern in order to write an equation.

- Only one column (usually the far-right column) should be labeled "Check." It holds a place for the desired result of all the computations.

- Use the pattern you develop in the table to write an equation, as modeled in LT-1.

- Check your answer in the original problem.

- To complete the solution, record these three important pieces of information:

    what the *variable* represents

    the *equation* itself

    your *answer* written in a complete sentence

For problems LT-3 and LT-4, you could assign each group just one of the problems. If time permits, have each group write its solution on an overhead transparency, and then present it to the whole class.

For problems LT-3 and LT-4, follow the "Guide for Using Guess-and-Check Tables to Write Equations" to solve the problem and write an equation.

LT-3. The perimeter of a triangle is 82 centimeters. The second side is half as long as the first side. The third side is 9 centimeters shorter than the second side. How long is each side? *Let x = length of the first side; then x + 0.5x + (0.5x − 9) = 82;   45.5 cm, 22.75 cm, 13.75 cm. Or let y = length of second side, so 2y + y + (y − 9) = 82.*

LT-4. Heather has twice as many dimes as nickels and two more quarters than nickels. The value of the coins is $5.50. How many quarters does she have? *Let n = the number of nickels; so $0.05n + 0.10(2n) + 0.25(n + 2) = $5.50, or (5¢)n + 10(2n) + 25(n + 2) = 550¢; 12 quarters*

Have all groups do LT-5 and LT-6. Problem LT-5 is a follow-up to SP-86, and prepares students to work through the guess-and-check table in LT-6.

LT-5. Write an expression that represents each amount. Refer to your work in SP-86 if you need help.

a. Hugh has $40,000 to invest in a high-risk stock and a safer mutual fund. If he invests $6438 in the high-risk stock, how much will he invest in the mutual fund? *$40,000 − $6438 = $33,562*

b. Nancee has 42 vacation days to spend on her trip to Uganda and Kenya. If she spends K days in Kenya, how many days will she spend in Uganda? *42 − K*

LT-6. Using the guess-and-check table started below, follow the "Guide for Using Guess-and-Check Tables to Write Equations" to solve the problem and write an equation.

At a football game, 2000 tickets were sold. General public tickets sold for $7.50 and student tickets for $5.00. The total revenue was $11,625. How many student tickets were sold? *5.00S + 7.50(2000 − S) = 11,625; 1350 student tickets*

Guess: Number of Student Tickets	Revenue from Student Tickets	Number of General Public Tickets	Revenue from General Public Tickets	Total Revenue from All Tickets	Check: Total Revenue = $11,625?
100	$5(100) = $500	2000 − 100 = 1900	$7.50(1900) = $14,250	$5(100) + $7.50(1900) =	$14,750 (high)
200	$5(200) = $1000	2000 − 200 = 1800	$7.50(1800) = $13,500	$5(200) + $7.50(1800) =	$14,500 (high)
500	$5(500) = $2500	2000 − 500 = 1500	$7.50(1500) = $11,250	$5(500) + $7.50(1500) =	$13,750 (high)
S	5S	2000 − S	7.5(2000 − S)	5S + 7.5(2000 − S) =	11,625

Variable: Let S = the number of student tickets.
Equation: 5.00S + 7.50(2000 − S) = 11,625
Answer: There were 1350 student tickets sold.

> **EXTENSION AND PRACTICE**

> For each of problems LT-7 through LT-9, follow the "Guide for Using Guess-and-Check Tables to Write Equations" to solve the problem and write an equation.

LT-7. Mark sold 105 tickets for the basketball game. Each adult's ticket cost $2.50 and each child's ticket cost $1.10. Mark collected $221.90. How many of each kind of ticket did he sell? (*Hint:* When you write the algebraic expression for the number of children's tickets, think about each of the numerical calculations you did in the Number of Children's Tickets column.) *Let A = the number of adults' tickets; $2.50A + 1.10(105 - A) = 221.90$; 29 children's and 76 adults' tickets*

LT-8. On a 520-mile trip, Chloë and Maude shared the driving. Chloë drove 80 miles more than Maude drove. How far did each person drive? *Let M = distance Maude drove; $M + (M + 80) = 520$; Maude, 220 mi; Chloë, 300 mi*

Note that LT-9 is a condensed version of OP-84. The equation for LT-9 is a quadratic; students will learn how to solve quadratic equations in Chapter 6.

LT-9. The borders of the state of Wyoming form a rectangle, and the area of the state is about 97,109 square miles. If the length of Wyoming is 92 miles more than its width, about how long are the borders? *Let w = width; $(w + 92)w = 97,109$; about 269 by 361 mi. (Hint: equation is nonlinear; you can write it, but you haven't learned the tools to solve it algebraically yet.)*

LT-10. Use the distributive property to rewrite each of the following products.

a. $6(x + 2)$ *$6x + 12$*   **b.** $x(3 + 5)$ *$3x + 5x$, or $8x$*

c. $9(3 - x)$ *$27 - 9x$*   **d.** $y(6 + 4)$ *$6y + 4y$, or $10y$*

LT-11. For each part in LT-10 you should now have two expressions: the original one and the rewritten one.

a. Evaluate each of the four expressions in LT-10a and b by replacing x with 2.   *a. 24, b. 16*

b. Evaluate each of the four expressions in LT-10c and d by replacing the variable x with -3.   *c. 54, d. -30*

LT-12. In this problem you will compare the expressions $5(x + 4)$ and $5x + 5(4)$.

a. Evaluate each of the following expressions for $x = 3$:

$$5(x + 4) \quad 35 \quad 5(x) + 5(4) \quad 35$$

b. Evaluate each of the following expressions for $x = -7$:

$$5(x + 4) \quad -15 \quad 5(x) + 5(4) \quad -15$$

c. Explain why the expressions $5(x + 4)$ and $5(x) + 5(4)$ both have the same value when a number is substituted for x.   *By the distributive property, the expressions are equal for all values of x.*

LT-13. Oscar and Felix are working with exponents again.

a. In OP-55, Oscar and Felix used the meaning of exponents to figure out what $(7^2)^3$ means. Look back at your work from OP-55, and then explain what $(y^2)^3$ means.   *$(y^2)^3 = y^2 \cdot y^2 \cdot y^2$*

b. Rewrite the product $(y^2)^3$ using a single exponent, without parentheses.   *y^6*

c. Explain what $(x^3)^2$ means, and then rewrite $(x^3)^2$ using a single exponent.   *$(x^3)^2 = x^3 \cdot x^3 = x^6$*

d. Explain what $(m^2)^4$ means, and then rewrite $(m^2)^4$ using a single exponent.   *$(m^2)^4 = m^2 \cdot m^2 \cdot m^2 \cdot m^2 = m^8$*

e. Look for a pattern in parts a through d. Then, without explaining the meaning, use the pattern to rewrite $(x^{31})^7$ using a single exponent.   *$x^{31 \cdot 7} = x^{217}$*

LT-14. Although the following two expressions are similar, they are fundamentally different. Explain how they are different, and then rewrite each expression as a power of x.

a. $x^{31} \cdot x^7$ *$x^{31} \cdot x^7 = x^{31+7} = x^{38}$*   **b.** $(x^{31})^7$ *$(x^{31})^7 = x^{31 \cdot 7} = x^{217}$*

$x^{31} \cdot x^7$ means to multiply 31 factors of x by seven factors of x, whereas $(x^{31})^7$ means to use x^{31} as a factor seven times.

LT-15. Derek's scientific calculator displays $\boxed{1.3^{\,03}}$. Explain what this represents. *(Hint: refer to OP-77.)* $1.3 \cdot 10^3 = 1300$

LT-16. Copy the following table, and put your calculator in scientific mode. Use your calculator to compute each result and complete the table.

Computation	Result in Scientific Notation	Result in Standard Form
$(2.1 \cdot 10^5)(3.25 \cdot 10^5) =$	$6.825 \cdot 10^{10}$	68,250,000,000
$(2.1 \cdot 10^5)+(3.25 \cdot 10^5) =$	$5.35 \cdot 15^5$	535,000

LT-17. In this problem you'll start to examine how graphing and solving equations are related.

a. Sketch a graph of the rule $y = 3(x - 5)$ using the input values $x = -4, -2, \ldots, 8, 10$. (You will need to make a table first.) Don't forget to label the important parts of your graph.

x	-4	-2	0	2	4	6	8	10
y	-27	-21	-15	-9	-3	3	9	15

b. If you connect the points to form a continuous graph, what are the coordinates of the x-intercept? What is the y-value at this point? *(5, 0); y-value is 0.*

c. Find a value for x that makes the equation $3(x - 5) = 0$ true. If you are unsure how to solve the equation algebraically, use a guess-and-check table as in SP-112. *$x = 5$*

d. How do your answers to parts b and c compare? *They should be about the same.*

LT-18. Suppose you have a standard deck of 52 playing cards.

a. What is the probability of randomly drawing a 5 of spades? $\frac{1}{52}$

b. What is the probability of randomly drawing a 5? $\frac{4}{52}$ or $\frac{1}{13}$

c. Suppose you draw a 5 and do not put the card back. What is the probability of randomly drawing another 5? $\frac{3}{51}$ or $\frac{1}{17}$

d. Write each of the probabilities in parts a through c as a percent. $\frac{1}{52} \approx 1.9\%; \frac{4}{52} \approx 7.7\%; \frac{3}{51} \approx 5.9\%;$

3.2 COMPOSITE RECTANGLES AND ALGEBRAIC EXPRESSIONS FOR AREA

SECTION MATERIALS

SETS OF ALGEBRA TILES (SEE P. 62)

ALGEBRA TILES FOR THE OVERHEAD PROJECTOR FOR LT-19 AND LT-20

LT-20 COMPOSITE RECTANGLES AND AREA RESOURCE PAGE

LT-26 RESOURCE PAGE

LT-27 RESOURCE PAGE

Head Problems: Spend about 10 minutes at the start of class on Head Problems. After students share results for Head Problem 7, pick "x" to represent a generic number, and go through the problem again. Record the algebra as shown in Section 3.1 for Head Problems 5 and 6. If time allows, repeat the process with Head Problem 8.

Head Problem 7. "Think of a number. Subtract 2 from it. Multiply the difference by 6. Divide the product by 3. Add 4, and then divide by 2. What is your result?" your original number

Head Problem 8. "Start with the number of seconds in a minute. Add a number of your own choice. Multiply the result by 2. Now subtract 40. Compute half of the difference. Subtract your number, and then add 20. What is your result?" 60

Algebra tiles for the overhead projector are useful for introducing LT-19 and having volunteers show composite rectangles for LT-20.

LT-19. Another Algebra Tile For this problem you'll need a set of three kinds of algebra tiles: the rectangles and small squares which you have already used, and an additional kind, the large squares.

a. Recall that the small square has a side of length 1, and the rectangle has a side of unknown length x. To form larger rectangles using just small squares and rectangles, the tiles must be aligned with their sides of common length—namely, their sides of length 1, put together. Similarly, to form larger rectangles using just large squares and rectangles, the tiles must be aligned with their sides of common length put together. Use this information to find the lengths of the sides and the area of the large square. *The large square has sides of length x and area x^2.*

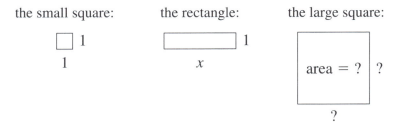

b. Trace the large square tile in your Tool Kit. Mark the dimensions along each side, and clearly label it with its area, since algebra tiles are often referred to by their areas.

LT-20. Composite Rectangles and Area

a. Determine whether it is possible to build a composite rectangle with each set of tiles indicated in the following chart. If so, on your LT-20 Composite Rectangles and Area Resource Page, sketch a composite rectangle you could build; if not, answer "not possible."

b. For each composite rectangle, label each tile with its area as shown, and then label the length and width of the rectangle.

c. Use what you have done to write the area of each set of tiles as a sum of the areas of the pieces, and then as a product of the length and the width, if possible.

Number of x^2 Tiles	Number of $1x$ Tiles	Number of 1 Tiles	Is a composite rectangle possible?	Sketch	Algebraic Expression for the Area as a Sum	Algebraic Expression for the Area as a Product
1	3	2	yes		$x^2 + 3x + 2 = (x + 1)(x + 2)$	
1	4	4	yes		$x^2 + 4x + 4 = (x + 2)(x + 2)$	
2	3	1	yes		$2x^2 + 3x + 1 = (2x + 1)(x + 1)$	
0	3	6	yes		$3x + 6 = (3)(x + 2)$	

(table continued on next page)

Number of x^2 Tiles	Number of 1 Tiles	Number of $1x$ Tiles	Is a composite rectangle possible?	Sketch	Algebraic Expression for the Area as a Sum	Algebraic Expression for the Area as a Product
1	6	4	no		$x^2 + 6x + 4$	not possible
1	4	0	yes	x wide by $x+4$ long: $x^2 \mid 1x\ 1x\ 1x\ 1x$	$x^2 + 4x = (x)(x+4)$	
2	7	6	yes	$(2x+3)$ by $(x+2)$ arrangement with x^2, $1x$, and 1 tiles	$2x^2 + 7x + 6 = (2x+3)(x+2)$	
1	7	10	yes	$(x+2)$ by $(x+5)$ arrangement with x^2, $1x$, and 1 tiles	$x^2 + 7x + 10 = (x+2)(x+5)$	

LT-21. Polynomials In the previous problem, you saw that the sum $x^2 + 3x + 2$ is equal to the product $(x + 1)(x + 2)$; both expressions represent the same area. The expression

$$x^2 + 3x + 2$$

is an example of a *polynomial* (from the Greek words *poly* meaning "many," and *nomos* meaning "part").

A **polynomial** is a **sum** of *monomials* or *terms*; in this case, $x^2 + 3x + 2$ is the sum of the monomials x^2, $3x$, and 2. The expressions $3x^2$, $-15x$, x, and -7 are also examples of monomials (from the Greek words *monos* meaning "single," and *nomos* meaning "part").

A *bi*nomial has *two* terms, such as $3x + 2$, or $7 - x^2$, and a *tri*nomial has *three* terms, such as $x^2 + 4x - 3$.

Certain polynomials can also be rewritten as **products** *of factors*. For example, the sum $x^2 + 3x + 2$ can be written as the product of the two factors $x + 1$ and $x + 2$. We say we have **factored** the polynomial $x^2 + 3x + 2$ as "the quantity $x + 1$ times the quantity $x + 2$."

Copy the following polynomial expressions without changing them. Then identify each expression as either a sum or a product.

a. $2x + 1$ sum

b. $(x + 5)(x + 2)$ product

c. $x^2 + 7x + 5$ sum

d. $x(2x + 5)$ product

e. $(x + 1)(x + 4)$ product

f. $5x^3 + 8x^2 + 10x + 13$ sum

g. $3x + 7$ sum

h. $2x(x^2 - 3x + 5)$ product

There are two answers to LT-22. When groups have finished the problem, have them share different sketches and solutions on the board or overhead projector.

LT-22. For the following problem, follow the "Guide for Using Guess-and-Check Tables to Write Equations" to solve the problem and write an equation.

A rectangular goat pen is enclosed by a barn on one side and by a total of 100 feet of fence on the three other sides. The area of the pen is 912 square feet. Draw a top-view diagram of the goat pen and then find its dimensions. $x(100 - 2x) = 912$; 12 by 76 ft or 38 by 24 ft

> **EXTENSION AND PRACTICE**

For Problems LT-23 through LT-25, follow the "Guide for Using Guess-and-Check Tables to Write Equations" to solve the problem and write an equation.

LT-23. Mr. Jordan keeps coins for paying the toll crossings on his commute to and from work. He presently has three more dimes than nickels and two fewer quarters than nickels. The total value is $5.40. Find the number of each type of coin. *Let n = the number of nickels; 0.05n + 0.10(n + 3) + 0.25(n − 2) = 5.40; 14 nickels, 17 dimes, 12 quarters*

LT-24. The length of Linnea's rectangular garden is 3 meters longer than twice its width. The perimeter is 51 meters. Find the dimensions of the garden. *Let w = the width of the garden; 2w + 2(2w + 3) = 51; 7.5 m wide, 18 m long*

LT-25. A rectangular sign is twice as long as it is wide. Its area is 722 square centimeters. What are the dimensions of the sign? *Let w = the width; w(2w) = 722; 19 by 38 cm*

LT-26. On your LT-26 Resource Page, write an algebraic expression for the area of each of the following composite rectangles in two different ways—first as a product, and then as a sum. Use "x^2" to represent each large square, "x" to represent each rectangular tile, and "1" to represent each small square.

Example: For this composite rectangle

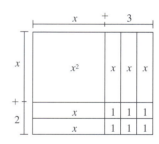

we can write

$$(x + 2)(x + 3) = x^2 + 5x + 6$$

area as a product area as a sum ▲

a.

$(x + 3)(x + 4) = x^2 + 7x + 12$

b.

$(x + 1)(x + 1) = x^2 + 2x + 1$

c.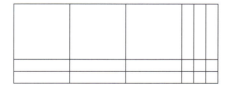

$(x + 2)(3x + 3) = 3x^2 + 9x + 6$

d.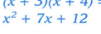

$x(x + 3) = x^2 + 3x$

e.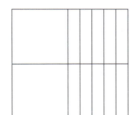

$2x(x + 5) = 2x^2 + 10x$

LT-27. Derek was so proud of the way he and Deanne tiled their lap pool (see SP-20) that he offered to tile the bottom surface of his mother-in-law's three large swimming pools. (She owns a gym.) Using his same old 10-foot-long measuring tape, he measured the lengths of the sides. Help Derek find the area of each pool in two ways:

1. by adding the areas of the parts, and
2. by multiplying the length of the whole rectangle by the width of the whole rectangle.

Example:

Area as a sum:

100 sq ft + 40 sq ft + 30 sq ft + 12 sq ft = 182 sq ft

Area as a product:

(10 ft + 3 ft)(10 ft + 4 ft) = (13 ft)(14 ft) = 182 sq ft ▲

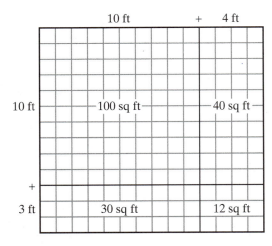

Follow the example and use your LT-27 Resource Page to complete parts a through c.

a.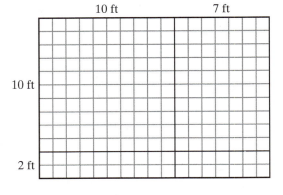

100 + 70 + 20 + 14 = 204 sq ft,
(10 + 2)(10 + 7) = 204 sq ft

b.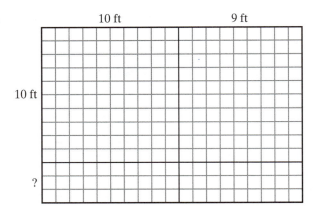

100 + 90 + 30 + 27 = 247 sq ft,
(10 + 3)(10 + 9) = 247 sq ft

c.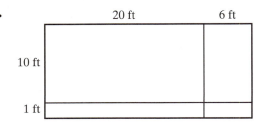

200 + 60 + 20 + 6 = 286, (10 + 1)(20 + 6) = 286

LT-28. To evaluate an expression like 8(30 + 2), you could add first and then multiply by 8, using the order of operations. The distributive property allows you to multiply first, and then add.

a. Without using a calculator, evaluate 8(30 + 2) in two ways, first by following the order of operations, and then by using the distributive property. *8(30 + 2) = 8(32) = 256, or, 8(30 + 2) = 240 + 16 = 256*

b. Without using a calculator, evaluate 5(18 + 2) in two ways, first by following the order of operations, and then by using the distributive property. *5(18 + 2) = 5(20) = 100, or, 5(18 + 2) = 90 + 10 = 100*

c. Look at your work from parts a and b. Which method made it easier for you to compute mentally? When is it more efficient to use the distributive property first, and when is it better to simply follow the order of operations? *For 8(30 + 2), the distributive property is more efficient. For 5(18 + 2), using order of operations is easier.*

LT-29. Using the distributive property, rewrite each of the following products as a sum, and rewrite each sum as a product:

a. $2(x + 4)$ *2x + 8*   b. $x(x + 4)$ *$x^2 + 4x$*

c. $x(x-2)$ *$x^2 - 2x$*   d. $3x + 6$ *3(x + 2) (Hint: 3x + 6 = 3·x + 3·2)*

e. $4y - 8$ *4(y – 2)*   f. $m^2 + 5m$ *m(m + 5)*

LT-30. Rewrite each of the following exponential expressions with a single base and a single exponent.

a. $(2^3)(2^4)$ *2^7*   b. $(x^2)(x^5)$ *x^7*

c. $(2^3)^4$ *2^{12}*   d. $(x^2)^5$ *x^{10}*

LT-31. Use the meaning of integer exponents to rewrite each of the following expressions using a single exponent:

a. $4^3(x^2)^3$ *$64x^6$*   b. $(4x^2)^3$ *$64x^6$*

c. $7^2(x^5)^2$ *$49x^{10}$*   d. $(7x^5)^2$ *$49x^{10}$*

e. $3^2(x)^2$ *$9x^2$*   f. $(3x)^2$ *$9x^2$*

g. Compare your results in parts a and b. Compare your results in parts c and d. Compare your results in parts e and f. What do you notice? *Each pair of expressions has the same result: they are equal.*

h. Make up another pair of expressions that will follow the same pattern you observed. *Answers vary; e.g., $(8x)^3 = 8^3(x)^3$*

LT-32. Follow the instructions below for each of these equations:

$$(1)\ y = x \quad (2)\ y = x + 2$$

a. Make a table with eight input values for x between -3 and 3, and find the corresponding y-values.

x	-3	-2	-1	0	0.5	1	2	3
$y = x$	-3	-2	-1	0	0.5	1	2	3
$y = x + 2$	-1	0	1	2	2.5	3	4	5

b. Use your tables to graph each equation. Your graphs are incomplete without appropriate labels.

c. List two similarities and one difference you see in the graphs. *Both are straight lines; the lines are parallel, but $y = x + 2$ is "above" $y = x$.*

LT-33. As part of their advertising campaign, Chevy's restaurant chain claims, "Our fresh tortillas are hot off 'El Machino™' every 53 seconds, flat!" As a footnote, Chevy's adds, "Hey! that's 900 tph (tortillas per hour!)."

a. If the first part of the claim means that "El Machino™" makes one tortilla every 53 seconds, how many tortillas does the machine make in an hour? *about 68*

b. Do you think the footnote claim is true? If so, show why. If not, how many tortillas must the machine make in 53 seconds in order to make the claim true? *untrue; about 13 tortillas every 53 sec*

LT-34. Copy and complete this table to evaluate each of the following polynomials for $x = 2$ and for $x = -3$:

Polynomial	Value when $x = 2$	Value when $x = -3$
Example: $x^2 - 5x - 7$	$(2)^2 - 5(2) - 7 = -13$	$(-3)^2 - 5(-3) - 7 = 17$
a. $x^2 - 3x + 8$	$(2)^2 - 3(2) + 8 = 6$	$(-3)^2 - 3(-3) + 8 = 26$
b. $-3x^2 + x$	$-3(2)^2 + 2 = -10$	$-3(-3)^2 + (-3) = -30$
c. $x^2 + x - 6$	$2^2 + 2 - 6 = 0$	$(-3)^2 + (-3) - 6 = 0$

LT-35. Make each of the following equations true by inserting one or more sets of grouping symbols (such as parentheses or brackets):

a. $2 \cdot 3^2 + 4 \cdot 3 - 1 = 47$ $(2 \cdot 3)^2 + 4 \cdot 3 - 1 = 47$

b. $2 \cdot 3^2 + 4 \cdot 3 - 1 = 80$ $[(2 \cdot 3)^2 + 4] \cdot (3 - 1) = 80$

3.3 SOLVING LINEAR EQUATIONS AT THE CANDY FACTORY

Section Materials

For class demonstration:

- **Two-pan balance scale**
- **About 30 jelly beans and several small resealable ("snack size") bags**
- **Small square algebra tiles for overhead projector**
- **Four film canisters**
- **Transparency of Two-Pan Balance Scale Resource Page**

For each pair of students in LT-36:

- **About 20 jelly beans and six small resealable bags (or 20 small square algebra tiles and six film canisters)**
- **Two-Pan Balance Scale Resource Page**
- **LT-36 Using Packets and Jelly Beans to Solve Linear Equations Resource Page**
- **LT-47 Representing Tiles with Polynomials Resource Page**

About the Materials for the Class Demonstration: A memorable way to start off this section is to show how a two-pan balance scale can be balanced with bags of jelly beans and some loose beans while a student reads the introduction to the section. The two-pan balance scale helps demonstrate the concept of balance between two sides of an equation, and its use is especially beneficial for students who may have never seen or used this kind of scale. Its use also helps students see that when equal amounts are added to (or subtracted from) both sides of a balanced scale (or equation), the balance (and equality) are maintained. However, if something is added to (or removed from) just one side of a balanced scale, then the balance is not maintained.

Once students see how a balanced scale could represent an equation, we switch to the overhead projector to demonstrate how to solve linear equations using bags, jelly beans, and a transparency of the Two-Pan Balance Scale Resource Page. The overhead projector makes it easier for students to see the demonstration and lets us avoid inaccuracies that arise with the balance scale. We use small square overhead algebra tiles in place of the beans, and film canisters to represent the sample packets. Students follow along as they work with partners, using bags and jelly beans (or small square algebra tiles and film canisters to represent the jelly beans and the sample packets of jelly beans) and their copies of the Two-Pan Balance Scale Resource Page.

Sara works summers as a tour guide at a jelly bean factory. During busy periods, while people wait in the reception area for their groups' turn to tour the facility, Sara entertains them with riddles and brain teasers. She's made up a whole series of number puzzles using an old two-pan balance scale that's on display and sample packets of jelly beans. She describes the situation and states the riddle:

> Here's a riddle to figure out how many jelly beans are in a sample packet. Suppose all the sample packets have the same number of jelly beans in them (but we don't know and can't see how many there are), that all the jelly beans have the same weight, and that the weight of the bag is too small to matter.
>
> On the two-pan balance scale, 16 loose jelly beans and 19 sample packets in one pan balance eight sample packets and 302 loose jelly beans in the other pan. How many jelly beans are in a sample packet?

Sara then shows the two-pan scale balanced with all the packets and jelly beans. Some people use a guess-and-check approach to solve the puzzle, while others use the balanced scale. Using the scale leads to developing an algebraic approach for solving similar problems.

Model Example 1 as described and shown here. Proceed slowly with the demonstration; let the students suggest what to do at each step. Record the corresponding algebraic step each time you move the packets (canisters) and individual jelly beans (tiles), and have students do the same.

Example 1

Description of Process	Overhead Display (Model with Packets and Beans)	Algebraic Record (Model with Variables)
Step 1. Present the problem. Use canisters and beans on the overhead projector. "Four packets and six jelly beans balance one packet and 12 jelly beans. Each packet has the same number of jelly beans. We want to find out what that number is." Write an equation to model the balanced scale.		Let P represent the number of beans in each packet. $4P + 6 = 1P + 12$
Step 2. Ask: "What could we do to both sides to eliminate some packets or jelly beans while keeping the balance?" Students will suggest that you remove two jelly beans, or one packet, or two jelly beans and one packet from each side. Follow suggestions, but do only one step at a time. Record the algebraic process corresponding to each step.	Remove / Remove or Remove / Remove	$\begin{aligned} 4P + 6 &= P + 12 \\ -6 & \quad -6 \\ \hline 4P &= P + 6 \end{aligned}$ or $\begin{aligned} 4P + 6 &= P + 12 \\ -P & \quad -P \\ \hline 3P + 6 &= 12 \end{aligned}$
Step 3. Continue: "Now four packets balance one packet and six jelly beans. How could we eliminate some packets while keeping the balance?" Remove one packet from each side. Record the algebraic process. or "Now three packets and six jelly beans balance 12 jelly beans. How could we eliminate some jelly beans while keeping the balance?" Remove six beans from each side. Record the algebraic process.	Remove / Remove or Remove / Remove	$\begin{aligned} 4P &= P + 6 \\ -P & \quad -P \\ \hline 3P &= 6 \end{aligned}$ or $\begin{aligned} 3P + 6 &= 12 \\ -6 & \quad -6 \\ \hline 3P &= 6 \end{aligned}$

Example 1 (continued)

Description of Process	Overhead Display (Model with Packets and Beans)	Algebraic Record (Model with Variables)
Step 4. "Now three packets balance six jelly beans. While maintaining the balance, what can we do to both sides to find out how many jelly beans there are in one packet?" After students suggest dividing each side by 3, rearrange the packets and beans into three balanced rows, and ask, "Why did we divide by 3?" Record the algebraic process.		$3P = 6$ $$\frac{3P}{3} = \frac{6}{3}$$
Step 5. Ask, "How many jelly beans are in a packet?" Record the answer.		$P = 2$
Step 6. Check the solution. Put two jelly beans in each packet, and check the balance to verify that the solution $P = 2$ is correct. Record the check algebraically.		$4(2) + 6 = 2 + 12$ ▲

After the demonstration, have students practice in pairs with $3P + 4 = P + 12$ while you circulate and check their work. Make sure they record the equations associated with each step of the display or process.

Next, model solving equations with fractional solutions, as in Example 2.

Example 2

Description of Process	Overhead Display (Model with Packets and Beans)	Algebraic Record (Model with Variables)
Step 1. Present the problem. Use canisters and beans on the overhead projector. Also sketch the display on the board. "Four jelly beans and three packets balance one packet and nine jelly beans. Each packet has the same number of jelly beans. We want to find out what that number is."		Let P represent the number of beans in each packet.
Write an equation to model the balanced scale.		$4 + 3P = P + 9$

Example 2 (continued)

Description of Process	Overhead Display (Model with Packets and Beans)	Algebraic Record (Model with Variables)
Step 2. Ask: "What could we do to both sides to eliminate some packets or jelly beans while keeping the balance?" Students will suggest that you remove four jelly beans, or one packet, or four jelly beans and one packet from each side. Follow suggestions, but do only one step at a time. Record the algebraic process corresponding to each step.	*Remove four jelly beans from each side* or *Remove one packet from each side*	$4 + 3P = P + 9$ $\underline{-4 \qquad\quad -4}$ $3P = P + 5$ or $4 + 3P = P + 9$ $\underline{-P \quad\; -P}$ $4 + 2P = 9$
Step 3. Continue: "Now three packets balance one packet and five jelly beans. How could we eliminate some bags while keeping the balance?" Remove one packet from each side. Record the algebraic process. or "Now two packets and four jelly beans balance nine jelly beans. How could we eliminate some jelly beans while keeping the balance?" Remove four beans from each side. Record the algebraic process.	*Remove one packet from each side* or *Remove four beans from each side*	$3P = P + 5$ $\underline{-P \quad\; -P}$ $2P = 5$ or $4 + 2P = 9$ $\underline{-4 \qquad\quad -4}$ $2P = 5$

Section 3.3 Solving Linear Equations at the Candy Factory **161**

Example 2 (continued)

Description of Process	Overhead Display (Model with Packets and Beans)	Algebraic Record (Model with Variables)
Step 4. Say to the students, "Look at the balanced scale—what question does it suggest now?" Pause for students to respond: "If two packets balance five jelly beans, how many jelly beans does one packet balance?" Ask the students what to do next, and then paraphrase their response: "Right. We must divide the five jelly beans between the two bags." Ask students, "Why do we divide by 2? Why not divide by 5?"		$2P = 5$
Step 5. Say, "The division process is represented algebraically by writing fractions."		$2P = 5$ $\dfrac{2P}{2} = \dfrac{5}{2}$
Step 6. Display one packet and two and a half jelly beans to show the solution. Then record the solution.		$P = \dfrac{5}{2},$ or $P = 2\dfrac{1}{2}$
Step 7. Check the solution. Put $2\dfrac{1}{2}$ jelly beans in each packet, and check the balance to verify that the solution $P = 2\dfrac{1}{2}$ is correct. Record the check algebraically.		$4 + 3P = P + 9$ $4 + 3\left(\dfrac{5}{2}\right) = \dfrac{5}{2} + 9$

▲

162 Chapter 3 Lions, Tigers, and Emus: Writing and Solving Equations

After the demonstration, provide a practice problem, such as 3x + 4 = 6 or 4x + 3 = 4, for students to do in pairs while you circulate and check their work. Make sure they record the equations associated with the display or process.

To give your students a historical perspective of this way of representing a linear equation, you could explain to them how in ancient Egypt the notion of a heap (aha) was used for the unknown amount—analogous to our use of a packet. Solving a problem for them meant finding how many grains or kernels were in a heap.*

*See, for example, Carl B. Boyer's *A History of Mathematics*, 2nd ed. (New York: Wiley, 1991).

If you have time now, and feel your students would be comfortable with the use of positive and negative integer tiles (as in Appendix A), you could model the use of negative jelly beans to solve equations shown in Example 3. However, it is not necessary to do so at this time—you could wait to deal with negative numbers using the "doing and undoing" ideas of Section 3.4. Since a negative number of jelly beans doesn't make sense, you might have students imagine "cosmic jelly beans" that act like charged particles; a positively charged bean and a negatively charged bean neutralize each other for a net charge of zero. Use two different colors of beans (tiles) to distinguish positives from negatives. In this example, the ⊞ and ⊟ represent the charged particles and the circles represent the film canisters.

Example 3

Description of Process	Overhead Display (Model with Packets and Beans)	Algebraic Record (Model with Variables)
Step 1. Present the problem. Use canisters and beans on the overhead projector. Also sketch the display on the board. "Six negatively charged beans and three packets balance one packet and two positively charged jelly beans. Each packet has the same number of jelly beans. We want to find out what that number is." Write an equation to model the balanced scale.		Let P represent the number of beans in each packet $-6 + 3P = P + 2$

Section 3.3 Solving Linear Equations at the Candy Factory **163**

Example 3 (continued)

Description of Process	Overhead Display (Model with Packets and Beans)	Algebraic Record (Model with Variables)
Step 2. *"We can't remove an equal number beans of the same charge from each side, but we could add an equal number beans of the same charge to both sides. What could we add?"* Add either six positive beans or two negative beans to each side. Record the algebraic process, and sketch the resulting display on the board.		$-6 + 3P = P + 2$ $+6 \qquad\qquad +6$ ──────────── $3P = P + 8$
Step 3. Continue: *"Now three packets balance one packet and eight jelly beans. How could we eliminate some bags while keeping the balance?"* Remove one packet from each side. Record the algebraic process, and sketch the resulting display on the board.		$3P = P + 8$ $-P \quad -P$ ──────── $2P = 8$
Step 4. *"Now two packets balance eight jelly beans. While maintaining the balance, what can we do to both sides to find out how many jelly beans there are in one packet?"* After students suggest dividing each side by 2, rearrange the packets and beans into two balanced rows, and ask, *"Why did we divide by 2?"* Record the algebraic process.		$2P = 8$ $\dfrac{2P}{2} = \dfrac{8}{2}$ $P = 4$
Step 5. Check the solution. Put four jelly beans in each packet, and check the balance to verify that the solution $P = 4$ is correct. Record the check algebraically.		$-6 + 3P = P + 2$ $-6 + 3(4) = 4 + 2$

Now provide practice problems for students to do in their groups. For example, you could use $-5 + 3x = 4$ and $-3 + x = 7 + 4x$.

Have students read LT-36 up to the table (which includes answers) in part a, then close their texts and follow the instructions on the LT-36 Resource Page. After completing the Resource Page, students can check their work against the table.

LT-36. Using Packets and Jelly Beans to Solve Linear Equations In this problem you'll see how using a balanced scale to solve Sara's jelly beans riddles can serve as a model for solving linear equations. But, rather than solving Sara's puzzle directly, you'll work through the following example, which has smaller numbers:

Jelly Bean Riddle: On the two-pan balance scale, two loose jelly beans and three sample packets in one pan balance one packet and 14 loose beans in the other pan. How many beans are in a packet? (Assume that all the packets have the same number of jelly beans in them, that all the beans have the same weight, and that the weight of the packet is too small to matter.)

a. Follow the steps in this example, putting your packets and beans right on the Two-Pan Balance Scale Resource Page. (The two sides of the page represent the two pans of the scale.) Record each step on the LT-36 Using Packets and Jelly Beans to Solve Linear Equations Resource Page.

Situation: All packets have the same number of jelly beans. Two jelly beans and three packets balance one packet and 14 jelly beans. How many jelly beans are in a packet?

	Description of Process	Overhead Display (Model with Packets and Beans)	Algebraic Record (Model with Variables)
STEP 1.	Draw a diagram to represent the balanced scale. Then write an equation to model the situation.		Let P represent the number of beans in each packet. $$2 + 3P = P + 14$$
STEP 2.	Ask yourself, "What could I do to both sides to eliminate some packets or some jelly beans while keeping the balance?" You might remove one packet from each side, as shown. Record the algebraic process.		$$\begin{array}{r} 2 + 3P = P + 14 \\ -P -P \\ \hline 2 + 2P = 14 \end{array}$$
STEP 3.	Continue. You could now remove two jelly beans from each side and keep the balance. Record the algebraic process.		$$\begin{array}{r} 2 + 2P = 14 \\ -2 -2 \\ \hline 2P = 12 \end{array}$$
STEP 4.	Now two packets balance 12 jelly beans. Arrange the diagram so you can see how many beans would balance one packet. Record the algebraic process.		$$2P = 12$$ $$\frac{2P}{2} = \frac{12}{2}$$
STEP 5.	Answer the question, "How many jelly beans are in a packet?" Record your answer.		$$P = 6$$
STEP 6.	Check your answer in the original diagram and equation.		$$2 + 3P = P + 14$$ $$2 + 3(6) = 6 + 14$$

b. For Step 2 above, describe another way to eliminate some packets or some jelly beans while keeping the balance. Then write algebraic steps that correspond to the process you described. *Remove two beans from each side; 2 + 3P − 2 = P + 14 − 2, 3P = 12*

LT-37. Build each equation with packets and beans, and then solve it as demonstrated in LT-36. Be sure to record each of the moves you make in solving an equation by making a sketch *and* writing algebraically what you did at each step. Don't forget to show your check!

a. $2P + 3 = 11$
  P = 4, 2(4) + 3 = 11

b. $5P + 4 = 3P + 10$
  P = 3, 5(3) + 4 = 3(3) + 10

c.

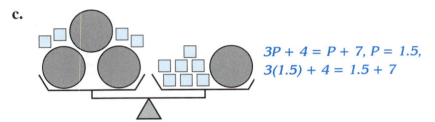

3P + 4 = P + 7, P = 1.5,
3(1.5) + 4 = 1.5 + 7

LT-38. Solve each of the following equations. Use the idea of keeping a scale balanced. For each equation, show your check.

a. $7x + 52 = 12x + 17$ *x = 7, 7(7) + 52 = 12(7) + 17*

b. $x + 8 = 2 + 5x$ *x = 1½, 1½ + 8 = 2 + 5(1½)*

c. $x + 6x + 31 = 15 + 39x$ *x = ½, ½ + 6(½) + 31 = 15 + 39(½)*

LT-39 could be done quickly as a whole-class, oral exercise.

LT-39. Some equations can be solved in your head without using pencil-and-paper techniques. Inspect each of the following equations, and, using only mental math, find a value of x that makes it true:

a. $x + 7 = 2$ *−5*

b. $4x − 7 = 1$ *2*

c. $5 = 4\frac{1}{3} + x$ *⅔*

d. $4x = x$ *0*

LT-40. Use a guess-and-check table to write an equation that models the following problem. Then use algebra to solve the equation. Record what the variable represents, the equation, and your answer. Show how you checked your answer.

Last year's pie-eating champion, Andy, trained seven consecutive days for this year's pie-eating contest. Each day he ate two more pies than the day before. Andy ate a total of 133 pies while in training. How many pies did he eat on the first day of training? *Let p = number of pies on first day; p + (p + 2) + (p + 4) + (p + 6) + (p + 8) + (p + 10) + (p + 12) = 133; 13 pies; 13 + (13 + 2) + (13 + 4) + (13 + 6) + (13 + 8) + (13 + 10) + (13 + 12) = 133*

EXTENSION AND PRACTICE

LT-41. Use your pattern-detection skills to write an equation represented by following the guess-and-check table. To do this, use x as a guess and then fill in the columns with expressions containing x. Define the variable, and write the equation. Then write a word problem that fits. (There are many possible word problems.)

Guess: First Number	Second Number	Product	Check: Product = 2176?
10	90	$10 \cdot 90 =$	90 (too low)
25	75	$25 \cdot 75 =$	1875 (too high)
40	60	$40 \cdot 60 =$	2400 (too high)
35	65	$35 \cdot 65 =$	2275 (too high)
32	68	$32 \cdot 68 =$	2176 *just right*

Variable: Let x represent the first number.

Equation: $x(100 - x) = 2176$

Answer: The numbers are 32 and 68.

Your word problem: Answers vary. One possibility: Find two numbers whose sum is 100 and whose product is 2176.

LT-42. Solve each of the following equations by any method you choose. For some problems, you might choose to use algebra and the idea of packets and beans on a balanced scale. It may be faster to solve others mentally. Regardless of the method you choose, show your check.

a. $7P + 15 = P + 81$

11, $7(11) + 15 = 11 + 81$

b. $9\frac{2}{5} + P = 10$

$\frac{3}{5}$, $9\frac{2}{5} + \frac{3}{5} = 10$

c. $53 = 4x + 22$

7.75, $53 = 4(7.75) + 22$

d. $1 + 4x = 3 + x$

$\frac{2}{3}$, $1 + 4\left(\frac{2}{3}\right) = 3 + \frac{2}{3}$

LT-43. Write and solve an equation for the jelly-bean riddle posed by Sara at the beginning of this section. Show how you checked your answer.
$16 + 19P = 8P + 302$, $P = 26$; $16 + 19(26) = 8(26) + 302$.

For each of problems LT-44 and LT-45, follow the "Guide for Using Guess-and-Check Tables to Write Equations" to solve the problem and write an equation.

LT-44. A large cake is cut into four pieces so that each piece is twice as heavy as the preceding one. The entire cake weighs 5 pounds. How many *ounces* does each piece weigh? (Hint: How many ounces does the cake weigh?) Check your answer. $x + 2x + 4x + 8x = 80$; $5\frac{1}{3}$, $10\frac{2}{3}$, $21\frac{1}{3}$, $42\frac{2}{3}$ oz; $5\frac{1}{3} + 2(5\frac{1}{3}) + 4(5\frac{1}{3}) + 8(5\frac{1}{3}) = 80$

LT-45. When Ellen started with Regina's favorite number and tripled it, the result was 12 more than twice the favorite number. Define a variable, then write an equation and use it to find Regina's number. Check your answer. *x = favorite number; 3x = 12 + 2x, x = 12; 3(12) = 12 + 2(12)*

LT-46. The two line segments below have equal lengths.

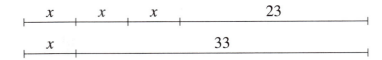

Copy the diagram, and then write an equation that represents it. Solve your equation to find the length x. Show how to check your answer on the diagram. *3x + 23 = x + 33; x = 5; 3(5) + 23 = 5 + 33*

Students should be thinking of the big square as having area x^2, the 1-by-x rectangles as having area x, and the small squares as having area 1.

LT-47. Representing Tiles with Polynomials On the Resource Page for this problem, represent each of the tile collections in parts a through c with an algebraic expression, as shown in the example. Represent the tile collection in part d with an equation.

Example: This collection of tiles

can be represented with the polynomial $x^2 + 2x + 3$. ▲

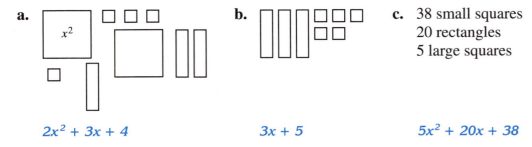

a. $2x^2 + 3x + 4$

b. $3x + 5$

c. 38 small squares
20 rectangles
5 large squares
$5x^2 + 20x + 38$

d.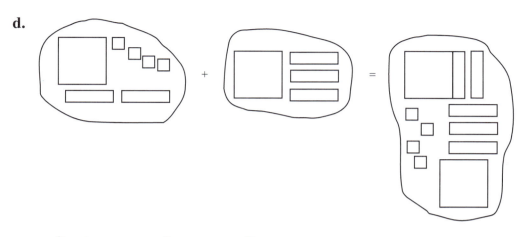

$(x^2 + 2x + 4) + (x^2 + 3x) = 2x^2 + 5x + 4$

LT-48. Suppose Tanya has two big squares, three rectangles, and one small square on her desk and Irena has one big square, five rectangles, and eight small squares. You decide to put all the tiles together on one desk. Write an algebraic equation that represents this situation. Your equation should look like this:

tiles on Tanya's desk + tiles on Irena's desk = all tiles together on one desk. $(2x^2 + 3x + 1) + (x^2 + 5x + 8) = 3x^2 + 8x + 9$

LT-49. Bob, Chris, Janelle, and Pat are in a group. Bob, Chris, and Janelle have algebra tiles on their desks. Bob has two big squares, four rectangles, and seven small squares; Chris has one big square and five small squares; and Janelle has 10 rectangles and three small squares. Pat's desk is empty. The group decides to put all of the tiles from the three desks onto Pat's desk. Write an algebraic equation that represents this situation.
$(2x^2 + 4x + 7) + (x^2 + 5) + (10x + 3) = 3x^2 + 14x + 15$

LT-50. Use the meaning of integer exponents to rewrite each of the following expressions more simply; that is, using as few exponents as possible.

a. $6^{34} \cdot 6^{25}$ 6^{59}
b. $(6^{34})^{25}$ 6^{850}
c. $(6x^2)^4$ $1296x^8$
d. $(xy)^3$ x^3y^3
e. $(2x^7)(3x^4)$ $6x^{11}$

LT-51. Molly knows that x^5 means $x \cdot x \cdot x \cdot x \cdot x$.

a. Explain to Molly how to rewrite $(2x^3)^5$ with a single exponent.
$(2x^3)^5 = 2x^3 \cdot 2x^3 \cdot 2x^3 \cdot 2x^3 \cdot 2x^3 = 2^5 \cdot x^{15} = 32x^{15}$

b. Without explaining the meaning, rewrite $(6x^4)^{11}$ with a single exponent.
$6^{11}x^{44} = 362{,}797{,}056 x^{44}$

LT-52. Write an algebraic expression for the area of each of the following composite rectangles in two different ways—first as a product, and then as a sum. For example, for this composite rectangle

you can write

$$(x + 1)(x + 2) = x^2 + 3x + 2$$
area as a product   area as a sum

a. 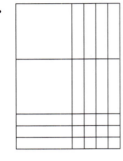   b.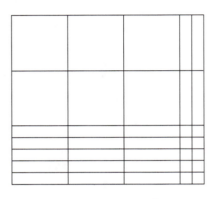

$(2x + 3)(x + 4) = 2x^2 + 11x + 12$ $(2x + 5)(3x + 2) = 6x^2 + 19x + 10$

LT-53. Using the distributive property, rewrite each of the following products as a sum, and rewrite each sum as a product.

a. $4(y - 7)$ $4y - 28$ b. $7y + 56$ $7(y + 8)$

c. $3z(2z - 4)$ $6z^2 - 12z$ d. $m^2 + m$ $m(m + 1)$

Problems LT-54, LT-72, LT-96 and LT-97 are a series of problems that introduce the zero product property in preparation for solving quadratic equations later on. Problems LT-126 and LT-127 summarize the series.

LT-54. The equation $(x + 1)(x - 2) = 0$ has two solutions; that is, there are two values of x that will make the equation $(x + 1)(x - 2) = 0$ true. One way to find the two solutions is to make a guess-and-check table like this:

Guess: Value of x	Value of $x + 1$	Value of $x - 2$	Product $(x + 1)(x - 2)$	Check: Product = 0?
10	11	8	$11 \cdot 8 =$	88 (too big)
1	2	-1	$2(-1) =$	-2 (too small)
-3	-2	-5	$(-2)(-5) =$	10 (too big)

a. Copy and complete the guess-and-check table to find the two values of x that will make the equation $(x + 1)(x - 2) = 0$ true. $x = -1, 2$

b. The equation $(x + 4)(x - 5) = 0$ also has two solutions. Set up and use a guess-and-check table to find two values that will make the equation true. $x = -4, 5$

LT-55. In this problem you'll compare the graphs of two equations that look similar:

$$(1)\ y = 2x \qquad (2)\ y = x^2$$

a. For the equation $y = 2x$, make a table with eight values for x between -4 and 4 and find the corresponding y-values.

x	-3	-2	-1	0	0.5	1	2	3
$y = 2x$	-6	-4	-2	0	1	2	4	6

b. Use your table in part a to graph the equation $y = 2x$. Be sure to label your graph with all the appropriate labels.

c. Repeat part a for the equation $y = x^2$.

x	-3	-2	-1	0	0.5	1	2	3
$y = x^2$	9	4	1	0	0.25	1	4	9

d. Use your table in part c to graph the equation $y = x^2$. Be sure to label your graph with all the appropriate labels.

e. Both of the equations $y = 2x$ and $y = x^2$ have "2" in them. What is it that makes one of the graphs a straight line and the other a curve (in this case a parabola) ?   *When an equation has an x-squared term, it will be a curve.*

LT-56. Copy and solve these Diamond Problems:

a.
b.
c.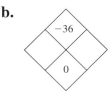
d.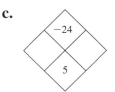
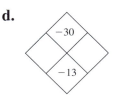

N = −27, S = −1.5   *6 and −6*   *8 and −3*   *−15 and 2*

172 Chapter 3 Lions, Tigers, and Emus: Writing and Solving Equations

3.4 DOING AND UNDOING

SECTION MATERIALS
A WRAPPED PACKAGE (FOR LT-57)
ALGEBRA TILES (FOR LT-61 AND LT-64)

This section extends the process of keeping a balance to solve more complicated linear equations using the idea of "doing and undoing."

We've successfully introduced the notion of "doing and undoing" by having students analyze the processes of wrapping and unwrapping a gift as in LT-57. Have groups share their lists of steps in wrapping a package from part a. Next, choose one list to write on the board, writing each step in abbreviated form. Then ask the class what steps are needed to systematically unwrap the package. Working back up the list, record each reverse step. Emphasize the idea that unwrapping is the reverse of the process of wrapping: Wrapping makes the gift look more elaborate, but the item inside the package remains unchanged; unwrapping makes the gift visible again.

This exercise is more memorable if you display an actual wrapped package to the students. You could also follow their directions for unwrapping the package to verify that the instructions work. One instructor who did this demonstration used graded quizzes as the "gift."

Much of mathematics consists of the closely related processes of doing and undoing. In many cases, to solve a problem we must "undo" something that has been "done." This is especially true when we solve equations. We can think of a complicated equation as the result of a series of processes that have been done to a simple equation. To uncomplicate the equation—and find its solution—we must reverse the series and undo each of the processes. The notion of undoing something is part of the problem-solving strategy of working backward.

LT-57. One way to think about the mathematical processes of doing and undoing is to imagine wrapping a gift for someone special. Imagine you've bought your grandmother an expensive crystal vase.

a. List four steps you could follow to wrap the gift.

b. Oops—you forgot to remove the price tag before you wrapped the gift! List the steps you'll need to take to systematically unwrap the vase so that you can remove the price tag and then rewrap the gift with the original paper and ribbon.

	Wrap		Unwrap
↓	1. *Put gift in box.*	↑	4. *Take out the gift.*
↓	2. *Cover with tissue paper.*	↑	3. *Remove the tissue paper.*
↓	3. *Wrap with fancy paper*	↑	2. *Unwrap the fancy paper*
↓	4. *Tie with ribbon.*	↑	1. *Untie the ribbon.*

The following Head Problem may help students make the connection between algebra and the wrapping and unwrapping processes.

Head Problem-9. *"Pick a number. Add 3. Multiply by 2. Subtract 1. (Pause.) Now add 1. Divide by 2. And subtract 3. What is your result?"* the original number

Instructor's Directions	Recording of Discussion	
Pick a number.	x	⎫
Add 3.	$x + 3$	⎬ doing
Multiply by 2.	$2(x + 3)$	
Subtract 1.	$2(x + 3) - 1$	⎭
[Pause]		
Add 1.	$2(x + 3) - 1 + 1$	⎫
Divide by 2.	$2(x + 3) \div 2$	⎬ undoing
Subtract 3.	$x + 3 - 3$	
Result:	x	⎭

Once you have recorded the steps of the Head Problem on the board, ask students to decide which operations were "undoing" the effect of the previous ones. Then label your list with "doing" and "undoing" as shown. In LT-58, students will use the technique of making a "do" list and "undo" list to help them decide what steps to take to solve linear equations.

In LT-58 first, have the students examine each equation and then list what has been done to x. Next have them make a list of the steps they will use to undo the operations they have described. Insist that students write words for the operations, since this will slow them down, promote understanding, and make their work far more accurate. Although students may not like doing the extra writing, it will help them in the long run.

LT-58. Some equations require only one step to solve; that is, undoing what has been done to the variable is a one-step process. Others are more complicated, requiring many steps. To become good at solving equations, it helps to be able to look at an equation and identify what has been done. Then you can figure out how to undo the process, just as you did with the wrapped gift in LT-57. When analyzing equations, it is always useful to keep the order of operations in mind. Complicated equations are solved one step at a time.

Example: For the equation $\frac{3x + 9}{2} - 4 = 1$, make a list to describe what is done to x and then list what you could do to undo each step. Show how to solve the equation for x by using your undo list. Finally, check your solution.

Do	Undo
↓ 1. Start with x.	↑ 5. The result is x.
↓ 2. Multiply by 3.	↑ 4. Divide by 3.
↓ 3. Add 9.	↑ 3. Subtract 9.
↓ 4. Divide by 2.	↑ 2. Multiply by 2.
↓ 5. Subtract 4.	↑ 1. Add 4.

After you make the two lists, use the undo list from bottom to top to solve the equation. Solve by undoing as follows:

$$\frac{3x + 9}{2} - 4 = 1$$

1. Add 4:
$$\frac{+4 \quad +4}{\frac{3x + 9}{2} = 5}$$

2. Multiply by 2:
$$2\left(\frac{3x + 9}{2}\right) = 2(5)$$
$$3x + 9 = 10$$

3. Subtract 9:
$$\frac{-9 \quad -9}{3x = 1}$$

4. Divide by 3:
$$\frac{3x}{3} = \frac{1}{3}$$

5. The result is x:
$$x = \frac{1}{3}$$

Finally, check your result:

$$\frac{3(\frac{1}{3}) + 9}{2} - 4 = 1 \quad \blacktriangle$$

For each of the following equations, make a list to describe what is done to x, and then make a list to describe how you could undo what is done. Solve each equation for x. Finally, check your solution by replacing x with your solution to see if the two sides of the equation balance.

a. $\dfrac{2(x-5)+4}{10} = 7$

Do list:
 subtract 5, multiply by 2, add 4, divide by 10.
Undo list:
 multiply by 10, subtract 4, divide by 2, add 5; $x = 38$

b. $2(\tfrac{5}{6}x - 8) + 1 = -20$

Do list:
 multiply by $\tfrac{5}{6}$, subtract 8, multiply by 2, add 1.
Undo list:
 subtract 1, divide by 2, add 8, divide by $\tfrac{5}{6}$; $x = -3$

c. $\dfrac{9\left(\tfrac{x}{3} - 7\right)}{4.5} = 100$

Do list:
 divide by 3, subtract 7, multiply by 9, divide by 4.5.
Undo list:
 multiply by 4.5, divide by 9, add 7, multiply by 3; $x = 171$

Problem LT-59 develops understanding of the content and construction of equations and provides practice with order of operations and arithmetic skills. After groups have finished both parts, have representatives from one or two groups put their creations on the board. Another way to do the problem is to have groups write their complicated equations on the board after completing part a. Then have the groups try to solve each other's equations after they complete part b. Alternatively, they could write their complicated equations on a fresh piece of paper and trade with another group to see if they can find the starting value for x.

LT-59. Complicating Equations Some linear equations, such as $x = 5$, are very simple. Others, such as

$$\frac{6(x+1)+7x-8}{10} = \frac{15-4x}{3},$$

are more complicated. One of your goals in this course is to develop skills in solving linear equations of all levels of difficulty. To help you do this, we extend the idea of "doing and undoing" to the idea of *complicating* a simple equation and then *uncomplicating* it—that is, solving it—by undoing the complicating process. (This is one way you could create a Head Problem with a specific answer.) Here's an example of the complicating process.

Example: Start with this simple equation:

$$x = 5$$

Now complicate the equation:

Add 1 to each side: $\quad x + 1 = 6$
Multiply both sides by 7: $\quad 7(x+1) = 42$
Subtract $4x$ from each side: $\quad 7(x+1) - 4x = 42 - 4x$
Subtract 8 from each side: $\quad 7(x+1) - 4x - 8 = 34 - 4x$

Divide both sides by 3: $\quad \dfrac{7(x+1) - 4x - 8}{3} = \dfrac{34 - 4x}{3}$ ▲

The whole idea of solving a complicated equation is to reverse the complicating process in order to obtain an equation of the form "$x = ?$"

a. With your group, choose a value for x and write a simple equation, like $x = \frac{2}{3}$. Use at least four steps to complicate the simple equation. Describe and record each step as shown in the example.

b. Copy your complicated equation. To uncomplicate the equation, describe and record the steps you could take to reverse the complicating process you did in part a.

LT-60. Solve each of the following equations for x. Check each solution by replacing x in the equation with your solution.

a. $x + \frac{2}{3} = 1\frac{1}{2}$ $\frac{5}{6}$
b. $5x + 0.5 = 0.2x + 1.5$ $\frac{5}{24} \approx 0.21$
c. $\frac{2}{5}(x + 3) - 4 = 5$ 19.5
d. $\frac{3}{4}x - 5 = 2x - 15$ 8
e. $0.3x - 1.2 = x + 8.4$ $-13\frac{5}{7} \approx -13.71$

Problems LT-61, LT-64, LT-84, LT-105, and LT-128 form a sequence of problems leading to multiplying binomials and other polynomials.

LT-61. If you know the dimensions of a composite rectangle, you can use algebra tiles to build the complete rectangle. And once you know what tiles to use to build a composite rectangle, you can write its area as a sum.

Example: Suppose the given dimensions of a composite rectangle are $x + 1$ and $x + 4$:

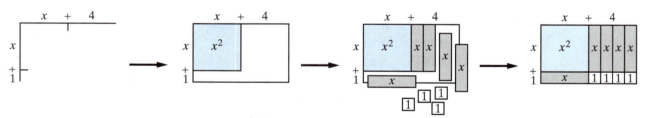

Then the composite rectangle shows

$$(x + 1)(x + 4) = x^2 + 5x + 4.$$
area as a product   area as a sum ▲

For each of parts a through c, the dimensions of a composite rectangle are given. Use your algebra tiles to construct the indicated rectangle. On your paper, draw the complete composite rectangle and then write its area as a product and as a sum.

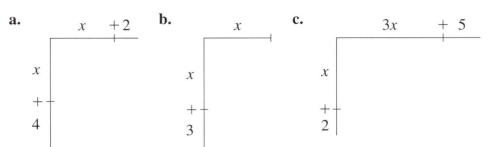

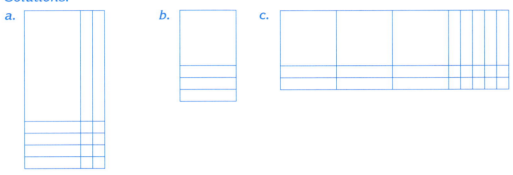

$(x+4)(x+2) = x^2 + 6x + 8 \quad (x+3)(x) = x^2 + 3x \quad (x+2)(3x+5) = 3x^2 + 11x + 10$

EXTENSION AND PRACTICE

LT-62. For each of the following equations, make a list to describe what is done to x, and then make a list to describe how you could undo what is done. Solve each equation for x. Finally, check your solution by replacing x with your solution to see if the two sides of the equation balance.

a. $\dfrac{7(x+3) - 5}{2} = -104$

Do list:
 add 3, multiply by 7,
 subtract 5, divide by 2.
Undo list:
 multiply by 2, add 5,
 divide by 7, subtract 3, $x = -32$

b. $\dfrac{\frac{2}{3}x + 14}{4} - 6 = 12$

Do list:
 multiply by $\frac{2}{3}$, add 14,
 divide by 4, subtract 6.
Undo list:
 add 6, multiply by 4, subtract 14,
 divide by $\frac{2}{3}$; $x = 87$

LT-63. Solve each equation below by any method you choose. Check your answer by replacing x with the value you find.

a. $\frac{7}{2}x + 6 = x + 12$ $\frac{12}{5}$

b. $-2x - 4 = 4x + 8$ -2

c. $\frac{7}{2} + x + 6 = 7x - 14$

$\frac{47}{12} \approx 3.92$

d. $10.5 = 7 - 4\left(\dfrac{x - 15}{9}\right)$

7.125

LT-64. For each of parts a through c, the dimensions of a composite rectangle are given. Use your algebra tiles to construct the indicated rectangle. On your paper, draw the *complete* composite rectangle (show each tile), and then write its area as a product and as a sum.

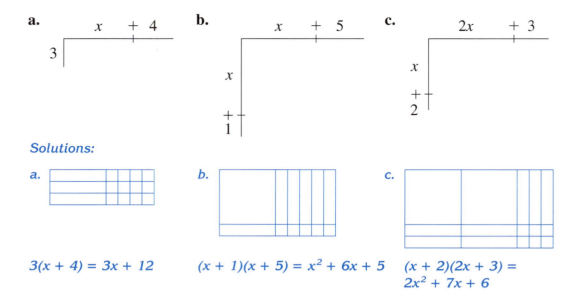

Solutions:

$3(x + 4) = 3x + 12$

$(x + 1)(x + 5) = x^2 + 6x + 5$

$(x + 2)(2x + 3) = 2x^2 + 7x + 6$

LT-65. Solving Equations Containing Parentheses Suppose you'd like to solve the equation $5(2x + 7) = x + 71$. The feature that makes this equation different from equations you've solved before are the parentheses. The standard order of operations require that you deal with parentheses first. Now, the $2x + 7$ inside the parentheses can't be combined; however, you can use the distributive property to rewrite the product $5(2x + 7)$ as the sum $5(2x) + 5(7)$. The following example shows one way you might proceed.

Example:

$$5(2x + 7) = x + 71$$

STEP 1 Use the distributive property to rewrite the left-hand side of the equation:

$$5(2x) + 5(7) = x + 71$$

Now simplify the left side to get an equation that looks like many you have seen before:

$$10x + 35 = x + 71$$

STEP 2 Subtract 35 from each side.

$$10x = x + 36$$

STEP 3 Subtract x from each side.

$$9x = 36$$

STEP 4 Divide both sides by 9.

Answer $\qquad\qquad\qquad x = 4$

a. Copy the example onto your paper.

b. How else might you proceed in Step 2? Try it! *You could subtract x from both sides.*

LT-66. You could mimic the process used in LT-65 to solve the equation $9(3x - 8) = 36$. Copy the problem shown below onto your paper. Then fill in lines b and c to explain what to do to obtain each equation below the line.

$$9(3x - 8) = 36$$

a. Distribute the 9.

$$27x - 72 = 36$$

b. *Add 72 to each side.*

$$27x = 108$$

c. *Divide each side by 27.*

$$x = 4$$

LT-67. Here's another way to solve the equation $9(3x - 8) = 36$. Copy the problem shown below onto your paper. Fill in lines a, b, and c to explain what to do to obtain each equation below the line.

$$9(3x - 8) = 36$$

a. *Divide each side by 9.*

$$3x - 8 = 4$$

b. *Add 8 to each side.*

$$3x = 12$$

c. *Divide each side by 3.*

$$x = 4$$

LT-68. You've used two different equation-solving methods to solve equations containing parentheses.

a. Describe how the equation-solving methods in LT-66 and LT-67 are different. In what ways are the two methods alike? *In LT-66, we used the distributive property in the first step. In LT-67, we divided first. Both ways led to the same solution.*

b. Explain why the method shown in problem LT-67 might be inconvenient for solving the equation $3(2x - 1) = 7$. *If you divide first, you'll get a fraction.*

c. Explain why the method shown in problem LT-66 might be inconvenient for solving the equation $1250(657x - 1324) = 1250$. *If you distribute first, you'll get large numbers.*

LT-69. In this problem you'll solve the equation $4(x - 183) = 20$ in two different ways.

a. Use the "distribute first" method illustrated in LT-66, and then use the "divide first" method shown in LT-67 to solve the equation.

$$4(x - 183) = 20$$

Distribute first:
$4(x - 183) = 20$
$4x - 732 = 20$
$4x = 752$
$x = 188$

Divide first:
$4(x - 183) = 20$
$x - 183 = 5$
$x = 188$

b. Which of the two methods was more convenient for solving this particular equation? Explain why you think so. *Answers vary: distributing takes one more step and is more difficult to do mentally.*

c. Add information about using the "distribute first" method and the "divide first" method for solving equations to your Tool Kit.

For each of problems LT-70 and LT-71, write an equation and solve it. Be sure to define the variable you use, record the equation you use, and write your answer in a complete sentence.

LT-70. Thinking of a Number I'm thinking of a number. If I add 3 to the number, multiply the sum by 2, and then subtract 5, the result will be 4 more than 3 times my original number. What is my number? Define a variable, write an equation, and then solve it. Check your answer in the original problem. *(Hint: a guess-and-check table is a good way to start.)* *Let x = unknown number; $2(x + 3) - 5 = 4 + 3x$; $x = -3$*

LT-71. Suppose the coins in a piggy bank are all dimes, nickels, and quarters. There are twice as many dimes as quarters, and two more nickels than quarters. If the total value of the coins is $7.60, how many coins are in the piggy bank? *Let q = number of quarters; $0.25q + 0.10(2q) + 0.05(q + 2) = 7.60$; 62 coins.*

LT-72 is another problem that prepares students for the zero product property. In checking their solutions to $(2x - 5)(x + 3) = 0$ in part c, students may be inclined to short-cut the process by judiciously substituting both values of x at the same time:

$$(2x - 5)(x + 3) = (2 \cdot 2.5 - 5)(-3 + 3) = (0)(0) = 0.$$

LT-72b shows the correct technique.

LT-72. Solving Equations with Two Solutions

a. The equation $(2x - 5)(x + 3) = 0$ has two solutions. Use a guess-and-check table as modeled in LT-54 to find the two values for x that make the equation true. *$x = 2.5, -3$*

b. In LT-54b you used a guess-and-check table to find two values for x that make the equation $(x + 4)(x - 5) = 0$ true. To check that $x = -4$ is a solution, you substituted -4 for x and calculated the product:

$$(x + 4)(x - 5) = (-4 + 4)(-4 - 5) = (0)(-9) = 0.$$

Show that $x = 5$ is also a solution to the equation $(x + 4)(x - 5) = 0$ by substituting 5 for x. $(x + 4)(x - 5) = (5 + 4)(5 - 5) = (9)(0) = 0$

c. Using the same method as you did in part b to show that each value you found for x in part a is indeed a solution to the equation $(2x - 5)(x + 3) = 0$. $(2x - 5)(x + 3) = (2(2.5) - 5)(2.5 + 3) = (0)(5.5) = 0$ and $(2x - 5)(x + 3) = (2(-3) - 5)(-3 + 3) = (-11)(0) = 0.$

LT-73. Sketch a composite rectangle described by each of the following expressions, and then find each product by writing each area as the sum of the parts.

a. $(10 + 4)(10 + 3)$

b. $(10 + 1)(20 + 5)$

Solutions:

a.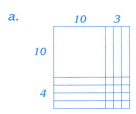

$(10 + 4)(10 + 3) = 100 + 30 + 40 + 12 = 182$

b.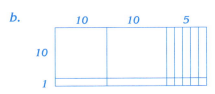

$(10 + 1)(20 + 5) = 200 + 50 + 20 + 5 = 275$

LT-74. You can use the meaning of integer exponents to rewrite complicated exponential expressions.

a. Write the meaning of the expression $x^4(x^2)^3$, and then rewrite it in a simpler exponential form. x^{10}

b. Write the meaning of the expression $(2x^3y^2)^4$, and then rewrite it in a simpler exponential form. $16x^{12}y^8$

LT-75. Exponent Practice Use what you know about exponents to write each of the following expressions more simply. Don't forget the order of operations!

a. $(5x)(2x^2)$ $10x^3$

b. $(5x \cdot 2x)^2$ $100x^4$

c. $(5x + 2x)^2$ $49x^2$

d. $(5x)(2x)^2$ $20x^3$

LT-76. Use your pattern-detection skills to fill in the missing column descriptions, write an equation, and create a word problem represented by the following guess-and-check table. There are many word problems you could write. Write only one, but be creative!

Guess:			Check: = 1974?
30	35	30 · 35 =	1050 (too low)
40	45	40 · 45 =	1800 (too low)
43	48	43 · 48 =	2064 (too high)
42	47	42 · 47 =	1974 *just right*

Variable: Possible answer: Let x = the first number.

Answer: The "number" is 42.

Equation: $x \cdot (x + 5) = 1974$

Your word problem: Possible answer: One number is 5 more than another. The product of the numbers is 1974. Find the smaller number.

LT-77. Alberto collects information about the speed of a roller-coaster compared to the location of the roller coaster along its track. Can Alberto use his data to draw a continuous graph, or should it be discrete? Explain your answer. The graph should be continuous.

Problem LT-78 is preparation for working with ratios in Chapter 4.

LT-78. Connie gets two weeks paid vacation each year. The comparison of Connie's paid vacation time to the number of weeks in a year can be expressed as a ratio in several ways:

$$\frac{2}{52} \quad \text{or} \quad \frac{1}{26} \quad \text{or} \quad 2:52 \quad \text{or} \quad 1:26.$$

Write a ratio for each of the following descriptions:

a. eight minutes to one day (in minutes) $\frac{8}{1440}$, or 1:180

b. one gallon to one pint (in pints) $\frac{8}{1}$, or 8:1

3.5 SOLVING LITERAL EQUATIONS

A **literal equation** is one that contains more than one variable. When there is a mathematical relationship between two or more quantities, you can write the relationship using a *literal equation,* or *formula.* For example, the relationship between temperature F measured in degrees Fahrenheit and temperature C measured in degrees Celsius is

$$F = \tfrac{9}{5}C + 32.$$

Section 3.5 Solving Literal Equations **183**

The formula for the perimeter P of a rectangle is another literal equation:
$$P = 2l + 2w,$$
where l represents the length of the rectangle and w represents the width. In this section, you will see how to solve a literal equation for one of its variables.

A quick review of the distributive property is a useful warm-up exercise for this section. Have students rewrite expressions such as

$$9(3x - 8) \quad -2(3 + 4x) \quad -3(2 - 5y)$$

without parentheses.

LT-79. Solve each of the following equations for x. Show the steps you use.

a. $5(x - 4) = x + 25$ $\quad \frac{45}{5} = 11.25$ **b.** $7(2x - 4) + 3 = 31$ *4*

c. $3(7x + 9) = -15$ *-2*

The focus in the next set of problems is on solving a linear equation for a specified variable in preparation for Chapter 5, where the students will be solving general linear equations, of the form $ax + by = c$, for y. We use patterning to show the students that this is similar to what they've done before. We have sequenced these problems to show that the same procedure works in literal linear equations as in linear equations of a single variable.

LT-80. Copy the following examples onto your paper. Fill in the blanks in parts b, c, and d to explain what to do to obtain the equation below it.

Example 1: Solve the equation $2(x + 4) = 13$.

$$2(x + 4) = 13$$

a. Distribute the 2.

$$2x + 8 = 13$$

b. *Subtract 8 from each side.*

$$2x = 5$$

c. *Divide each side by 2.*

$$x = \frac{5}{2}$$

Example 2: Solve the equation $2(x + w) = 13$ for x.

One Way:

$$2(x + w) = 13$$

a. Distribute the 2.

$$2x + 2w = 13$$

b. <u>Subtract 2w from each side.</u>

$$2x = 13 - 2w$$

c. <u>Divide each side by 2.</u>

$$x = \frac{13 - 2w}{2}$$

A Slight Variation:

$$2(x + w) = 13$$

a. Distribute the 2.

$$2x + 2w = 13$$

b. <u>Subtract 2w from each side.</u>

$$2x = 13 - 2w$$

c. <u>Multiply each side by $\frac{1}{2}$.</u>

$$x = \frac{1}{2} \cdot (13 - 2w)$$

d. <u>Distribute the $\frac{1}{2}$.</u>

$$x = 6.5 - w \quad \blacktriangle$$

LT-81. Solve each of the following equations for x.

a. $2x + 3 = 13$ $x = 5$

b. $2x + b = 13$ $x = \frac{(13 - b)}{2}$

c. $5x + 4 = x + 20$ $x = 4$

d. $5x + d = x + 20$ $x = \frac{(20 - d)}{4}$

e. $\frac{x}{2} + 5 = 9$ $x = 8$

f. $\frac{x}{2} + e = 9$ $x = 2(9 - e) = 18 - 2e$

LT-82. Look back at problem LT-81. Compare how you solved the equation in part b with how you solved the equation in part a. Write one or two sentences to explain how the processes you used to solve each pair of equations were alike, and how they were different. *The processes are the same—you treat all the variables (except the one you are solving for) as if they are numbers.*

LT-83. Solve the equation $5x + 3c = 17$ for x, and then solve it for c.
$x = \frac{(17 - 3c)}{5}$; $c = \frac{(17 - 5x)}{3}$

Problem LT-84 is part of a sequence of problems leading to multiplying binomials.

LT-84. We can make our work drawing tiled rectangles easier by not filling in the whole picture. That is, we can show a **generic rectangle** by using an outline instead of drawing in all the dividing lines for the rectangular tiles and unit squares. For example, we can represent the rectangle whose dimensions are $x + 3$ by $x + 4$ with the generic rectangle shown below:

Example using the Tiles:

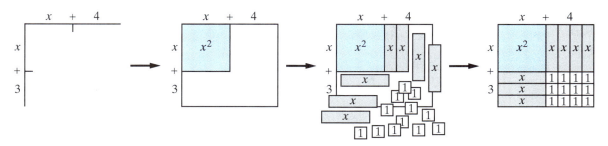

Same Example using a Generic Rectangle:

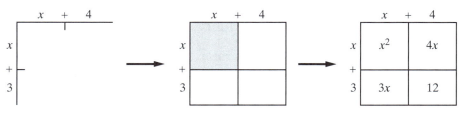

$(x + 3)(x + 4) = x^2 + 4x + 3x + 12 = x^2 + 7x + 12$
area as a product   area as a sum

In the last step, we found the area of each part of the generic rectangle by multiplying its length and width, and then we recorded the area on the part. Note that a generic rectangle just helps you organize a problem—it does not have to be drawn accurately to scale. ▲

Complete each of the following generic rectangles; that is, complete the outline of the parts without drawing in all the dividing lines for the individual tiles. Find and record the area of each part. For each completed generic rectangle, write an equation of the form

area as a product = area as a sum.

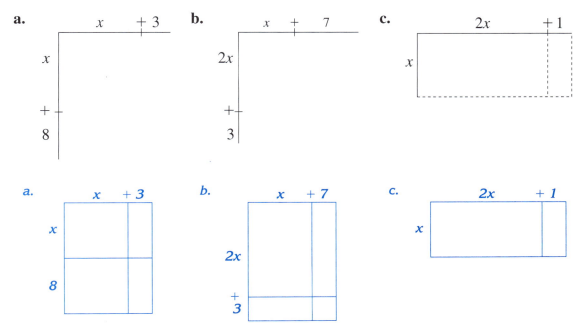

$(x + 8)(x + 3) = x^2 + 11x + 24$ $(2x + 3)(x + 7) = 2x^2 + 17x + 21$ $x(2x + 1) = 2x^2 + x$

EXTENSION AND PRACTICE

LT-85. Solve each of the following equations for x:

a. $3x - 2 = 10$ $x = 4$

b. $bx - 2 = 10$ $x = \dfrac{12}{b}$

LT-86. Solve each of the following equations for the indicated variable.

a. Solve $3x + 91 = 43 - x$ for x. $x = -12$

b. Solve $3x + c = 43 - x$ for x. $x = \dfrac{(43 - c)}{4}$

c. Solve $3x + c = 43 - x$ for c. $c = 43 - 4x$

LT-87. Depreciation The value, V (in dollars), of a refrigerator t years after you buy it is given by the equation

$$V = 550 - 30.5t.$$

a. How much is the refrigerator worth at the time of purchase? after 10 years? *$550; $245*

b. After how many years will the refrigerator be worth $150? When will it be worth nothing? *about 13 yr; about 18 yr*

c. Solve the equation $V = 550 - 30.5t$ for t. $t = \dfrac{(V - 550)}{-30.5}$

LT-88. Complete each of the following generic rectangles; that is, complete the outline of the parts without drawing in all the dividing lines for the rectangular tiles and small squares. Then find and record the area of each part. For each completed generic rectangle, write an equation of the form "area as a product = area as a sum."

a.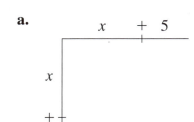
b.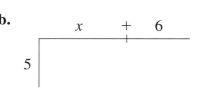
c.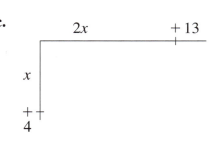

a.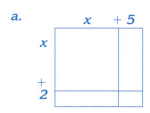
b.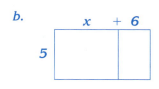
c.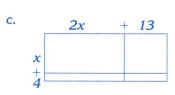

$(x + 2)(x + 5) = x^2 + 7x + 10$ $\qquad$ $5(x + 6) = 5x + 30$ $\qquad$ $(x + 4)(2x + 13) = 2x^2 + 21x + 52$

LT-89. Copy and solve these Diamond Problems:

 a. b. 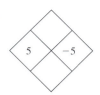 c. 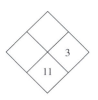 d.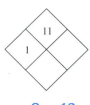

   $N = 14,$ $N = -25,$ $N = 24,$ $S = 12,$
   $S = -9$ $S = 0$ $W = 8$ $E = 11$

If you assign LT-90, a good way to start the next section is to have students share and solve each other's "thinking of a number" problems.

LT-90. Write your own "thinking of a number" problem like LT-70 that will give you a complicated linear equation. Copy your problem on a separate piece of paper, or an index card, so that you can share it with your group tomorrow.

In LT-91 and LT-92, write an equation for each problem and solve it. Be sure to define the variables you use, and write your final answer in a complete sentence.

LT-91. At the Santa Rosalita Senior Center's annual New Year's Eve Party, the band and its guests were given 15 free tickets, and 750 tickets were sold. Tickets for women were $5 each, but men paid $7.50 each. The center took in $175 more on men's tickets than it did on women's tickets. How many men's tickets were sold?   *Let m = number of men's tickets; 7.50m − 5(750 − m) = 175; 314 men's tickets*

LT-92. In honor of the 27th anniversary of Zintha's Department Store, everything was 27 percent off. Jarlene bought a jacket at the sale for $45.26. What was the original price of the jacket?   *Let x = price of original jacket; x − 0.27x = 45.26, or 0.73x = 45.26; $62*

LT-93. Solve each of the following equations:

  a. $3(4x + 1) = 163$ *$\frac{40}{3} \approx 13.33$*   **b.** $6(3x + 2) - 18 = 7$ *$\frac{13}{18} \approx 0.72$*

  c. $4 - 8(3x - 5) = 92$ *−2 (Hint: Remember that 4 − 8(3x − 5) means 4 + (−8)(3x − 5), and the order of operations implies the multiplication is done first.)*

LT-94. Sleepy Time An equation used to relate the age of a person to the number of hours of sleep required each day is

$$H = \frac{34 - A}{2},$$

where H represents the number of hours of sleep required, and A represents the age in years.

a. Answer the following questions in complete sentences:

Does this formula work for you? Does it work for babies? Does it work for older people?

For what ages does the formula seem to work best?

According to the formula, at what age should 6 hours of sleep per day be sufficient? *It works for babies, but past about age 20 it seems not to work as well. A 22-year-old should only require 6 hours of sleep according to the formula.*

b. Solve the equation $H = \dfrac{34 - A}{2}$ for A.   *A = 34 − 2H*

c. How could the equation you wrote in part b be helpful for answering the questions in part a?   *It expresses the age A in terms of the number of hours of sleep.*

LT-95. Use your pattern-detection skills to write an equation and a word problem represented by the following guess-and-check table. (There are many word problems you could write.)

Guess: First Number	Second Number	Sum	Check: Sum = 22?
4	9	4 + 9 =	13 (too low)
5	11	5 + 11 =	16 (too low)
9	19	9 + 19 =	28 (too high)
7	15	7 + 15 =	22 just right

Variable: *Let x = the first number.*

Answers: *The two numbers are 7 and 15.*

Equation: *x + (2x+1) = 22*

Your word problem: *Possible answer: One number is 1 more than twice the other. The sum of the numbers is 22. Find the numbers.*

Problem LT-96 leads directly to the zero product property in LT-97.

LT-96. In this problem, you'll examine some expressions and equations containing zero.

a. Find each of the following products:

   i. $(5.75)(8.33)(4.724)(0)$ *0*   ii. $(5.75)(0)(8.33)(4.724)$ *0*

   iii. $(3.27 \cdot 10^{-2})(0)$ *0*   iv. $(6^3)(4^2 - 1)(x)(0)(y)$ *0*

   v. $(0)(x^2)(0)$ *0*   vi. $(x)(0)$ *0*

b. If $(35)(x)(4)(7) = 0$, what must be the value of x?   *x = 0*

c. If A and B are two numbers, and $A \cdot B = 0$, what can you conclude about the possible values of A and B?   *At least one of them must be 0.*

d. If the product of several numbers is zero, what can you conclude about the numbers?   *At least one of them must be 0.*

In LT-97a watch for students who check their answers by incorrectly substituting both solutions at the same time, as in $(-30 + 30)(17 - 17) = 0(0) = 0$.

LT-97. The Zero Product Property In the preceding problem you used a property of numbers called the zero product property:

If the product of two or more numbers is zero, then at least one of the numbers must be zero.

a. Consider the equation $(x + 30)(x - 17) = 0$. Use the zero product property to determine value(s) of x that will make the equation a true statement. Demonstrate that each of your solutions works by substituting it for x as you did in LT-74.   *x = −30, 17. To check: (−30 + 30)(−30 − 17) = 0(−47) = 0 and (17 + 30)(17 − 17) = 47(0) = 0.*

b. Add notes about the *zero product property* to your Tool Kit.

LT-98. In this problem you will continue to examine how graphs and equations are related.

a. Sketch a graph of the rule $y = 0.75x - 6$ using input values x so that $-1 \leq x \leq 11$. (You will need to make a table first.) Be sure to label the important parts of your graph.

b. If you connect the points to form a continuous graph, what are the coordinates of the x-intercept? What is the y-value at this point?   *(8, 0); y-value is 0.*

c. Find a value for x that makes the equation $0.75x - 6 = 0$ true. Solve the equation algebraically, or use the guess-and-check strategy as in SP-112.   *x = 8*

d. How do your answers to parts b and c compare?   *They should be the same.*

LT-99. Simplifying Exponential Expressions

a. Write the meaning of the expression $5(2x^3y^2)^4$, and then write it in simplified exponential form.   *$80x^{12}y^8$*

b. Write the meaning of the expression $(2y^2)^3 \cdot y^5$, and then write it in simplified exponential form.   *$8y^{11}$*

LT-100. Harold was practicing using his calculator's exponentiation key to evaluate exponential expressions. Molly came along and asked him what would happen if the exponent were a negative integer, so Harold decided to check it out. He remembered from OP-21 that in order to enter a negative number, he needed to use either the change-sign ⌊+/−⌋ key or the negation ⌊(−)⌋ key, but *not* the subtraction ⌊−⌋ key. Harold and Molly tried the following examples:

- Harold's calculator has a change-sign key, so to evaluate the expression 3^{-2}, he entered a "3," pressed ⌊y^x⌋, entered "2," pressed ⌊+/−⌋, and then pressed ⌊=⌋.

- Molly's calculator has a negation key, so to evaluate the expression 3^{-2}, she entered a "3," pressed ⌊y^x⌋, then pressed ⌊(−)⌋, entered "2," and finally pressed ⌊=⌋.

a. Next Harold made two lists of exponential expressions. Copy Harold's table, and use a calculator to compute a value for each of the exponential expressions in which the exponent is a negative integer.

Expressions with Positive Exponents	Expressions with Negative Exponents
3^2	$3^{-2} \approx 0.11111$
5^3	2^{-3} *0.125*
4^3	4^{-3} *0.015625*
4^1	4^{-1} *0.25*

b. Molly decided to plot each of calculator values on this number line:

Harold claims that he can predict where each expression from the table will be plotted before Molly can finish her graph: some will lie between 0 and 1, while the others will lie to the right of 1, and none will be to the left of 0. Decide whether Harold's claim is correct or not. If it is correct, describe the expressions that would be plotted in each region. If Harold's claim is not correct, explain why not. *The claim is correct; the expressions in the list with positive exponents lie to the right of 1; those with negative exponents will lie between 0 and 1.*

c. Add information to your Calculator Tool Kit about using a calculator to *evaluate exponential expressions* in which the *exponent is a negative integer*.

LT-101. Pat found a die that had been poorly made: it had two 6s on it and no 1s. If Pat rolls the die, how likely is it to come up …

a. a 2? $\frac{1}{6}$ b. a 6? $\frac{2}{6}$ or $\frac{1}{3}$ c. a 5 or 6? $\frac{3}{6}$ or $\frac{1}{2}$

3.6 MODELING SITUATIONS WITH LINEAR EQUATIONS

SECTION MATERIALS
SUMMARY DRAFT
 TEMPLATE RESOURCE
 PAGE (FOR LT-123)

If you assigned LT-90, start this section by having students share and solve each other's "thinking of a number" problems as you tour the room and check their work.

So far in this chapter you've seen how algebra can be used as a tool to model and solve problems. You've practiced solving linear equations, both simple and complicated. The power of using an algebraic approach to solving equations lies in the relative ease and efficiency that it provides. However, most "real-life" problems don't occur in the form of equations, let alone linear equations. And the process of writing an equation to model a situation accurately can be extremely challenging. By using the problem-solving strategies of organizing a guess-and-check table and looking for patterns, you've been learning a way to take the relatively formidable task of writing an equation to model a situation and turn it into a relatively simple one.

Once a problem has been expressed as a linear equation, we can use the ideas of "doing and undoing" and "maintaining balance on a two-pan scale" to solve the equation. The two key problems of this section give you an opportunity to apply your skills to an unusual situation (LT-102) and to a scenario similar to one you may have encountered in your own life (LT-103 and LT-104).

LT-102. Lions, Tigers, and Emus Use a guess-and-check table to write an equation that you could use to solve the following problem. Then use algebra to solve the equation. Record your answer along with the equation and what the variable represents. Show how you checked your answer.

At Wild Animal Park, an expansive reserve modeled on animal preserves in Africa and Australia, zoologists keep track of lions, tigers, and emus. When Sydney, a ranger at the park who is known for his odd sense of humor and love of puzzles, took his turn at the official animal count in January, his report was unusual. Sydney noted that the three types of animals had combined totals of 152 heads and 562 legs, and there were twice as many tigers as lions. How many of each type of animal were in the park in January?   *If L is the number of lions, $4L + 4(2L) + 2(152 - 3L) = 562$; 43 lions, 86 tigers, 23 emus*

In Problem LT-103, students first write and graph an equation to solve a problem, and then solve it algebraically. Although the graph in this problem is not continuous, it is useful to view it as a linear trend and draw in the line.

LT-103. Kayak Rental Outdoor Adventures, the student-run nonprofit organization that offers weekend trips for students and rents a variety of equipment, wants to encourage more students to get into kayaking. The kayaks are rented on a daily basis, and the cost includes an $11.50 maintenance fee plus $6.50 per day. Because the rates are so low, there are no discounts for extended rentals.

a. How much would it cost to rent a kayak from Outdoor Adventures for three days? for 10 days? for two weeks? *$31.00, $76.50, $102.50*

b. Write an equation to model the relationship between the cost, *y* in dollars, of renting a kayak and the number of days, *x*, for which it is rented. Use the pattern from part a to help you. *y = 11.50 + 6.50x*

c. What kinds of numbers make sense for *x*, the number of days? Could *x* be negative? 0? 1? a fraction? What about *y*, the cost in dollars?
The number of days, x, is a positive integer, and the cost, y, is a positive integer or fraction.

d. Now set up coordinate axes on graph paper. Scale the horizontal axis so that each tick mark represents 1 day and the vertical axis so that one tick mark represents $2. Be sure to label the axes with their units. Then, explain why the graph of the kayak-rental equation is discrete, and finally, graph its trend line. *Kayaks are rented on a daily basis, so the x-coordinates of points of the graph would be positive integers and the graph would consist of disconnected points.*

e. Use your equation for part b to find the *y*-intercept of the trend line. What would the *y*-intercept represent in terms of renting a kayak? *(0, 11.5); Kayak rental would cost $11.50 before even putting the kayak in the water.*

LT-104. Imagine you're planning to rent a kayak from Outdoor Adventures for an upcoming outing. Unfortunately, you can afford to spend at most $50 on the cost of renting a kayak. How long a trip would your budget limitation allow you to take?

a. Explain and show how to use the graph in LT-103 to find an answer.
Find the integer that precedes the x-coordinate of the point on the line associated with height y = 50.

b. Write and solve an equation to find how long could the outing be.
50 = 11.50 + 6.50x; x ≈ 5.9, so at most 5 days.

LT-105. You can use generic rectangles to find products of binomials, (polynomials with two terms) even when it's not possible to use algebra tiles.

Example: To find $(x - 3)(2x + 5)$, think of $x - 3$ as $x + (-3)$. After labeling the dimensions of a generic rectangle, you can write in the area of each part, and then add:

	2x	+ 5
x		
+		
−3		

	2x	+ 5
x	$2x^2$	$5x$
+		
−3	$-6x$	-15

$(x - 3)(2x + 5) = 2x^2 - x - 15$
area as a product   area as a sum

Find each of the following products of binomials by drawing and labeling a generic rectangle. Not all of the generic rectangles will have the same number of parts.

a. $(3x + 2)(x - 8)$ $3x^2 - 22x - 16$ b. $2x(x + 3)$ $2x^2 + 6x$

c. $(x - 7)(x + 7)$ $x^2 - 49$ d. $(x + 7)(x + 7)$ $x^2 + 14x + 49$

Solutions:

a. 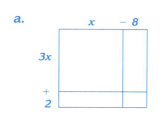 b. 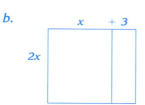 c. 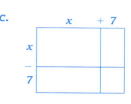 d.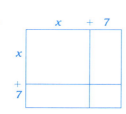

LT-106. Rewrite the expression $(2x + 3)(x - 6) - x(x + 5)$ as a polynomial by following the order of operations and using generic rectangles when needed. $x^2 - 14x - 18$

EXTENSION AND PRACTICE

For LT-107 through LT-109, write an equation for each problem and solve it. Be sure to define the variables you use. Write your answer in a complete sentence.

LT-107. Each side of a square garden is increased by 3 meters. The perimeter of the new garden is 50 meters. What was the length of the original square garden? Draw a diagram and label the sides.   *Let s = length of original side; 4(s + 3) = 50; 9.5 m*

LT-108. Mrs. Agnos keeps cats and canaries. The animals have a combined total of 45 heads and 104 legs. How many canaries does Mrs. Agnos have?   *If x is the number of canaries, then 2x + 4(45 − x) = 104; 38 canaries*

LT-109. Sunrise Sand and Gravel charges $4.50 for each cubic yard of sand plus $6.00 to deliver the sand. How much sand could you have delivered for $100.00?   *Let x = cubic yards of sand; 4.5x + 6 = 100; about 20.89 cu yd*

LT-110. Suppose in class you were working with one big square algebra tile, seven rectangles, and six small squares when you accidentally knocked the big square and five of the small squares onto the floor. Write an equation to represent the tiles you had, what fell off the desk, and the tiles that remained.   *$(x^2 + 7x + 6) - (x^2 + 5) = 7x + 1$*

LT-111. Suppose you were minding your own business using your algebra tiles—two big squares, four rectangles, and three small squares—when Kim came in and took the four rectangles and two of the small squares. Write an equation to represent the tiles you started with, what Kim took, and the tiles you had after Kim left.   *$(2x^2 + 4x + 3) - (4x + 2) = 2x^2 + 1$*

LT-112. Solve each of the following equations by any method you choose. Show how you check your answer.

a. $\dfrac{x}{1.5} = 2$ 3

b. $\tfrac{1}{2}x + 4 = -\tfrac{1}{4}x + 7\tfrac{1}{2}$ $\dfrac{14}{3} \approx 4.67$

c. $m + 0.63 = 1.56$ 0.93

d. $1.2 - w = 0.8w - 1.2$ $\dfrac{4}{3} \approx 1.33$

LT-113. There are two values of x that make the equation $(x - 2)(x + 3) = 0$ true.

a. Is $x = 2$ a solution to the equation $(x - 2)(x + 3) = 0$? Substitute $x = 2$ for both x's in the equation to justify your response.   yes; $(2 - 2)(2 + 3) = 0(5) = 0$.

b. Is $x = 0$ a solution to the equation $(x - 2)(x + 3) = 0$? Justify your response.   no; $(0 - 2)(0 + 3) = (-2)(3) = -6$

c. Is $x = -3$ a solution to the equation $(x - 2)(x + 3) = 0$? Justify your response.   yes; $(-3 - 2)(-3 + 3) = (-5)(0) = 0$

LT-114. Find each of the following products by drawing and labeling a generic rectangle. Not all of the generic rectangles will have the same number of parts.

a. $(2x - 5)(2x + 5)$ $4x^2 - 25$
b. $(7x + 2)(x - 1)$ $7x^2 - 5x - 2$
c. $15x(2x - 1)$ $30x^2 - 15x$
d. $(12 - x)(3x + 4)$ $-3x^2 + 32x + 48$

LT-115. Mary Beth wanted to multiply 14 by 32 without using a calculator, so she drew the following rectangle:

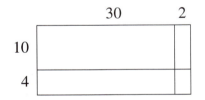

a. Explain why the area of the large composite rectangle is equal to $14 \cdot 32$.   area of rectangle = length $\cdot$ width = $(10 + 4)(30 + 2) = (14)(32)$

b. Find a value for $14 \cdot 32$ by finding the areas of the four smaller rectangles and summing. $300 + 20 + 120 + 8 = 448$

LT-116. Rewrite each of the following expressions in a simpler exponential form, if possible. Otherwise, write an expression in standard form.

a. $3^4 \cdot 3^5$ 3^9
b. $3^4 + 3^5$ $81 + 243$, or 324
c. $(x^3)^4$ x^{12}
d. $(10x^3y^2)^5$ $10^5 x^{15} y^{10}$ or $100{,}000\, x^{15} y^{10}$
e. $(5x^3)^2$ $25x^6$
f. $5x(2x)^3$ $40x^4$

LT-117. Use any method you like to solve each of the following equations. Show how you check your answer.

a. $8 = 4(y + 7) - 20$ *0*
b. $\frac{7x}{4} + 1 = 6 + x$ *$\frac{20}{3}$*
c. $1308(x + 2) = 0$ *-2*
d. $\frac{5m - 3}{2} + 12 = m$ *7*
e. $3 + 4\left(\frac{1}{2}R + 1\right) = 159$ *76*
f. $(x - 12)(x + 1) = 0$ *12, -1*

LT-118. Solve each of the following equations twice: once using the "distribute first" method of LT-66, and then using the "divide first" method of LT-67. Show how you check your answers.

a. $5(4x + 3) = 75$ *3*
b. $-3(8x - 4) = 18$ *$\frac{-1}{4}$*
c. Which method was easier for you? Explain why.   *In both cases, we can divide first to work with smaller numbers.*

LT-119. Slides: A Way to Represent Change Many situations we encounter involve quantities that change in relation to another quantity. One way to represent the relationship between two changing quantities is with a *slide*. A slide is a way to record change graphically, as shown in this example.

Example: Chauncey likes to play the slot machines in Lake Tahoe. Last Tuesday, he started at 8 o'clock in the morning with $20. Lady Luck was with him, and Chauncey played steadily for five hours. On counting his money, he realized that he had made a gain of $40!

We can represent the change in Chauncey's pocket money with a graph, as follows:

Chauncey started the day on Tuesday with $20.

Start at point (0, 20): Tuesday

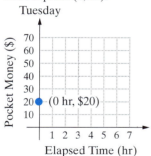

Five hours later, he was ready to count his winnings.

Slide 5 units to the right:

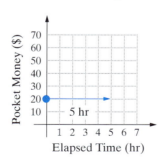

He counted, and realized he had gained $40.

Slide 40 units up:

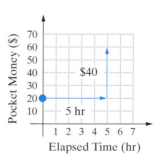

Chauncey stopped after 5 hours, with $60 in his pocket.

Slide 5 units to the right:

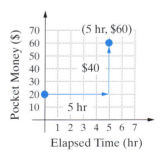

▲

The following graphs represent how Chauncey's pocket money changed on Wednesday. Describe what happened.

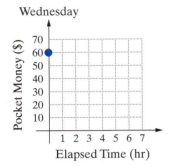
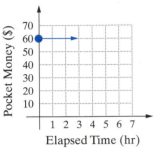
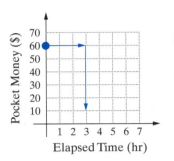
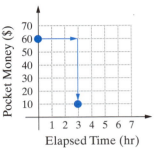

Chauncey started with $60. After 3 hrs, he had lost $50. He stopped after 3 hrs with $10 in his pocket.

LT-120. As you saw in LT-119, a **slide** is a horizontal move together with a vertical move on a graph.

  a. Start with a point P at $(-1, -5)$. Slide P to the right 3 units and then up 4 units. What are the coordinates of P's new position? *(2, −1)*

  b. How could you describe the slide that was used to move point Q from $(-3, 1)$ to $(4, -2)$? *move 7 units right and 3 units down.*

  c. Add to your Tool Kit information about how to describe a move from one point to another using *slide*.

LT-121. Write a ratio for each of the following descriptions:

  a. forty-five minutes to two hours (in minutes) $\frac{45}{120}$, *or 3:8*

  b. one pound to twelve ounces (in ounces) $\frac{16}{12}$, *or 4:3*

LT-122. For each of the following equations, make a table with entries for values of x, and compute the corresponding values for y. Graph each equation on a separate set of axes. Be sure everything is labeled.

  a. $y = x^2 - 3x + 2$ for $-1 \le x \le 4$. Explain why it is important to use 1.5 as an input value. *a parabola with vertex: (1.5, −0.25); (1, 0), (2, 0), the low point of the parabola occurs when x = 1.5.*

  b. $y = 8 - x$

LT-123. Chapter 3 Summary Rough Draft In this chapter you've extended your use of guess-and-check tables from solving problems to being able to write equations to model situations. The ability to model a problem with an equation and to solve the equation algebraically is a tremendous strength. Because you can now solve linear equations in one variable—for example, $3 - x = 6(2x + 1)$—you don't need to rely on guessing and checking to solve problems. The distributive property and the zero product property are useful tools for solving equations.

These are the four main ideas of Chapter 3:

- You can use guess-and-check tables to write equations that model situations.
- You can use the distributive property to rewrite expressions.
- To solve linear equations, it helps to think of the processes of "maintaining a balance on a two-pan scale" and of "undoing what has be done to a variable x."
- The zero product property is useful for solving equations of the form $A \cdot B = 0$.

a. Make photocopies of the Summary Draft Template Resource Page (or follow the same format on your own paper). Use a page to complete the following steps for each of the four main ideas of the chapter:

- State the main idea.
- Select and recopy a problem that is a good example of the idea and that you did well.
- Include a completely worked-out solution to the selected problem.
- Write one or two sentences that describe what you learned about the idea.

As in previous chapters, you will have the opportunity to revise your work later, so your focus now should be on the content of your summary rather than its appearance. Be ready to discuss your responses with your group at the next class meeting.

b. Which problem in this chapter did you like best and what did you like about it?

c. Which were the most difficult parts of this chapter? List sample problems, and discuss the hard parts.

3.7 SUMMARY AND REVIEW

In case you have more than one class period to spend on this section, you might start with a Head Problem. Otherwise, have groups discuss their summaries, and then move on to LT-125 through LT-130. Assign selected problems, as needed for review.

Head Problem 10: "Start with the number of doughnuts in a baker's dozen. Subtract any number of your choice between 5 and 10. Multiply the difference by the largest number of dots on a side of a standard fair six-sided die. Add twice the fifth prime. Divide the sum by 2. Subtract half of 10^2, then divide by your number. What is your result?" −3

LT-124. Chapter 3 Summary Group Discussion Take out the rough-draft summary you completed in LT-123. Use some class time to discuss your work; use homework time to revise your summaries as needed.

For each of the four main ideas of the chapter, listed in LT-123, choose one member of the group to lead a short discussion. When it is your turn to be the discussion leader, you should do the following:

- Explain the problem you chose to illustrate your main idea.
- Explain why you chose that particular problem.
- Tell which problem you liked best, and what you liked about it.
- Tell what you thought were the most difficult parts of this chapter.

This is also your chance to make sure your summary is complete, to update your Tool Kit, and to work together on problems you may not be able to solve yet.

LT-125. Susan wants to use up an extra 80 square feet of lawn sod by planting a border of sod around a rectangular flower bed. If Susan follows her plan, what will be the dimensions of the entire flower bed, including the border? All dimensions are in feet.

Plan for New Flower Bed

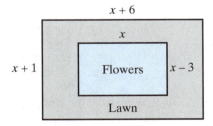

Use the diagram to write an equation, then solve the equation, and check the answer. Write you answer in a complete sentence.
$(x + 1)(x + 6) - x(x - 3) = 80$; $x = 7.4$; The flower bed, including border, would be 8.4 by 13.4 ft.

LT-126. As you saw in LT-97, the zero product property can be used to solve some types of equations.

a. If $(x - 7)(x + 45) = 0$, what must be true about either the factor $x - 7$ or the factor $x + 45$?   At least one factor must equal 0.

b. Use what you wrote in part a to find two values for x that make equation $(x - 7)(x + 45) = 0$ true. $x = 7$ or -45

c. Solve the equation $(2x + 12)(x + 5) = 0$. $-6, -5$

d. Solve the equation $(3x + 4)(2x - 3) = 0$. $-\frac{4}{3}, \frac{3}{2}$

LT-127. Using the Zero Product Property to Solve Equations

a. Write an equation that could be solved with the direct use of the zero product property.   *An example: $(2x + 5)(x - 7) = 0$*

b. Write an equation for which the zero product property is of no help.   *An example: $x + 5 = 0$*

c. Compare the equations you wrote in parts a and b, and describe a fundamental difference.   *Equation (a) is nonlinear, and is a product of two linear expressions; equation (b) is not of the form "product equals zero."*

LT-128. Finding the Area of a Rectangle In LT-84 you were introduced to the use of a generic rectangle to find the area of a rectangle given its dimensions. The idea that

$$\text{area as a product} = \text{area as a sum}$$

can help when you want to rewrite a product of factors in the form of a polynomial.

Examples: Using a generic rectangle, we could rewrite the product of a monomial and a binomial as a polynomial:

$$\underset{\text{product of factors}}{4(x + 3)} = \underset{\text{polynomial}}{4x + 12}$$

We could also write the product of two binomials as a polynomial:

$$\underset{\text{product of factors}}{(x + 2)(x + 8)} = \underset{\text{polynomial}}{x^2 + 10x + 16}$$ ▲

a. How are the distributive property and the use of a generic rectangle related? Make up your own example and draw a generic rectangle. Explain how the picture helps you rewrite the product of factors as a polynomial.

b. Write examples and explanations in your Tool Kit about how to use a generic rectangle to write the product of two expressions as a polynomial.

For each of problems LT-129 and LT-130, write an equation and solve it. Be sure to define the variables you use. Write your answer in a complete sentence.

LT-129. Moe started washing dishes at Larry's Café. He washed nine dishes per minute. Fifteen minutes later, Curley joined Moe and washed 16 dishes per minute. They washed a total of 760 dishes before stopping to rest. How long did Curley work? How many dishes did Moe wash?
Let x = minutes Curly worked; 9(15) + 9x + 16x = 760; 25 min; 360 dishes

LT-130. On an algebra test, each question in Part A was worth three points and each question in Part B was worth five points. Mabel answered 19 questions correctly and had a score of 81. How many questions on each part of the test did she answer correctly? *Let A = number of questions Mabel answered correctly in Part A; 3A + 5(19 − A) = 81; 7 questions on Part A and 12 on Part B*

EXTENSION AND PRACTICE

LT-131. Consider the following three exponential expressions:

    i. $x^5 \cdot x^2$ **ii.** $(x^5)^2$ **iii.** $x^5 + x^2$

a. Although the three expressions have some similarities, they are fundamentally different. In complete sentences, explain how they are different. *One is a product of powers of x, another is a power of a power of x, and the third is a sum of powers of x.*

b. Use the meaning of exponents to rewrite each of the expressions using a single exponent, if possible. If not possible, explain why not.
$x^5 \cdot x^2 = (x \cdot x \cdot x \cdot x \cdot x) \cdot (x \cdot x) = x^7$ and $(x^5)^2 = (x^5)(x^5) = x^{10}$

c. Explain to a student in another algebra class how to rewrite $x^{47} \cdot x^{21}$ with a single exponent without writing out all the factors of x.
$x^{47} \cdot x^{21} = x^{47+21} = x^{68}$

d. Explain to a student in another algebra class how to rewrite $(x^{103})^6$ with a single exponent without writing out all the factors of x.
$(x^{103})^6 = x^{103 \cdot 6} = x^{618}$

e. Update your Tool Kit by adding explanations and examples about *how to rewrite exponential expressions*.

LT-132. Tool Kit Check In addition to your Tool Kit entries from Chapters 1 and 2, your Algebra Tool Kit should also include entries for at least the following topics:

Algebra

- large algebra tile, labeled with its dimensions and area   *LT-19*
- equations containing parentheses: "divide first" or "distribute first"?   *LT-69*
- zero product property   *LT-97*
- sliding to move from one point to another   *LT-120*
- using a generic rectangle to rewrite a product as a sum   *LT-128*
- how to rewrite exponential expressions   *LT-131*

Calculator

- how to enter negative integer exponents   *LT-100*

You are not limited to topics on these lists. If there are other ideas you'd like help remembering, add them to your Tool Kit as well.

a. To be sure that your Tool Kit entries are clear and accurate, exchange Tool Kits with a classmate, and read each other's entries.

b. Make any necessary revisions, clarifications, or updates to your Tool Kit.

LT-133. If one side of a square is increased by 12 feet and an adjacent side is decreased by 3 feet, a rectangle is formed whose perimeter is 64 feet. Draw diagrams of the original square and the new rectangle, and find the length of the side of the original square. *Let s = length of side of original square; 2(s + 12) + 2(s − 3) = 64; 11.5 ft*
Solution:

LT-134. In this program, you will work with an equation that relates degrees Fahrenheit F with degrees Celsius C.

a. For the equation $5F - 9C = 160$, solve for F.
$F = \dfrac{9c + 160}{5}$ or $\dfrac{9}{5}C + 32$

b. Solve the same equation for C. $C = \dfrac{(5F - 160)}{9}$

LT-135. Solve each of the following equations for x. Show how you check your answer.

a. $3 - 2(x - 5) = 15$ *−1*

b. $\dfrac{2(x + 3) - 5}{7} = 0$ $\dfrac{-1}{2}$

c. $\dfrac{2}{5}x + 14 = 20$ *15*

d. $7\dfrac{3}{4} + 0.38x = \dfrac{3}{10} + x$ ≈ 12.02

e. $0.78x - 2 = 0.8x + 8.4$ *−520*

f. $0.38x(1.24 - x) = 0$ *0, 1.24*

g. $\dfrac{9}{14} = \dfrac{-3}{7}x$ $\dfrac{-3}{2}$

h. $8 = -4 + 9\left(\dfrac{11}{2}x + 5\right)$ $\dfrac{-2}{3}$

LT-136. Solve each of the following equations for the indicated variable:

a. $3r = b$ for r $\dfrac{b}{3}$

b. $A = \dfrac{1}{2} \cdot b \cdot h$ for h $\dfrac{2A}{b}$

c. $\dfrac{x}{2} + 5 = 2r$ for x
$2(2r - 5)$ or $4r - 10$

d. $\dfrac{2}{3}P - 6 = 4S$ for P
$\dfrac{3}{2}(4S + 6)$, or $6S + 9$

For each of problems LT-137 through LT-141, write an equation and solve it. Be sure to define the variables you use. Write your answer in a complete sentence.

LT-137. Lassa's favorite cookies at the Cookie Company, the double chocolate chip cookies, cost $0.75 each. Her brother Trevor bought some for Lassa's birthday and spent an extra $0.15 for a gift box. If Trevor spent a total of $12.90, how many cookies did he get? *Let x represent the number of cookies; 0.75x + 0.15 = 12.90; 17 cookies*

LT-138. A 112-centimeter board is cut so one piece is three times as long as the other. How long is each piece? *Let x represent the length of the shorter piece; x + 3x = 112; 28 cm, 84 cm*

LT-139. The perimeter of a triangle is 33 centimeters. The second side is twice as long as the first side, and the third side is 8 centimeters longer than the second. How long is each side? *Let x represent the length of the first side; x + 2x + (2x + 8) = 33; 5 cm, 10 cm, 18 cm*

LT-140. A stick 152 centimeters long is cut up into four short pieces, all the same length, and two longer pieces (both the same length). A long piece is 10 centimeters longer than a short piece. How long is each piece? *Let s represent the length of a short piece; 4s + 2(s + 10) = 152; 22 cm, 32 cm*

LT-141. Orlando's savings account has a yield of 8%. After one year of not adding or withdrawing from the account the balance is $1350.00. What amount was in the account at the start of the year? *Let x represent the amount in Orlando's account at the start; x + 0.08x = 1350; $1250*

LT-142. Copy and solve these Diamond Problems:

a.

N = 30,
S = 11

b.

N = 6,
S = −5

c.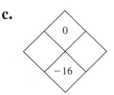

0, −16

d.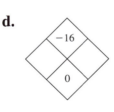

4, −4

LT-143. Graph each of the following equations on the same set of axes. Remember to label the axes and each graph.

a. $y = x$ b. $y = x - 3$ c. $y = 3x$

d. Compare the graphs in parts a and b: How did the "3" in the equation in part b change the graph from part a? *The line was moved down 3 units.*

e. Compare the graphs in parts a and c: How did the "3" in the equation in part c change the graph from part a? *The line is steeper.*

LT-144. Examine each graph, and then write a story or description about what each graph shows. Your story may be different from those of others in your group.

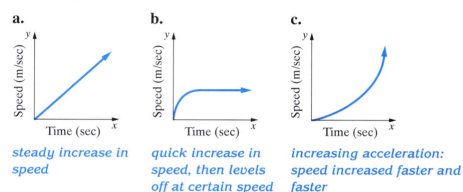

a. *steady increase in speed*

b. *quick increase in speed, then levels off at certain speed*

c. *increasing acceleration: speed increased faster and faster*

LT-145. The two line segments below have equal lengths. Find the length x. *$x = 6$*

LT-146. Solve each of the following equations for x:

a. $\frac{3}{4}x - 6 = 12$ *24*

b. $\frac{3x - 6}{4} = 12$ *18*

c. $8 = 4(x - 7) + 20$ *0*

d. $3x + c = 10$ *$\frac{10 - c}{3}$*

e. $2x + 3 = 0$ *$\frac{-3}{2}$*

f. $(2x + 3)(x + 5) = 0$ *$\frac{-3}{2}, -5$*

g. $14 - 3(2x + c) = 15$ *$\frac{-(1 + 3c)}{6}$*

h. $xc - 5 = y$ *$\frac{y + 5}{c}$*

LT-147. Rewrite each of the following expressions in simpler exponential form, if possible.

a. $x^2 \cdot x^5$ *x^7*

b. $(x^3 y^5)^2$ *$x^6 y^{10}$*

c. $(10x^6 y^3)^4$ *$10^4 x^{24} y^{12}$*

d. $x^2 + x^3$ *Not possible*

e. $(10x^2)(3x^5)^2$ *$90x^{12}$*

f. $10x(5x^2)$ *$50x^3$*

LT-148. Sketch a graph of the relationship between your test scores and the amount of homework that you do.

LT-149. Consider this number trick:

"Pick a number. Multiply it by 2. Add 6. Multiply the sum by 2. Divide by 4, then subtract 3. The result equals your starting number."

a. Choose a number, and use it to show that the trick works.

b. Show that the trick works if you start with x. *$(2(2x + 6) \div 4) - 3 = (4x + 12) \div 4 - 3 = (x + 3) - 3 = x$*

LT-150. Rewrite each of the following products, and sketch a composite rectangle to illustrate each product:

a. $5(z + 4)$ *$5z + 20$*
b. $x(3 + 4)$ *$3x + 4x$, or $7x$*
c. $x(x + 3)$ *$x^2 + 3x$*
d. $(y - 2)(y + 7)$ *$y^2 + 5y - 14$*

LT-151. Imagine you are busy working on a problem with algebra tiles. You have three big squares, five rectangles, and 10 small squares on your desk when your friend leans over and borrows two big squares, two rectangles, and four small squares. Write an algebraic equation that represents the tiles you have to begin with, what your friend takes, and the tiles you have left. *$(3x^2 + 5x + 10) - (2x^2 + 2x + 4) = x^2 + 3x + 6$*

LT-152 One way to demonstrate that $2x$ does not usually equal x^2 is to use two rectangles and one large square, as shown here.

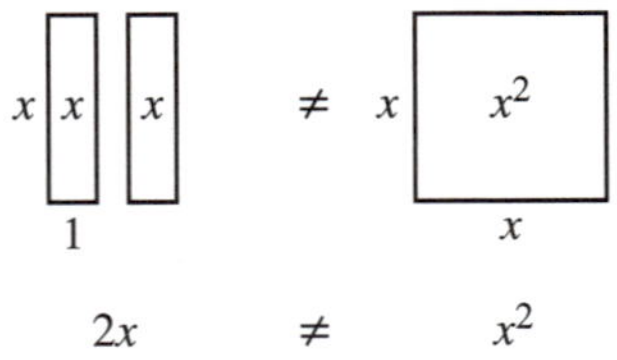

We say "$2x$ *does not usually* equal x^2" because *sometimes* it does. Check specifically to see that if $x = 0$, or if $x = 2$, then $2x = x^2$.

a. Sketch the tiles you would use to show that $3x + x$ does not usually equal $3x^2$.

Solution:

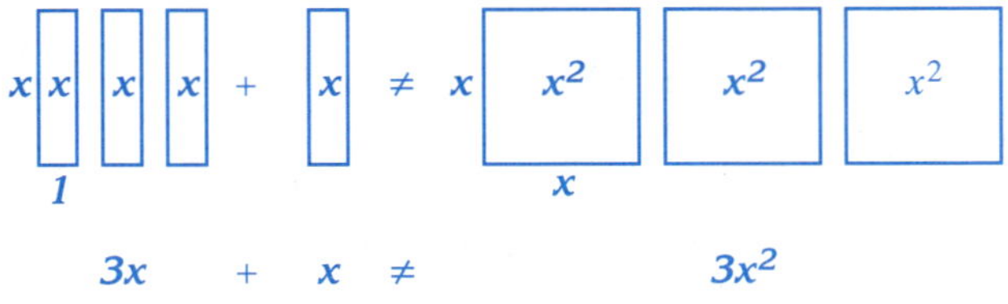

$3x + x \neq 3x^2$

b. Make a sketch of the tiles you would use to show that $2x - x$ does not usually equal 2.

Solution:

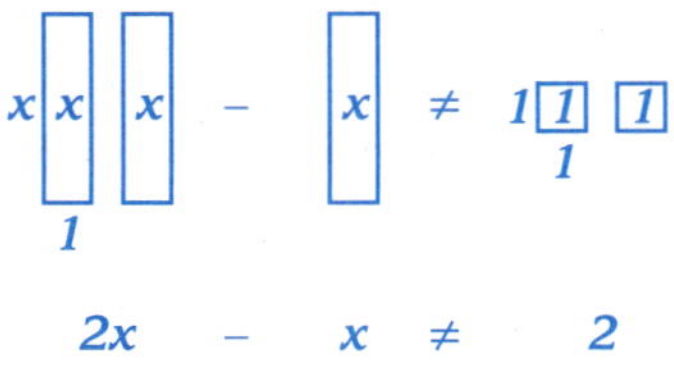

$2x - x \neq 2$

LT-153. Chapter 3 Summary Revision This is the final summary problem for Chapter 3. Using your rough draft from LT-123 and the ideas you discussed in your groups from LT-124, spend time revising and refining your Chapter 3 Summary. Your presentation should be thorough and organized, and on separate pages from the rest of your homework.

MATERIALS

- At least 6 pages of dot paper copied from the template in the Resource Pages
- Algebra tiles
- Protractors and metric rulers
- Poster-size graph for EF-112 Fruit Punch
- For each group—a 3-oz paper cup (to use for a net) and a strong lunch-size paper sack (to use for a lake)
- For distribution among the groups' lakes (paper sacks)—about five pounds of dried beans, such as kidney, pinto, or baby lima (to represent fish), plus one pound of the same size beans in a contrasting color (to represent the tagged fish).
- Chart for recording class data
- Resource Pages for EF-2, EF-3, EF-4, EF-6, EF-21, EF-22 and 23, EF-63, EF-64, EF-96, EF-124, EF-130, and EF-134; also Summary Draft Template (optional)

CHAPTER FOUR—FOR THE INSTRUCTOR

ESTIMATING FISH POPULATIONS
Numeric, Geometric, and Algebraic Ratios

The concept of ratio is one of the most fundamental and widely used concepts in mathematics. In this chapter, we explore ratios from several perspectives and tie the various approaches together so students can see relationships among them.

We begin with an informal look at similarity by examining ratios from a geometric perspective. The problems reinforce students' previous experience with numeric ratios and familiarize them with what happens to both linear measure (perimeter) and area measure when a simple lattice polygon ("dot-grid figure") is enlarged or reduced. We then formally introduce the concept of similar triangles. Similar right triangles acquaint students with the idea of slope in preparation for their work in Chapter 5.

Throughout the chapter, we integrate and extend more familiar ratio concepts: numeric ratios, percents as ratios with a denominator of 100, and equations that involve equivalent ratios. Students will continue to develop their facility with using ratios to solve problems throughout the course. We don't expect a sense of mastery of these concepts until later in Chapters 5 and 6.

A second focus of this chapter is the problem-solving strategy of breaking larger problems into more manageable parts called **subproblems**. *Students find areas of complicated regions, manipulate algebraic expressions (such as factoring polynomials geometrically with algebra tiles), and solve equations involving fractions (using "fraction busters")—all in the context of solving subproblems. Additional applications of the subproblem strategy will appear throughout the text.*

You will probably have to emphasize that **identifying** *the subproblems is important. Expect serious resistance; most students will just want to get on to the business of solving the problem rather than spend time thinking about how they will do so. Point out the usefulness of identifying and solving subproblems each time they attempt to solve complicated problems. If students truly develop the ability to identify and solve subproblems, it will be one of the most useful skills they learn in this course.*

CHAPTER FOUR

ESTIMATING FISH POPULATIONS
Numeric, Geometric, and Algebraic Ratios

In this chapter you will be exploring the concept of ratio, an idea that is useful in many aspects of daily life. One of the goals in this chapter is to develop skills using ratios to solve a variety of problems, including ones like this:

> To protect wild strains and population numbers, marine biologists need to keep track of salmon in the waters they monitor. They might want to know, for example, how many chinook salmon there are in the river systems above San Francisco Bay. This number changes throughout the year as fish move in and out of the rivers to spawn, so biologists need a way to gather current data fairly quickly and inexpensively. How do you think a marine biologist might estimate the number of salmon on spawning grounds in the river systems?

Another goal is to learn to solve complicated problems by breaking them into simpler problems, or *subproblems*. In this chapter, you will use the idea of subproblems to find the areas of complex geometric figures. In later chapters, you will use them to "uncomplicate" complicated algebraic processes. To do this, you have to be able to step back from the whole problem and identify useful subproblems before you begin.

The ability to identify subproblems could be one of the most useful skills you learn in this course!

IN THIS CHAPTER YOU WILL HAVE THE OPPORTUNITY TO:

- explore "similar" triangles and dot-grid figures—and see how the ratios of corresponding sides, perimeters, and areas are related;
- use the strategy of identifying, writing, and solving subproblems to solve a larger problem;
- use a generic rectangle to rewrite a polynomial as a product of two factors;
- represent ratio problems by the graph of a line passing through the origin, and represent the rate at which a quantity changes as a ratio;
- use equivalent ratios to write equations to solve problems;
- use "fraction busters" to convert an equation with fractions into an equivalent equation without fractions.

Chapter 4 Estimating Fish Populations: Numeric, Geometric, and Algebraic Ratios

MATERIALS	CHAPTER CONTENTS	PROBLEM SETS
EF-2 Enlargement Ratios Resource Page; EF-3 Subproblems and EF-4 Resource Page; EF-6 Reducing Figures Resource Page; dot paper (for EF-11)	**4.1** Enlarging and Reducing Geometric Figures *Subproblems*	EF-1 – EF-20
EF-21 Enlargement and Reduction Ratios Summary Resource Page; dot paper (for EF-22, EF-23, and EF-26); EF-22 and EF-23 Testing Your Enlargement and Reduction Ratios Conjecture Resource Page	**4.2** More Enlarging, Reducing, and Ratios	EF-21 – EF-40
Algebra tiles (for EF-43)	**4.3** Area and Subproblems: Geometric Factoring	EF-41 – EF-62
Protractors; metric rulers; EF-63 Similar Triangles Resource Page; EF-64 Enlargement and Reduction Ratios and Ratios of Side Lengths Resource Page; dot paper (for EF-64)	**4.4** Similar Triangles: More Geometric Ratios	EF-63 – EF-83
Overhead transparency and pen for each group (optional); EF-96 Sums and Products: Part I Resource Page	**4.5** More Similarity: Right Triangles and Missing Sides *Fraction Busters*	EF-84 – EF-108
Poster-size graph (for EF-112); EF-124 Sums and Products: Part II Resource Page	**4.6** Fruit Punch: Using Equivalent Ratios to Graph	EF-109 – EF-129
For EF-130: about 5 lb of dried beans, such as pinto, plus 1 lb of beans of the same size in contrasting color; 3-oz plastic cups; strong, lunch-size paper sacks; EF-130 Estimating Fish Populations Resource Page; class-data chart; EF-134 Sums and Products: Part III Resource Page; Chapter Summary Draft Resource Page	**4.7** Estimating Fish Populations: A Simulation	EF-130 – EF-142
Dot paper (for EF-151)	**4.8** Summary and Review	EF-143 – EF-166

Problem Solving

Graphing

Writing Equations

Ratios

Solving Equations

Symbol Manipulation

Reasoning & Communication

4.1 ENLARGING AND REDUCING GEOMETRIC FIGURES

SECTION MATERIALS
- EF-2 ENLARGEMENT RATIOS RESOURCE PAGE
- EF-3 AND EF-4 RESOURCE PAGES
- EF-6 REDUCING FIGURES RESOURCE PAGE
- DOT PAPER (COPIED FROM THE DOT PAPER TEMPLATE RESOURCE PAGE, FOR EF-11)

Make sure students understand EF-1 before they try EF-2 in groups. You will probably need to explain what "corresponding sides" means, so be prepared to give an informal definition.

EF-1. Enlarging Geometric Figures Architects or graphic artists sometimes need to make a small-scale drawing from a larger one, or a larger drawing from a small blueprint or plan. Cartographers also use the ideas of enlarging or reducing when making maps of various sizes. In this section, you will see how geometric ratios can be used to make **enlargements** or **reductions** of geometric figures.

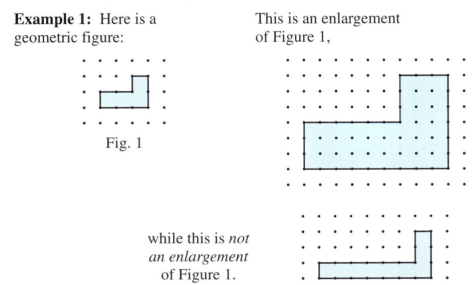

Example 1: Here is a geometric figure:

Fig. 1

This is an enlargement of Figure 1,

while this is *not an enlargement* of Figure 1.

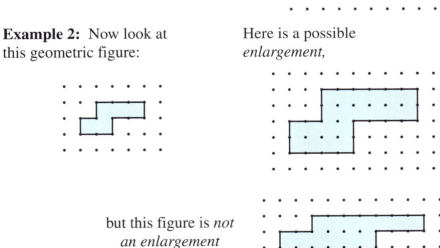

Example 2: Now look at this geometric figure:

Here is a possible *enlargement,*

but this figure is *not an enlargement* of Figure 2.

Write one or two sentences to explain why the third figure in Example 2 is not an enlargement of Figure 2.

Enlargements have to be the same shape. In this figure some sides are the same as the original, some are doubled, and some tripled. To keep the shape the same you have to multiply each side by the same number. If students simply say "increase each side by the same number," be sure to question them as to what they mean by "increase."

You may want to guide the class in EF-2 by using an overhead of Lissa's goal-post-shaped figure. Emphasize that each of the three ratios must be labeled clearly, and that students are going to be gathering data so that they can look for a pattern and make a conjecture about the ratios.

Be sure to check the groups' diagrams while you circulate. Watch out for students who compute perimeters by counting dots or squares rather than segments.

EF-2. Enlargement Ratios Lissa saw a dot grid figure and decided to enlarge it. She made the corresponding sides of the new figure *twice* as long as those in the original figure. In other words, the enlargement ratio for the sides was $\frac{2}{1}$. Here are Lissa's results:

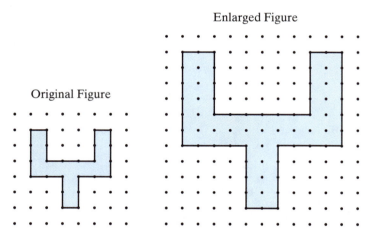

Select two of the following figures. Then follow these steps for each figure you selected:

1. Copy the original figure on the EF-2 Enlargement Ratios Resource Page, and then enlarge the figure by making the corresponding sides of the new figure *twice* as long as those in the original figure. In other words, use a side enlargement ratio of $\frac{2}{1}$ to enlarge your copy of the original figure.

2. Compute the perimeters and areas of both the original figure and its enlargement.

3. Finally, for each pair of figures, find and simplify the following ratios:

$$\frac{\text{length of } Side \text{ of } new \text{ figure}}{\text{length of } corresponding \text{ } Side \text{ of } original \text{ figure}}$$

$$\frac{Perimeter \text{ of } new \text{ figure}}{Perimeter \text{ of } original \text{ figure}} \quad \text{and} \quad \frac{Area \text{ of } new \text{ figure}}{Area \text{ of } original \text{ figure}}$$

a. 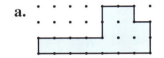 b. 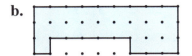 c.  d.

Solutions:

Figure	$\dfrac{S_{new}}{S_{orig}}$	$\dfrac{P_{new}}{P_{orig}}$	$\dfrac{A_{new}}{A_{orig}}$
(a)	$\dfrac{2}{1}$	$\dfrac{40}{20} = \dfrac{2}{1}$	$\dfrac{48}{12} = \dfrac{4}{1}$
(b)	$\dfrac{2}{1}$	$\dfrac{52}{26} = \dfrac{2}{1}$	$\dfrac{88}{22} = \dfrac{4}{1}$
(c)	$\dfrac{2}{1}$	$\dfrac{48}{24} = \dfrac{2}{1}$	$\dfrac{80}{20} = \dfrac{4}{1}$
(d)	$\dfrac{2}{1}$	$\dfrac{32}{16} = \dfrac{2}{1}$	$\dfrac{40}{10} = \dfrac{4}{1}$

In EF-3 the notion of subproblem shouldn't be interpreted too narrowly. It doesn't mean that the region must be chopped into smaller subregions. Indeed, one could compute the area of a 7 by 11 rectangle and subtract from it the areas of the "missing corner" rectangles: $7 \cdot 11 - 2 \cdot 2 - 4 \cdot 5 = 53$.

Identifying subproblems is an important and useful problem-solving strategy that we will revisit throughout this chapter and also in later chapters. Most students at this level have some experience with the strategy of identifying and solving subproblems even if they don't know it by that name. The advantage of giving the strategy a name, however, is that students can call the technique explicitly to mind whenever they need to use it.

EF-3. Subproblems Most problems that you will meet in this course, in other mathematics courses, and in life are made up of smaller problems whose solutions you can put together to solve the original problem. These smaller problems are called **subproblems**. For example, suppose you want to find the area of the shaded region in Figure 1. The area is equal to the number of square units, so you could simply count each individual square, as shown in Figure 1, to get 53 square units. However, counting each individual square can be tedious.

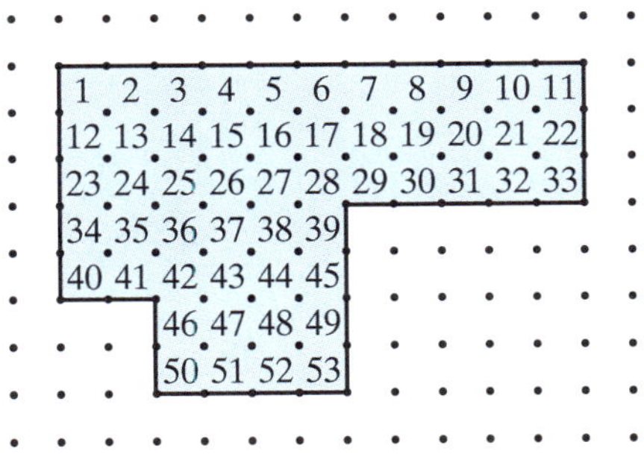

In this problem, you will explore two ways you could use subproblems to find the area of the shaded region more efficiently. Subproblems are usually not described explicitly, so you must first identify them.

Fig. 1

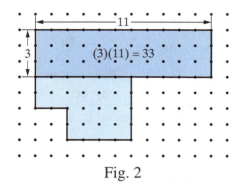

Fig. 2

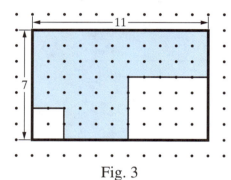

Fig. 3

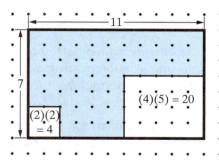

Fig. 4

Subdivide: One way to find the area of the shaded region in Fig. 1 is to subdivide it into shapes whose areas are easy to calculate. You could break this region into three rectangular pieces. Then finding the area of a subregion would be a *subproblem* of the original problem.

One subproblem is to find the area of the rectangle at the top of Figure 2. Since the length is 11 and the width is 3, the area of that rectangle is $3 \cdot 11 = 33$ square units.

The rest is left to you. On your EF-3 Subproblems Resource Page identify two more subproblems you could use to subdivide the shaded region in Figure 2 to finish computing the total area. Solve each of your subproblems, and then check to see that the total shaded area is 53 square units. *One possibility: break the remaining region into a 2-by-2 square and a 4-by-4 square; the area of the region is $2(2) + 4(4) = 20$ sq units. Another possibility: break the remaining region into a 2-by-6 rectangle and a 2-by-4 rectangle; the area of the region is $2(6) + 2(4) = 20$ sq units. Check: the total area is $33 + 20 = 53$ sq units.*

Surround and Subtract: A second way of using subproblems to find the area of the shaded region in Fig. 1 is to first draw a rectangle that just fits around the original figure. The first subproblem is to find the area of this large rectangle (see Figure 3). What is it? *$(7)(11) = 77$*

Two more subproblems are solved in Figure 4. What are they? *The areas of the two "missing" rectangles are $(2)(2) = 4$ and $(4)(5) = 20$, respectively.*

Using the solutions to the three subproblems, you can now subtract to find the area of the original region:

$$77 - 4 - 20 = 53 \text{ square units.}$$

 EF-4. As demonstrated in EF-3, there is often more than one way to solve a complicated problem. On your EF-4 Resource Page use the idea of subproblems to find the area of the shaded region. Show all of your work by stating each subproblem you use.

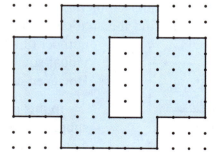

One possible method: $(9)(12) - 4(6) - 10 = 74$ square units

EF-5. Look back at EF-4 and compare your work with your group. How many different ways of finding the shaded areas did your group find? How did you divide up the larger problem into subproblems? Was one way more efficient than the other ways? Be ready to share your methods with the rest of the class.

In EF-6 point out to the students that the "H" inside of the third figure is part of the perimeter.

EF-6. Reducing Figures Follow the directions listed below to apply what you know about enlarging figures to making copies that are smaller:

1. Choose *two* of the following shaded figures. On your EF-6 Reducing Figures Resource Page, reduce each figure you chose so that the corresponding sides are *half* the size of the original. In this case, the reduction ratio for the sides is $\frac{1}{2}$.

2. Compute the perimeters and areas of the original figures and their reductions.

3. Then compute and reduce the following ratios. Be sure to label your answers!

$$\frac{\text{length of } Side \text{ of } new \text{ figure}}{\text{length of } corresponding \; Side \text{ of } original \text{ figure}}$$

$$\frac{Perimeter \text{ of } new \text{ figure}}{Perimeter \text{ of } original \text{ figure}} \quad \text{and} \quad \frac{Area \text{ of } new \text{ figure}}{Area \text{ of } original \text{ figure}}$$

a. 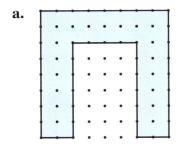 b. 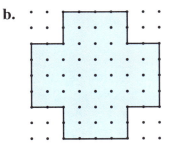 c.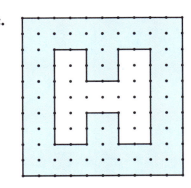

Solutions	Figure	$\frac{S_{new}}{S_{orig}}$	$\frac{P_{new}}{P_{orig}}$	$\frac{A_{new}}{A_{orig}}$
	(a)	$\frac{1}{2}$	$\frac{22}{44} = \frac{1}{2}$	$\frac{10}{40} = \frac{1}{4}$
	(b)	$\frac{1}{2}$	$\frac{16}{32} = \frac{1}{2}$	$\frac{12}{48} = \frac{1}{4}$
	(c)	$\frac{1}{2}$	$\frac{36}{72} = \frac{1}{2}$	$\frac{18}{72} = \frac{1}{4}$

214 Chapter 4 Estimating Fish Populations: Numeric, Geometric, and Algebraic Ratios

Note: You will need to make at least six copies of the Dot Paper Template Resource Page for use with some of the problems in this chapter, including Problem EF-11 in this section.

EXTENSION AND PRACTICE

EF-7. Notation To save time and effort, the notation will be changed

from $\dfrac{\text{length of } Side \text{ of } new \text{ figure}}{\text{length of } corresponding \text{ } Side \text{ of } original \text{ figure}}$ to $\dfrac{S_{new}}{S_{orig}}$

from $\dfrac{Perimeter \text{ of } new \text{ figure}}{Perimeter \text{ of } original \text{ figure}}$ to $\dfrac{P_{new}}{P_{orig}}$

and from $\dfrac{Area \text{ of } new \text{ figure}}{Area \text{ of } original \text{ figure}}$ to $\dfrac{A_{new}}{A_{orig}}$.

a. Suppose a 4-ft by 5-ft rectangle is enlarged so that the ratio of corresponding sides is $\frac{4}{1}$. Make a rough sketch of the original rectangle and the enlargement. What are the dimensions of the new rectangle?   *16 ft by 20 ft*

b. Compute the perimeter and the area of original rectangle and the new, enlarged rectangle.
   P_{orig} = 18 ft, A_{orig} = 20 sq ft, P_{new} = 72 ft, A_{new} = 320 sq ft

c. Compute the following ratios:

$$\frac{S_{new}}{S_{orig}}, \frac{P_{new}}{P_{orig}}, \frac{A_{new}}{A_{orig}} \quad \frac{4}{1}, \frac{4}{1}, \frac{16}{1}$$

EF-8. Max, the inquisitive third grader, just took his first spelling test and got 8 out of a possible 10 points. When he got his paper back, "80%" was written at the top of the page, and Max wondered what it meant. Explain it to him in one or two complete sentences.   *80% means a ratio of 80 to 100, or $\frac{80}{100}$. This is equivalent to $\frac{8}{10}$.*

EF-9. Writing Ratios as Percents To rewrite each of the following ratios as a percent, use mental math to solve each equation for x:

a. $\dfrac{43}{100}$ means *what* percent? $\dfrac{43}{100} = \dfrac{what}{100}$ *43*

b. To rewrite $\dfrac{7}{100}$ as a percent, solve mentally: $\dfrac{7}{100} = \dfrac{x}{100}$ *7*

c. To rewrite $\dfrac{6}{10}$ as a percent, solve mentally: $\dfrac{6}{10} = \dfrac{x}{100}$ *60*

d. To rewrite $\frac{3}{50}$ as a percent, solve mentally: $\frac{3}{50} = \frac{x}{100}$ 6

e. Explain why each ratio on the right-hand side of the equation in parts a through d is called a percent.   *The ratios show number per hundred, and "percent" means number per hundred; 7% means 7 hundredths.*

EF-10. Ratios are often used to compare two quantities.

a. Derek took three and a half hours to do his homework. Dean took 15 minutes. Find a ratio of Derek's time to Dean's time in terms of minutes. Be careful. $\frac{210}{15}$ or $\frac{14}{1}$ or 210:15 or 14:1

b. The ratio of Derek's age to Dean's age is 2 to 1. Explain what this means in plain English.   *Derek is twice as old as Dean.*

EF-11. In this problem, you'll use the idea of subproblems to find the area of a shaded figure.

a. Copy the figure on dot paper. State subproblems you can use to find the area of the shaded region. Then solve the suproblems to find the area. Show all of your work by stating each subproblem you use.

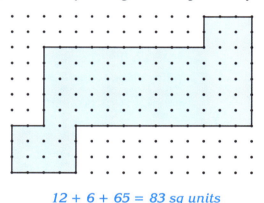

12 + 6 + 65 = 83 sq units

b. Add information about the *strategy of using subproblems* to your Tool Kit.

EF-12. Suppose a bucket filled with water weighs 8.4 kilograms.

a. If the water by itself weighs five times as much as the bucket, what is the weight of the bucket? Solve the problem, and write an equation for it.   *b + 5b = 8.4, 1.4 kg*

b. What is the ratio of the weight of the bucket to the weight of the water?   *5:1*

EF-13. Solve each of the following equations:

a. $0.24(2x + 7) = 108$ *x = 221.5*

b. $\frac{18(19x - 4)}{2} = 3x$ *x = $\frac{3}{14}$ ≈ 0.21*

c. $7x + 15 - (2x + 7) = 14$ (*Hint:* To combine the two polynomials on the left-hand side, imagine using algebra tiles. Why does $-(2x + 7) = -2x - 7$?) $x = \frac{6}{5} = 1.2$

d. $0.25(3n) + 0.10(68 - 4n) + 0.05(n) = 11.60$ $n = 12$

EF-14. According to the *University of California at Berkeley Wellness Letter*, if you exercise strenuously, you can easily sweat away more than 4 cups of water per hour.*

*"Wellness Made Easy," *University of California at Berkeley Wellness Letter*, 13, no. 11 (August 1997), p. 8

Consider the graph shown at the left.

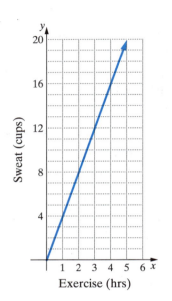

a. Why does the graph go through the point (2, 8)? *If you exercise for 2 hrs, you'll sweat 8 cups.*

b. The equation for this graph is $y = 4x$. Some athletes reportedly can sweat at a rate of nearly *a gallon* an hour. Would the graph of their line be steeper or flatter? Explain your answer, and make a sketch of such an athlete's sweat graph. (There are 16 cups in a gallon.) *Steeper because they will produce more cups of sweat per hour; the line will go up at a rate of 16 to 1: the graph will be just below $y = 16x$.*

c. A certain couch potato sweats at a rate of 0.5 cup per hour when he walks to the TV and back. Would the line for this graph be steeper or flatter? Sketch this couch potato's sweat graph. *Flatter because he would produce fewer cups of sweat per hour: the graph would be $y = 0.5x$.*

EF-15. In this problem you'll examine the graph of $y = x^2 - 3$.

a. Graph the curve $y = x^2 - 3$ by first making a table of values. Choose values for x so that $-3 \leq x \leq 3$.

b. Estimate the points where the parabola in part a crosses the x-axis. *at $(\pm\sqrt{3}, 0) \approx (\pm 1.73, 0)$*

Students need to be able to recognize fractional forms equivalent to 1, especially those of the form $\frac{ax + b}{ax + b}$, where a and b are not both zero. We "simplify" algebraic fractions by finding factors to make forms equivalent to 1. Avoid referring to the process as "canceling," which students confuse with subtracting to get 0. If you talk about "canceling like terms" you risk having students write nonsense like $\frac{x + 6}{x} = 6$. If they do, suggest that they pick a value to substitute for x and check it.

EF-16. Simplifying Fractions You can "simplify" fractions by finding factors of the numerator and the denominator to make forms that are equivalent to 1, such as $\frac{2}{2}$ or $\frac{139}{139}$. The trick is to find as many fractional forms of 1 as you can.

Section 4.1 Enlarging and Reducing Geometric Figures **217**

Simplify each fraction in column A by first finding an equivalent product in column B. Then find an equivalent product with a "1" in column C. The first one is done for you as an example.

Example: a. $\dfrac{5 \cdot 3}{2 \cdot 3} = \dfrac{5}{2} \cdot \dfrac{3}{3} = \dfrac{5}{2} \cdot 1 = \dfrac{5}{2}$ ▲

fractional form of 1

	Column A	Column B	Column C
a.	$\dfrac{5 \cdot 3}{2 \cdot 3}$	$\dfrac{5}{3} \cdot \dfrac{20}{20}$	$1 \cdot \dfrac{5}{6}$
b.	$\dfrac{5 \cdot 5 \cdot 5}{5 \cdot 5 \cdot 6}$	$\dfrac{5}{5} \cdot \dfrac{1}{5^2}$	$\dfrac{5}{3} \cdot 1$
c.	$\dfrac{5}{5^3}$	$\dfrac{5}{2} \cdot \dfrac{3}{3}$	$\dfrac{3}{5} \cdot 1$
d.	$\dfrac{100}{60}$	$\dfrac{5}{5} \cdot \dfrac{x}{x} \cdot \dfrac{x}{3}$	$1 \cdot \dfrac{1}{5^2}$
e.	$\dfrac{36}{60}$	$\dfrac{3}{5} \cdot \dfrac{12}{12}$	$1 \cdot 1 \cdot \dfrac{x}{3}$
f.	$\dfrac{5x^2}{15x}$	$\dfrac{5 \cdot 5}{5 \cdot 5} \cdot \dfrac{5}{6}$	$\dfrac{5}{2} \cdot 1$

(b) $\dfrac{5 \cdot 5 \cdot 5}{5 \cdot 5 \cdot 6} = \dfrac{5 \cdot 5}{5 \cdot 5} \cdot \dfrac{5}{6} = 1 \cdot \dfrac{5}{6} = \dfrac{5}{6};$ (c) $\dfrac{5}{5^3} = \dfrac{5}{5} \cdot \dfrac{5}{5^2} = 1 \cdot \dfrac{5}{5^2} = \dfrac{5}{5^2};$

(d) $\dfrac{100}{60} = \dfrac{5}{3} \cdot \dfrac{20}{20} = \dfrac{5}{3} \cdot 1 = \dfrac{5}{3};$ (e) $\dfrac{36}{60} = \dfrac{3}{5} \cdot \dfrac{12}{12} = \dfrac{3}{5} \cdot 1 = \dfrac{3}{5};$

(f) $\dfrac{5x^2}{15x} = \dfrac{5}{5} \cdot \dfrac{x}{x} \cdot \dfrac{x}{3} = 1 \cdot 1 \cdot \dfrac{x}{3} = \dfrac{x}{3}$

EF-17. Evelyn tried to complete the matching in EF-16, but she still does not understand why $\dfrac{3 \cdot 4}{3 \cdot 9}$ is the same as $\dfrac{3}{3} \cdot \dfrac{4}{9}$. Explain to Evelyn why the two expressions are equivalent, and how the second form helps you simplify the fraction. *To multiply fractions, you multiply the numerators and multiply the denominators. So, $\dfrac{3}{3} \cdot \dfrac{4}{9}$ must be the same as $\dfrac{3 \cdot 4}{3 \cdot 9}$. To simplify $\dfrac{3 \cdot 4}{3 \cdot 9}$, you notice that $\dfrac{3}{3} = 1$, and 1 times $\dfrac{4}{9}$ is $\dfrac{4}{9}$.*

EF-18. Exponent Practice Use what you know about exponents to rewrite each of the following expressions with a single exponent.

a. $(1.4x^2)^3$ *2.744x^6*   **b.** $(1.4x^2)(2.5x^3)$ *3.5x^5*

c. $\dfrac{5y^2}{20y}$ *$\dfrac{y}{4}$*   **d.** $\dfrac{5^2}{5^3}$ *$\dfrac{1}{5}$*

EF-19. Copy and solve these Diamond Problems:

a.

b. [diamond: top $10z^2$, left $2z$, right ___]
$E = 5z$,
$S = 7z$

c. [diamond: top ___, left z, right $6z$]
$N = 6z^2$,
$S = 7z$

For (a): [diamond: left 100, right 0.1]
$N = 10$,
$S = 100.1$

EF-20. The area of the following rectangle is $6x^2 + 12x$. Factor the expression $6x^2 + 12x$ three different ways by finding three different possibilities for the length and the width of the rectangle.

[rectangle divided into two parts labeled $6x^2$ and $12x$]

There are many possibilities: $x(6x + 12)$, $3x(2x + 4)$, $2x(3x + 6)$, $6x(x + 2)$, $2(3x^2 + 6x)$, $3(2x^2 + 4x)$, $6(x^2 + 2x)$

4.2 MORE ENLARGING, REDUCING, AND RATIOS

SECTION MATERIALS

EF-21 ENLARGEMENT AND REDUCTION RATIOS SUMMARY RESOURCE PAGE

DOT PAPER (COPIED FROM THE DOT PAPER TEMPLATE RESOURCE PAGE, FOR EF-22, EF-23, AND EF-26)

EF-22 AND EF-23 TESTING YOUR ENLARGEMENT AND REDUCTION RATIOS CONJECTURE RESOURCE PAGE

In EF-21b, some students may need more concrete questions, such as "If $N = 2$ or $N = 3$, we know the answers. What if $N = 10$? or $N = 100$?" Remind students that a conjecture is just an educated guess and that we're basically looking for a pattern. Stronger students may be able to justify why the area ratio is N^2, but for most of them it's too soon.

EF-21. Enlargement and Reduction Ratios Summary In Section 4.1, you reduced and enlarged several figures and then computed ratios. In this problem you will organize the ratios your group found to look for a pattern and make a conjecture about enlargement and reduction ratios.

a. **Organize your data and look for a pattern.** On the EF-21 Resource Page ("Enlargement and Reduction Ratios Summary") compile your group's data from problems EF-2, EF-6, and EF-7. For each figure, record the enlargement or reduction ratio, and the three ratios

$$\frac{S_{new}}{S_{orig}}, \quad \frac{P_{new}}{P_{orig}}, \quad \text{and} \quad \frac{A_{new}}{A_{orig}}$$

in the appropriate columns. Look for patterns in your chart: When you enlarge (or reduce) a figure on dot paper, what relationships are there among the ratios for the figure? *The ratio of perimeters equals the ratio of the sides and the ratio of areas is the square of the ratio of the sides.*

b. **Make some conjectures.** What generally happens when you enlarge (or reduce) a figure on dot paper? Without actually drawing any figures, use the patterns you saw on the Summary Page to write what you *think* the ratios $\frac{P_{new}}{P_{orig}}$ and $\frac{A_{new}}{A_{orig}}$ would be if the following were true:

   i. The ratio $\frac{S_{new}}{S_{orig}}$ is $\frac{2}{1}$.   *$\frac{2}{1}, \frac{4}{1}$*

   ii. The new figure has corresponding side lengths three times the original figure's.   *$\frac{3}{1}, \frac{9}{1}$*

   iii. The new figure has corresponding side lengths 10 times the original figure's.   *$\frac{10}{1}, \frac{100}{1}$*

c. Now suppose N represents a positive integer. If the ratio $\frac{S_{new}}{S_{orig}}$ were $\frac{N}{1}$, what do you think the ratios $\frac{P_{new}}{P_{orig}}$ and $\frac{A_{new}}{A_{orig}}$ would be? Write your conjecture in a complete sentence. You will be testing your conjecture in the next two problems.   *$\frac{N}{1}, \frac{N^2}{1}$*

EF-22. Testing Your Enlargement and Reduction Ratios Conjecture
In this problem, you will test the conjecture you made about geometric ratios in the previous problem. To do this, you will draw a rectangle and then draw a new, reduced rectangle. Then you will decide whether the ratios of the perimeters and areas fit the pattern you predicted.

a. Draw a rectangle with a perimeter of 24 units and an area of 32 square units on a piece of dot paper.   *rectangles will be 4 by 8 or 8 by 4*

b. To test your conjecture from problem EF-21, have each member of your group complete *one* of the following two tests:

   Test 1: Reduce the rectangle from part a so that the ratio of corresponding sides is $\frac{1}{2}$. Then compute the ratios

   $$\frac{S_{new}}{S_{orig}}, \quad \frac{P_{new}}{P_{orig}}, \quad \text{and} \quad \frac{A_{new}}{A_{orig}}. \qquad \frac{1}{2}, \frac{1}{2}, \frac{1}{4}$$

Test 2: Reduce the rectangle from part a so that the ratio of corresponding sides is $\frac{1}{4}$. Then compute the ratios

$$\frac{S_{new}}{S_{orig}}, \frac{P_{new}}{P_{orig}}, \text{ and } \frac{A_{new}}{A_{orig}}. \quad \frac{1}{4}, \frac{1}{4}, \frac{1}{16}$$

c. Compile your group's data on the Resource Page for EF-22 and EF-23 ("Testing Your Enlargement and Reduction Ratios Conjecture").

EF-23. More Tests Test your conjecture with this figure.

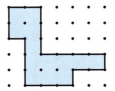

a. Copy the given figure on a sheet of dot paper.

b. Have each member of your group complete *one* of the following tests:

Test 1: Enlarge the original figure by making each corresponding side three times as long as in the original. Compute and reduce the ratios

$$\frac{S_{new}}{S_{orig}}, \frac{P_{new}}{P_{orig}}, \text{ and } \frac{A_{new}}{A_{orig}}. \quad \frac{3}{1}, \frac{66}{22} = \frac{3}{1}, \frac{117}{13} = \frac{9}{1}$$

Test 2: Enlarge the original figure, but this time make each corresponding side five times as long as in the original. Compute and reduce the ratios

$$\frac{S_{new}}{S_{orig}}, \frac{P_{new}}{P_{orig}}, \text{ and } \frac{A_{new}}{A_{orig}}. \quad \frac{5}{1}, \frac{110}{22} = \frac{5}{1}, \frac{325}{13} = \frac{25}{1}$$

c. Add your group's data to the EF-22 and EF-23 Testing Your Enlargement and Reduction Ratios Conjecture Resource Page.

EF-24. Revising Your Enlargement and Reduction Ratios Conjecture

You've made a conjecture about geometric ratios and tested it. Was your conjecture from EF-21c correct, or do you need to revise it? If necessary, rewrite or revise your conjecture about what the ratios $\frac{P_{new}}{P_{orig}}$ and $\frac{A_{new}}{A_{orig}}$ will be if the ratio $\frac{S_{new}}{S_{orig}}$ is $\frac{N}{1}$, where N represents a positive integer. Use complete sentences, and add the information to your Tool Kit. *If the side ratio is $\frac{N}{1}$, then the ratio of perimeters will be $\frac{N}{1}$, and the ratio of areas will be $\frac{N^2}{1}$.*

EF-25. Using the Conjecture about Geometric Ratios to Make Predictions You probably noticed that when the side ratio is $\frac{N}{1}$, the perimeter ratio will also be $\frac{N}{1}$ and the area ratio will be $\frac{N^2}{1}$. We will not prove that this conjecture is correct, but in this case, the pattern you noticed is indeed true for any value of N you choose (including values that are not integers). You can use this idea to solve problems. Use it to solve the following problem.

Sean created a figure on dot paper with a perimeter of 60 units and an area of 100 square units. He is wondering whether he could calculate the area and perimeter of a reduction without actually drawing the reduced figure. (Sean is always looking for shortcuts.)

a. If Sean's reduction ratio is $\frac{1}{4}$, what will be the perimeter ratio? What will be the area ratio? *perimeter ratio = $\frac{1}{4}$ and area ratio = $\frac{1}{16}$.*

b. To find the new perimeter without drawing the reduced figure, Sean wrote the following equation using equivalent ratios, but he is not sure what to do next.

$$\frac{P_{new}}{P_{orig}} = \frac{1}{4}$$

Explain to Sean what to do first in order to use this equation to solve for the new perimeter. Then solve the equation to find the perimeter. *Substitute 60 for P_{orig}; and then solve the resulting equation to get $P_{new} = 15$ units.*

c. Write an equation you could use to find the area of the reduced figure, and then solve it. $\frac{A_{new}}{A_{orig}} = \frac{1}{16}$; $A_{new} = 6.25$ sq units.

EXTENSION AND PRACTICE

EF-26. Copy each of the following triangles onto dot paper. Then use the idea of "surround and subtract" to find the area of each triangle.

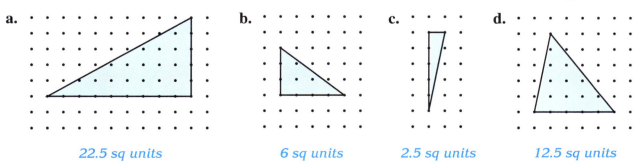

a. 22.5 sq units   b. 6 sq units   c. 2.5 sq units   d. 12.5 sq units

EF-27. Look again at the triangles in EF-26.

a. If you enlarged the triangle in EF-26b using an enlargement ratio of $\frac{4}{1}$, what would be the area of the enlarged triangle? *16(6) = 96 sq units*

b. If you reduced the triangle in EF-26a using an side ratio of $\frac{1}{3}$, what would be the area of the reduced triangle? $\left(\frac{1}{9}\right)(22.5) = 2.5$ *sq units*

If students ask for help with EF-28, suggest that they apply what they already know about undoing. Some may remember and use the process of cross-multiplying, which is just a way of combining two steps in the undoing process.

EF-28. Here are some equations to solve. They look a bit different from the equations you've already solved, but you can use the same process to solve them that you've used before. The key is to focus on *undoing*.

a. $\frac{y}{3} = 10$ *30*

b. $\frac{2x}{5} = \frac{9}{5}$ *4.5*

c. $\frac{m}{3} = \frac{6}{7}$ $\frac{18}{7} \approx 2.57$

d. $\frac{3z}{2} = \frac{12}{5}$ $\frac{8}{5} = 1.6$

e. $\frac{5w}{8} = \frac{w-4}{4}$ $-\frac{8}{3} \approx -2.67$

Problem EF-29 is really a word problem like those in Chapter 3 except that it involves ratios. Students can use guess-and-check tables to write an equation and solve the problem. There are, of course, several ways to solve this type of problem. For example, in EF-29, instead of writing and solving an equation with equivalent ratios, $\frac{1}{5} = \frac{x}{30-x}$, we could solve an equation that involves ratios implicitly, $x + 5x = 30$.

EF-29. Helena used a guess-and-check table to start to solve the following problem. Look at Helena's table of guesses, and use it to write an equation. Then solve the equation using algebra and answer the question.

Two numbers are in a ratio of 1 to 5. If their sum is 27, what are the numbers?
4.5 and 22.5

Guess: First Number	Second Number	Ratio of the Numbers	Check: Ratio = $\frac{1}{5}$?	
10	17	$\frac{10}{17}$	$\neq \frac{1}{5}$	no
3	24	$\frac{3}{24} = \frac{1}{8}$	$\neq \frac{1}{5}$	no
6	21	$\frac{6}{21} = \frac{2}{7}$	$\neq \frac{1}{5}$	no
x	?	?		

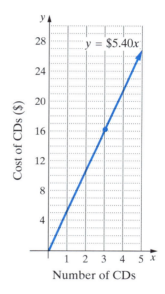

EF-30. At Ralph's Recycled Records, compact discs (CDs) cost $5.40 each. To find the cost of x CDs, you would multiply by $5.40. Thus a formula for the cost, y, of x CDs is $y = \$5.40x$. The graph of this cost equation is shown.

a. Use the graph to estimate the cost of four CDs. *About $22.*

b. Explain why (3, 16.20) is a point on the graph. *3 CDs cost $16.20.*

c. For the situation described by the point (3, 16.20), write the ratio of the total cost to the number of CDs. *$\frac{\$16.20}{3}$*

d. What point would correspond to the purchase of five CDs? Write the ratio of the total cost to the number of CDs. *(5, $27.00); $\frac{\$27.00}{5}$*

e. What point would correspond to the purchase of one CD? Write the ratio of the cost to the number of CDs for this point. *(1, $5.40); $\frac{\$5.40}{1}$*

f. Reduce each of the ratios in parts c and d, and compare the reduced ratios to the ratio in part e. What do you notice? What do the reduced ratios tell you about the CDs? *They're all equal to $\frac{\$5.40}{1}$. They tell the cost per CD.*

EF-31. While reading the newspaper, Laci noticed a whole-page ad for Zinntha's Department Store that declared all merchandise to be half price. At the bottom of the page, the ad concluded, "Hurry! Every item 50% of its usual price!" Laci wondered why "half price" and "50% price" meant the same thing. Explain to her in one or two complete sentences why $\frac{1}{2} = 50\%$. *Fifty percent means 50 out of 100, and $\frac{50}{100}$ equals $\frac{1}{2}$.*

EF-32. Express each of the following ratios as a percent. For example, to express $\frac{1}{5}$ as a percent, solve the equation $\frac{1}{5} = \frac{x}{100}$. To do this mentally, notice that $\frac{1}{5} \cdot \frac{20}{20} = \frac{20}{100}$, which means 20 percent.

a. $\frac{1}{4} = \frac{?}{100}$ $\frac{25}{100} = 25\%$

b. $\frac{3}{10}$ $\frac{30}{100} = 30\%$

c. $\frac{17}{25}$ $\frac{68}{100} = 68\%$

d. $\frac{9}{5}$ $\frac{180}{100} = 180\%$

EF-33. Copy and solve these Diamond Problems:

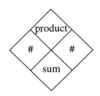

a.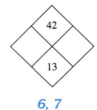

6, 7

b.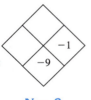

$N = 8$,
$W = -8$

EF-34. Symbol "| |" We will use the symbol "| |" to denote the length of a line segment. For example, "$|AB|$" means "the length of segment AB." In the following diagram we see that the segment AB has length 3, so we write $|AB| = 3$. We also see that segment BD has length $x + 2$, so we write $|BD| = x + 2$.

$$\underset{ABCD}{\overset{3x2}{\vdash\!\!-\!\!\!-\!\!\!-\!\!\!-\!\!\!-\!\!\!\dashv}}$$

a. Copy the information about "| |" into your Tool Kit.

b. Write statements using the symbol "| |" for the length of segment AC and the length of segment AD. $|AC| = x + 3$ and $|AD| = x + 5$

c. If $x = 7$, write the ratio of $|AB|$ to $|AD|$ in simplified form. $\frac{3}{12}$, or $\frac{1}{4}$

EF-35. Rewrite each of the following expressions using only one base and one exponent. Then check your work for the numeric ratios in parts a and c by using your calculator.

a. $\frac{2^4}{2^3}$ 2

b. $\frac{x^5}{x^2}$ x^3

c. $\frac{4^9}{4^6}$ 4^3

d. $\frac{y^{18}}{y^5}$ y^{13}

EF-36. Look for a pattern in your work from EF-35. Could you rewrite an expression like $\dfrac{x^{79}}{x^{50}}$ without writing out all the factors of x first? Using complete sentences, explain how to rewrite expressions like those in EF-35. You might want to make up an example to use in your explanation. *When dividing exponential expressions with the same base, you subtract the exponents. For example, $\dfrac{x^{79}}{x^{50}} = x^{79-50} = x^{29}$.*

EF-37. Exponent Practice Use what you know about exponents to rewrite each of the following expressions more simply:

a. $3.1 \cdot 2 \cdot 10^6 \cdot 10^9$ *$6.2 \cdot 10^{15}$*

b. $(3.1 \cdot 10^6)(2 \cdot 10^9)$ *$6.2 \cdot 10^{15}$*

c. $10^5 \cdot 7^4 \cdot 10^3 \cdot 7^8$ *$7^{12} \cdot 10^8$*

d. $\dfrac{8^6}{8^4}$ *$8^2 = 64$*

e. $(\dfrac{1}{4}x^{30})(84x^{20})$ *$21x^{50}$*

f. $\dfrac{-42y^2}{6y^3}$ *$\dfrac{-7}{y}$*

EF-38. Barbie Trivia According to a July 2, 1996 article from the Davis Enterprise, the typical American girl between the ages of 3 and 11 owns an average of eight Barbie dolls. Barbie's popularity contributes to sales: in 1995, Mattel sold $1.4 billion worth of Barbie-related merchandise, up from sales of $430 million in 1987.

a. Find the difference in sales from 1987 to 1995, and write the result in both standard notation and scientific notation.   *Mattel made $970,000,000, or $9.7 \cdot 10^8$, more in 1995.*

b. Find the ratio of 1995 sales to 1987 sales, and then reduce it. Explain what this ratio means in a complete sentence.   *$\dfrac{\$1.4 \cdot 10^9}{\$430 \cdot 10^6} = \dfrac{\$1.4 \cdot 10^9}{\$4.3 \cdot 10^8} \approx 0.326 \cdot 10^1 = 3.26$. In 1995, sales (in dollars) were about 3.26 times as much as in 1987.*

EF-39. René collected information from his fellow basketball players and made these two graphs:

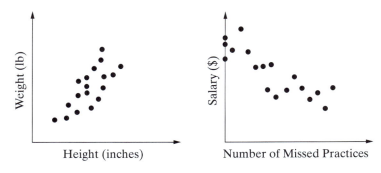

Describe what you notice about the data René plotted by copying and completing these sentences:

In general, as height increases, _____. *weight increases.*

In general, as the number of missed practices increases, _____. *salary decreases.*

EF-40. Use the following graph to answer each of the following questions:

a. What are the coordinates of point A? *(2, 5.5)*

b. What are the coordinates of point B? *(−6, −2.5)*

c. If the x-coordinate of a point on the graph is −8, what is its y-coordinate? *$y \approx 5.5$*

d. If the y-coordinate of a point on the graph is 3, what is its x-coordinate? *$x \approx 1.5$ or -7.5*

e. What are the coordinates of the x-intercepts? *$\approx (0.8, 0)$ and $\approx (-6.8, 0)$*

f. What are the coordinates of the y-intercept? *(0, −2.5)*

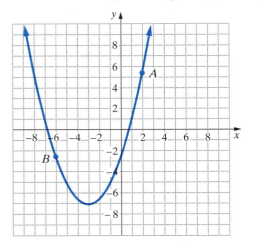

4.3 AREA AND SUBPROBLEMS: GEOMETRIC FACTORING

SECTION MATERIALS

ALGEBRA TILES (FOR EF-43)

Most students can already identify subproblems in easy cases but don't realize they're doing so. To exploit the power of this strategy, however, they must step back from a problem and ask, "Are there some subproblems here that I can solve?" and then, "How do I put these solutions together to solve the original problem?" Caution students that it is sometimes possible to choose a set of subproblems that does not lead to a solution of the original problem.

In EF-3 you used the idea of identifying and solving subproblems to solve a large problem by breaking it into smaller problems whose solutions you could put together to solve the original problem. In this section you'll apply the subproblems strategy to find the dimensions of a rectangle with known area.

Get students started on EF-41 by encouraging them to look for and clearly state subproblems—the process of identifying subproblems is the main point here. You might be tempted to focus on how to do this problem efficiently without a calculator, but then students might see efficiency as the goal, when that is not the case.

EF-41. Imagine that you want to tile your 6-foot by 8-foot rectangular bathroom floor with 3-inch by 3-inch square tiles. How many tiles should you buy? Identify a subproblem you could use to get started, and clearly state each of the subproblems you use to solve the problem.   *Possible subproblems: Figure out the number of 3-in square tiles that would fit along a 6-ft wall, or how many 3-in by 3-in tiles fit on one sq ft. Number of tiles = 768.*

Problems EF-42 through EF-46 develop factoring quadratics starting with algebra tiles and then moving on to Diamond Problems and generic rectangles. The Diamond Problems here are easy compared with others students have done, but they are here so students can make the connection between Diamond Problems and factoring quadratics with a leading coefficient of 1.

EF-42. For each of the following rectangles, first write the area as a product. Then draw and label a generic rectangle and use it to rewrite the area as a sum.

a. a rectangle with length $x + 4$, and width $x + 7$
   $(x + 4)(x + 7) = x^2 + 11x + 28$

b. a rectangle with length $2x - 6$, and width $3x + 10$.
   $(2x - 6)(3x + 10) = 6x^2 + 2x - 60$

c. a rectangle with length $x - 25$, and width $x + 25$.
   $(x - 25)(x + 25) = x^2 - 625$

Solution:

	x	+	7
x		x^2	$7x$
+			
4		$4x$	28

As you circulate, ask students if they can find more than one answer for EF-43c. As a follow-up, you might have a whole-class discussion of the multiple ways to factor this expression.

EF-43. Geometric Factoring What if you knew the area of a rectangle and you wanted to find the dimensions? You would have to work backward. For a given area expression, say $x^2 + 5x + 6$, you can use algebra tiles to find out whether or not it is possible to form a rectangle.

Example: You tried this with $x^2 + 5x + 6$ before, and it worked:

		x	+	3		
x		x^2		$1x$	$1x$	$1x$
+						
2		$1x$		1	1	1
		$1x$		1	1	1

You can see that the rectangle has area $x^2 + 5x + 6$, and its dimensions are $x + 2$ and $x + 3$, so you can write $x^2 + 5x + 6 = (x + 2)(x + 3)$. ▲

The process of rewriting a polynomial as a product is called **factoring** the polynomial. We say that $x + 2$ and $x + 3$ are **factors** of $x^2 + 5x + 6$.

Use algebra tiles to build rectangles with each of the following areas. Draw the composite rectangle and write its dimensions (length and width) algebraically as in the example above. Then write an equation to illustrate "the area as a sum = the area as a product." Check your work with your group.

a. $x^2 + 6x + 8$ $x^2 + 6x + 8 = (x + 4)(x + 2)$

b. $x^2 + 5x + 4$ $x^2 + 5x + 4 = (x + 1)(x + 4)$

c. $2x^2 + 8x$ $2x^2 + 8x = x(2x + 8)$ or $2x(x + 4)$

d. $2x^2 + 5x + 3$ $2x^2 + 5x + 3 = (2x + 3)(x + 1)$

e. Add information about *factoring a polynomial* to your Tool Kit.

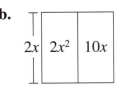

EF-44. Suppose that each of the generic rectangles shown here represents a composite rectangle made with algebra tiles. (The parts are not necessarily drawn to scale.) Copy each composite rectangle and find its dimensions, as shown in the example. Then write the area equation:

area as a sum = area as a product.

Example:

	x^2	$3x$
	$2x$	6

$$\longrightarrow$$

	x	$+$	3
x	x^2		$3x$
$+$			
2	$2x$		6

So, $x^2 + 5x + 6 = (x + 2)(x + 3)$ ▲

a.

x^2	$5x$
$3x$	15

$x^2 + 8x + 15 = (x + 3)(x + 5)$

b.

$2x$	$2x^2$	$10x$

$2x^2 + 10x = 2x(x + 5)$

c.

$2x^2$	$8x$
$3x$	12

$2x^2 + 11x + 12 = (2x + 3)(x + 4)$

d.

x^2	$5x$
$2x$	10

$x^2 + 7x + 10 = (x + 2)(x + 5)$

EF-45. Copy and solve each of the following Diamond Problems.

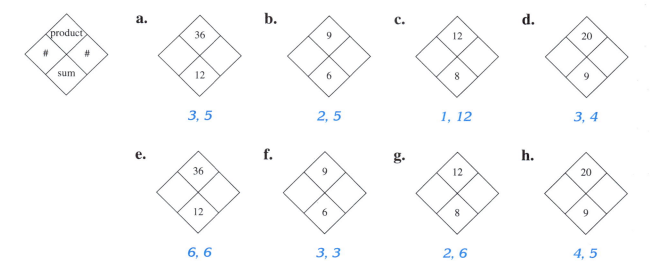

EF-46. Factoring with Generic Rectangles Let's go back to the problem of finding the dimensions of a rectangle when you know its area. You could try to replace the tiles with a generic rectangle and then work backward. For example, you could start with $x^2 + 8x + 12$, draw a generic rectangle as a sort of frame, and then fill in the parts you know:

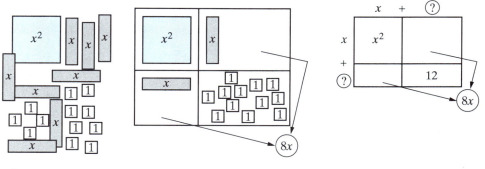

Start with $x^2 + 8x + 12$.

Imagine forming a rectangle by placing the x^2 tile and the 12 "1" tiles first, without knowing how you will fit the eight x tiles yet.

Draw a generic rectangle to represent the rectangle you imagined.

In doing this, the first subproblem is to figure out just how to place the 12 "1" tiles. You know they must form a small rectangle with area 12 in the lower corner. But what should be the dimensions of that small rectangle? There are several ways to arrange the "1" tiles to get an area of 12:

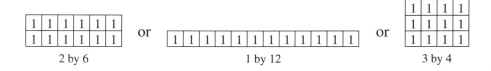

Rather than sketching the details—remember, you're working in a generic rectangle—imagine how the dimensions of the small rectangle will affect the shape of your generic rectangle.

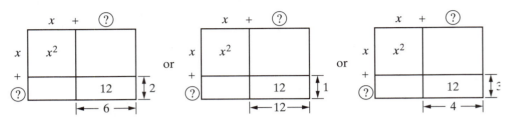

There is additional information that will help you decide which shape is correct: the numbers that go in the two circled ?'s must have a sum of 8, because there are 8 of the x tiles.

a. Find a Diamond Problem that you could have used to help you solve the subproblem of arranging the 12 "1" tiles. *product = 12; sum = 8*

b. Copy and complete the generic rectangle. Label its dimensions to show how to factor the polynomial $x^2 + 8x + 12$. *(x + 2)(x + 6)*

EF-47. Using generic rectangles and the Diamond Problems from problem EF-45, work backward to factor each of the following quadratic polynomials as described in EF-46. Draw and label a generic rectangle for each polynomial, and then write an algebraic equation of the form

$$\text{area as a sum} = \text{area as a product}.$$

a. $x^2 + 7x + 10$ (First think of factor pairs for 10, and then find the pair of factors whose sum is 7.) *(x + 5)(x + 2)*

b. $x^2 + 7x + 12$ *(x + 3)(x + 4)*

c. $x^2 + 13x + 12$ *(x + 1)(x + 12)*

d. $x^2 + 6x + 9$ *(x + 3)(x + 3)*

EXTENSION AND PRACTICE

EF-48. Draw and label a generic rectangle for each polynomial as you did in EF-47. Then write an equation of the form "area as a sum = area as a product."

a. $x^2 + 9x + 20$ *(x + 4)(x + 5)*

b. $x^2 + 24x + 63$ *(x + 3)(x + 21)*

EF-49. Suppose that each generic rectangle shown here represents a composite rectangle made with algebra tiles. (The parts are not necessarily drawn to scale.) Copy each composite rectangle, and find its dimensions as you did in EF-47. Then write the area equation

$$\text{area as a sum} = \text{area as a product}.$$

a.

x^2	$6x$
$3x$	18

b.

$4x^2$	$10x$
$10x$	25

$x^2 + 9x + 18 = (x + 6)(x + 3)$

$4x^2 + 20x + 25 = (2x + 5)(2x + 5)$

EF-50. You've practiced factoring trinomials (polynomials with three terms) by drawing generic rectangles with four parts. You can also use generic rectangles to factor binomials. In that case, the generic rectangle will have just two parts.

Example: You could factor $2x^2 + 10x$ as $2x(x + 5)$ by drawing the following generic rectangle:

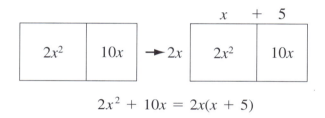

$2x^2 + 10x = 2x(x + 5)$

Alternatively, you could factor $2x^2 + 10x$ as $x(2x + 10)$ by drawing this generic rectangle:

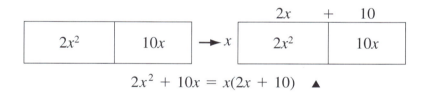

$2x^2 + 10x = x(2x + 10)$ ▲

Factor each of the following binomials; in other words, find the dimensions of a generic rectangle with each given area. Each generic rectangle in this problem has only two parts. Draw and label each generic rectangle as shown in the example above.

a. $x^2 + 7x$ $x(x + 7)$

b. $3x^2 + 6x$ $3x(x + 2)$ or $3(x^2 + 2x)$ or $x(3x + 6)$

c. $3x + 6$ $3(x + 2)$

d. There is more than one way to factor $3x^2 + 6x$. Compare your answer in part b with those of your group mates and write down all the solutions your group found.

EF-51. You have already seen that you can use generic rectangles to find certain products, whether or not it is possible to model with algebra tiles. For example, you can find the product $(x + 2)(6x + 2y - 4)$, even though you don't have any tiles of length y or of length -4, by using the patterns you established with the tiles.

Example: Use a generic rectangle to rewrite the product $(x + 2)(6x + 2y - 4)$ as a sum.

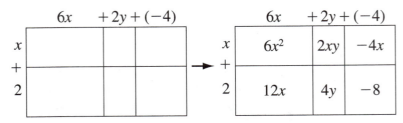

$(x + 2)(6x + 2y - 4) = 6x^2 + 2xy - 4x + 12x + 4y - 8 = 6x^2 + 2xy + 8x + 4y - 8$
area as a product $\qquad\qquad\qquad=\qquad\qquad\qquad$ area as a sum. ▲

Remember, a generic rectangle just helps you organize a problem. It does not have to be drawn accurately to scale. Some students see the generic rectangle as a table, like a small multiplication table. In the example above, you can see that the part representing $6x$ is not as big as it would really be compared with the part representing x.

Find each of the following products by drawing and labeling a generic rectangle. In some of the problems, the generic rectangles will have only two parts; some will have three or six parts.

a. $(2x - y)(x + 3y)$
   $2x^2 + 5xy - 3y^2$

b. $(3x^2)(5x - 4)$
   $15x^3 - 12x^2$

c. $5(2t^2 - 3t + 1)$
   $10t^2 - 15t + 5$

d. $(x + 3)(x^2 - 5x + 4)$
   $x^3 - 2x^2 - 11x + 12$

e. $(x - 50)(3y + 20)$
   $3xy + 20x - 150y - 1000$

EF-52. The perimeter of a circle is usually called its circumference. A circle of radius r units has a circumference of length $2\pi r$ units, and it has an area of πr^2 square units.

$$\text{circumference } C = 2\pi r \text{ units}$$
$$\text{area } A = \pi r^2 \text{ square units}$$

a. Use the π key on your calculator to get an approximate decimal value, up to hundredths, for the irrational number π. $\pi \approx 3.14$

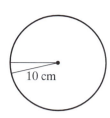

Example: Suppose you want to find the area of the circle shown at left. Since the radius of the circle is 10 cm, then

$$\text{area} = \pi r^2 = \pi \cdot (10)^2 = 100 \cdot \pi \approx (100) \cdot (3.14) = 314 \text{ square cm.}$$

The area is exactly 100π square cm, or approximately 314 square cm. ▲

b. Find the exact circumference of a circle with radius 10 cm. Then use the $\boxed{\pi}$ key on your calculator to find an approximate value, to the nearest tenth, for the circumference.
circumference = $2\pi r = 2 \cdot \pi \cdot 10 = 20 \cdot \pi \approx 62.8$ cm

c. Add to your Tool Kit information about the *circumference* and *area of a circle*.

EF-53.

a. The diameter of a circle is 40 inches. What is the circumference of the circle in terms of π? Find an approximate value of the circumference to the nearest tenth. $40\pi \approx 125.7$ in.

b. The circumference of a circle is 40 inches. What is the diameter of the circle in terms of π? Find an approximate value of the diameter to the nearest tenth. $\dfrac{40}{\pi} \approx 12.7$ in.

EF-54. Use the formula for the area of a circle to find the area of each shaded region. Identify each of the subproblems you use.

a. 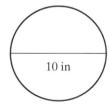   b. 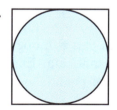   c. 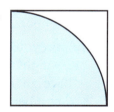   d.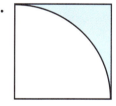

(a) Subproblem: find the radius; $25\pi \approx 75.8$ sq in.
(b) Subproblem: find the side of the square and the radius; $16\pi \approx 50.3$ sq in.
(c) Subproblem: find area of whole circle; $16\pi \approx 50.3$ sq in.
(d) Subproblem: find area in part c; $64 - 16\pi \approx 13.7$ sq in.

For problem EF-55, do not give a formal presentation on computing probabilities. Instead of saying "Find the probability of..." you could ask, "What chance is there of...?" or "What is the likelihood that...?" It would be helpful to have a deck of cards on hand, in case students are not familiar with a standard deck.

EF-55. Two aces have been removed from a deck of 52 playing cards.

a. Find the probability of drawing a king from the remaining deck.
$\frac{4}{50}$, or 0.08

b. Find the probability of drawing an ace from the remaining deck.
$\frac{2}{50}$, or 0.04

c. Write each of the probabilities as a percent.   8%, 4%

EF-56. Allen forgot to study for the last history quiz, so he scored only 7 out of 25 points. Explain why solving the equation $\frac{x}{100} = \frac{7}{25}$ will help Allen to know what percentage of the points he got right.

EF-57. Larry weighs 240 pounds and decides to go on a miracle diet. The diet claims he will lose 20 pounds in three days.

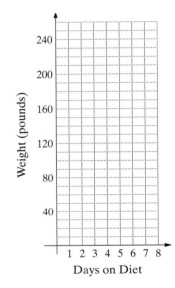

a. Copy the coordinate grid shown here onto your graph paper. On your grid use a *slide* to show the change Larry could expect if the diet's claim is true. Describe the slide in a sentence. You might want to look back at your work with slides in LT-119 first.   *down 20, right 3*

b. What would Larry's weight be after three days if the claim is correct?
220 lb

c. Predict Larry's weight in six days.   *200 lb*

d. Predict Larry's weight in 12 days.   *160 lb*

e. Suppose Larry stuck to the diet and kept losing weight at the constant rate of 20 pounds every three days. When would Larry disappear completely?   *36 days*

f. As you figured in part e, it doesn't make sense for Larry to keep losing weight at a constant rate. Express the rate at which Larry is losing weight as a ratio of pounds to days, and then write this ratio as a mixed number. Realistically, what would happen to this ratio as time passed?
Ratio: 20 lb per 3 days ≈ 6.7 lb/day. Realistically, the ratio would get smaller.

EF-58. Which Method Is Correct? Keisha and Rudy were arguing about how to compute the expression $\frac{2^4}{2^7}$. Keisha used the meaning of exponents and the idea of finding fractional forms of 1 (as in problem EF-16) to show Rudy that

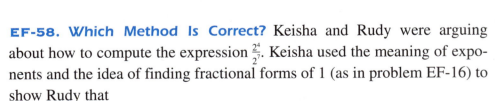

$\frac{2^4}{2^7}$ means $\frac{2 \cdot 2 \cdot 2 \cdot 2}{2 \cdot 2 \cdot 2 \cdot 2 \cdot 2 \cdot 2 \cdot 2} = \frac{2 \cdot 2 \cdot 2 \cdot 2}{2 \cdot 2 \cdot 2 \cdot 2} \cdot \frac{1}{2 \cdot 2 \cdot 2} = 1 \cdot \frac{1}{2^3} = \frac{1}{2^3}.$

Rudy says that since the bases are the same, it's easier to just subtract the exponents to find how many factors of 2 are left in the denominator, so

$$\frac{2^4}{2^7} \text{ should equal } 2^{4-7} = 2^{-3}.$$

a. Find the value of $\frac{2^4}{2^7}$ by using your calculator and the order of operations.
0.125

b. Use your calculator to find the values of $\frac{1}{2^3}$ and 2^{-3} (using the exponentiation key). *Both are 0.125*

c. What do you observe in parts a and b? *The values are the same.*

d. Write a fraction with a numerator of 1 that you think is equal to 2^{-5}. Check your conjecture using your calculator. *$2^{-5} = \frac{1}{2^5} = 0.03125$*

e. Repeat part d for 4^{-3}. *$4^{-3} = \frac{1}{4^3} = 0.015625$*

EF-59. Exponent Practice Rewrite each of the following expressions with a single base and a single exponent:

a. $\frac{2^4}{2^9}$ *2^{-5}*

b. $\frac{6^{10}}{6^3}$ *6^7*

c. $\frac{x^2}{x^9}$ *x^{-7}*

d. $\frac{m^4}{m^5}$ *m^{-1}*

EF-60. More Exponent Practice Use what you know about exponents to rewrite each of the following expressions more simply:

a. $\frac{x^3 x^5}{x^6}$ *$\frac{x^8}{x^6} = x^2$*

b. $\frac{x^2 y^5}{x^4 y}$ *$\frac{y^4}{x^2}$*

c. $(2 \cdot 10^5)(x \cdot 10^4)$ *$2x \cdot 10^9$*

EF-61. Cindy and Emily took the red-eye flight from Miami to Las Vegas. When they got close to Las Vegas, the pilot spoke to the passengers,

"Ladies and gentlemen, we are now about 80 miles from Las Vegas, and we should be there in 20 minutes, at 3:30 a.m. local time. The temperature is about 65°F. Enjoy your stay and thank you for flying MathAir."

What was the average speed of the plane during the last 20 minutes of the flight? *240 mph*

EF-62. For the following problem, define a variable, and then write an equation and solve it.

Cassandra is planning to deposit money into two simple interest accounts. She has already decided to put $5000 into an account that pays an annual simple interest rate of $4\frac{1}{2}\%$. How much should she deposit in a second account which earns 6% annual simple interest so that both accounts earn the same amount of money in interest?
0.06x = 0.045(5000); $3750

4.4 SIMILAR TRIANGLES: MORE GEOMETRIC RATIOS

SECTION MATERIALS
PROTRACTORS
METRIC RULERS
EF-63 SIMILAR TRIANGLES RESOURCE PAGE
EF-64 SIMILAR TRIANGLES AND RATIOS OF SIDE LENGTHS RESOURCE PAGE
DOT PAPER (COPIED FROM THE DOT PAPER TEMPLATE RESOURCE PAGE, FOR EF-69)

In this section we introduce the notion of similar triangles and examine the ratios of corresponding sides of similar triangles. Students may need help with the convention that angles marked with the same symbol have equal measures.

Some students may have difficulty seeing that the corresponding angles are equal in size. Suggest that they trace and cut out a copy of the smaller triangle to place on top of the larger. They can then shift the smaller triangle from vertex to vertex inside the larger triangle to see that the corresponding angles are the same size.

To tie the ideas presented here with the notion of enlargements, you might ask students to predict whether or not doubling the lengths of the sides of a triangle will produce a similar triangle.

As you circulate, check to see that students are using their protractors correctly and accurately. Some instructors also use a class chart on the board where groups can record angle measurements and then compare their measurements with other groups' results.

In this section you'll focus on an important geometric relationship—similarity—in the case of triangles. You'll extend the idea of enlargement and reduction ratios of dot-grid figures to ratios of corresponding sides of similar triangles. By working carefully through the problems, you'll develop a way to write equations using equivalent (same size) ratios.

Ratio and similarity are big ideas and, as with any new situation, you can help yourself by slowing down and reading carefully.

EF-63. Definition We say two triangles are **similar** if their corresponding angles are the same size (that is, they have the same measure). For example, the two triangles in each pair shown next are similar since all the angles marked in the same way are the same size. Here, $\triangle ABC$ is similar to $\triangle DEF$, and $\triangle ZIP$ is similar to $\triangle GUM$. Notice that each triangle is named by stating the vertices in the same order as corresponding angles.

Section 4.4 Similar Triangles: More Geometric Ratios **237**

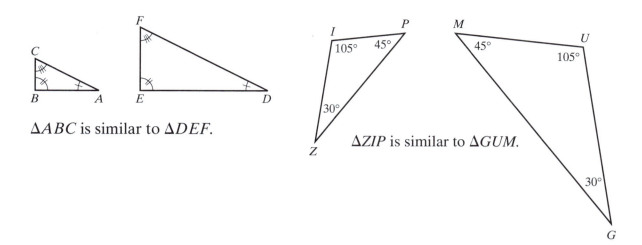

△ABC is similar to △DEF.

△ZIP is similar to △GUM.

a. Copy the definition for *similar triangles* (include the drawings) into your Tool Kit.

b. On your Similar Triangles Resource Page, use a protractor to measure the angles in the following triangles. Since there are 15 angles to measure, it will be more efficient if you share the work with your group. Label each angle with its measure as shown in △ZIP and △GUM.
$m\angle H = m\angle R = 55°$, $m\angle U = m\angle T = 90°$, $m\angle M = m\angle A = 35°$, $m\angle C = m\angle B = 38°$, $m\angle O = m\angle E = 123°$, $m\angle G = m\angle Z = 19°$, $m\angle N = 57°$, $m\angle I = 75°$, $m\angle X = 48°$

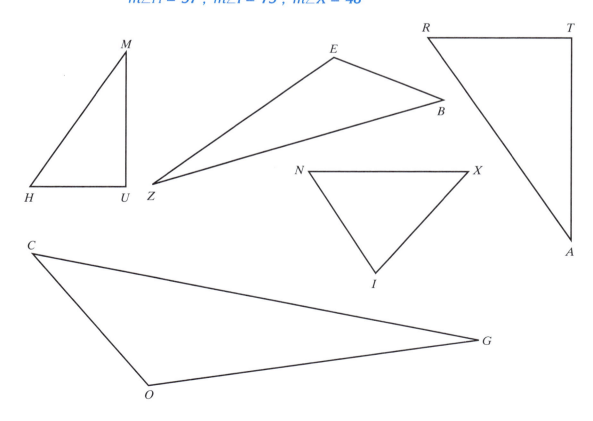

c. State which triangles are similar to each other. Be sure that you name each triangle by stating the vertices in the same order as corresponding angles. △HUM~△RTA, △BEZ~△COG

EF-64. Similar Triangles and Ratios of Side Lengths When cartographers make an enlargement of a map, distances between cities must be proportional, and shapes must be similar. For example, each of the following two maps shows the direct air-delivery route—from Sacramento to Los Angeles, then to Reno and back to Sacramento—for ExpressAir, a small air-freight company in California.

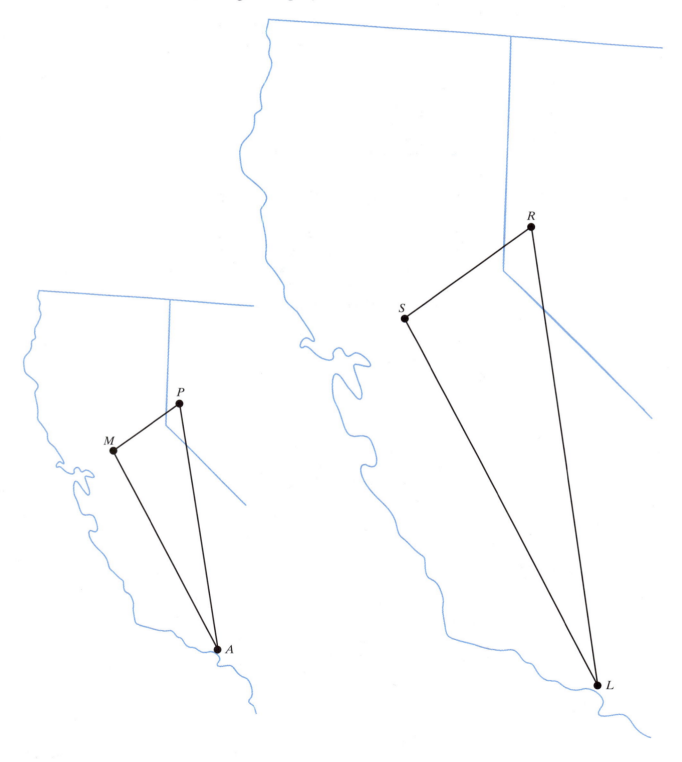

a. On your EF-64 Resource Page ("Similar Triangles and Ratios of Side Lengths"), use a protractor to measure each of the six angles to the nearest degree. Label each angle with its measure. **$m\angle M = m\angle S = 99°$; $m\angle P = m\angle R = 59°$; $m\angle A = m\angle L = 22°$**

b. Are $\triangle MAP$ and $\triangle SLR$ similar triangles? Explain how you know. **Yes, because their corresponding angles are equal.**

c. Use a ruler to measure the lengths of each of the six sides of the two triangles to the nearest millimeter, and label the lengths on your Resource Page. **For $\triangle MAP$: 27, 60, and 70 mm. For $\triangle SLR$: 51, 112, and 129 mm.**

d. Using your measurements from part c, find each of the following ratios. Then use a calculator to write each ratio as a decimal to the nearest tenth:

$$\frac{|SL|}{|MA|} \qquad \frac{|LR|}{|AP|} \qquad \frac{|RS|}{|PM|}$$

$\frac{112}{60} \approx 1.9, \quad \frac{129}{70} \approx 1.9, \quad \frac{51}{27} \approx 1.9$

e. What do you notice about the three ratios in part d? **They are all approximately equal to 1.9. We want the students to see that the ratios of corresponding sides of similar triangles are equal. In this case the ratio is 1.9:1.**

f. Joe thinks $\triangle SLR$ just *looks* like an enlargement of $\triangle MAP$ but really isn't. Do you think it is an enlargement? If not, explain why not. If so, find the enlargement ratio. (In other words, find the ratio $\frac{S_{new}}{S_{orig}}$. Use a calculator to write your ratio as a decimal to the nearest tenth.) **The enlargement ratio is 1.9 to 1.**

EF-65. In EF-64, you compared only certain ratios of the sides of the two triangles; that is, you compared ratios of **corresponding sides**.

a. What is meant by "corresponding sides"? **There are many possible ways to answer this. Look for some sense that when the triangles are positioned so the equal angles are matched, the corresponding sides are aligned or parallel.**

b. What did you observe about the ratios of the corresponding sides of two similar triangles? Write your observation in a complete sentence. Add your observation to your Tool Kit. **The ratios of corresponding sides of similar triangles are equal.**

EF-66. Triangle *ABC* and △*DEF* are similar triangles.

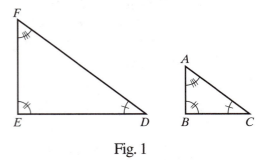

Fig. 1

a. Use Figure 1 to find two more ratios of sides that are equal to $\frac{|EF|}{|BC|}$. $\frac{|ED|}{|BA|}, \frac{|DF|}{|AC|}$

b. Use the lengths of the sides given in Figure 2 and the ratios from part a to write an equation you can solve for *x*, the length of *EF*. Then solve your equation to find *x*. $\frac{x}{3} = \frac{8}{5}; x = \frac{24}{5} = 4.8$

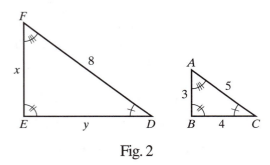

Fig. 2

c. Follow the process of parts a and b to write an equation to solve for *y*, the length of *ED*. Then solve your equation for *y*. $\frac{y}{4} = \frac{8}{5}; y = \frac{32}{5} = 6.4$

EF-67. The two triangles shown here are similar. Triangle *DEF* is an enlargement of △*ABC*.

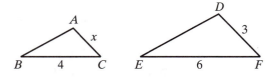

a. Find *x*, the length of *AC*. *x* = 2

b. What is the enlargement ratio? $\frac{3}{2}$

c. What do you think the ratio of the perimeter of △*DEF* to the perimeter of △*ABC* is? Explain why. $\frac{3}{2}$; because it would be consistent with EF-25.

d. What do you think the ratio of the area of △*DEF* to the area of △*ABC* is? Explain why. $\frac{9}{4}$; because it would be consistent with EF-25.

(**EXTENSION AND PRACTICE**)

EF-68. Notation When naming the angles of a triangle, we use the symbol "∠" to mean "angle." For example, rather than writing, "the measure of angle G is about 29°," we could write the shorthand, "the measure of ∠G ≈ 29°" or "m∠G ≈ 29°."

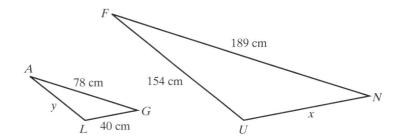

a. Add information about the symbol "∠" to your Tool Kit.

b. Using a protractor, find the measure of each of the following angles:

measure of ∠G ≈ ____29°____

measure of ∠L ≈ ____130°____

measure of ∠N ≈ ____29°____

measure of ∠U ≈ ____130°____

c. How can you be sure that $\triangle ALG$ is similar to $\triangle FUN$? Explain.
Measure ∠A and ∠F and then check that all corresponding angles are equal. Some students might mention that m∠F = m∠A = 21° because the sum of the angles of a triangle is 180°.

d. Write equations you could use to find the length of each of the missing sides, UN and AL. Then find the lengths of the missing sides.
$\frac{y}{154} = \frac{78}{189}$, $y = 63.6$ cm; $\frac{x}{40} = \frac{189}{78}$, $x = 96.9$ cm

EF-69. More Ratios of Sides of Similar Triangles In problem EF-64, you found a relationship between the ratios of corresponding sides of similar triangles. In this problem, you will look for more equivalent ratios involving similar triangles.

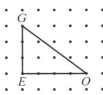

a. Copy $\triangle GEO$ on a piece of dot paper, and then enlarge it by making the corresponding sides of the new triangle four times as long. Label your new triangle $\triangle JAM$. What is the enlargement ratio? (In parts c and d you will be considering ratios other than the enlargement ratio.) $\frac{4}{1}$

b. Is △GEO similar to △JAM ? Explain how you know. *Yes: since it's an enlargement, it's the same shape, so its corresponding angles are equal.*

c. For △GEO, find each of the following ratios:

$$\frac{\text{Short side}}{\text{Longest side}} \quad \frac{\text{Short side}}{\text{Middle-length side}} \quad \frac{\text{Longest side}}{\text{Middle-length side}}.$$

$\frac{3}{5} = 0.6$; $\frac{3}{4} = 0.75$; $\frac{5}{4} = 1.25$

(Hint: To find $|GO|$, mark a segment of length GO on a scrap of paper, and then use it to count how many units long GO would be if it were drawn vertically or horizontally on the dot paper.)

d. For △JAM, find each of the following ratios:

$$\frac{\text{Short side}}{\text{Longest side}} \quad \frac{\text{Short side}}{\text{Middle-length side}} \quad \frac{\text{Longest side}}{\text{Middle-length side}}$$

$\frac{12}{20} = 0.6$; $\frac{12}{16} = 0.75$; $\frac{20}{16} = 1.25$

e. What do you notice about the ratios in parts c and d? *Corresponding ratios are equal.*

EF-70. Ginger measured the lengths of the sides of the two similar triangles shown here, △ABC and △DEF:

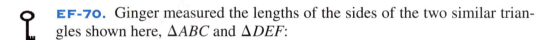

a. What is the enlargement ratio from △ABC to △DEF? Write all three ratios of corresponding sides that are equal to the enlargement ratio.

enlargement ratio $= \frac{3}{1}$; $\frac{|DE|}{|AB|} = \frac{|EF|}{|BC|} = \frac{|DF|}{|AC|} = \frac{3}{1}$.

b. Using the ideas from problem EF-69, write as many statements showing other equivalent ratios as you can.

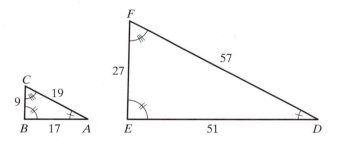

c. Describe in your own words the difference between the ratios in part a and the ratios in part b. *The ratios in part a are all ratios of sides from two triangles; the ratios in part b are ratios of sides from one triangle.*

EF-71. Suppose you are given these two similar triangles:

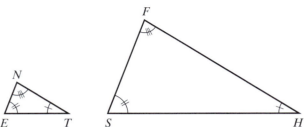

Look back at your results from problems EF-64 and EF-69. Use what you learned to make two conjectures about what is true about ratios of corresponding sides of $\triangle NET$ and $\triangle FSH$. Write your conjectures carefully and completely. *Conjecture #1: $\frac{|NE|}{|FS|} = \frac{|NT|}{|FH|} = \frac{|ET|}{|SH|}$; ratios of corresponding sides from two similar triangles are equal. Conjecture #2: $\frac{|NE|}{|ET|} = \frac{|FS|}{|SH|}, \frac{|NT|}{|ET|} = \frac{|FH|}{|SH|}$, and $\frac{|NT|}{|NE|} = \frac{|FH|}{|FS|}$; ratios of sides from one triangle are equal to ratios of corresponding sides from a similar triangle.*

EF-72. Write an equation and use it to solve the following problem:

Two numbers are in a ratio of 2 to 3. If their sum is 50, find the numbers. *(Hint: problem EF-29 could help students get started.) Possible equations: $\frac{x}{50-x} = \frac{2}{3}$, or $\frac{50-x}{x} = \frac{2}{3}$, or $2x + 3x = 50$; 20 and 30*

EF-73. Copy each of the following generic rectangles, and find its dimensions. Then write an equation of the form

area as a sum = area as a product,

as you did in EF-47.

a.
$3x^2$	$4x$
$15x$	20

b.
$2x^2$	$3x$
$12x$	18

$3x^2 + 19x + 20 = (x + 5)(3x + 4)$ *$2x^2 + 15x + 18 = (x + 6)(2x + 3)$*

EF-74. Draw and label a generic rectangle for each polynomial as you did in EF-47. Then, for parts a-c, write an equation of the form

area as a sum = area as a product.

a. $x^2 + 12x + 36$ $(x + 6)(x + 6)$ b. $x^2 + 16x + 28$ $(x + 2)(x + 14)$

c. $2m^2 + 7m + 3$ $(2m + 1)(m + 3)$

EF-75. Before completing each of the following problems, identify a subproblem you plan to solve.

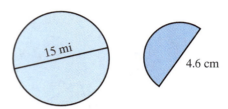

a. Find the area and the circumference of the circle.
Subproblem: Find the radius. Area = $56.25\pi \approx 176.7$ sq mi; circumference = $15\pi \approx 47.1$ mi.

b. Find the area and the circumference of the semicircle.
Subproblem: Find the radius, then find area and circumference of the whole circle, and then compute half. Area = $2.645\pi \approx 8.3$ sq cm; circumference = $4.6 + 2.3\pi \approx 11.83$ cm

EF-76. Maya and Mike are 60 miles apart and traveling toward each other. Maya travels at 20 miles per hour on her bike, while Mike can manage only 30 miles per hour in his car. In how many hours will they meet? *In $\frac{6}{5}$ hours, or 1 hour 12 minutes*

EF-77. Negative Exponents In EF-58, Keisha and Rudy discussed how to deal with quotients of exponential expressions such as $\frac{2^4}{2^7}$, and you made an observation about expressions with negative exponents such as 2^{-5} and 4^{-3}. The following example pulls together what you observed.

Example: The exponential expression 4^{-3} means $\frac{1}{4^3}$. Using the meaning of positive integer exponents you can rewrite this in "expanded" form as $\frac{1}{4 \cdot 4 \cdot 4}$. You can then divide using a calculator to find the decimal value, so $\frac{1}{4 \cdot 4 \cdot 4} = \frac{1}{64} = 1 \div 64 = 0.015625$. Putting this all together, you have

$$4^{-3} = \frac{1}{4^3} = \frac{1}{4 \cdot 4 \cdot 4} = \frac{1}{64} = 0.015625. \blacktriangle$$

a. Add information and examples about the meaning of *negative integer exponents* to your Tool Kit.

b. Copy and complete the following table by rewriting each of the numbers in exponential form, fraction form, expanded form, and decimal form. The first two have been done for you.

Exponential Form	Fraction Form	Expanded Form	Decimal (Standard) Form
2^{-3}	$\frac{1}{2^3}$	$\frac{1}{2 \cdot 2 \cdot 2}$	0.125
4^{-2}	$\frac{1}{4^2}$	$\frac{1}{4 \cdot 4}$	0.0625
2^{-4}	$\frac{1}{2^4}$	$\frac{1}{2 \cdot 2 \cdot 2 \cdot 2}$	0.0625
5^{-3}	$\frac{1}{5^3}$	$\frac{1}{5 \cdot 5 \cdot 5}$	0.008
3^{-2}	$\frac{1}{3^2}$	$\frac{1}{3 \cdot 3}$	≈ 0.111
4^{-1}	$\frac{1}{4^1}$	$\frac{1}{4}$ or $\frac{1}{2 \cdot 2}$	0.25
5^{-2}	$\frac{1}{5^2}$	$\frac{1}{5 \cdot 5}$	0.04

EF-78. Write each of the following expressions in fraction form and then in standard (decimal) form.

Example: $2^{-3} = \frac{1}{2^3} = \frac{1}{8} = 0.125$

fraction form ↗   ↑ decimal form ▲

a. 4^{-3} $4^{-3} = \frac{1}{4^3} = 0.015625$ b. 2^{-5} $2^{-5} = \frac{1}{2^5} = 0.03125$

c. 3^{-1} $3^{-1} = \frac{1}{3^1} \approx 0.3333$ d. 5^{-2} $5^{-2} = \frac{1}{5^2} = 0.04$

EF-79. Rewrite each of the following expressions using each base only once.

a. $3x^3y \cdot 4x^2y^2$ $12x^5y^3$ b. $3x(x^2)^3$ $3x^7$

c. $(3x^2)^3$ $27x^6$ d. $\frac{8x^2y^3}{2x^2y}$ $4y^2$

EF-80. Sometimes an understanding of exponents is more useful than a calculator.

a. Use a calculator to compute $(1.4 \cdot 10^{98}) \cdot (2.3 \cdot 10^5)$.

b. Explain what happened.   *overflow because product is so large*

c. Now use what you know about exponents to compute $(1.4 \cdot 10^{98}) \cdot (2.3 \cdot 10^5)$. $3.22 \cdot 10^{103}$

EF-81. Solve each of the following equations for x.

a. $3x + c - 5 = -4$ $\quad \frac{1-c}{3}$

b. $3c + x - 5 = -4$ $\quad 1 - 3c$

c. $7(x - 2) = m + 3x$ $\quad \frac{m+14}{4}$

EF-82. Monique and her classmates were working with algebra tiles to factor the binomial $2x^2 + 10x$, and they came up with the following four possible figures to help them factor the expression. Study the four figures, and then answer the following questions.

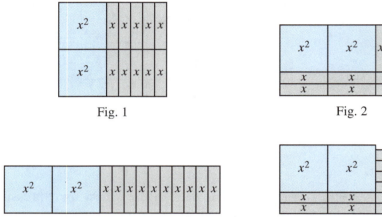

Fig. 1   Fig. 2   Fig. 3   Fig. 4

a. Which of the figures represents a correctly factored form of the expression $2x^2 + 10x$? Explain how you know. *Figures 1 and 3 are the only figures that are composite rectangles.*

b. For each composite rectangle from part a, write an equation of the form

    area as a sum = area as a product.

Figure 1: $2x^2 + 10x = 2x(x + 5)$; Figure 3: $2x^2 + 10x = x(2x + 10)$

At the next class meeting, ask students if they were surprised that the result in EF-83 is the same as that in EF-41. Then have them share their explanations from part b.

EF-83. Here's another floor-tiling problem.

a. How many 6-inch by 6-inch square tiles would be needed to cover a 12-foot by 16-foot rectangular floor? *768 tiles*

b. What's an easy way to see that the answer to part a is the same as the answer to EF-47? *All lengths, of both of the rectangle and the tiles, are doubled.*

4.5 MORE SIMILARITY: RIGHT TRIANGLES AND MISSING SIDES

SECTION MATERIALS

AN OVERHEAD TRANSPARENCY AND PEN FOR EACH GROUP (OPTIONAL, FOR EF-84)

EF-96 SUMS AND PRODUCTS: PART I RESOURCE PAGE

In this section students revise their conjectures about ratios of corresponding sides of similar triangles and apply them to similar right triangles. Working with similar right triangles helps prepare students for understanding the concept of the slope of a line, which is the heart of Chapter 5. Here, students look at slopes of lines in context, as rates of change. The formal definition of slope comes in Section 5.1, so we don't use formal notation or vocabulary until then.

Students also see (in EF-90) how to solve equations that contain fractions by using subproblems and what they know about solving equations.

In problems EF-95 through EF-97, they start to connect the Diamond Problems that they have been solving since Chapter 1 to generic rectangles and factoring quadratic polynomials with a lead coefficient of 1. The work continues in the remaining sections of the chapter in problems EF-123 through EF-125, and EF-134, and culminates in EF-148. Some students will see the sum-product pattern easily. Others will need to see it demonstrated several times; if possible, let those students hear it from their peers.

In EF-84, encourage students to slow down and recall the pairs of sides that were compared in EF-64 and EF-71. Stress that each group needs to agree on and write two conjectures. Each group member should copy them for later reference. You also could have each group write their conjecture on an overhead transparency for presentation, or have a member from each group put the group's conjecture on the board. This will give everyone a chance to see different ways of describing their observations.

EF-84. Suppose you are given these two similar triangles:

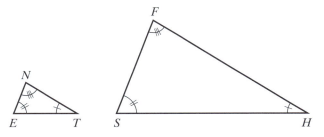

In EF-71, you wrote two conjectures about ratios of corresponding sides of $\triangle NET$ and $\triangle FSH$. Share your conjectures with the rest of your group's members. Compare the conjectures, and then revise them to make two conjectures that satisfy everyone in the group. Write the group's two conjectures carefully and completely, and be ready to share them with the rest of the class.

In this section, you will use these ratio relationships for similar triangles to find the length of an unknown side.

EF-85. Missing Sides In this problem, $\triangle NET$ is similar to $\triangle FSH$, and, the lengths of some of the sides are as given.

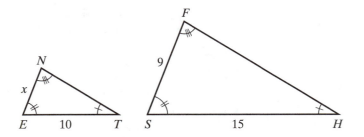

a. Use your group's two conjectures from EF-84 to write two different equations that could be used to solve for the length x of side NE.
$\frac{x}{10} = \frac{9}{15}, \frac{x}{9} = \frac{10}{15}, \frac{10}{x} = \frac{15}{9}$, or $\frac{9}{x} = \frac{15}{10}$

b. Find the length x using one of your equations. **6**

EF-86. Definitions A **right angle** is an angle that measures 90°. We indicate a right angle with a small square, this way:

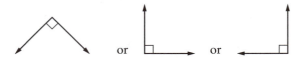

A **right triangle** is a triangle with exactly one right angle. For example, $\triangle ABC$ is a right triangle with its right angle at B.

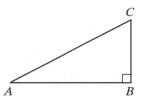

a. Copy the definitions of *right angle* and *right triangle*, including the diagrams, into your Tool Kit.

b. The following sketch contains four right triangles. Draw each right triangle separately, and write the name of each one next to its drawing.
$\triangle ARG, \triangle ATB, \triangle ETB,$ and $\triangle EXH$

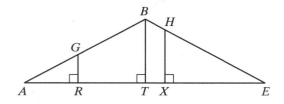

EF-87. Look at right triangles *MAT* and *TCH,* shown here:

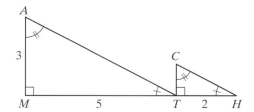

a. How can you tell that they are similar? **Their corresponding angles are equal.**

b. Find $|TC|$ by writing and solving an equation. (Hint: Use the variable x to represent the unknown length $|TC|$.) $|TC| = \frac{6}{5} = 1.2$

c. Write another equation using ratios that you could have used to solve for $|TC|$. $\frac{x}{2} = \frac{3}{5}$ or $\frac{x}{3} = \frac{2}{5}$ or $\frac{2}{x} = \frac{5}{3}$ or $\frac{3}{x} = \frac{5}{2}$

Problem EF-88 introduces a strategy for dealing with situations students will encounter throughout the remainder of the course, namely problems that involve a constant rate of change. There's no need for a rigorous proof of similarity in part a. Students might "justify" the congruence of the third pair of angles in a number of ways, including tracing and comparing, cutting and comparing, measuring, and using the fact that the sum of the measures of the angles of a triangle is 180°. Let students come up with their own methods. The important point is that students be able to recognize a pair of similar right triangles and use them to write equations to model the situation.

EF-88. Look at the right triangles in the following diagram:

a. Explain how you can tell that the two triangles are similar, even though the measures of all the angles are not given. **The right angles are equal, the smallest angles are equal since they coincide, and so the third angles must also be equal.**

b. To write an equation based on the diagram, you need to be able to see two separate triangles. One technique for seeing the two triangles more easily is to draw a separate sketch of each one and include the dimensions of each as shown in part a.

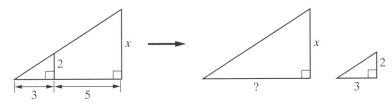

Copy the original diagram and the separated sketch on your paper. What is the length of the longer leg of the large triangle? Label the separated sketch. **3 + 5 = 8 units**

c. Write *two* different equations you could use to solve for the length *x*.
$\frac{8}{3} = \frac{x}{2}$ or $\frac{2}{3} = \frac{x}{8}$ or $\frac{3}{8} = \frac{2}{x}$ or $\frac{3}{2} = \frac{8}{x}$

d. Use one of your equations to find the length *x*. $x = \frac{16}{3} \approx 5.33$

EF-89. Jason is making a skateboard ramp with two vertical supports. What length of board should he cut to make the center support? **x = 9.6 ft.**

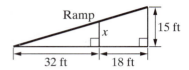

In EF-90 and EF-91 we expand the notion of "undoing while maintaining a balance" to include "fraction busters," an approach to solving equations that contain fractions in which the subproblem of "undoing the denominators" is identified. Because it simply combines strategies students have already successfully used to solve problems, it's not really a new method at all.

EF-90. Fraction Busters In this course you've been solving equations that contain algebraic fractions. Some of the equations took several steps to solve. These steps applied the idea of "undoing" and "keeping the balance." Here's a more efficient approach, based on the same ideas, where the "undoing" is accomplished by multiplying by a "fraction buster."

Example: Suppose you want to solve the equation $\frac{x}{3} + \frac{x}{5} = 2$.

The hard part of this problem is dealing with the fractions. You could add them by first writing them in terms of a common denominator, ... *but there is an easier way.*

You can avoid dealing with fractions by "eliminating" or "undoing" the denominators. The first subproblem is to *find a common denominator*. Then you'll *multiply both sides of the equation by that common denominator*. In this case the least common denominator is 15, and you could multiply both sides of the equation by 15 (don't forget to use the distributive property):

$\frac{x}{3} + \frac{x}{5} = 2$

The least common denominator of $\frac{x}{3}$ and $\frac{x}{5}$ is 15.

$15 \cdot \left(\frac{x}{3} + \frac{x}{5}\right) = 15 \cdot 2$

$15 \cdot \frac{x}{3} + 15 \cdot \frac{x}{5} = 15 \cdot 2$

The number you use to eliminate the denominators is called a **fraction buster**. Here we used "15" to bust the fractions.

Section 4.5 More Similarity: Right Triangles and Missing Sides

The result is a valid equation *without fractions*: It looks like many you have seen before and can be solved in the usual way.

$$5x + 3x = 30$$
$$8x = 30$$
$$x = \frac{30}{8} = \frac{15}{4} = 3.75$$

You will, of course, always want to check the answer!

Check:
$$\frac{3.75}{3} + \frac{3.75}{5} = 2$$
$$1.25 + 0.75 = 2 \quad \blacktriangle$$

Copy the following problem on your paper. Fill in each of the lines labeled a through f to explain how the statement or equation to its right was obtained from the equation above it.

Solve the equation: $\quad \frac{4}{x} + \frac{3}{2x} = \frac{11}{6}$

a. *The least common denominator for x, 2x, and 6 is 6x.*   A fraction buster is $6x$.

b. *Multiply both sides by 6x, the fraction buster.* $\quad 6x\left(\frac{4}{x} + \frac{3}{2x}\right) = 6x \cdot \left(\frac{11}{6}\right)$

c. *Use the distributive property.* $\quad 6x \cdot \left(\frac{4}{x}\right) + 6x \cdot \left(\frac{3}{2x}\right) = 6x \cdot \left(\frac{11}{6}\right)$

d. *Reduce the fractions.* $\quad 24 + 9 = 11x$

e. *Add 24 and 9.* $\quad 33 = 11x$

f. *Divide both sides by 11.* $\quad 3 = x$

EF-91. Solve each of the following equations. In each case, find the least common denominator first, as in problem EF-90.

a. $\frac{3x}{5} + 2 = \frac{x+2}{7}$ *-3.75*   b. $\frac{1}{y} + \frac{1}{2y} = 3$ *0.5*

c. Add information to your Tool Kit about using *fraction busters* to solve equations.

EXTENSION AND PRACTICE

EF-92. Solve each of the following equations. In each case, find the least common denominator first, as in problem EF-90.

a. $\frac{4}{z} + 1 = 9$ *0.5*   b. $\frac{y}{2} + \frac{y}{3} - 4 = \frac{5}{6}$ *$\frac{29}{5} = 5.8$*

c. $x + \frac{x}{5} + \frac{x}{3} = 69$ *45*

EF-93. Write and solve an equation using ratios to find $|BY|$.

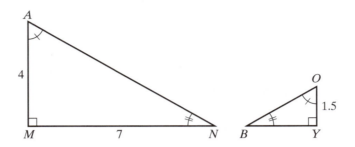

For example, $\dfrac{|BY|}{1.5} = \dfrac{7}{4}$, $|BY| = \dfrac{21}{8} \approx 2.63$

EF-94. Dan wants to buy the smallest round pizza he can and still get at least 100 square inches of pizza. What should be the diameter of the pizza that Dan orders? Be sure to draw pictures and identify any subproblems. *≈11.28 inches, or a 12-in. pizza*

Problems EF-95 through EF-97 are part of the series of problems in which students connect Diamond Problems, generic rectangles, and factoring quadratic polynomials with a lead coefficient of 1.

EF-95. Use a generic rectangle to factor (rewrite as a product) each of the following quadratic polynomials.

a. $x^2 + 10x + 16$ *(x + 2)(x + 8)*

b. $x^2 + 8x + 16$ *(x + 4)(x + 4)*

c. $m^2 - 5m + 6$ *(m − 3)(m − 2)*

The following list includes positive and negative sums and products and prepares students to use generic rectangles to factor quadratics with integer coefficients. Note that the integer pairs are not ordered pairs. The problem is based on an idea developed by teacher Steve Legé. Students will get more practice and consolidate the patterns in the homework of the following two sections.

EF-96. Sums and Products: Part I

a. Use your Resource Page for EF-96 ("Sums and Products: Part I") to record integer pairs whose sums and products are indicated in the table. For each sum-product pair, find two integers whose sum is the number in the left-hand column and whose product is the number in the right-hand column.

Example: Find two integers whose sum is 5 and whose product is 6.

Solution: The integers 2 and 3 work, since $2 + 3 = 5$ and $2 \cdot 3 = 6$. ▲

Note: There may be some sum-product pairs for which no pair of integers works.

Sum	Product	Integer Pair	Sum	Product	Integer Pair
5	6	2, 3 or 3, 2	−5	6	*−3, −2*
7	6	*1, 6*	−7	6	*−6, −1*
13	12	*1, 12*	−11	30	*−6, −5*
10	16	*2, 8*	−8	16	*−4, −4*
6	12	*none*	−10	16	*−8, −2*
8	16	*4, 4*	7	12	*3, 4*
10	0	*10, 0*	−10	24	*−6, −4*

b. We can use a line in the Sums and Products table in part a to write a quadratic polynomial and its factored form.

For example, $x^2 + 5x + 6 = (x + 2)(x + 3)$.

Which line in the Sums and Products table is related to each of the quadratic polynomials in problem EF-95 ? Look back at your work, circle the related lines in your Sums and Products table, and then complete a table like the one below.

	Sum	Product	Integer Pair	Polynomial area as a sum	Factored Form area as a product
Example:	5	6	2, 3	$x^2 + 5x + 6$	$= (x + 2)(x + 3)$
				$x^2 + 10x + 16 =$	
				$x^2 + 8x + 16 =$	
				$m^2 - 5m + 6 =$	

EF-97. Copy and solve these Diamond Problems:

a.
12
13
1, 12

b.
16
10
2, 8

c.
6
−5
−3, −2

d.
12
7
3, 4

EF-98. Write an equation that represents the following problem, and then solve it.

Find three consecutive odd numbers such that the sum of the smallest number and seven times the largest number is 68. *$x + 7(x + 4) = 68$; 5, 7, 9*

EF-99. Adding Fractions

a. Add: $\frac{5}{8} + \frac{1}{3}$ $\frac{23}{24}$

b. Describe what you did in part a to find the sum of the two fractions.

c. Now find the sum of this pair of fractions: $\frac{5}{x} + \frac{2}{3}$. $\frac{15 + 2x}{3x}$

EF-100. Find each sum without using a calculator. Leave your results in fractional form.

a. $\frac{3}{7} + \frac{4}{5}$ $\frac{43}{35}$

b. $\frac{3}{7} + \frac{x}{7}$ $\frac{3 + x}{7}$

c. $\frac{3}{z} + \frac{4}{5}$ $\frac{15 + 4z}{5z}$

d. $\frac{w}{3} + \frac{w + 1}{7}$ $\frac{10w + 3}{21}$

EF-101. Describe a *slide* you could use to move a point from (3, 0) to (5, 2). *Slide up 2 units and then to the right 2 units.*

EF-102. Lora loves going to Reno to play the slot machines. On her last trip, she played all day long on Friday, Saturday, and Sunday. Each day, she gathered a purseful of silver dollars, started at 6:00 A.M., and played until there were no more silver dollars left in her purse. These graphs show her results for the three days:

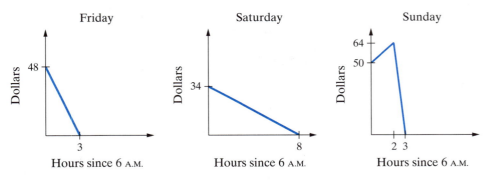

a. Describe how much money Lora started with each day, and state what time she ran out of money and had to stop gambling. *On Friday, she started with $48 and stopped at 9 A.M. On Saturday, she started with $34 and stopped at 2 P.M. On Sunday, she started with $50 and stopped at 9 A.M.*

b. For each graph, write a ratio to describe the rate at which Lora lost (or gained) money. For Sunday, you will need to write two different ratios. *On Friday, $\frac{-\$48}{3}$ hr = −$16 per hour.
On Saturday, $\frac{-\$34}{8}$ hr = −$4.25 per hour.
On Sunday, she gained at + $\frac{\$14}{2}$ hr = + $7 per hour for the first two hours, then she lost at $\frac{-\$64}{1}$ hr = $64 per hour during the last hour.*

EF-103. The mass of the sun is about 330,000 times the mass of the Earth. If the mass of the Earth is about $6 \cdot 10^{24}$ kilograms, what is the approximate mass of the sun?   *approximately $1.98 \cdot 10^{30}$ kg*

EF-104. Kari bought a box of beads in 13 different colors. There were four beads of each color (red, blue, green, orange, violet, ...). Two red beads fell out of the box as Kari was going home.

a. Find the probability of drawing a green bead from the beads that remain in the box. $\frac{4}{50}$

b. Find the probability of drawing a red bead from the beads that remain in the box. $\frac{2}{50}$

c. Write each of the probabilities in parts a and b as a percent.   *8%, 4%*

d. Compare your work in this problem with your work in EF-55. What do you notice?   *The two problems are essentially the same.*

Problem EF-105 provides an opportunity to show how the use of precise language helps us communicate clearly: "three times the square of x" versus "the square of 3x."

EF-105. Does $3x^2 = (3x)^2$? That is, does "three times the square of x" equal the "square of three times x"? Draw a picture to justify your answer.   *no*

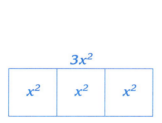
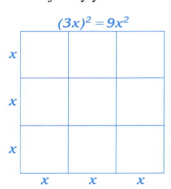

EF-106. Rewrite each of the following exponential forms in both fraction form and decimal form.

a. 4^{-3} $\frac{1}{64} = 0.015625$

b. 7^0 1

c. 5^{-2} $\frac{1}{25} = 0.04$

d. 8^{-1} $\frac{1}{8} = 0.125$

EF-107. Exponent Practice Use what you know about exponents to rewrite each of the following expressions more simply.

a. $\frac{10^8}{10^6}$ 10^2

b. $\frac{n^5}{n^{13}}$ $\frac{1}{n^8}$, or n^{-8}

c. $\frac{3^5}{3^7}$ $\frac{1}{3^2}$, or 3^{-2}

d. $(x^4 \cdot 10^8) \div (x^5 \cdot 10^8)$ $\frac{1}{x}$, or x^{-1}

EF-108. Tina has been skipping every problem she thought might cause an argument between Keisha and Rudy, but now she is hopelessly behind. Help Tina catch up by writing her a letter to explain how to simplify fractions in which the numerator and denominator have the same base, but different exponents. Include examples such as

$$\frac{x^5}{x^3}, \frac{x^3}{x^5}, \text{ and } \frac{x^4}{x^4},$$

or make up your own examples.

4.6 FRUIT PUNCH: USING EQUIVALENT RATIOS TO GRAPH

SECTION MATERIALS

A POSTER-SIZE GRAPH (FOR EF-112)

EF-124 SUMS AND PRODUCTS: PART II RESOURCE PAGE

The main goal of this section is to tie together the numeric and the geometric views of ratios. To do this, after introducing equations of horizontal and vertical lines in problems EF-109 through EF-111, we have the students plot some straight lines through the origin (in the context of a punch recipe) and then observe that by dropping perpendicular lines to the x-axis, we can form similar right triangles. For this chapter, we use only lines with positive slopes; later we'll extend the work to include lines with negative slopes.

In EF-112 we suggest that you model parts a through c one step at a time using a poster-size graph. You can help students see the dimensions of the two triangles if you "pull them apart" by sketching each one separately, as shown in EF-88. (We have found it helpful to leave this large graph posted in the room for future reference.) Let the students finish the problem in their groups.

In EF-112f, you may need to guide the students back to the definition of similar triangles given in EF-63. Students should recognize that the two triangles are right triangles that share a common angle, as in EF-88a. There's no need for rigorous justifications here. There are several ways that students might "justify" the congruence of the third pair of angles, including tracing and comparing, cutting and comparing, measuring, and using the fact that the sum of the measures of the angles of a triangle is 180°. Let students come up with their own methods.

EF-109. Make a table of values and graph each of the following lines on the same coordinate system. Label each line with its equation. For example, a table you could use for the line in part a is

x	-2	-1	0	1	2
y	4	4	4	4	4

since y must be equal to 4, and there is no restriction given for x.

a. $y = 4$ *horizontal line through (0, 4)*

b. $x = -2$ *vertical line through (−2, 0)*

c. $y = 0$ *x-axis*

d. $x = 0$ *y-axis*

EF-110. *Without graphing,* write a sentence to describe each of the following lines so that someone who is absent today could graph the lines from your descriptions.

a. $y = 6$ horizontal line through (0, 6) and (n, 6)

b. $x = 3$ vertical line through (3, 0) and (3, n)

EF-111. Use what you learned in EF-109 and EF-110 to complete this problem about horizontal and vertical lines.

a. Marci graphed an equation and got a vertical line that contained the point (1, 5). Assuming that the graph was correct, what equation did Marci graph? $x = 1$

b. When Marc graphed his equation, he got a horizontal line that contained the point (1, 5). Assuming that his graph was correct, what was Marc's equation? $y = 5$

EF-112. Fruit Punch Lori's favorite punch recipe uses 3 cups of sugar for 5 gallons of punch. She wants to make 7 gallons of punch for her daughter's gymnastics club party. She knows she could double the recipe to make 10 gallons, but she doesn't want to make any extra punch. In this section, you will explore how geometric and algebraic ratios are related, and look at several ways to solve the problem of how much sugar to add.

a. What is the sugar to punch ratio for Lori's recipe? Write this fraction as a decimal. What does this tell you about the recipe?   *3 cups to 5 gallons; 0.6 cup of sugar for each gallon of punch*

b. Make a graph of the line that represents this situation with "gallons of punch" on the horizontal axis. Explain why the points (0, 0) and (10, 6) lie on this line.   *(0, 0) means no punch, no sugar; (10, 6) represents doubling the recipe.*

c. Form a right triangle by drawing the vertical line $x = 10$. One side of the triangle is on the *x*-axis. Label the lengths of the vertical and horizontal sides of the triangle.   *The vertical side has length 6; the horizontal side has length 10.*

d. Form a second right triangle within the first one by drawing the vertical line $x = 7$. Label the length of the horizontal side of the triangle. Since you don't know the length of the vertical side, label it *y*. What does *y* represent in the context of this problem?   *number of cups of sugar needed for 7 gal of punch*

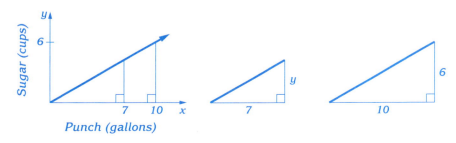

258 Chapter 4 Estimating Fish Populations: Numeric, Geometric, and Algebraic Ratios

e. Use your graph to estimate how much sugar is needed for 7 gallons of punch. ≈ **4 cups; actual answer is 4.2 cups.**

f. To get a more accurate answer, you can use an algebraic approach. First, explain why the two triangles from parts b and c are similar. Then, write an equation using equivalent ratios to find an exact value for y.
They're similar because they share an angle and both are right triangles; $\frac{y}{7} = \frac{6}{10}$; y = 4.2 cups

EF-113. More Fruit Punch

a. Explain how you could use your graph from EF-112 to estimate the amount of sugar needed for 18 gallons of Lori's punch. What estimate do you get using your graph? **Start at 18 on the x-axis, draw a vertical line up to the graph, then a horizontal line over to the y-axis. About 11 cups**

b. Write an equation using equivalent ratios and solve the problem in part a algebraically. $\frac{y}{18} = \frac{6}{10}$ or $\frac{y}{18} = \frac{3}{5}$; **10.8 cups**

Circulate around the room to check students' graphs—when the vertical distance is given and the horizontal is unknown, as in EF-114, students may have trouble seeing how to draw the two similar triangles.

EF-114. Marta has 8.5 cups of sugar in the pantry. If she uses it all in the same sugar-to-punch ratio as in Lori's recipe, how much punch can she make?

a. Use your graph to estimate the number of gallons of punch. **A little more than 14 gal**

b. Write an equation using equivalent ratios and solve the problem in part a algebraically. $\frac{8.5}{x} = \frac{3}{5}$; ≈ **14.17 gal**

In parts a and b of EF-115, students use the problem-solving techniques of making a table and looking for a pattern to develop the equation

$$\text{\# gallons used} = \frac{\text{\# miles driven}}{\text{rate of gasoline usage}}.$$

EF-115. C.J.'s car gets 20 miles per gallon of gas.

a. Copy and complete the following table of data for C.J.'s car:

# Miles Traveled	# Gallons of Gas Used
20	1
100	5
40	2
10	0.5
0	0
m	$\frac{m}{20}$

b. Use the table you made in part a to write an equation relating g, the number of gallons used, to m, the number of miles driven. A good way to check your equation is to answer the question: Does your equation make sense for $m = 20$ miles? $g = \dfrac{m}{20}$

c. Graph your equation with m, the number of miles, on the horizontal axis, and g, the number of gallons, on the vertical axis. Label both axes. Scale the horizontal axis so one unit represents 10 miles. Scale the vertical axis so one unit represents 1 gallon.

d. Use your graph to estimate the value for g when $m = 72$.   *between 3 and 4 gal*

e. Now form two similar triangles by drawing the vertical lines $m = 20$ and $m = 72$. Explain why these triangles are similar, and label the lengths of the horizontal and vertical sides of each triangle. (You will have to use a variable for one of the lengths.)

f. Use your triangles to write an equation using equivalent ratios, and solve it to find out how much gas C.J. used in driving 72 miles. Compare your solution to your estimate in part d.   *3.6 gal*

g. What is the gallons-to-miles ratio for C.J.'s car? Write this ratio as a decimal. What does this tell you about the car?   *1 gal to 20 mi; the car uses 0.05 gal of gas for each mile driven.*

EXTENSION AND PRACTICE

EF-116. Identify a variable, then write an equation and use it to solve the following problem.

Kim noticed that 100 vitamins cost $1.89.

a. At this rate, how much should 350 vitamins cost? $\dfrac{100}{1.89} = \dfrac{350}{x}$; *$6.62*

b. Write a ratio to represent the cost per vitamin. $\dfrac{\$1.89}{100} \approx$ *1.9¢ per vitamin*

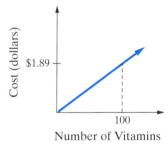

EF-117. Identify a variable, then write an equation and use it to solve the following problem:

Joe came to bat 464 times in 131 games. At this rate, how many times may he expect to come to bat in a full season of 162 games? $\dfrac{464}{131} = \dfrac{x}{162}$; *573 times*

EF-118. In a city of 3 million, 3472 people were surveyed. Of those surveyed, 28 of them admitted that they watched *Gilligan's Island*.

a. What fraction of the people surveyed admitted watching the program? $\frac{28}{3472}$

b. What percentage of the people surveyed admitted watching the program? ≈ 0.8%

c. If the survey did represent the city's TV viewing habits, how many people in the city would admit watching the program? ≈ 24,194, or ≈ 24,000 if students round to 0.8%

EF-119. A 5-inch by 7-inch photograph is enlarged so its width is 17.5 inches.

a. What is the length of the enlarged photograph? (Note that the width of the original photo is 5 inches.) 24.5 in.

b. What is the enlargement ratio? $\frac{17.5}{5} = \frac{3.5}{1} = \frac{7}{2}$

EF-120. James is making a skateboard ramp with two vertical supports. He cannot find a saw to cut the center support, but he did find a board 4 feet long. How far from the end should he put the 4-foot board so that the ramp remains flat? $x = 15.75$ ft

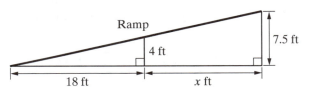

EF-121. In this problem you will examine what happens when you perform the same slide repeatedly.

a. Describe a slide that could be used to move a point from $(-7, -5)$ to $(-4, -3)$. *Slide up 2 units, and then to the right 3 units*

b. On a piece of graph paper, plot the points $(-7, -5)$ and $(-4, -3)$. Starting at the point $(-4, -3)$, perform the slide you found in part a. What are the coordinates of the stopping point? $(-1, -1)$

c. Perform the slide three more times, plotting your stopping point each time, and starting the next slide from the previous stopping point. What do you notice about the six points you have plotted? *They lie on a straight line.*

EF-122. Find the area of the shaded region in each of the following figures. Identify the subproblems you use.

a. 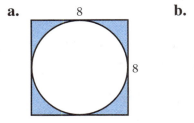   b.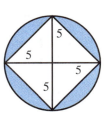

$64 - 16\pi \approx 13.73$ sq units $25\pi - 50 \approx 28.54$ sq units

Problems EF-123 through EF-125 are part of the series of problems in which students connect generic rectangles, factoring quadratic polynomials with a leading coefficient of 1, and Diamond Problems. It is helpful for students briefly to share their results for EF-124b and c with the whole class before they do EF-134, which is part the Extension and Practice for Section 4.7.

EF-123. Use a generic rectangle to factor each of the following quadratic polynomials.

a. $x^2 - 5x - 6$ *(x − 6)(x + 1)*

b. $x^2 + 11x + 30$ *(x + 5)(x + 6)*

c. $y^2 + 2y - 35$ *(y − 5)(y + 7)*

EF-124. Sums and Products: Part II

a. Use your Resource Page for EF-124 ("Sums and Products: Part II") to record integer pairs whose sums and products are indicated in the following table. For each sum-product pair, find two integers whose sum is the number in the left-hand column and whose product is the number in the right-hand column, as you did in EF-96.

Sum	Product	Integer Pair	Sum	Product	Integer Pair
15	56	*7, 8*	5	0	*5, 0*
−3	−28	*−7, 4*	−9	18	*−6, −3*
5	−6	*−1, 6*	−5	−6	*−6, 1*
2	−35	*−5, 7*	−2	−35	*−7, 5*
1	−72	*−8, 9*	11	30	*5, 6*
0	−16	*−4, 4*	−1	−12	*−4, 3*
3	−10	*−2, 5*	0	−25	*−5, 5*

b. Cindy says the entries in the Sums and Products table in part a are just Diamond Problems in disguise. Explain what she means.   *Diamond Problems are just sums and products in a different arrangement.*

c. We can use a line in the Sums and Products table to write a quadratic polynomial and its factors, such as $x^2 - 5x - 6 = (x - 6)(x + 1)$. Find which line in the Sums and Products table above is related to each of the quadratic polynomials in problem EF-123. Circle the related lines in your Sums and Products table. On your Resource Page, complete a table like the one started below, and then add two additional examples to your table.

	Sum	Product	Integer Pair	Polynomial area as a sum	Factored Form = area as a product
Example:	−5	−6	−6, 1	$x^2 - 5x - 6$	$= (x - 6)(x + 1)$

EF-125. Copy and solve these Diamond Problems and write a quadratic polynomial and its factored form for each as in the example.

Example: For this Diamond Problem, West and East must be 5 and 6, since $5 \cdot 6 = 30$, and $5 + 6 = 11$.

So the corresponding quadratic polynomial and its factored form are $x^2 + 11x + 30 = (x + 5)(x + 6)$. ▲

a.   b.   c.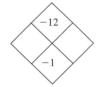

(a) $-2, 5$; $x^2 + 3x - 10 = (x - 2)(x + 5)$;
(b) $7, -5$; $x^2 + 2x - 35 = (x - 5)(x + 7)$;
(c) $-4, 3$; $x^2 - x - 12 = (x - 4)(x + 3)$

EF-126. Solve each of the following equations.

a. $\dfrac{x}{3} = \dfrac{5}{7}$ $\dfrac{15}{7}$ b. $\dfrac{5}{y} = \dfrac{2}{y+3}$ -5 c. $\dfrac{1}{x} = \dfrac{5}{x+1}$ $\dfrac{1}{4}$

d. Draw two similar triangles for which the equation in part a can be used to find a side of length x.   *One possibility: two similar right triangles, one with legs 5 and 7 units, and the other with legs x and 3 units.*

EF-127. The moon is about $3.8 \cdot 10^5$ kilometers from the earth. If light travels at about 300,000 kilometers per second, about how long does it take light reflected from the moon to reach Earth?   *about 1.27 sec*

EF-128. Melissa knows that 4^{-3} means $\dfrac{1}{4 \cdot 4 \cdot 4}$, or $\dfrac{1}{64}$.

a. Explain to her how to rewrite x^{-3} as a fraction.
x^{-3} means $\dfrac{1}{x \cdot x \cdot x}$, or $\dfrac{1}{x^3}$.

b. Rewrite $(xy)^{-2}$ as a fraction. $(xy)^{-2}$ means $\dfrac{1}{xy \cdot xy}$, or $\dfrac{1}{x^2y^2}$

EF-129. Scientists use beakers of many shapes for holding and measuring liquids. Imagine that you add water at a steady rate to each of the beakers illustrated below. As water is added the height of the water in the beaker will rise. The graph shows how the height of the water increases as water is added at a steady rate to beaker A.

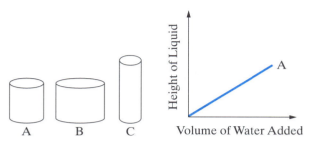

a. Copy the beaker shapes and the graph for A. Sketch lines to show how beakers B and C fill up as water is added. Label each line with the letter of the corresponding beaker.

b. Explain why you drew the lines where you did.

Solutions: (a)

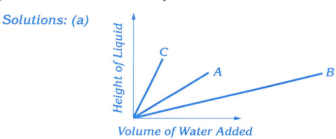

(b) The wider the cylinder, the slower the water will rise. So graph B rises more slowly than A, and C goes up faster.

4.7 ESTIMATING FISH POPULATIONS: A SIMULATION

SECTION MATERIALS

FOR EF-130: ABOUT 5 POUNDS OF DRIED BEANS (TO REPRESENT FISH)

ABOUT 1 POUND OF THE SAME SIZE BEANS AS ABOVE BUT IN A CONTRASTING COLOR

3-OZ PLASTIC CUPS (TO USE FOR NETS), STRONG LUNCH-SIZE PAPER SACKS (TO USE FOR LAKES)

EF-130 ESTIMATING FISH POPULATIONS RESOURCE PAGE

CHART FOR RECORDING CLASS DATA

EF-134 SUMS AND PRODUCTS: PART III RESOURCE PAGE

CHAPTER SUMMARY DRAFT RESOURCE PAGE (OPTIONAL)

About the Materials: The class chart for EF-130 should look like this:

Group	1	2	3	4	5	6	7	8	9
Guess (before sampling)									
Final population estimate (after all samples are completed)									
Total cost of sampling									
Actual population									
Error									
Percent error									

To prepare for the Estimating Fish Populations problem (EF-130), you'll need to stock each "lake" with "fish": fill each of the paper sacks (lakes) with about one-third to one-half pound of one color of beans such as pinto, kidney, or pink beans (about 400 to 600 "fish" per lake works well.) Each group also needs a small supply (a little more than one 3-oz plastic cup full) of beans of the same size but in a contrasting color—the "tagged fish"—in a resealable plastic bag, and a 3-oz plastic cup (net).

Before starting the class work for this section, it would be helpful to have students briefly share their observations for EF-124b and c with the whole class. This will make sure they're on track before they do EF-134, which is part of the Extension and Practice for this section. In addition to EF-134 and EF-142 Chapter 4 Summary Rough Draft, select some problems from EF-131 through EF-141 for practice and review.

About EF-130: *The Estimating Fish Populations investigation is a simulation of a procedure used by marine biologists to estimate the number of fish in a lake at a given time. We suggest that you first read the student version of the investigation (EF-130) before reading the hints and suggestions given here.*

Introduce the investigation by having a volunteer read the introduction of the problem to the class. Before students tag any fish or take any samples, have each group make a "30-second guess" as to how many fish are in their lake, and record the guess in the first line in the class chart. (We want students to see that the goal of the problem is to estimate the number of fish in the lake, but not to spend too much time making up their own ways of estimating.)

We have found that the simulation goes more smoothly if we discuss with the class how the catch-and-release method works before turning groups loose to tag their fish. Some instructors find it motivating to make a competition with "cash" prizes for the groups with the most accurate estimations—providing an incentive to make the most accurate estimate but at the same time to be cautious about spending money to take extra samples. As you circulate, you will probably need to check to see that groups understand the difference between tagging the fish and taking a sample, and that they are sharing the counting work efficiently and writing their ratio equations correctly.

To start the simulation, each group will net a sample of fish (beans) from their lake and then replace them with the other color of beans to represent the tagged fish. (Alternatively, rather than replacing netted fish with beans of a contrasting color, the students could "tag" the netted beans on both sides with marking pens. This process is slower and messier than the replacement process, however.)

Once the tagged fish are back in the lake, students should shake the sack a little to disperse them among the other fish and then use the net to draw a new sample. The students will count both kinds of fish in the sample and use the information along with the original number of tagged fish to estimate the total number of fish in the lake.

If students need help with setting up the ratios in part f, prompt their memory with a reference to earlier ratio exercises when they compared $\frac{part}{whole}$. In this case, to make an estimate for the total number of fish in the lake (the "unknown" in this problem), they'll compare

$$\frac{\text{\# tagged fish in sample}}{\text{total \# fish in sample}} \text{ to } \frac{\text{total \# tagged fish}}{\text{total \# fish in lake}}.$$

The estimates for the total number of fish will be more accurate if students net sufficiently many fish in each sample and if all samples are about the same size. We've found that a sample size of three-fourths to one full 3-oz cup of beans works well. After all the data have been recorded on the class chart, we like to have each group give a brief report on their procedure. We have the class consider reasons for the various degrees of accuracy in the estimates by asking them questions such as, "What factors might have affected the sampling?" and "How accurate do you think we can expect this procedure to be?" If there's time we have the class compute and discuss each group's relative (percent) error.

Before they leave class, have groups remove and bag all tagged fish from their lakes. (If beans were marked to indicate tagging, you could leave them in the lakes and use a different color of pen to tag the fish the next time one of your classes does the simulation.)

EF-130. Estimating Fish Populations This investigation is a simulation of a process used by marine biologists to mathematically estimate the number of fish in a lake.

Biologists net some fish, count them and tag each one, and then return the tagged fish to the lake. After allowing enough time for the tagged fish to disperse among the other fish in the lake, the biologists net another sample of fish. In the new sample, they note the ratio of the number of tagged fish to the number of fish netted, and use this ratio to estimate the total fish population of the lake.

Your group, acting as a team of research biologists, will have a lake (paper sack) full of fish (beans), and a net (a small cup). Like marine biologists, you will be trying to estimate the number of fish in your lake by netting samples and using ratios. You will record your estimates and calculations on a copy of the EF-130 Estimating Fish Populations Resource Page.

a. Guess Before you start your mathematical investigation, *guess* how many fish are in your lake. Record your guess on the class chart to compare later with the actual number.

b. Tag Some Fish In your first outing, you need to tag some fish. Use your "net" to take out a cupful of fish in order to "tag" them.

c. Count the number of fish you netted in your initial outing. To "tag" the netted fish, replace each one with a bean of a different color. Record the number of tagged fish as the "total number of tagged fish."

d. Put all the tagged fish back into the lake. (Put the beans they replaced in a "holding pond.") Be careful not to let any of the fish jump out onto the floor. Gently shake the bag to thoroughly mix all the fish in the lake. Try not to bruise them.

e. **First Sample (to Make an Estimate)** Use your net to take out another cupful of fish. Count the number of tagged fish in the sample and record this number on the Resource Page as "number of tagged fish in sample." Also, count and record the "total number of fish in sample."

f. You now have three pieces of information: the *total number of tagged fish in the lake*, the *number of tagged fish in the sample*, and the *total number of fish in the sample*. Use this information to write two equivalent ratios. Use the two equivalent ratios to write an equation, and then solve your equation to find a mathematical estimate of the total number of fish in the lake.

g. **Second Sample (to Make another Estimate)** Return your sample of fish to the lake, gently mix the fish, and take another sample. Repeat the counting procedure of part e, and the use of ratios in part f, to get another estimate of the lake's fish population.

h. It is important to get an accurate count of the fish population, but each time you take the boat out to net fish, taxpayers must pay for your time and equipment. Assume that it costs $1000 to net and tag the fish initially, and each time you net and count a sample it costs an additional $800. Therefore, your first three outings have cost $2600. If you feel your estimate at this point is accurate, record it on the class chart with your cost. If you think you should try another sample for better accuracy, do the same steps as before. Draw as many samples as you feel you need, but remember, each sampling costs $800.

i. After your data are recorded on the class chart, count the fish in your lake to find the actual population. (Biologists can't actually do this!) Record the actual population on the class chart.

j. How close was your estimate? Calculate the percentage error. What might have thrown your estimate off? What do you think of this method for estimating fish populations?

Be sure each group member records all the group's data, calculations, and conclusions.

EXTENSION AND PRACTICE

EF-131. At the Movies It is now 7:51 P.M. The movie that you've waited three weeks to see starts at 8:15. Standing in line for the movie, you count 146 people ahead of you. Nine people can buy their tickets in 70 seconds.

To determine whether you will be buying your ticket before the movie starts, write an equation using equivalent ratios and solve it. *Yes; you'll buy your ticket in about 19 minutes, or at 8:10 P.M. If x = number of minutes until you buy your ticket, an equation is $\frac{9}{70} = \frac{146}{x}$. Another way to look at it: there is enough time, since 185 people buy tickets in the 24 min. This is found by letting n = number of people buying tickets and solving the equation $\frac{9}{70} = \frac{n}{24 \cdot 60 \text{ sec}}$.*

EF-132. A 70-foot length of wire weighs 23 pounds.

a. How much does zero feet of wire weigh? *0 lb*

b. Draw a graph showing the weight of the wire (vertical axis) compared with the length of the wire (horizontal axis). Label the axes with appropriate units.

c. Use your graph to *estimate* the weight of a 50-foot length of wire. Then, to find a more accurate answer, write and solve an equation using equivalent ratios to find the weight of a 50-foot length of wire. *a possible equation: $\frac{23 \text{ lb}}{70 \text{ ft}} = \frac{x \text{ lb}}{50 \text{ ft}}$; ≈ 16.43 lb*

d. Use your graph to estimate the length of 15 pounds of wire. Then, to find a more accurate answer, write and solve an equation using equivalent ratios to find the length of a piece of wire that weighs 15 pounds. *a possible equation: $\frac{70 \text{ ft}}{23 \text{ lb}} = \frac{x \text{ ft}}{15 \text{ lb}}$; ≈ 45.65 ft*

e. How close were your estimates to the actual values you found?

f. Write the weight-to-length ratio for a piece of wire, and then use your calculator to write it as a decimal. What does this ratio tell you about the wire? *$\frac{23}{70}$ ≈ 0.33 lb per foot of wire*

EF-133. Write an equation that represents the following problem, and then solve it:

Admission to the football homecoming dance was $3 in advance and $4 at the door. There were 30 fewer tickets sold in advance than at the door, and the ticket sales totaled $1590. How many of each kind of ticket were sold? *3a + 4(a + 30) = 1590; or 3(d − 30) + 4d = 1590; 210 in advance and 240 at the door*

Problem EF-134 is part of the series of problems in which students connect Diamond Problems, generic rectangles, and factoring quadratic polynomials with a leading coefficient of 1. The students will summarize their work in EF-148.

EF-134. Sums and Products: Part III

a. Use your Resource Page for EF-134 ("Sums and Products: Part III") to record integer pairs whose sums and products are indicated in the table. For each sum-product pair, find two integers whose sum is the number in the left-hand column and whose product is the number in the right-hand column, as you did in problem EF-96.

Note: There may be some sum-product pairs for which no pair of integers works.

Sum	Product	Integer Pair	Sum	Product	Integer Pair
−11	28	−4, −7	2	−120	12, −10
14	0	0, 14	18	−63	−3, 21
0	−100	10, −10	−10	−25	none
0	−32	none	−7	−30	3, −10
23	−24	−1, 24	−1	−42	−7, 6
10	−24	−2, 12	−12	−13	−13, 1
20	99	11, 9	13	−30	15, −2

    b. We can use a line in the sum-product table to write a quadratic polynomial and its factored form. For example,

$$x^2 - 11x + 28 = (x - 4)(x - 7).$$

Give three more examples.
    There are many possibilities: $x^2 + 14x = x(x + 14)$;
    $x^2 - 100 = (x + 10)(x - 10)$; $x^2 + 23x - 24 = (x - 1)(x + 24)$

    c. Use the table to write a quadratic polynomial that has *no* factors.
    $x^2 + 0x - 32$ or $x^2 - 10x - 25$

EF-135. Use a generic rectangle to factor each of the following quadratic polynomials.

    a. $x^2 - x - 30$ b. $x^2 - 100$ c. $3m^2 + 16m + 5$
    $(x + 5)(x - 6)$ *$(x - 10)(x + 10)$*   *$(3m + 1)(m + 5)$*

    d. $y^2 + 2y - 80$ e. $z^2 + 10z - 24$
    $(y - 8)(y + 10)$ *$(z - 2)(z + 12)$*

EF-136. Solve each of the following equations. If possible, use a fraction buster.

    a. $\frac{w}{3} = \frac{w+1}{7}$ *$\frac{3}{4}$*   b. $\frac{3}{z} = \frac{4}{5}$ *$\frac{15}{4}$*

    c. $\frac{3}{y+1} = \frac{8}{y}$ *$\frac{-8}{5}$*   d. $(3m + 2)(m - 17) = 0$ *$\frac{-2}{3}$, 17*

    e. $\frac{4}{5}x - 3(x - 6) = 6 - x$ *10*   f. $\frac{2x-5}{3} + 1 = \frac{x}{4} + \frac{x+2}{3}$ *16*

EF-137. Solve the following problem by writing an equation and solving it.

The Second Time Around Theatre shows old movies. Student tickets sell for $4.00, while general public tickets are $6.50. Last Saturday night's show sold out, and the 588-seat theater was full. If the proceeds from the student tickets matched the sales of the general public tickets, how many of each type of ticket was sold?   *4.00x = 6.50(588 − x); 364 student tickets and 224 general public tickets*

EF-138. Solve each of the following problems by writing an equation and then solving it:

a. A 23-ounce box of raisin bran costs $2.89. At that rate, what should a 2-pound, 5-ounce box cost? $\frac{2.89}{23} = \frac{x}{37}$; *$4.65 for the larger box.*

b. A biscuit recipe calls for $\frac{1}{2}$ teaspoon of baking powder for $\frac{3}{4}$ cup flour. How much baking powder is needed for 3 cups of flour? $\frac{0.5}{0.75} = \frac{x}{3}$; *2 teaspoons*

EF-139. Use what you know about exponents to rewrite each of the following expressions:

a. $3x^2 \cdot x$ $3x^3$

b. $\dfrac{n^{12}}{n^3}$ n^9

c. $(x^3)^2$ x^6

d. $(-2x^2)(-2x)$ $4x^3$

e. $\dfrac{-8x^6y^2}{-4xy}$ $2x^5y$

f. $(2x^3)^3$ $8x^9$

EF-140. L.J.'s car has a gas tank that holds 19 gallons.

a. If L.J. used 8 gallons to drive 200 miles, does the car have enough gas to go another 250 miles?   *Yes, it can go 275 mi.*

b. What assumptions did you make in your solution to part a?   *The car uses gas at a constant rate, and the gas tank was full.*

c. At what rate, in gallons per mile, does L.J.'s car use gas? $\frac{1}{25}$ *gallon per mile, or 0.04 gal/mi*

EF-141. Jermaine is really hungry, but he is also on a budget. He wants to buy a lot of pizza but is also trying to get a good deal. At Pietro's Authentic Italian Pizza, he could order a 12-inch pepperoni pizza for $8.99, or he could order the Special of the Week—half of a 16-inch pepperoni pizza for $8.49.

A 12-inch pizza   Half of a 16-inch pizza

a. Which pizza is bigger, the whole 12-inch pizza or the Special of the Week? Find the exact area of each pizza to help Jermaine decide. *The area of the Special is 32π sq in. The 12-in. whole pizza, with area 36π square inches, is larger.*

b. How much bigger is the larger of the two orders? Find the difference in areas, and then on your paper sketch and label a rectangle that would have the same approximate area as the extra piece. *The difference is 4π, or about 12.6, square inches. One possible rectangle would be 3.14-in. by 4-in.*

c. Remember, Jermaine is on a budget. How much would it cost for 1 square inch of the 12-inch pizza? How much would it cost for 1 square inch of the 16-inch half pizza? *The 12-in. pizza costs $\frac{\$8.99}{36\pi} \approx \$0.079$, about \$0.08, per square inch. The 16-in. half pizza costs $\frac{\$8.49}{32\pi} \approx \$0.084$, also about \$0.08, per square inch.*

d. Which pizza should Jermaine order? *The 12-in. pizza—it's slightly cheaper (about half a cent) per square inch, and he'll get more.*

EF-142. Chapter 4 Summary Rough Draft In this chapter you've worked with one of the most fundamental and widely used concepts in mathematics, the concept of the "ratio" of two quantities. You explored ratios from a *geometric* perspective by enlarging and reducing dot-paper figures and by working with similar triangles. You worked with ratios from an *algebraic* perspective by writing equations using equivalent ratios. In simplifying fractions by finding factors of the numerator and the denominator to make forms that are equivalent to 1, and by writing ratios as percent, you strengthened your familiarity with *numeric* ratios. All three perspectives were tied together when you modeled a situation with a graph and then used the graph to write an equation to solve a problem.

This chapter also introduced one of the most frequently used problem-solving strategies in mathematics, namely the process of identifying, writing, and solving subproblems. Throughout the remainder of the course, you'll have opportunities to use the important ideas introduced in this chapter.

These are the main ideas of Chapter 4:

- The enlargement and reduction of shapes involves geometric ratios. The ratios of corresponding sides, the ratios of perimeters, and the ratios of areas of a figure and it enlargement (or reduction) are related. The ratios of corresponding sides of similar right triangles are equal.

- We can use the strategy of identifying, writing, and solving sub-problems to solve a larger problem in several contexts, including areas of circles.

- We can use a generic rectangle to rewrite a polynomial as a product of two factors.

- We can represent a ratio problem with the graph of a line passing through the origin, and we can use a ratio to represent the rate at which a quantity changes.

- We can use equivalent ratios to write equations to solve problems, including percent problems.

- We can use a fraction buster to convert an equation with fractions into an equivalent equation without fractions.

a. Make photocopies of the Chapter Summary Draft Resource Page (or follow the same format on your own paper). Use a page to complete the following steps for each of the six main ideas of the chapter:

- State the main idea.

- Select and recopy a problem that is a good example of the idea and that you did well in.

- Include a completely worked-out solution to the selected problem.

- Write one or two sentences that describe what you learned about the idea.

As in previous chapters, you will have the opportunity to revise your work, so your focus should be on the content of your summary rather than its appearance. Be ready to discuss your responses with your group at the next class meeting.

b. If you had to choose a favorite problem from the chapter, which would it be? Why?

c. Find a problem that you still cannot solve or that you are unsure you would be able to solve on a test. Write out the question and as much of the solution as you can. Then explain what you think is keeping you from solving the problem. Be clear and precise in describing the hard part.

4.8 SUMMARY AND REVIEW

SECTION MATERIALS
DOT PAPER (COPIED FROM THE DOT PAPER TEMPLATE RESOURCE PAGE, FOR EF-151)

Problems EF-143 and EF-166 complete the chapter summary. Select some problems from EF-150 through EF-165 for review.

EF-143. Chapter 4 Summary Group Discussion Take out the rough-draft summary you completed in EF-142. Use this class time to discuss your work; use homework time to revise.

For each of the six main ideas of the chapter, choose one member of the group to lead a short discussion. When it is your turn to be the discussion leader, you should do the following:

- Explain the problem you chose to illustrate your main idea.
- Explain why you chose that particular problem.
- Explain, as far as possible, the problem you wrote about in EF-142c that you still cannot solve.

This is also your chance to make sure your summary is complete and to work together on problems you may not be able to solve yet.

EF-144. For each of the following problems, write an equation and use it to solve the problem.

a. The ratio of gas to oil for Alonzo's motorcycle should be 16 to 1. He just put 56 ounces of gasoline in his tank. Assuming that what was already in the tank was in the right ratio, how many ounces of oil should Alonzo add? $\frac{gas}{oil} = \frac{16}{1} = \frac{56}{x}$; 3.5 oz

b. On another occasion, Alonzo's girlfriend Tina had 36 ounces of gas and 0.5 ounce of oil in her motorcycle. How much oil does she need to add to make the ratio of gas to oil 16 to 1 ? $\frac{gas}{oil} = \frac{16}{1} = \frac{36}{0.5 + x}$; 1.75 oz

For students who are having trouble getting started in EF-145, suggest that drawing a picture and identifying subproblems might help.

EF-145. Suppose a right triangle with sides of lengths 5, 12, and 13 centimeters is similar to a right triangle whose shortest side is 15 centimeters long.

a. What is the perimeter of the larger triangle? *90 cm*

b. What is the ratio of the perimeter of the smaller triangle to the perimeter of the larger triangle? *1 to 3*

c. How does the ratio in part b above compare with the ratio of the lengths of corresponding sides of the triangles? *They're the same.*

d. What is the ratio of the area of the smaller triangle to the area of the larger triangle? *1 to 9*

EF-146. Carrie Ann's car gets 20 miles per gallon. For each question below, use equivalent ratios to write an equation, then solve it.

a. How far will Carrie Ann's car go on 8 gallons of gas? *160 mi*

b. If Carrie Ann drives 80 miles, how much gas will be used?
$\frac{g}{80} = \frac{1}{20}$; *4 gal*

c. If she drives 118 miles, how much gas will be used? $\frac{g}{118} = \frac{1}{20}$; *5.9 gal*

EF-147. Mario's car needs 12 gallons of gas to go 320 miles.

a. How much gas is needed to go 0 miles? *0 gallons*

b. Draw a graph that shows g, the number of gallons of gas used (vertical axis), compared with m, the number of miles driven (horizontal axis).
The graph is a line from (0, 0) through (12, 320).

c. From your graph, *estimate* how much gas will be needed to go 70 miles.
about 2.5 gal

d. From your graph, *estimate* how many miles can be driven with 10 gallons of gas. *about 270 mi*

e. Use equivalent ratios to write equations to find the exact answers for parts c and d. Solve your equations. (c) $\frac{12}{320} = \frac{g}{70}$; *2.625 gal*; (d) $\frac{12}{320} = \frac{10}{m}$; *266 $\frac{2}{3}$ mi*

EF-148. In the series of Sums and Products problems (EF-95 through EF-97, EF-123 through EF-125, and EF-134) you saw how Diamond Problems, generic rectangles, and factoring quadratic polynomials can be related.

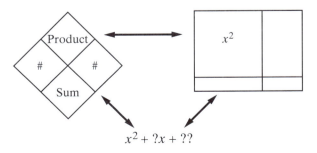

To illustrate the relationships you observed, follow these instructions as directed in parts a through d:

   i. Draw and complete a Diamond Problem.

   ii. Draw and label a generic rectangle.

   iii. Write and factor a quadratic polynomial.

a. Do tasks i and ii for the equation
$$x^2 - 19x + 48 = (x - 3)(x - 16).$$

b. Do tasks ii and iii for this Diamond Problem:

(ii) Upper left x^2, lower right 24, lower left $-3x$, upper right $-8x$ or lower left $-8x$, upper right $-3x$; (iii) $(x - 3)(x - 8) = x^2 - 11x + 24$ or $(x - 8)(x - 3) = x^2 - 11x + 24$

c. Do tasks i and iii for this generic rectangle:

(i) N = 42, S = 17; either E = 3, W = 14 or E = 14, W = 3; (iii) $(x + 3)(x + 14) = x^2 + 17x + 42$

d. Do tasks i and ii for the equation
$$x^2 + (A + B)x + A \cdot B = (x + A)(x + B),$$

where A and B are integers. *(i) N = A·B, S = A + B; E = A and W = B, or E = B and W = A; (ii) Parts (clockwise from upper left) are x^2, Bx, A·B, and Ax.*

EF-149. Tool Kit Check In addition to your Tool Kit entries from earlier chapters, your Tool Kit should also include entries for at least the following topics:

- writing a percent as a ratio   *EF-9*
- breaking a problem into subproblems   *EF-3, 11*
- predicting perimeter and area ratios in enlarged or reduced figures   *EF-24*
- symbol for the length of a line segment   *EF-34*
- factoring a polynomial   *EF-43*
- finding the area and circumference of a circle   *EF-52*
- definition of "similar triangles"   *EF-63*
- using ratios of corresponding sides of similar triangles to write equations   *EF-65, 66*
- the "∠" symbol for denoting an angle   *EF-68*

- rewriting expressions with negative exponents **EF-77**
- definition of "right angle" and "right triangle" **EF-86**
- how a fraction buster can help you solve an equation **EF-90, 91**

You are not limited to the topics on this list. If there are other ideas you'd like help remembering, add them to your Tool Kit as well.

a. To be sure that your Tool Kit entries are clear and accurate, exchange Tool Kits with members of your group, and read each other's entries.

b. At home, make any necessary clarification, revisions, or updates to your Tool Kit.

EXTENSION AND PRACTICE

EF-150. A certain line contains the points (3, 1) and (6, 2). Find the coordinates of five other points with integer coordinates on the line. *..., (−3, −1), (0, 0), (9, 3), (12, 4), (15, 5), (18, 6), ...*

EF-151. Copy this figure onto dot paper:

a. Enlarge this figure on dot paper by making each side of the new figure twice as long as its corresponding side in the original figure.

b. Write the enlargement ratio of the corresponding sides. *2 to 1*

c. For the figures in part a, write and reduce these ratios

$$\frac{\text{length of } Side \text{ of } new \text{ figure}}{\text{length of } Side \text{ of } original \text{ figure}}$$

$$\frac{Perimeter \text{ of } new \text{ figure}}{Perimeter \text{ of } original \text{ figure}}, \text{ and } \frac{Area \text{ of } new \text{ figure}}{Area \text{ of } original \text{ figure}}.$$

$\frac{P_{new}}{P_{orig}} = \frac{40}{20} = \frac{2}{1}$, $\frac{A_{new}}{A_{orig}} = \frac{52}{13} = \frac{4}{1}$

EF-152. Find $|DE|$ and $|DF|$. *$|DE| = \frac{25}{9} \approx 2.78$, and $|DF| = \frac{20}{3} \approx 6.67$*

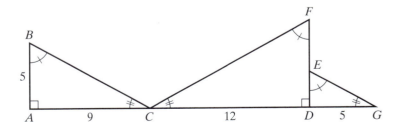

EF-153. Use a generic rectangle to factor each of the following quadratic polynomials.

a. $10x^2 + 14x$
   $2x(5x + 7)$

b. $m^2 - 9m + 18$
   $(m - 3)(m - 6)$

c. $y^2 - 16y + 64$
   $(y - 8)(y - 8)$

d. $2x^2 + x - 10$
   $(2x + 5)(x - 2)$

EF-154. Solve the following problem by writing an equation and solving it.

The Monochrome Movie House shows only black-and-white movies. Tickets for children under age 10 are only $2.50, while adult tickets cost $4.00. When the Monochrome showed a Charlie Chaplin film last Sunday, 144 tickets were sold. The revenue from adult tickets was twice that of the sale of children's tickets. How many of each type of ticket were sold? *4.00A = 2(2.50)(144 − A); 64 children's tickets and 80 adult tickets*

EF-155. A decorative window design uses four semicircles, each with a diameter of 22 inches, so that one semicircle is placed on each side of a square with a side length of 22 inches.

a. Sketch a picture of the window, and describe the subproblems that must be solved in order to compute the total area of the window. Then find the area of the window. *Find the area of a circle with radius 11 in., double it, and then add the area of the square; total area = 484 + 242π sq in. ≈ 1244 sq in.*

b. List the subproblems needed to calculate the perimeter of the window. Then find the perimeter. *Find the circumference of a circle with a 22-in. diameter and then double it; perimeter = 44π in. ≈ 138 in.*

c. If you decided to enlarge the window using an enlargement ratio of $\frac{3}{1}$, what would be the perimeter and the area of the enlarged window? *New perimeter = 3(44π) in. ≈ 415 in. New area = 9(484 + 242π) sq in. ≈ 11,198 sq in.*

EF-156. Two numbers are in a ratio of 3 to 7. Their sum is 700. Find the two numbers. *210 and 490*

EF-157. In a recent survey, seven out of 10 basketball players recommended Itch-Gone athlete's foot powder.

a. If 150 players were surveyed, how many recommended Itch-Gone foot powder? *105*

b. What is the probability that a player did not recommend the product? $\frac{3}{10}$

EF-158. Solve each of the following equations for x:

a. $\frac{x}{3} = \frac{x+4}{5}$ 6
b. $\frac{1}{x} = \frac{3}{x+5}$ $\frac{5}{2}$
c. $\frac{17}{x} = \frac{25}{100}$ 68
d. $\frac{2x+5}{3} + 3 = \frac{x}{5}$ -10

EF-159. For each of the following questions, write an equation and solve it.

a. What percent of 60 is 45? **75%**

b. What is 45% of 60? **27**

c. Sixty percent of what number is 45? **75**

EF-160. Write each of the following expressions in a simpler form:

a. $x^2y^3 \cdot x^3y^4$ x^5y^7
b. $(-3x^2) \cdot (4x^3)$ $-12x^5$
c. $(x^3)^4$ x^{12}
d. $(2x^2)^3$ $8x^6$
e. $\frac{6x^2y^3}{2xy}$ $3xy^2$
f. $(x^3y)^2(2x)^3$ $8x^9y^2$

EF-161. Copy and complete the following table by rewriting each of the following numbers in exponential form, fraction form, expanded form, and decimal form.

Exponential Form	Fraction Form	Expanded Form	Decimal (Standard) Form
3^{-1}	$\frac{1}{3^1}$	$\frac{1}{3}$	≈ 0.333
8^{-2}	$\frac{1}{8^2}$	$\frac{1}{8 \cdot 8}$	0.015625
10^{-2}	$\frac{1}{10^2}$	$\frac{1}{10 \cdot 10}$	0.01
5^{-3}	$\frac{1}{5^3}$	$\frac{1}{5 \cdot 5 \cdot 5}$	0.008
10^{-4}	$\frac{1}{10^4}$	$\frac{1}{10 \cdot 10 \cdot 10 \cdot 10}$	0.0001

Discuss why the results in EF-162c don't check.

EF-162. In this problem you'll examine the graph of $y = 2^x$.

a. Graph the equation $y = 2^x$ for $-3 \le x \le 3$. Scale the vertical axis so that each tick mark represents $\frac{1}{2}$ unit.

b. Using your graph, estimate to the nearest tenth the x-value for which the y-value is 5. $x \approx 2.3$

c. Use a calculator to check whether $5 = 2^x$ for the value of x you found in part b. Then use guess and check to see if you can find a value of x closer to the solution of $2^x = 5$. $2^{2.3} \approx 4.92; x \approx 2.322$

d. Using your graph, estimate to the nearest tenth a solution to the equation $0.75 = 2^x$.   *Hint: Find a value of x for which $y = 0.75$. $x \approx -0.4$*

e. Use a calculator to check whether $0.75 = 2^x$ for the value of x you found in part d. Then use guess and check to see if you can find a value of x closer to the solution of $2^x = 0.75$. $2^{-0.4} \approx 0.7579; x \approx 2.415$

EF-163. What is the area of the shaded region? Identify the subproblems you use.   *Shaded area = area of smaller rectangle + area larger rectangle − 2·area of overlap = 36 sq units*

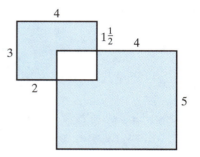

EF-164. A closet floor can be covered with 40 square tiles that are 9 inches on a side. If instead the floor is covered with square tiles that are 3 inches on a side, how many tiles would be needed?   *360*

EF-165. Here's another beaker problem. The graph shows how the height of the water increases as water is added at a steady rate to beaker A.

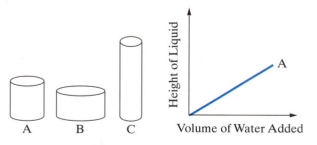

a. Copy the beaker shapes and the graph for A. Plot lines to show how beakers B and C fill up as water is added. Label each line.

b. Explain why you drew the lines where you did.

Solution:

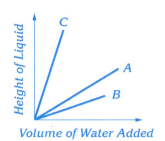

EF-166. Chapter 4 Summary Revision This is the final summary problem for Chapter 4. Using your rough draft from EF-142 and the ideas you discussed in your groups from EF-143, spend time revising and refining your Chapter 4 Summary. Your presentation should be thorough and organized, and should be done on a separate piece of paper.

CHAPTER FIVE—FOR THE INSTRUCTOR

MANAGING WATER RESOURCES
Graphs, Slope Ratios, and Linear Equations

MATERIALS

- Graph paper and straight-edges (needed throughout the chapter)
- Dot paper transparencies for the overhead projector (optional)
- Transparency of the WR-57 Resource Page (optional)
- Graphing calculators (one per pair of students) if using WR-103
- Resource Pages for WR-27, WR-28, WR-45, WR-57, and either WR-103 or WR-104; also, Dot Paper and Summary Draft Templates (optional)

In this chapter we revisit the concept of ratio to build a strong understanding of the slope of a line. Students see that slope is a ratio and relate slope computations to similar right triangles. Students also learn to interpret the slope as a rate of change in the context of modeling linear situations such as the Cricket Thermometers Problem.

We use kinesthetic exercises twice: first to reinforce the relationship between the numeric slope and the geometric steepness of a line, and again to reinforce the connection between the slope-intercept form of a line and the graph. We also recommend using graphing calculators to help students strengthen their understanding of the relationships between the slope-intercept form of an equation of a line and its graph.

Students learn to write an equation of a line by estimating the slope and the y-intercept. Do not expect mastery of writing equations at this point. (Students will learn in Chapter 6 to write equations of lines given either any point and a slope, or two points.)

By the end of this chapter, students will be able to describe a slide from one given point to another and then be able to give the coordinates of the next point if the slide is repeated. They'll be able to describe the shape of the graph formed if a slide were made repeatedly, and to use units to interpret the slope of a line as a rate of change in the context of a situation. They'll graph equations of the form $y = mx + b$ more efficiently, and given a linear graph, they'll estimate its y-intercept and slope, and use them to write an equation.

CHAPTER FIVE

MANAGING WATER RESOURCES
Graphs, Slope Ratios, and Linear Equations

In particularly wet years, water must be released from mountain river reservoirs to make room for water from the snowpack and spring rains. People managing the water resources must figure out the rates of water release over certain periods of time to prevent flooding and to assure enough water supplies for the dry season. It might be that water is released from a reservoir at a constant rate over a period of time. In this case, it would be important to know how long water could be released this way before the reservoir is drained.

You may have worked on a problem similar to this one in the Burning Candle Investigation from Chapter 2, where you estimated how long it would take for the candle to extinguish by collecting data and then graphing the data points. In the Sierra River Reservoir problem you'll use data points and what you learn about the graphs of linear relationships to examine the following water management situation:

> As Sierra River Reservoir approached its capacity level, water releases were begun. Water was released at a constant rate, and after 10 weeks, the water level measured at the spillway was 123.50 feet. After 3 more weeks of releasing water at that same rate, the reading was 116.75 feet. If possible, hydrologists would like to model this situation with an equation for use in making decisions.

In this chapter you will be extending your understanding of the concept of ratio to learn techniques for writing equations of lines from data or graphs. First you'll review the concept of ratio, and then you'll look at some applications of ratios.

IN THIS CHAPTER, IN ADDITION TO REVIEWING RATIOS, YOU WILL HAVE THE OPPORTUNITY TO:

- develop a good understanding of what the slope of a line represents and estimate the slopes of arbitrary lines;
- understand that slope is a ratio and relate computing slopes to similar right triangles;
- use units to interpret the slope as a rate of change in the context of a physical situation;
- sketch graphs using the slope-intercept form of a line;
- use a graph to write an approximate equation for a line given either a point and a slope, or two points.

Chapter 5 Managing Water Resources: Graphs, Slope Ratios, and Linear Equations

MATERIALS		CHAPTER CONTENTS	PROBLEM SETS
Graph paper and straightedges (needed throughout the chapter)	5.1	Ratios and the Slope of a Line	WR-1–WR-25
WR-27 and WR-28 Graph Comparisons Resource Page; WR-45 The Bike Race Resource Page	5.2	Graphing Linear Equations Using Two Points and a Slide	WR-26–WR-55
Dot paper and straightedges for WR-56; dot paper transparencies (optional); WR-57 Slopes of Segments Resource Page; transparency of WR-57 Slopes of Segments Resource Page (optional)	5.3	Slopes of Parallel Lines and the Slope-Intercept Pattern in $y = mx + b$	WR-56–WR-82
	5.4	Using Graphs to Investigate Linear Relationships	WR-83–WR-102
Either: WR-103 Patterns with Lines and Their Equations Resource Page and graphing calculators (one per pair of students); or WR-104 Patterns with Lines and Their Equations Resource Page	5.5	Practice with $y = mx + b$	WR-103–WR-121
Chapter Summary Draft Resource Page (optional, for WR-141)	5.6	Using Graphs to Write Equations of Lines	WR-122–WR-141
Poster paper or overhead transparencies (optional, for WR-164)	5.7	Summary and Review	WR-142–WR-164

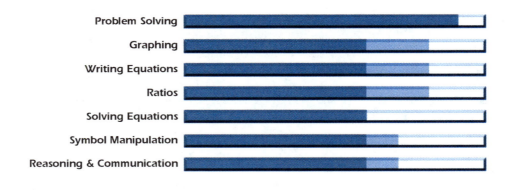

5.1 RATIOS AND THE SLOPE OF A LINE

SECTION MATERIALS
GRAPH PAPER (NEEDED THROUGHOUT THE CHAPTER)
STRAIGHTEDGES (NEEDED THROUGHOUT THE CHAPTER)

After groups have done problems WR-1 through WR-3, have them share their responses and methods with the entire class. Guide students to solve the problems without drawing the line but instead by using a slide; the goal is to have students state their intuitive understanding that points on a line can be determined by using ratios.

WR-1. A line contains the point $(3, -2)$ and the origin. Plot the two points on graph paper, but *do not draw in the line*. Use your picture to find four other points with integer coordinates that are on the line.
Some possibilities: $(6, -4)$, $(9, -6)$, $(-3, 2)$, $(-6, 4)$

WR-2. The two points $(3, 1)$ and $(4, 3)$ lie on a line. Describe a slide you could use to find five other points with integer coordinates on the line, and then find them! *Some possibilities: $(5, 5)$, $(6, 7)$, $(2, -1)$, $(1, -3)$*

WR-3. Imagine that a classmate who was absent today has called you for help. Given a line through two points—say $(0, 1)$ and $(-2, -3)$—she'd like to know how to find other points on the line that have integer coordinates. She also wants to be able to find these other points without using a straightedge. Use what your group learned in solving WR-1 and WR-2 to explain to your classmate a method for using two points on a line for determining more points on the line. *A slide can be used to find other points on the line.*

In problems WR-4 to WR-8, students extend the intuitive notions of slope as steepness, direction, and rate to a formal definition of slope by noticing that it can be written as a ratio from a graph.

WR-4. Fruit Punch Revisited In EF-112, you made a graph to represent Lori's favorite punch recipe. In this problem, you will use ratios to compare three recipes shown in the graph: Lori's, Elena's, and Phil's.

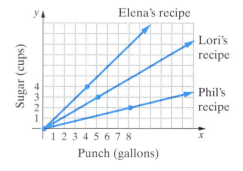

Lori's recipe uses 3 cups of sugar for every 5 gallons of punch. Elena's 4-gallon recipe uses 4 cups of sugar. Phil adds 2 cups of sugar to make 8 gallons of punch.

a. You already have seen that the points $(0, 0)$ and $(5, 3)$ lie on the graph for Lori's recipe. What does each of these points represent? Describe a slide you could use to find three other points on the Lori's line. *$(5, 3)$ represents that 5 gallons of punch uses 3 cups of sugar; $(0, 0)$ means that the recipe uses no sugar if you make no punch; move up 3 and to the right 5.*

b. What is the ratio of sugar to punch for Lori's recipe? How is this ratio related to the slide you found in part a? Write this ratio as a decimal. What does the decimal tell you about the recipe? **3 cups to 5 gal; it's the same as the ratio of "up" to "over" in the slide. The recipe requires 0.6 c sugar for each gallon of punch.**

c. Look at the graph of the three punch recipes, and decide which one you think makes the sweetest punch. Explain why you think so. Then, write the ratio of sugar to punch for Elena's recipe and Phil's recipe. Which line has the largest sugar-to-punch ratio? How is the sugar-to-punch ratio related to the steepness of the lines? **Elena's recipe makes the sweetest punch because its graph is the steepest. Elena's: $\frac{4\text{ c sugar}}{4\text{ gal of punch}}$; Phil's: $\frac{2\text{ c sugar}}{8\text{ gal punch}}$. Elena's recipe has the largest sugar-to-punch ratio. The larger the sugar-to-punch ratio, the steeper the line.**

d. In mathematics, a ratio that describes the steepness of a line is called the **slope of the line**. The slope of Lori's line is $\frac{3\text{ cups}}{5\text{ gallons}}$. (We could also write this as "$\frac{3}{5}$ cup per gallon" or "0.6 cup per gallon.") What is the slope ratio for Elena's line? What is the slope ratio for Phil's line? **Elena's: $\frac{4\text{ c}}{4\text{ gal}} = 1$ c/gal; Phil's: $\frac{2\text{ c}}{8\text{ gal}} = \frac{1}{4}$ c/gal**

e. How is the slide you found in part a related to the slope ratio in part d? **The vertical change is the numerator of the slope ratio; the horizontal change is the denominator.**

In WR-5, for lines with positive slope, we purposely draw the slope triangles above the lines to encourage students to think in terms of "up and over" while making a slide from left to right. On the other hand, for lines with negative slopes, we draw the slope triangles below the lines to encourage the sense of "down and over" from left to right. This way students are thinking of the vertical change and the horizontal change in the same order as the numerator and the denominator, respectively, of the slope ratio. This seems to lead to fewer mistakes when students write the slope ratio.

WR-5. By comparing the graphs in this problem, you will find a way to determine the slope ratio for a line from a slope triangle.

Example: The three graphs in Figure 1 describe the distances traveled by three participants in a 40-mile bicycle ride, while the two graphs in Figure 2 describe the total amount of oil remaining in each of two leaky oil tanks, a smaller tank and a larger tank. Below the graph of each line there is a ratio that describes how steep it is. This ratio is the *slope of the line*. The triangle drawn on the graphs is called the **slope triangle**.

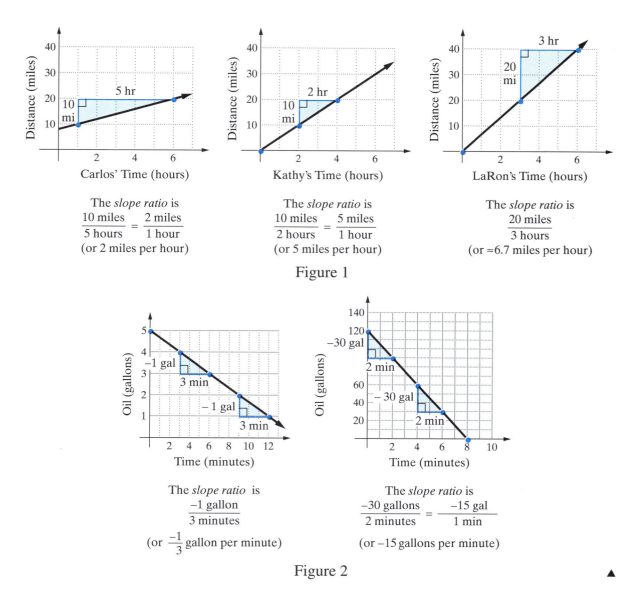

Figure 1

Figure 2

Compare all five graphs to see how the slope of each line was determined. Discuss with your group how you could find the slope of any line before answering parts a, b, and c.

a. Describe how to find the slope of a line. If your group found more than one method, describe each method. Your descriptions should be clear enough for other groups to understand. *Possible answer: Find a slide that goes from one point on the line to another point on the line. The vertical change (up or down) is the numerator of the slope, and the horizontal change (right or left) is the denominator.*

b. What does the slope ratio tell you about the graph of the line?
It describes the steepness, and whether it is up- or downhill from left to right.

c. What does the slope ratio tell you about the situation the line is modeling?
It tells about the rate of change of the situation.

WR-6. Drawing a Slope Triangle Irene and Crystal are arguing about where a slope triangle should be drawn. They decide to draw three different triangles, as shown here, and compare the slope ratios they get using each triangle.

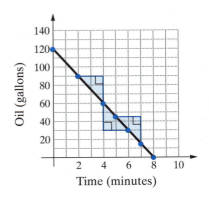

a. Write the slope ratio for this line by using each of the three triangles. *from left to right: $\frac{-3}{2}, \frac{-3}{2}, \frac{-3}{2}$*

b. What do you notice? Does it matter whether a slope triangle is drawn above the line or below the line? Explain why or why not. *No, it doesn't matter. The vertical change and horizontal change are the same in both cases.*

WR-7. Sketch a slope triangle and then state the slope of each of the lines below. Remember to describe each slope with a positive or negative ratio. Then write a complete sentence and use units to interpret the slope as a rate.

a. *Larry's Diet:*

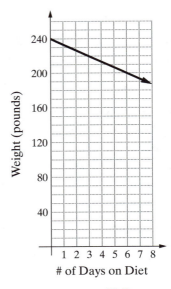

Possible answer: $\frac{-40\ lb}{6\ days}$. Larry loses weight at a rate of $6\frac{2}{3}$ lb/day.

b. *Bicycle Race:*

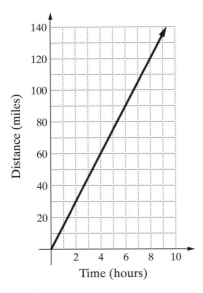

Possible answer: $\frac{60\ mi}{4\ hr}$. The racer rides at a rate of 15 mi/hr.

c. *Leaky Vessel:*

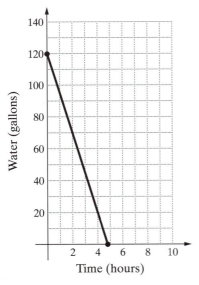

Possible answer: $\frac{-50\ gal}{2\ hr}$. Water is leaking out at a rate of 25 gal/hr.

Section 5.1 Ratios and the Slope of a Line **287**

Have groups share their responses to WR-7, and discuss what different slope ratios might mean in terms of steepness and rate before groups do WR-8.

WR-8. The Slope of a Line As stated in WR-4, the slope of a line is a number—a ratio—that measures the steepness and upward/downward direction (from left to right) of the line. One way to find the slope of a line is first to find a slide that goes from one point on the line to another point farther to the right on the line. Then describe the slide as a ratio of the vertical distance (or change) to the horizontal distance (or change) from the point on the left to the point on the right.

Definition: The **slope of a line** is defined as the ratio

$$\frac{\text{vertical change}}{\text{horizontal change}} = \frac{\text{change in } y\text{-values}}{\text{change in } x\text{-values}}.$$

Note that a line that goes upward from left to right (↗) has a **positive slope ratio**, while a line that goes downward from left to right (↘) has a **negative slope ratio**. The slope of a line is commonly represented by the letter m.

To calculate the slope of a line, pick two points on the line and use the segment between them as the longest side of a right triangle.

Example: If the two chosen points are $(-2, 3)$ and $(3, 5)$, then

$$m = \text{slope} = \frac{\text{change in } y\text{-values}}{\text{change in } x\text{-values}} = \frac{2}{5}.$$

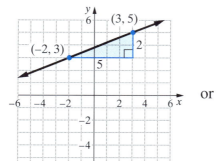 or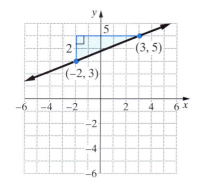

Plot each of the following pairs of points on graph paper, and then determine the slope of the line passing through each pair by using a slope triangle and counting to find the vertical and horizontal changes.

a. $(3, 8)$ and $(5, 6)$ *slope = −1*   **b.** $(-4, 5)$ and $(5, 4)$ *slope = $\frac{-1}{9}$*

c. $(-6, 8)$ and $(-4, 5)$ *slope = $\frac{-3}{2}$*   **d.** $(3, 0)$ and $(0, -6)$ *2*

WR-9. Interpreting the Slope of a Line In this problem you will see how units can help you interpret the meaning of the slope ratio in the context of the situation.

Example: This graph shows information from the practice session of two students in typing class. At the start of the session, Dalia had already typed 20 words. During the first 5 minutes, she typed 80 words (for a total of 100 words), and she continued typing at that same rate.

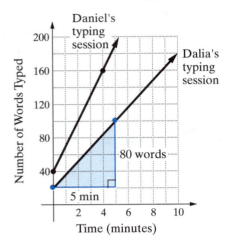

To find the slope of Dalia's line, we draw a slope triangle and label the vertical and horizontal sides. In this case, the slope ratio is $\frac{80}{5}$ since the vertical change is 80 and the horizontal change is 5. Thus, the slope ratio is $\frac{80}{5}$, or $\frac{16}{1}$.

To see how this numeric ratio relates to Dalia's practice session, we use words to state what the vertical and horizontal changes represent. In this case, the vertical change is "80 *words*" and the horizontal change is "5 *minutes.*"

Including units with the slope will help you to interpret the slope as a rate. Here the slope ratio is $\frac{80 \text{ words}}{5 \text{ minutes}}$, or $\frac{16 \text{ words}}{1 \text{ minute}}$, which means Dalia is typing at a constant rate of 80 words every 5 minutes, or 16 words per minute. This is an example of what we mean by **interpreting the slope of the line**. ▲

a. Daniel started the practice session with 40 words already typed, and after 4 minutes had passed, he counted 160 words on his page. Find the slope of Daniel's line by drawing a slope triangle and labeling it.
$$slope = \frac{120}{4} = \frac{30}{1}$$

b. Now use units to interpret Daniel's slope ratio as a rate.
$$slope = \frac{120 \text{ words}}{4 \text{ mins}} = 30 \text{ words/min}$$

EXTENSION AND PRACTICE

If you assign WR-10 for homework, be sure to have students discuss their responses to part d in class the following day.

WR-10. You know that you can use a pair of points to determine the slope ratio of a line. In this problem, you will examine what happens when you use two different pairs of points from the same line.

a. Graph the line $y = 1.6x + 2$.

b. Find the *y*-coordinate of the point on the line whose *x*-coordinate is 1. Find the *y*-coordinate of the point on the line whose *x*-coordinate is 4. Plot these two points and use them to sketch a slope triangle.

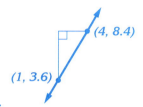

The points are (1, 3.6) and (4, 8.4).

c. Use the slope triangle you drew in part b to determine the slope of the line, and write this slope ratio as a decimal.   *slope = $\frac{4.8}{3}$ = 1.6*

d. Repeat parts b and c for the two points whose *x*-coordinates are 10 and 16. What do you notice?   *(10, 18); (16, 27.6); slope = $\frac{9.6}{6}$ = 1.6; the slopes are the same.*

WR-11. Ginger got "1" for the slope of the line through points (1, 2) and (4, −1).

a. What mistake might Ginger have made? Explain to her how to correctly find the slope of the line.   *Ginger may have made a slope triangle whose vertical and horizontal sides were 3 units each, and forgotten that the line goes downhill from left to right. The slope is actually $\frac{-3}{3} = -1$.*

b. Copy the definition of the *slope of a line* and information about *how to find the slope* of a line into your Tool Kit.

WR-12. Fran's Pharmacy claims that because their prices are so low, there is no discount for buying in volume. This graph shows how much the pharmacy charges depending on how many Mega-Vitamins you buy.

a. Explain why it makes sense that the point (0, 0) is on the line.   *Zero vitamins cost $0.*

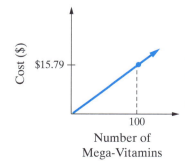

b. How much does it cost to buy 100 Mega-Vitamins? How much does it cost to buy 50 Mega-Vitamins? How about for 200 Mega-Vitamins? *$15.79, $7.90, $31.58*

c. What is the slope of this line? $\text{slope} = \dfrac{\$15.79}{100 \text{ vitamins}} = \dfrac{\$0.1579}{1 \text{ vitamin}} \approx \dfrac{\$0.16}{1 \text{ vitamin}}$

d. Explain what the slope represents. Include the units in your answer. *Mega-Vitamins are priced at a rate of about 16¢ per vitamin.*

WR-13. As of 6 A.M. on Tuesday, the volume of water storage of the Oroville Dam was 3,131,284 acre-feet. On Friday at 6 A.M., the storage was 3,166,328 acre-feet. What was the change in storage from Tuesday to Friday? During that time, what was the average rate of increase in storage in acre-feet per day? *The storage increased 35,044 acre-feet over the three days, for an average rate of about 11,681.3 acre-feet per day.*

WR-14. In parts a and b, $\triangle ABC$ is similar to $\triangle DEF$. For each part, write an equation with equivalent ratios to find the length $|BC|$.

a.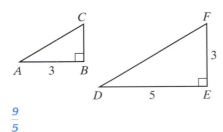

$\dfrac{9}{5}$

b.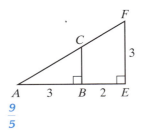

$\dfrac{9}{5}$

c. For the drawing in part b, if the coordinates of point A were $(0, 0)$, what would be the coordinates of points C and F? What would be the slope of the line through points A and C? *$C = (3, \dfrac{9}{5})$, $F = (5, 3)$; slope $= \dfrac{3}{5}$*

WR-15. Find each of the following sums. Write each answer in fraction form.

a. $\dfrac{2}{3} + \dfrac{1}{5}$ $\dfrac{13}{15}$

b. $\dfrac{2}{n} + \dfrac{1}{5}$ $\dfrac{10 + n}{5n}$

c. $\dfrac{2}{w} + \dfrac{5}{3}$ $\dfrac{6 + 5w}{3w}$

d. $\dfrac{3}{z} + \dfrac{5}{2z}$ $\dfrac{11}{2z}$

e. Use part d to solve the equation $\dfrac{3}{z} + \dfrac{5}{2z} = \dfrac{1}{6}$.   *$z = 33$*

WR-16. Solve each of the following equations. Show your steps, and check your answers.

a. $4x - 6(4 - 2x) = 68$ *x = 5.75*

b. $\dfrac{n}{1.6 \cdot 10^9} = \dfrac{4 \cdot 10^{12}}{3.75}$ *n ≈ 1.71·10²¹*

c. $(R - 6)(R - 2) = 0$ *R = 2, 6*

WR-17. Sari knows that $3C + 2 = 14$, and wants to find out what $5C + 1$ equals.

a. Identify a subproblem that Sari could use to help her, and then solve the subproblem.   *Find what C equals; C = 4.*

b. Solve the original problem: if $3C + 2 = 14$, what does $5C + 1$ equal?   *21*

WR-18. Simplifying Fractions Rewrite each of the following fractions using as few digits or variables as possible.

a. $\dfrac{7659 \cdot 10}{7659}$ *10*

b. $\dfrac{3x}{x}$ for $x \neq 0$ *3*

c. $\dfrac{x + 3}{x + 3}$ for $x \neq -3$ *1*

d. $\dfrac{16(x + 2)}{4}$ *4(x + 2) or 4x + 8*

WR-19. Solve each of the following equations for x. Write each answer in simplified fraction form.

a. $\dfrac{x}{6} = \dfrac{7}{10}$ *21/5*

b. $\dfrac{6}{x} = \dfrac{7}{10}$ *60/7*

c. $\dfrac{6}{x} + \dfrac{4}{7} = \dfrac{7}{3}$ *126/37*

WR-20. Show how to answer each of the following percent problems by writing an equation and solving it.

a. Percy has earned 237 points out of a possible 280 on all his algebra tests so far. What percent of the total test points has Percy earned?
$\dfrac{237}{280} = \dfrac{x}{100}$; ≈ 84.6%

b. If 92% of the 280 points are required for a grade of A, how many more points should Percy have earned in order to get an A at this time in the course?   *$\dfrac{92}{100} = \dfrac{x}{280}$; he needs a total of 257.6 points, so needs 21 more.*

c. Two hundred thirty-seven is 92% of what number?

$\frac{237}{x} = \frac{92}{100}; \approx 257.61$

d. Percy has just one more test to take, and it's worth 100 points. To earn a B in the algebra class, a student must earn 80% of the total test points. What is the lowest score Percy could get on the last test and still earn a B?

$\frac{237 + x}{380} = \frac{80}{100};$ 67 points

WR-21. Dart Boards Bill plays darts every night at the local pool hall. Last night there was a new square dart board with a circle inside. To win, Bill's dart must hit the circle. The area of the entire square is 81 square inches.

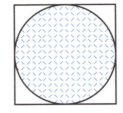

a. Find the area of the circle. $20.25\pi \approx 63.6$ sq in.

b. If Bill throws randomly, what is the probability that he will hit the circle and win the game? $\frac{20.25\pi}{81} \approx 0.785$

WR-22. Exponent Practice Write each of the following expressions with a single base and a single exponent.

a. $x^3 \cdot x^4$ x^7

b. $x^{-3} \cdot x^{17}$ x^{14}

c. $\frac{n^{12}}{n^{17}}$ n^{-5}

d. $\frac{n^5}{n^{-2}}$ n^7

WR-23. Rewrite each of the following expressions by multiplying the polynomials in parts a and b and factoring the polynomials in parts c through f.

a. $x(x^2 - 6x + 8)$ $x^3 - 6x^2 + 8x$

b. $(x + 11)(x - 11)$ $x^2 - 121$

c. $x^2 + 8x$ $x(x + 8)$

d. $6 - 2x^2$ $2(3 - x^2)$

e. $x^2 + 6x + 5$ $(x + 5)(x + 1)$

f. $3x^2 + 16x + 5$ $(3x + 1)(x + 5)$

Problem WR-24 is part of a sequence of problems that focuses on factoring differences of squares. Students will need to refer to their work from WR-24 to complete problem WR-71 in Section 5.3.

WR-24. In this problem you'll explore factoring relationships for special sum-and-product pairs. Copy and complete the following Sums and Products table. Then for each pair, draw and complete a Diamond Problem, draw and label a generic rectangle, and write and factor a quadratic polynomial.

	Sum	Product	Integer Pair	
a.	−4	0	−4, 0	$x^2 - 4x = x(x - 4)$
b.	0	−4	−2, 2	$x^2 - 4 = (x - 2)(x + 2)$
c.	−16	0	−16, 0	$x^2 - 16x = x(x - 16)$
d.	0	−16	−4, 4	$x^2 - 16 = (x - 4)(x + 4)$
e.	−10	0	−10, 0	$x^2 - 10x = x(x - 10)$
f.	0	−10	no solution	no solution with integers
g.	−25	0	−25, 0	$x^2 - 25x = x(x - 25)$
h.	0	−25	−5, 5	$x^2 - 25 = (x - 5)(x + 5)$

The East and West entries in each Diamond Problem are the integers from the integer pair. The dimensions of generic rectangles are the factors of the polynomials.

Problem WR-25 reinforces the students' understanding of the relationship between a linear equation and its graph. Section 5.2 opens with a follow-up problem so you can check students' understanding before moving on.

WR-25. It is often useful to know whether a particular point lies on a line without actually graphing the line.

a. Graph the equation $y = 3x + 2$ for $-3 \leq x \leq 3$.

b. On the same set of axes, graph and label the points (1, 5), (2, 6), (−2, −4), (0, 1) and (−1, −1).

c. Does the point (1, 5) make the equation $y = 3x + 2$ true or false? Explain how you know. *True; by substituting for x, we get $y = 3(1) + 2 = 5$.*

d. Now check whether each of the points (2, 6), (−2, −4), (0, 1) and (−1, −1) makes the equation $y = 3x + 2$ true or false. *F, T, F, T*

e. Look for a relationship between the points from parts b and d that lie on (or off) the line and those that make the equation true (or false). Describe what you observe. *The points that lie on the line make the equation true; those off the line make the equation false.*

5.2 GRAPHING LINEAR EQUATIONS USING TWO POINTS AND A SLIDE

SECTION MATERIALS

WR-27 AND WR-28 GRAPH COMPARISONS RESOURCE PAGE

WR-45 THE BIKE RACE RESOURCE PAGE

Problems WR-26 through WR-29 should be assigned together and will take a class period. They form an investigation, designed to convince the students of the following:

- *Only two points need to be plotted in order to sketch the graph of a linear equation.*

- One of the two points may be read easily from an equation in "y-form."
- Any linear equation can be transformed into "y-form."

Encourage students to do each problem thoughtfully and to graph accurately and neatly.

NOTE In the remainder of this chapter many of the graphs will require a full sheet of graph paper in order to make them clear, accurate, and large enough for you to analyze. Use a straightedge or ruler to draw the graphs of lines.

So far in this course you have graphed equations by first selecting several different values for x and then calculating the corresponding y-values. Each of the equations you've graphed has been written in **y-form**; that is, y is alone on the left-hand side of the equation while x and numbers are on the right. Problems WR-26 through WR-27 will help you develop a faster method for graphing linear equations.

Problem WR-26 is a follow-up problem for WR-25. Have groups work on it as you circulate and check the students' understanding of the relationship between a linear equation and its graph.

WR-26. Which, if any, of the following points are on the graph of the equation $y = 3x + 2$?

a. $(1, 3)$ *no*   b. $(0, 2)$ *yes*

c. $(2, 8)$ *yes*   d. $(-3, -3)$ *no*

e. Explain how you can tell without drawing the graph of the line $y = 10x - 1$ whether the point $(53, 527)$ lies on the line. *If substituting the x- and y-values into the equation results in a true statement, then the point lies on the line.*

WR-27. In this problem, you will see whether the graph of an equation depends on which x-values you use to calculate the points.

a. Use three different x-values to find three points that lie on the graph of the equation

$$y = 2x - 5.$$

The members of your group should all choose different x-values.

b. On the WR-27 and WR-28 Graph Comparisons Resource Page, use the three points you found in part a to graph the equation $y = 2x - 5$. Label the points you plotted with their coordinates, and label the line with its equation. Also label the x-intercept and the y-intercept with their coordinates.

c. Now stack your group's Resource Pages to compare your graphs. Does the graph of the equation $y = 2x - 5$ depend on which x-values were selected, or are your graphs of the equation $y = 2x - 5$ the same? *The graphs should be identical.*

WR-28. For this problem, you will use the same Resource Page as in WR-27 and will examine three equations and their graphs:

(1) $y = 2x - 5$ (2) $y = 2x + 6$ (3) $y = 2x - 1$

a. You've already graphed $y = 2x - 5$ in WR-27. On the same set of axes, carefully graph $y = 2x + 6$ and $y = 2x - 1$. Use -2, 0, and 2 as x-values for each equation. Label each graph with its equation.

b. For each of the three lines, find the coordinates of the x-intercept and the y-intercept, and determine the slope. Then copy and complete the following table:

y-Form of Equation	x-Intercept	y-Intercept	Slope
$y = 2x - 5$	$\approx (2.5, 0)$	$(0, -5)$	2
$y = 2x + 6$	$(-3, 0)$	$(0, 6)$	2
$y = 2x - 1$	$\approx (0.5, 0)$	$(0, -1)$	2

c. Look at the y-intercept column of your table. What do you notice about the x-coordinates of the y-intercepts? Write a sentence to describe what you observe. *The x-coordinate of the y-intercept is always 0.*

d. For each line, compare the y-intercept with the y-form of the line's equation. Write a sentence to describe what you observe. *The y-coordinate of the y-intercept is the constant of the y-form of the equation.*

WR-29. Look back at your Resource Page from WR-27 and WR-28. Discuss the following questions with your group, and then write a paragraph to answer them.

How are the graphs of $y = 2x - 5$, $y = 2x + 6$, and $y = 2x - 1$ alike? In what ways are they different?

How are the three equations alike? How are they different? *The graphs are all lines with slope 2. They have different x- and y-intercepts. The equations all are of the form "y = 2x ± a constant" for different constants.*

WR-30. Here are some linear equations in y-form. Use your observations in the previous problem to find the y-intercept of the graph of each equation. Write your answers in the form

"The y-intercept of the line _____ is _____."

a. $y = 2x + 3$ *(0, 3)*   b. $y = -x - 1$ *(0, -1)*
c. $y = 2x - 4$ *(0, -4)*   d. $y = -3x + 2$ *(0, 2)*

WR-31. The graph of the equation $x - 2y = 4$ is a line.

a. Use your Tool Kit and/or observations from WR-28c to find the coordinates of the y-intercept of the line. *(0, −2)*

b. Use your Tool Kit and/or observations from SP-62a to find the coordinates of the x-intercept of the line. *(4, 0)*

c. Use the x- and y-intercepts to graph the line.

d. Draw a slope triangle and then find the slope of the line. *slope = $\frac{1}{2}$*

WR-32. If you know the equation of a line, it is very easy to calculate the y-intercept of the line. Explain, in a way that a classmate who was absent today could understand, why this is so. *The x-coordinate of the y-intercept is 0, so all you need to do is substitute 0 for x in the equation. If the equation is already in the form $y = mx + b$, then the y-coordinate of the y-intercept is the constant b.*

WR-33. In problem WR-27, each member of your group graphed the equation $y = 2x - 5$ using three x-values. Although you all used different x-values, your graphs were the same line. In this problem you will develop another method for graphing lines.

a. Suppose you want to sketch a graph of a line. Would the graph be accurate if you plotted just two points that lie on the line? Explain why or why not. Would the graph be accurate if you plotted just one point that lies on the line? Explain why or why not. Using no other information about the line, what is the *least* number of points you need to plot in order to sketch a graph of a line? *Yes, if the points were plotted correctly; not necessarily, as many lines pass through a given point; we need to plot at least 2 points.*

b. Suppose you now want to graph the line $y = 2x - 7$. Use your observations from WR-30 and WR-32 to find one point on the graph of $y = 2x - 7$. *(0, −7).*

c. Now find a second point on the line by choosing *any x-value other than zero* and finding its corresponding y-value. *For example, (1, −5) or (4, 1).*

d. Plot the points you found in parts b and c on a piece of graph paper. Describe a slide you could use to find five other points on the line, and plot those points. Connect your points using a ruler or straightedge. *Slide up 2 and to the right 1.*

e. To check the accuracy of your graph, pick another x-value and find a point on the line by substituting for x in the equation $y = 2x - 7$. Does this point lie on the line you drew?

Section 5.2 Graphing Linear Equations Using Two Points and a Slide

 WR-34. Two-Point Slide Method for Graphing Lines In problem WR-33 you graphed a line by using the *y*-form of an equation to find two points on the line—its *y*-intercept and one other point. The two points determined a slide that you used to graph more points on the line. You then checked the accuracy of your graph by finding a third point that fit the equation and making sure it also was on the line formed by the first two points.

This method, called the **two-point slide method for graphing lines**, can be used to graph any linear equation. To use it, follow these steps:

- If the equation is not already written in *y*-form, rewrite it.

- Use the *y*-form of the equation to find two points on the line: the *y*-intercept and one other point.

- Use the slide to find more points, and then connect the points using a ruler or straightedge.

- Check your work by finding a third point that fits the equation and then plotting it.

Use the handy two-point slide method for graphing lines to graph the four equations in problem WR-30 on the same set of axes you used in WR-33. Be sure to label each line with its equation and to check the accuracy of your graphs.

 WR-35. It's often easy to graph equations that are written in *y*-form, but not all equations are written that way. For example, the equation

$$3x + y = -1$$

is not in *y*-form. However, you could rewrite it in *y*-form by solving it for *y* as shown in the following example.

a. Copy this example into your Tool Kit.

Example: To rewrite the equation $3x + y = -1$ in *y*-form, you just need to solve it for *y*. You can do this by subtracting $3x$ from each side of the equation:

$$\begin{array}{r} 3x + y = -1 \\ -3x \quad\quad -3x \\ \hline y = -1 - 3x \end{array}$$

Note that you could also write the equation $y = -1 - 3x$ as $y = -3x - 1$. ▲

b. Write the equation $3x + 2y = 2$ in *y*-form by solving it for *y*. Show each step. $y = \dfrac{2 - 3x}{2} = 1 - \dfrac{3x}{2}$ or $y = \dfrac{-3x + 2}{2} = \dfrac{-3x}{2} + 1$

In WR-36, the forms $y = mx + b$ and $y = b + mx$ are both acceptable.

WR-36. Write each of the following equations in *y*-form.

a. $6x + 2y = 10$ *$y = 5 - 3x$ or $y = -3x + 5$*

b. $2x - y = 3$ *$y = -3 + 2x$ or $y = 2x - 3$*

c. $\dfrac{y - b}{m} = x$ *$y = b + mx$ or $y = mx + b$*

WR-37. Use the two-point slide method to graph each of the following linear equations. You will need to change the equations into *y*-form first.

a. $3x - 6y = -24$ *$y = \frac{1}{2}x + 4$*   b. $6x + 2y = 7$ *$y = -3x + \frac{7}{2}$*

EXTENSION AND PRACTICE

WR-38. You've practiced graphing lines using two points and a slide in problems WR-34 through WR-37. Now add to your Tool Kit information about using the *two-point slide method for graphing lines*.

WR-39. Use the two-point slide method to graph each of the following linear equations. You will need to change the equations into *y*-form first.

a. $4x + y = 1$ *$y = -4x + 1$*   b. $2x - 4y = 5$ *$y = \frac{1}{2}x - \frac{5}{4}$*

WR-40. In problem WR-37 you used the two-point slide method to graph the equation $3x - 6y = -24$. Choose one of the points you found using the slide and check that it makes the equation true. *For example, sliding "up 1, to the right 2" from (2, 5), we come to (4, 6). Substituting this into the equation, we get $3(4) - 6(6) = -24$.*

WR-41. The graph of the equation $2x - 3y = 7$ is a line.

a. Find the coordinates of the line's *x*-intercept.   *(3.5, 0)*

b. Find the coordinates of the *y*-intercept of the line.   *$(0, \approx -2.33)$*

c. If a point on the line given by $2x - 3y = 7$ has an *x*-coordinate of 10, what is its *y*-coordinate?   *$y = \frac{13}{3} \approx 4.33$*

WR-42. Can each of these equations be graphed using the two-point slide method? Explain why or why not. You do not need to graph the equations.

a. $y = 2x + 1$ *Yes, the equation is linear.*

b. $y = 2x$ *Yes, the equation is linear.*

c. $y = x^2$ *No, the equation is not linear.*

WR-43. Without drawing a graph, determine which, if any, of the listed points are on the graph of the equation $y = 3x - 4$.

a. $(-1, -4)$ *no* **b.** $(4, 3)$ *no*

c. $(1, -1)$ *yes* **d.** $(2, 2)$ *yes*

WR-44. Bike Rental It costs $1.50 per hour plus $5 to rent a bicycle. To describe the relationship between the cost, C, of renting a bike and the number of hours, H, Marta wrote the equation

$$C = 1.5H + 5.$$

a. Draw a graph to show this relationship.

b. Show how to use your graph to estimate the value of C when $H = 6$.

c. Show how to use your graph to estimate the value of H when $C = 20$.

In WR-45, students will need to extend both Cindy's and Dean's graphs to answer part c, and they'll need to extend both axes to answer part f.

WR-45. The Bike Race Cindy and Dean are engaged in a friendly 60-mile bike race that started at noon. The following graph represents their race. Notice that Dean, who doesn't feel a great need to hurry, stops to rest for a while.

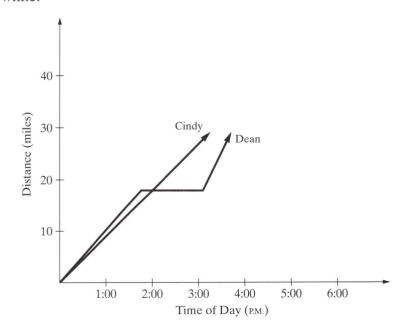

After reading through the questions in parts a through f, make marks on your copy of the WR-45 The Bike Race Resource Page to show how to use the graph to answer each question. Write your responses to the questions in complete sentences.

a. At what time (approximately) does Cindy pass Dean? *around 2:00 P.M.*

b. About how far has Cindy traveled when she passes Dean? *about 18 mi*

c. You can see that Dean will catch up to Cindy again if they both continue to ride at their same rates. At approximately what time will this happen? About how far will they each have traveled when they reach this point? *Around 4:00 P.M., about 37 mi*

d. After his rest, does Dean travel faster or slower than before? Why do you think so? *He rides faster after his rest. The graph is steeper.*

e. About how long did Dean rest? On the graph, show how you know. *about $1\frac{1}{4}$ hr*

f. If Cindy continues at a steady pace, about how long will it take her to ride the entire race? Why do you think so? *It will take Cindy about $6\frac{2}{3}$ hours in all. If her graph is continued, it intersects the horizontal line $y = 60$ when $x \approx 6:40$ P.M.*

WR-46. Use the graph shown here to answer the following questions:

a. Which car is traveling at the greater rate? *A*

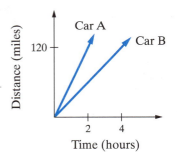

b. What is the average rate in miles per hour for Car A? for Car B? *60 mph, 30 mph*

c. What is the slope of the line representing Car A? What is the slope of the line representing Car B? *60 mph, 30 mph*

WR-47. Solve each of the following equations. Show your steps and check your answers.

a. $34 = 8 - 2w$
    $w = -13$

b. $8x - 72 = -8x - 40$
    $x = 2$

c. $\dfrac{2x - 3}{5} = \dfrac{x}{4} + 2$
    $x = \dfrac{52}{3} \approx 17.3$

d. $5(5y + 4) - (y - 3) = 5$
    $y = -0.75$

e. $(n + 5)(n + 2)(n - 1) = 0$ $n = -5, -2, 1$

WR-48. Arrange these tiles, which represent $2x^2$, $9x$, and 9 square units, into a composite rectangle.

a. Sketch and label the composite rectangle.

Solution:

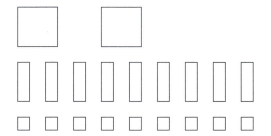

b. Write the area of the composite rectangle as a sum. $2x^2 + 9x + 9$

c. Write the area of the composite rectangle as a product. $(x + 3)(2x + 3)$

d. Is there more than one way to factor the polynomial $2x^2 + 9x + 9$? Explain your answer. *No, because there is only one way to create a composite rectangle using the given tiles.*

WR-49. Write one polynomial to represent each of the following sums or differences. You may want to draw or visualize algebra tiles to help you understand these problems. Note that it is possible to get negative results algebraically, even though they can't easily be represented by tiles.

a. $(2x^2 + 3x + 5) + (x^2 + 2x + 8)$ $3x^2 + 5x + 13$

b. $(3x^2 + 5x + 7) - (4x^2 + x + 1)$ $-x^2 + 4x + 6$

c. $(x^2 + 9x + 8) - (x^2 + 4x + 8)$ $5x$

d. $(7x^2 + x + 10) - (3x^2 + 12x + 12)$ $4x^2 - 11x - 2$

WR-50. Simplifying Fractions Use what you know about exponents to rewrite each of the following fractions in a simpler exponential form. Assume that $x \neq 0$.

a. $\dfrac{10}{10^3}$ $\dfrac{1}{10^2}$ or 10^{-2} b. $\dfrac{-6x^3}{2x}$ $-3x^2$

c. $\dfrac{8}{2x^3}$ $4x^{-3}$ d. $\left(\dfrac{3x^2y}{x}\right)^3$ $27x^3y^3$

e. $\dfrac{x^5}{x^5}$ x^0 or 1 f. $\dfrac{(-2x^2)^2}{2x}$ $2x^3$

WR-51. Exponent Practice Use guess and check to solve each of the following equations for n.

a. $a^4 \cdot a^n = a^{11}$ **7**

b. $\dfrac{6^n}{6^4} = 6^{-8}$ **−4**

c. $(x^n)^6 = x^{96}$ **16**

WR-52. Keisha showed Tina why she thinks it makes sense that $3^{-2} = \dfrac{1}{3^2}$. Unfortunately, a blob of pizza sauce later fell on the paper, so when Tina got home all she could read was

$$\dfrac{1}{3^2} = \dfrac{3^0}{3^2} = \bigcirc = 3^{-2}$$

a. Copy the work without the blob and fill in the step that is hidden.
$\dfrac{1}{3^2} = \dfrac{3^0}{3^2} = 3^{0-2} = 3^{-2}$

b. Explain why $\dfrac{1}{3^2} = \dfrac{3^0}{3^2}$. *We know $3^0 = 1$, and if two fractions have equal denominators and equal numerators, then the fractions are equal.*

WR-53. Rewrite each of the following expressions by multiplying the binomials in parts a and b and factoring the polynomials in parts c through f.

a. $(x + 10)(x + 10)$
   $x^2 + 20x + 100$

b. $(x + 2y)(x + 3y)$
   $x^2 + 5xy + 6y^2$

c. $9x^2 - 16$
   $(3x + 4)(3x - 4)$

d. $3x^2 - 21x$
   $3x(x - 7)$

e. $x^2 + 3x - 10$
   $(x - 2)(x + 5)$

f. $x^2 - 7x - 18$
   $(x + 2)(x - 9)$

Section 5.3 Slopes of Parallel Lines and the Slope-Intercept Pattern in $y = mx + b$ **303**

WR-54. A box contains 24 red cubes and B blue cubes.

a. If $B = 6$, what is the probability of randomly pulling out one blue cube? $\frac{6}{30}$ or $\frac{1}{5}$

b. Identify a subproblem you used to solve part a. *Find the total number of cubes.*

c. Suppose the probability of randomly pulling out one blue cube is $\frac{1}{3}$. To find B, make a guess-and-check table to help you write an equation, and then solve the equation. *Possible column headings: "Guess: # Blue Cubes," "Total # Cubes," "$\frac{B}{Total}$," "Check: $\frac{B}{Total} = \frac{1}{3}$?" Possible equation: $\frac{B}{24 + B} = \frac{1}{3}$; $B = 12$*

WR-55. Copy and solve these Diamond Problems:

a.

b.

$N = -35,$

$S = 11.5,$

$9, -2$

c.

$E = -15,$

$S = -14\frac{2}{3}$

d.

$N = \frac{1}{4},$

$S = \frac{13}{12}$

5.3 SLOPES OF PARALLEL LINES AND THE SLOPE-INTERCEPT PATTERN IN $y = mx + b$

SECTION MATERIALS

DOT PAPER AND STRAIGHTEDGES (FOR WR-56)

DOT PAPER TRANSPARENCIES (OPTIONAL FOR CLASS DISCUSSION OF WR-56)

WR-57 SLOPES OF SEGMENTS RESOURCE PAGE

TRANSPARENCY OF THE WR-57 SLOPES OF SEGMENTS RESOURCE PAGE (OPTIONAL)

As students work on WR-56, encourage them to check the accuracy of their work with other group members. In a class discussion after WR-56, try to elicit the ideas that the negative sign indicates the downward direction of the line, and that the sign can be associated with the numerator or the denominator, but not both.

In Section 5.2 you saw that having an equation of a line in y-form is useful for graphing. In this section you'll examine a relationship between a line's slope and an equation of the line.

WR-56. Use dot paper for this exercise. Label a point A in the middle of the paper. Use a straightedge to carefully draw lines through point A with each of the following slopes:

$$1, 2, 3, 5, -1, -4, \frac{3}{5}, \frac{1}{10}, \frac{-2}{3}, \text{ and } 0.$$

Next to each line, write its slope.

After students have finished problem WR-56, if time permits, begin these kinesthetic activities on slope:

1. For reference, students should have their completed work on problem WR-56 out on their desks **where they can see it**.
2. Have all the students stand and face forward, holding their left arms in front of them in an L shape with their elbows bent at a right angle so that their upper arms are parallel to the floor and their forearms are vertical.
3. The elbow will represent the y-intercept and the left forearm will indicate the given slope of a line.
4. While facing the class, use **your right** forearm to demonstrate slopes with values 0, 1, $\frac{1}{2}$, 2, -1, and $-\frac{1}{2}$. Have the students use their **left** forearms to practice along with you.
5. Choose volunteers to lead practice with other slopes, such as those in problem WR-56: 3, 4, 5, -2, -3, -4, $\frac{3}{5}$, $\frac{1}{10}$, $-\frac{2}{3}$, and 0.

If you have a few minutes at the end of the period, do more exercises where students try to estimate the slopes of various lines, either drawn on the chalkboard or in the air with their arms.

A transparency of the WR-57 Slopes of Segments Resource Page may be helpful. The problem doesn't examine the slope of vertical lines, but you could ask for the slope of segment *GI* if you want students to discuss this special case.

WR-57. Slopes of Segments Use your copy of the WR-57 Slopes of Segments Resource Page to draw each of the following segments: *DF*, *GH*, *CF*, *EF*, and *EG*.

a. Find the slope of each segment. For example, the slope of $DF = \frac{2}{2} = 1$.

$1, -2, -\dfrac{1}{12}, -\dfrac{2}{3}, \dfrac{17}{7}$

b. Now find the slope of each of the following segments: *FD*, *HG*, *FC*, *FE*, and *GE*. $1, -2, -\dfrac{1}{12}, -\dfrac{2}{3}, \dfrac{17}{7}$

c. Compare the slopes of the following pairs of segments: *DF* and *FD*; *GH* and *HG*; *CF* and *FC*; *EF* and *FE*; *EG* and *GE*. Does the slope of a segment seem to depend on the order in which endpoints of the segment are written?   *The slopes in each pair are equal; no.*

There are two purposes to WR-58 through WR-62: to help students understand that lines with the same slope are parallel, and to help them discover the significance of the m and the b in the equation $y = mx + b$.

After groups work on WR-58 and WR-59, call them together to discuss their answers to parts e and f of WR-59.

Section 5.3 Slopes of Parallel Lines and the Slope-Intercept Pattern in $y = mx + b$ **305**

WR-58. Consider the rule, "the output is half of the input." The equation for this rule is $y = \frac{1}{2}x$.

   a. Graph the line $y = \frac{1}{2}x$.

   b. Mentally verify that this line goes through the points (0, 0), (20, 10), (800, 400), and (2000, 1000).

   c. Compute the slope of the line $y = \frac{1}{2}x$. *slope* $= \dfrac{1}{2}$

   d. What is the y-intercept of the line $y = \frac{1}{2}x$? *(0, 0)*

WR-59. In this problem you will compare your graph from WR-58 with the graph of another line.

   a. Graph the line $y = \frac{1}{2}x + 3$ on the same coordinate axes as the graph in WR-58.

   b. Mentally verify that this line goes through the points (0, 3), (20, 13), (800, 403), and (2000, 1003).

   c. Compute the slope of the line $y = \frac{1}{2}x + 3$. *slope* $= \dfrac{1}{2}$

   d. What is the y-intercept of the line $y = \frac{1}{2}x + 3$? *(0, 3)*

   e. Do the two lines represented by $y = \frac{1}{2}x$ and $y = \frac{1}{2}x + 3$ ever intersect?

   No, they're parallel.

   f. Write an equation for another line that is parallel to these two lines.

   $y = \dfrac{1}{2}x \pm$ *constant*

   g. Do these lines slope upward or downward? *Upward*

Part e of WR-60 may be challenging for most students; they'll see the idea again later.

WR-60. Here's another line to graph. Use a new set of coordinate axes for this one.

a. Find two points on the graph of $y = -\frac{3}{2}x + 4$. *For example, (0, 4) and (2, 1)*

b. Find the slope of the line $y = -\frac{3}{2}x + 4$ using the two points you found on the graph in part a. *slope = $-\frac{3}{2}$*

c. Does the line slope upward or downward? *downward*

d. Where does the graph cross the y-axis? *(0, 4)*

e) Write an equation for a line that is parallel to the line $y = -\frac{3}{2}x + 4$.
$y = -\frac{3}{2}x \pm$ constant

Be sure that groups all get WR-61 right. Students should be able to find out for themselves that m is the slope and b is the y-intercept.

WR-61. Slope and y-intercept in $y = mx + b$ In the previous three problems you studied the lines given by $y = \frac{1}{2}x$, $y = \frac{1}{2}x + 3$, and $y = -\frac{3}{2}x + 4$. In this problem you will organize information from your work and look for general patterns to connect an algebraic representation of a line with its graph.

a. Copy the following table, and then use your work in problems WR-58 through WR-60 to fill in the missing information:

y-Form of Equation	y-Intercept	Slope
$y = \frac{1}{2}x$		
$y = \frac{1}{2}x + 3$		
$y = -\frac{3}{2}x + 4$		

b. Notice that the three equations

$$y = \frac{1}{2}x, \quad y = \frac{1}{2}x + 3, \quad \text{and} \quad y = -\frac{3}{2}x + 4$$

Section 5.3 Slopes of Parallel Lines and the Slope-Intercept Pattern in $y = mx + b$ **307**

are each written in the y-form $y = mx + b$, where m and b are numbers. For example, in the linear equation $y = \frac{1}{2}x + 3$, the coefficient of x is $\frac{1}{2}$, so $m = \frac{1}{2}$, and the constant term is 3, so $b = 3$. Look for patterns in your table. Then, with your group, state as clearly as you can a rule for a fast and sure way of knowing the slope of a line in the form $y = mx + b$.

The slope is m, the coefficient of x.

c. Now state a rule for finding the y-intercept of a line in the form $y = mx + b$. *The y-intercept is the constant b.*

d. An equation written in the form **$y = mx + b$** is said to be written in **slope-intercept form** of a linear equation. Add your rules for finding the slope and the y-intercept of a line in the form $y = mx + b$ to your Tool Kit.

WR-62. Use the observations you made in problem WR-61 to state the slope and y-intercept of each of the following lines:

a. $y = 2x - 5$ $m = 2; (0, -5)$

b. $y = \frac{3}{4}x + \frac{7}{5}$ $m = \frac{3}{4}; (0, \frac{7}{4})$

c. $2y = 4x - 5$ (*Hint:* First solve for y.) $m = 2; (0, -2.5)$

d. $4y = 3x + 6$ $m = \frac{3}{4} = 0.75; (0, \frac{3}{2}) = (0, 1.5)$

e. $y + 2x = 4$ $m = -2; (0, 4)$

f. Two pairs of lines in parts a through e are parallel. Write their equations. *a and c, b and d*

WR-63. Graph each of the following lines from WR-62:

a. $4y = 3x + 6$ *x-intercept is (−2, 0); y-intercept is (0, 1.5); $m = \frac{3}{4}$*

b. $y + 2x = 4$ *x-intercept is (2, 0); y-intercept is (0, 4); $m = -2$*

WR-64. In problem WR-33, two points were needed to graph a line. Now suppose you know the slope of a line you'd like to graph. Do you think you still need to know two points on the line, or could you sketch a graph if you know the slope and only one point?

a. Use the two-point slide method to graph the equation $y = \frac{2}{3}x - 4$.

b. State the slope and the y-intercept of $y = \frac{2}{3}x - 4$. How could you have used the slope to help you determine the slide?

c. Explain how you could have used only one point and the slope to graph the line $y = \frac{2}{3}x - 4$.

WR-65. Point-Slope Method for Graphing Lines The method you described in problem WR-64 is often called the **point-slope method for graphing lines**. To use it you simply plot one point that lies on the line (the y-intercept is an easy choice) and then use the slope to find additional points on the line. To check your work, substitute the coordinates of one of the additional points you found into the equation for the line.

a. Use the point-slope method to graph each of the following lines:

  i. $y = \frac{1}{3}x - 2$

  ii. a line with the slope -2 that passes through the point $(0, 1)$
  graph of $y = -2x + 1$

b. For each line in part a, select one of the additional points you found using the slope. Show that each selected point makes the equation of the line true by substituting its coordinates into the equation. *For example, starting with the y-intercept in part i we can use the slope to find $(3, -1)$. Substituting 3 for x in the equation, we get $y = \frac{1}{3}(3) - 2 = -1$.*

c. Add a description of the *point-slope method for graphing lines* to your Tool Kit.

EXTENSION AND PRACTICE

WR-66. Apply what you know about equations written in the form $y = mx + b$ to determine the slope and the y-intercept of the following lines. Then use the point-slope method to graph each line. Label each graph with its equation.

a. $y = \frac{2}{3}x - 4$ **b.** $2y + 3x = 8$ *$y = \frac{-3}{2}x + 4$*

c. $y = 7 - 3x$ **d.** $5(y - 2x) = 30$ *$y = 2x + 6$*

Time (hours)

WR-67. This graph shows the distances traveled by Cars A, B, and C over a certain time interval.

a. List the cars in order by speed, from greatest speed to least, and then explain how you know.   *Car A is fastest because its $\frac{distance}{time}$ slope ratio is greatest; cars B and C have the same speed because their lines have the same slope.*

b. Where did Cars A and B start? Explain how you know.   *Both started at Salt Lake. Their graphs contain (0, 0), which means that at time 0 hr their distance from Salt Lake was 0 mi.*

WR-68. A line has the equation $y = \frac{2}{3}x - 1$.

a. Verify that the points A (3, 1), B (6, 3), and C (15, 9) are all on the line.

b. Compute the slope of the line using points A and B. $m = \frac{2}{3}$

c. Compute the slope of the line using points A and C. $m = \frac{2}{3}$

d. Compute the slope of the line using points B and C. $m = \frac{2}{3}$

e. What do the results of parts b, c, and d tell you about using different points to calculate the slope of this line?   *No matter which points on the line are chosen, the slope is the same.*

WR-69. A line contains the points $(-3, 2)$ and $(2, 5)$. Margo thinks that the point $(12, 12)$ is also on this line. Do you agree? If so, tell why. If not, tell how you know it's not on the line.   *No, (12, 12) is not on the line; the point (12, 11) is.*

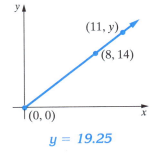

$y = 19.25$

WR-70. Use the graph and similar right triangles to write an equation. Solve the equation to find y.

WR-71. More Sums and Products Patterns In this problem you'll explore factoring relationships for the special sum-and-product pairs you examined in WR-24.

a. Look at just the Sum and the Product columns of the table in WR-24. How are the sum-and-product pairs alike? How are some of the sum-and-product pairs different from others?   *In every pair, one of the two numbers is 0 and the other is negative. In some pairs the sum is 0, in others the product is 0. In every pair except the ones with -10, the nonzero number is the negative of a perfect square. The pair in which -10 is the product is the only one with no solution.*

b. Now look at the sum-and-product table that you completed for WR-24. What patterns do you notice in the Integer Pair column? *If the product is 0, then one of the numbers is zero. If the sum is 0, then the integer pair is ± an integer, except when −10 is the product.*

c. How do the generic rectangles and factored polynomials of the entries in which the product is zero compare with the generic rectangles and factored polynomials of the entries in which the sum is zero? *If the product is 0, the rectangle and factored form are x by (x minus the sum). If the sum is zero, the rectangle and factored form are (x minus an integer) by (x plus the integer), except when −10 is the product.*

WR-72. Factor each of the following quadratic polynomials. Treat the set as a puzzle. If the numbers you try don't work for one expression, they may work for another.

a. $x^2 - 2x - 24$ *(x − 6)(x + 4)*   b. $x^2 + 11x + 24$ *(x + 3)(x + 8)*

c. $x^2 - 10x + 24$ *(x − 4)(x − 6)*   d. $x^2 + 5x - 24$ *(x + 8)(x − 3)*

Problem WR-73 focuses on the zero product property in preparation for WR-74 as well as for work with quadratics in Chapter 6.

WR-73. Margo and her group have been trying to remember whether an equation can have more than one solution.

a. Although Margo is certain that $T = -1002$ and $T = 471$ are both solutions to the equation

$$(T + 1002)(T - 471) = 0,$$

her group mates are skeptical. Show how Margo can convince her friends that $T = -1002$ and $T = 471$ make the equation true. *Hint: Substitute T = −1002 for both T's, and show that the product is zero. What happens if you do the same thing with T = 471?*

b. If $(x - 98)(x + 335) = 0$, what must be true about either the factor $x - 98$ or the factor $x + 335$? Explain how you know. *According to the zero product property, at least one factor must equal 0.*

Problem WR-74 is optional. As an extension of their work with factoring and the zero product property, it has students use both tools to solve quadratic equations. This not a major issue at this time—the problem serves as a preproblem for a focused development of solving quadratic equations by factoring in Chapter 6.

WR-74. Use factorizations you found in WR-72 and your observations in WR-73 to help you solve these equations:

a. $x^2 - 2x - 24 = 0$ *6, −4*   b. $x^2 - 10x + 24 = 0$ *4, 6*

WR-75. The length of a certain rectangle is five times its width. Use this information to sketch and label the rectangle. Write an expression to represent the perimeter. Find the rectangle's dimensions if its perimeter is 36 centimeters.   *3 by 15 cm*

WR-76. Driving on a country road, Henry travels 25 miles in 35 minutes.

a. If he continues at the same speed, how long will it take Henry to drive 90 miles?   *126 min, or 2 hr and 6 min*

b. Determine Henry's average speed in miles per hour.
$$\frac{25 \text{ mi}}{35 \text{ min}} = \frac{x \text{ mi}}{60 \text{ min}}; x = \frac{25 \cdot 60}{35} \approx 42.86 \text{ mph}$$

WR-77. Solve the following problem by writing and solving an equation. It may help to draw a simple diagram or make a guess-and-check table. Check your answer and then write it in a complete sentence.

Two cars, an Edsel and a Studebaker, are 635 kilometers apart. They start at the same time and are driven toward each other. The Edsel travels at a rate of 70 kilometers per hour, and the Studebaker travels at 57 kilometers per hour. In how many hours will the two cars meet?
70t + 57t = 635; 5 hr

WR-78. For the equation $2y - 4x = 6$, do the following:

a. Write the coordinates of the x- and y-intercepts.   *(0, 3), $(-\frac{3}{2}, 0)$*

b. Graph the intercepts, and draw the line.

c. Find the slope of the line.   *2*

WR-79. Exponent Practice Write each of the following expressions in a simpler exponential form:

a. $x^{-3} \cdot (x^4)^2$ *x^5*   b. $(x^3 \cdot y^7)^2$ *$x^6 y^{14}$*

c. $\frac{12}{4n^2}$ *$3n^{-2}$*   d. $\frac{(-2n^5)^3}{n^{-2}}$ *$-8n^{17}$*

e. $(99^2)(99^x)$ *99^{x+2}*   f. $\frac{1}{m^{12}} \cdot (m^6)^2$ *1 if m ≠ 0*

WR-80. Solve each of the following equations. Show your steps and how to check your answers.

a. $\dfrac{n+2}{6} = \dfrac{5n+6}{14}$ $\quad -\dfrac{1}{2}$

b. $\dfrac{4-w}{3} + \dfrac{5}{12} = \dfrac{2w+1}{18}$ $\quad \dfrac{61}{16} = 3.81215$

*Problem WR-81 outlines the "F.O.I.L. method" for multiplying binomials. Many students will have seen it before. We include it here only to show its connection to the use of generic rectangles, **not** because we believe students should memorize the algorithm.*

WR-81. Find each of following products of binomials:

a. $(2x+4)(3x+5)$ $\quad 6x^2 + 22x + 20$

b. $(5x-2)(x+3)$ $\quad 5x^2 + 13x - 6$

To multiply two binomials such as $(3x-2)$ and $(4x+5)$, you might use a generic rectangle, or you could use a method known as F.O.I.L. (first, outside, inside, last). The acronym F.O.I.L. is a reminder for finding the product of two binomials. Applying it to the product $(3x-2)(4x+5)$, you get

F multiply the **first** term from each binomial → $(3x)(4x) = 12x^2$

O multiply the **outside** terms → $(3x)(5) = 15x$

I multiply the **inside** terms → $(-2)(4x) = -8x$

L multiply the **last** term from each binomial → $(-2)(5) = -10$

This means $(3x-2)(4x+5) = 12x^2 + 15x - 8x - 10$

$= 12x^2 + 7x - 10.$

You can see how the F.O.I.L. method works if you look at a generic rectangle:

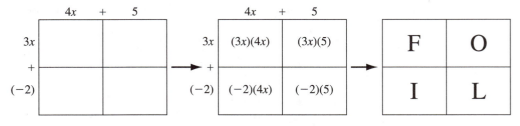

c. Use a generic rectangle or the F.O.I.L. method to find each of following products of binomials:

i. $(3x-1)(2x-5)$ $\quad 6x^2 - 17x + 5$

ii. $(5x+4)(x-2)$ $\quad 5x^2 - 6x - 8$

WR-82 More Dart Boards There's another new dart board at Bill's favorite pool hall. To win, a dart must hit the shaded part of the square and miss the circle. The area of the entire square is 144 square inches.

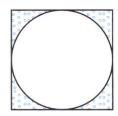

a. Find the area of the circle.   *36π ≈ 113.1 sq in.*

b. Find the area of the shaded part.   *144 − 36π ≈ 30.9 sq in.*

c. If he throws randomly, what is the probability that Bill will hit the shaded part and win the game? $\dfrac{144 - 36\pi}{144} \approx 0.2146$

5.4 USING GRAPHS TO INVESTIGATE LINEAR RELATIONSHIPS

NOTE Many of the problems in the remainder of this chapter require several steps in their solutions. Some problems are broken into several tasks and model a process you can use to solve other problems. Be careful to answer *all* questions and to complete *all* algebraic expressions, equations, and graphs as specified in each problem. In general, you should solve these problems first by referring to the graph and then by algebraic methods.

*Set groups to work right away on WR-83, a guided investigation. You may need to clarify why using a **trend line** makes sense when dealing with discrete data points. Suggest that students use WR-83 as a model if they need help with subsequent problems of this type.*

WR-83. Cricket Thermometers Biologists have observed that we can use the chirping of crickets to figure out the ambient temperature. An equation that relates the chirping rate and temperature is

$$T = \tfrac{1}{4}C + 37,$$

where C represents the number of chirps per minute and T is the temperature in degrees Fahrenheit.

A graph gives us an easy way to interpret this relationship. First we need to decide which variable goes on which axis. We usually choose axes so that the variable on the vertical axis *depends* on the variable on the horizontal axis. In this case, we are using chirping rate to calculate the current temperature, so the temperature we calculate (the output) *depends* on the chirping rate (the input).

a. Before you draw the coordinate axes, think about what kind of number makes sense for C. Could the number of chirps per minute be negative? zero? a fraction? Explain why or why not.   *No, yes, no; the number of chirps per minute is a whole number because it represents distinct items.*

b. Now draw a pair of coordinate axes on graph paper. Scale the horizontal axis so that each tick mark represents five chirps. Scale the vertical axis so that one tick mark represents 10°. Mark each axis up to 100. Be sure to label the axes with their units.

Because it only makes sense for the number of chirps to be a whole number, the Cricket Thermometers graph is *discrete*. It is also true that the Cricket Thermometers equation,

$$T = \tfrac{1}{4}C + 37,$$

is *linear*, so the points on the graph will all lie on a line. It is much easier to graph a line than it is to plot many discrete points, and we can use the **trend line** of the points to help answer questions related to the relationship modeled by the Cricket Thermometers equation.

c. Use the two-point slide method, or make a brief table of values, to graph the trend line of $T = \tfrac{1}{4}C + 37$.

d. Based on your graph, for what temperatures do you think the cricket indicator is not valid? Explain your answer. *when the temperature is less than 37° F*

e. What is the approximate temperature when a cricket chirps at a rate of 35 times per minute? *about 46° F*

f. You can represent the statement

"a cricket chirps 35 times per minute"

by drawing a vertical line. Draw the appropriate line so that it goes through your graph of the Cricket Thermometers equation.

g. The equation $T = 50$ means, "the temperature is 50°." Draw a horizontal line to represent this situation. How many chirps per minute would you expect when the temperature is 50°? Draw the appropriate line on the graph and answer the question. *about 52 chirps/min*

h. What does the equation $C = 72$ mean? How can it be represented on the graph? *There are 72 chirps per minute, which is represented by a vertical line.*

i. Write another equation similar to $T = 50$ or $C = 72$, and explain in writing what the equation means, just as you did in parts g and h.

WR-84. Look again at your graph of the Cricket Thermometers equation in WR-83.

a. Draw a slope triangle for the trend line. How many cricket chirps per minute correspond to a rise of one degree in temperature? What is the slope of the line?   *about 4; $m = \dfrac{1°F}{4 \text{ chirps/min}}$*

b. Where does the line intersect the vertical (T) axis?   *T-intercept is (0, 37)*

c. Compare the equation $y = \frac{1}{4}x + 37$ with the Cricket Thermometers equation $T = \frac{1}{4}C + 37$. Describe how they are alike and how they are different. How would you expect the graphs of the two equations to compare?   *Similarities: both graphs are linear, with the same y-intercepts and the same slopes. Difference: the graph of the equation $y = \dfrac{1}{4}x + 37$ is continuous, while the graph of the Cricket Thermometers equation is discrete; the two equations use different variables and have different sets of valid input and output values. We'd expect the graphs to be similar ($y = \dfrac{1}{4}x + 37$ is the trend line for $T = \dfrac{1}{4}C + 37$).*

d. Could you use the point-slope method to graph the line $y = \frac{1}{4}x + 37$? If so, explain what to do. If not, explain why not.   *You could, but it would be awkward using the given scaling of the axes.*

WR-85. The price the Village Bakery charges a supermarket for loaves of its sourdough bread depends on how many loaves the supermarket buys. If the supermarket buys 20 loaves or more, the price is $2 per loaf. Otherwise, the price per loaf is given by the relationship

$$y = \frac{24 - x}{2}$$

where y is the price in dollars and x is the number of loaves bought.

a. What numbers make sense for x and y in this situation?   *Both x and y are positive, and x is an integer; if $0 < x < 20$, then $2 < y < 12$ and if $x \geq 20$, then $y = 2$.*

b. After rewriting the equation $y = \dfrac{24 - x}{2}$ in the form $y = mx + b$, state the slope and y-intercept of the graph and its trend line.   *$y = -\dfrac{1}{2}x + 12$; $m = \dfrac{-1}{2}$; y-intercept is (0, 12)*

c. Using the point-slope method, graph the trend line for the equation you wrote in part b for $0 \leq x \leq 20$.

d. If the y-intercept of the trend line were a point on the cost graph, what would it represent? *If the relationship applied to x = 0, the cost would be $12 per loaf assuming 0 loaves were bought.*

e. When Bobbie graphed the trend line, she extended the graph until it intersected the x-axis. What is the x-intercept of the trend line? Why does the x-intercept not make sense in this situation? *(24, 0); the equation doesn't apply to x-values > 20. If it did, the cost would be $0 per loaf for 24 loaves, which doesn't make sense!*

EXTENSION AND PRACTICE

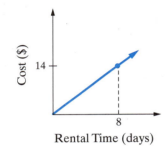

WR-86 Outdoor Adventures in a student-run nonprofit organization that offers weekend trips for students and rents a variety of equipment. All equipment is rented on a daily basis, and because their rates are so low, there are no discounts for extended rentals. This graph shows the cost to rent a backpacking stove.

a. Explain why it makes sense that the point (0, 0) is on the line.
 0 days cost $0

b. How much does it cost to rent a stove for 8 days? How much does it cost to rent a stove for 3 days? How about for 14 days? *$14, $5.25, $24.50*

c. What is the slope of the line, and what does it represent? Include the units in your answer. $slope = \dfrac{\$14.00}{8 \text{ days}} = \dfrac{\$1.75}{1 \text{ day}}$; *the rental rate for a stove is $1.75 per day.*

WR-87. Use your observations from problem WR-61 to state the slope and y-intercept of each of the following lines:

a. $y = -4x + 7$ *m = −4, b = 7*   b. $y = 4x$ *m = 4, b = 0*

c. $y = 4$ *m = 0, b = 4*   d. $y = x - \frac{3}{2}$ *m = 1, b = $-\frac{3}{2}$*

WR-88. Rewrite each of the following equations in y-form, and then graph them using the point-slope method:

a. $y - 4x = -3$ *y = 4x − 3*

b. $3x + 2y = 12$ *y = $-\frac{3}{2}$x + 6*

c. $3y - 3x = 7$ *y = x + $\frac{7}{3}$*

WR-89. Solve each of the following equations. Show your steps, and check your answers.

a. $5 + 4(x + 1) = 5$ -1

b. $4(2y - 1) = -10(y - 5)$ 3

c. $4(z + 5) - 3(z + 2) = 14$ 0

d. $\dfrac{2x}{7} = \dfrac{4}{5} + x$ $\dfrac{-28}{25} = -1.12$

WR-90. At 6 A.M. on Thursday, water was flowing into the lake behind Shasta Dam at a rate of 14,506 cubic feet per second, while it was being let out at a rate of 5287 cubic feet per second. At what rate was the water storage changing? If the water storage had continued to change at that rate, what would have been the overall change in the storage of Shasta Dam by 6 P.M. Thursday?   *The water storage was increasing at an average rate of 9219 cu ft/sec. At that rate, over a 12-hr period the storage would have increased by 398,260,800 cu ft.*

WR-91. Write and solve an equation in order to solve the following problem:

At 7:00 P.M. a red CRV started traveling north from Cheyenne at 50 miles per hour. At the same time, a Land Cruiser started traveling south at 40 mph. If the two vehicles keep traveling at these rates, at what time will the they be 225 miles apart?   *50t + 40t = 225; t = 2.5; at 9:30 P.M.*

WR-92. To make raisin-peanut gorp for his backpacking trip, Howard bought 8 pounds of raisins and 5 pounds of peanuts from the bulk bins at the co-op. The raisins cost $3.49 per pound, and the peanuts cost $6.30 per pound. Approximately how much did Howard spend per pound of gorp?

a. Identify and solve two subproblems that will help you solve the problem.   *How many pounds of gorp did Howard buy, and what was the total cost? 8 + 5 = 13 lb; 8($3.49) + 5($6.30) = $59.42*

b. Use your work from part a to find out about how much 1 pound of gorp cost Howard. $\dfrac{8(\$3.49) + 5(\$6.30)}{(8 + 5) \text{ lb}} = \dfrac{\$59.42}{13 \text{ lb}} \approx \$4.57/lb$

WR-93. Find x and y for each of the following pairs of similar triangles.

a. b.

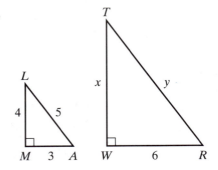 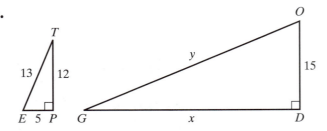

c. d.

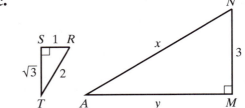 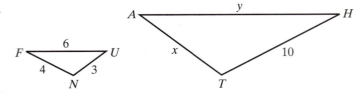

(a) $x = 8$, $y = 10$; (b) $x = 36$, $y = 39$; (c) $x = 6$, $y \approx 5.2$; (d) $x = 7.5$, $y = 15$

WR-94. Reggie bragged that he saved $12.60 by buying his new sneakers on sale at 30% off. How much did the sneakers cost originally?
$0.3x = 12.60$, so $x = \$42.00$

WR-95. Rewrite each of the following expressions by multiplying the binomials in parts a and b and factoring the polynomials in parts c and d.

a. $(2x + 3)(x + 5)$
 $2x^2 + 13x + 15$

b. $(4 - x)(3x + 10)$
 $-3x^2 + 2x + 40$

c. $3x^2 + 5x - 2$
 $(3x - 1)(x + 2)$

d. $x^2 + 8x - 20$
 $(x - 2)(x + 10)$

WR-96. Draw and complete a Diamond Problem for each of the following quadratic polynomials. Then use the Diamond Problem to factor the polynomial.

a. $x^2 - 0x - 9$ E, W = −3, 3; $x^2 - 0x - 9 = (x - 3)(x + 3)$

b. $m^2 - 0m - 36$ E, W = −6, 6; $m^2 - 0m - 36 = (m - 6)(m + 6)$

WR-97. Draw and label a generic rectangle for each of the following quadratic polynomials. Then use the generic rectangle to factor the polynomial.

a. $x^2 - 0x - 64$ Rectangle sides are $x - 8$ and $x + 8$;
 $x^2 - 0x - 64 = (x - 8)(x + 8)$

b. $w^2 - 0w - 1$ Rectangle sides are $w - 1$ and $w + 1$;
 $w^2 - 0w - 1 = (w - 1)(w + 1)$

WR-98. In this problem you will examine whether points on a parabola make the equation of the parabola true or false.

a. Graph the parabola given by the equation $y = x^2 - 3x$.

b. Use your graph to estimate the x- and y-intercepts of the parabola. Use substitution to check that the values you found make the equation $y = x^2 - 3x$ true. *y-intercept is (0, 0), x-intercepts are (0, 0) and (3, 0); $0^2 - 3(0) = 0$, $3^2 - 3(3) = 0$.*

c. Use your graph to find what value(s) x has if $y = 10$. *−2, 5*

d. Do the values for x that you found in part (c) make the equation $10 = x^2 - 3x$ true? Show how you know. *yes; $(-2)^2 - 3(-2) = 10$ and $5^2 - 3(5) = 10$*

WR-99. Keisha and Rudy were both working simplifying exponential expressions as you did in WR-50. To simplify $\dfrac{x^3}{x^3}$ for $x \neq 0$, Keisha wrote

$$\frac{x^3}{x^3} = 1.$$

Looking at Keisha's work, Rudy said, "But if we subtract exponents, we get

$$\frac{x^3}{x^3} = x^{3-3} = x^0.$$

In light of these two facts, what must be true about x^0 and 1? *They're equal.*

WR-100. Exponent Practice Use guess and check to solve each of the following equations for n:

a. $a^8 \cdot a^n = a^{-3}$ *n = −11*
b. $\dfrac{6^{2n}}{6^{n+1}} = 6^3$ *n = 4*
c. $a^{-5} \cdot a^n = 1$ *n = 5*
d. $a^{28} = \dfrac{1}{a^{4n}}$ *n = −7*

WR-101. Simplifying Fractions Rewrite each of the following fractions using as few digits or variables as possible:

a. $\dfrac{12x}{3x}$ for $x \neq 0$ *4*

b. $\dfrac{-2(x+3)}{x+3}$ for $x \neq -3$ *−2*

c. $\dfrac{(5-x)(x-1)}{(x-1)}$ for $x \neq 1$ *5 − x*

d. $\dfrac{(x+87)^2}{x+87}$ for $x \neq -87$ *x + 87*

WR-102. Feeding Frenzy According to Phil Gates and Gaden Robinson in *365 Days of Nature and Discovery*, piranhas have razor-sharp, interlocking teeth that can strip 15 cubic centimeters of flesh from their prey with each bite. Each piranha takes about five bites per minute. When piranhas gather in shoals of 20 or more, they can go on a feeding frenzy and quickly reduce their prey to nothing but bones.

An adult human body has about 50,000 cubic centimeters of flesh on it. If a shoal of 100 piranhas encountered a man swimming in the Orinoco River, Gates and Robinson claim that it would it take 6.25 minutes for the piranhas to reduce the swimmer to a skeleton.* Do you agree with the claim? Justify your response by showing your calculations, including units of measure.

*Phil Gates and Gaden Robinson, *365 Days of Nature and Discovery* (New York: Harry N. Abrams, Inc., 1994), p.32.

no; $\dfrac{50{,}000 \text{ cm}^3}{100 \cdot 5 \frac{\text{bites}}{\text{min}} \cdot 15 \frac{\text{cm}^3}{\text{bite}}} \approx 6.67$ min; it would take about 25 sec longer.

5.5 PRACTICE WITH $y = mx + b$

SECTION MATERIALS

EITHER:

WR-103 PATTERNS WITH LINES AND THEIR EQUATIONS RESOURCE PAGE

GRAPHING CALCULATORS (ONE PER PAIR OF STUDENTS, FOR WR-103)

OR

WR-104 PATTERNS WITH LINES AND THEIR EQUATIONS RESOURCE PAGE

Students should complete only two of the first three problems in this section. We suggest using graphing calculators to explore questions about slopes and y-intercepts of lines. Classes with graphing calculators should complete WR-103 and WR-105 (omitting WR-104). However, if graphing calculators are not available, have students complete WR-104 and WR-105 (omitting WR-103). Either combination of two problems will take a class period to complete.

If your students have graphing calculators, have them work in pairs on WR-103. The technology removes the tedium of plotting points and gives students the freedom to explore questions such as "What if the coefficient of x is 1? or 10? or 100? or 1000?... " and "What if the constant is 1? or 10? or 100? or 1000?... " (This activity could also be done, though less effectively, using a single computer with an overhead display attached, or with small groups taking turns using the computer while the rest of the class works on something else.)

If graphing calculators aren't available, begin with WR-104, which relates the calculator activities of WR-103 to graphs on paper. If students have done WR-103, they do not need to do WR-104.

If time permits **after** the students have completed their in-class work, do the following kinesthetic activities to review and summarize the equations $y = mx$ and $y = mx + b$:

a. In this activity, a forearm represents a line, and an elbow represents the y-intercept of the line.

b. Have students start with their left elbow in front of them at chest height; in this location, the elbow is at the origin.

NOTE *If you model along with the students, use your right elbow and forearm so the location and direction of the line are correct from the students' perspective.*

c. With their left elbows at the origin, ask students to use their left forearms to show the lines $y = x$, $y = -x$, $y = 2x$, $y = -2x$, $y = 10x$, $y = -10x$, $y = 100x$, and $y = -100x$.

To reinforce the idea that, as the slope of a line increases, its graph approaches a vertical line, ask students to describe how the line $y = 1000x$ would look.

d. Next ask the students to show the line $y = x + 1$ by first showing the line $y = x$ and then moving their elbows up one unit (about 2 to 3 inches) while maintaining the same slope.

Continue practicing by having students show the lines $y = x - 1$, $y = x + 3$, $y = x - 4$, $y = 2x + 1$, and $y = -x + 1$.

 WR-103. Patterns with Lines and Their Equations *This exercise is to be done with a graphing calculator.* You will also need your copy of the Resource Page for WR-103 ("Patterns with Lines and Their Equations").

Interesting patterns often occur when graphing several equations on one set of coordinate axes. The goal here is to find similarities among sets of graphs of lines and their equations. Try these examples, and then create some combinations on your own.

In your graphing calculator, enter each set of equations on the same set of coordinate axes. On your Resource Page, **sketch** the graphs on one set of axes, so you will have a record of them. Be sure to label each line with its equation. Look for differences and similarities among the graphs.

a. $y = x$

   $y = 2x$

   $y = 4.9x$

   What happens to the graph of $y = x$ when the coefficient of x is greater than 1—that is, when x is multiplied by a number greater than 1?
   It gets steeper.

b. $y = x$

   $y = \frac{1}{2}x$

   $y = \frac{1}{4}x$

   $y = \frac{1}{6}x$

   What happens to the graph of $y = x$ when the coefficient of x is between 0 and 1? *It gets less steep.*

c. $y = x$

   $y = \frac{-1}{4}x$

   $y = \frac{-1}{2}x$

   $y = -4x$

   $y = -2x$

   What happens to the graph of $y = x$ when the coefficient of x is negative?
   It slants downward from left to right.

d. $y = x$

$y = x + 2$

$y = x - 2$

$y = x + 4$

$y = x - 4$

What happens to the graph of $y = x$ when a constant (number) is added to x? *It shifts upward if the constant is positive, and downward if the constant is negative.*

e. How are all of the equations in parts a through d alike? *They're all linear and written in slope-intercept form.*

f. Identify the slope and the y-intercept for each of the following equations:

$y = x$	slope = __1__	y-intercept = __(0, 0)__
$y = 0.5x$	slope = __0.5__	y-intercept = __(0, 0)__
$y = 0.5x + 5$	slope = __0.5__	y-intercept = __(0, 5)__
$y = -0.5x$	slope = __−0.5__	y-intercept = __(0, 0)__
$y = 0.5x - 5$	slope = __0.5__	y-intercept = __(0, −5)__
$y = -0.5x - 5$	slope = __−0.5__	y-intercept = __(0, −5)__

WR-104. Patterns with Lines and Their Equations Interesting patterns often occur if you graph several equations on one set of coordinate axes. The goal here is to find similarities among sets of graphs of lines and their equations. Use your copy of the Resource Page for WR-104 ("Patterns with Lines and Their Equations") to complete this exercise.

a. Find the slope and y-intercept of each of the lines a, b, and c. Near each line, write its equation in slope-intercept form. *a: $y = x$; b: $y = 2x$; c: $y = 5x$*

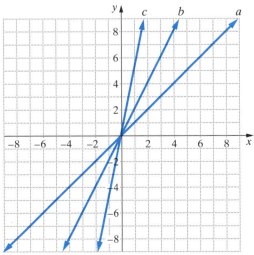

b. Look again at the lines in part a. What happens to the graph of $y = x$ when the coefficient of x is greater than 1—that is, when x is multiplied by a number greater than 1? *It gets steeper.*

c. Now repeat part a for the lines a, d, and e in the following graph:

a: $y = x$; d: $y = \dfrac{1}{2}x$; e: $y = \dfrac{1}{6}x$

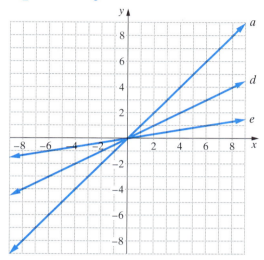

d. Look again at the lines in part c. What happens to the graph of $y = x$ when the coefficient of x is between 0 and 1? *It gets less steep.*

e. Repeat part a for the lines $a, f, g, h,$ and i in the following graph:

a: $y = x$; f: $y = \dfrac{-1}{6}x$; g: $y = \dfrac{-1}{2}x$; h: $y = -2x$; i: $y = -4x$

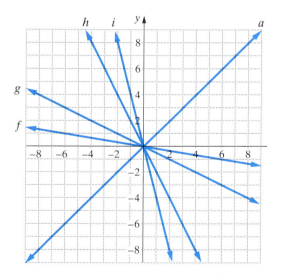

f. What happens to the graph of $y = x$ when the coefficient of x is negative? *It slants downward from left to right.*

g. Repeat part a for the lines a, j, k, l, and m in the following graph:
  a: $y = x$; j: $y = x + 4$; k: $y = x - 2$; l: $y = x + 2$; m: $y = x - 4$

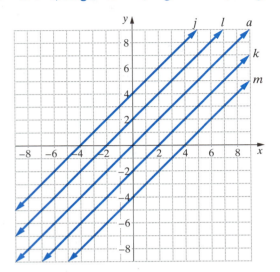

h. What happens to the graph of $y = x$ when a constant (number) is added to x? *It shifts upward if the constant is positive, and downward if the constant is negative.*

i. How are all of the equations in parts a through d alike? *They're all linear, of the form $y = mx + b$.*

j. Identify the slope and the y-intercept for each of the following equations:

equation	slope	y-intercept
$y = x$	1	(0, 0)
$y = 0.5x$	0.5	(0, 0)
$y = 0.5x + 5$	0.5	(0, 5)
$y = -0.5x$	-0.5	(0, 0)
$y = 0.5x - 5$	0.5	(0, -5)
$y = -0.5x - 5$	-0.5	(0, -5)

All students should complete WR-105.

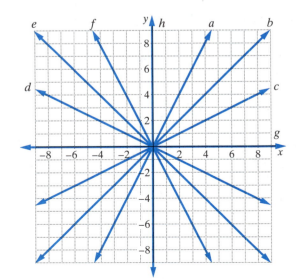

WR-105. On the following coordinate grid, there are eight lines: a through f, g (which is the x-axis), and h (the y-axis).

The slope-intercept equation of each line—
a: $y = 2x$; b: $y = x$; c: $y = \frac{1}{2}x$; d: $y = -\frac{1}{2}x$; e: $y = 2x$; f: $y = -2x$; g: $y = 0$.

a. Between which two of the eight lines would the graph of $y = 100x$ lie? Justify your answer. *It would lie between line a and the y-axis because its slope is greater than 2.*

b. Between which two of the lines would the graph of $y = \frac{2}{5}x$ lie? Justify your answer. *It would lie between line c and the x-axis because its slope is between 0 and $\frac{1}{2}$.*

c. Between which two of the lines would the graph of $y = \frac{3}{2}x$ lie? Justify your answer. *It would lie between lines a and b because its slope is between 1 and 2.*

EXTENSION AND PRACTICE

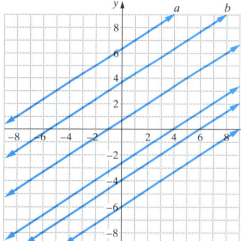

WR-106. Use the following graph to answer parts a through f.

a. Estimate the y-intercept of each of the lines *a* through *f*. *a: $(0, 6\frac{1}{3})$; b: $(0, 3\frac{2}{3})$; c: $(0, \frac{2}{3})$; d: $(0, -2\frac{2}{3})$; e: $(0, -4)$; f: $(0, -6)$*

b. What do lines *a* through *f* have in common? *They are all parallel and have the same slope; namely $\frac{2}{3}$.*

c. Between which two lines would the graph of $y = \frac{2}{3}x + 2$ lie? Justify your answer. *b and c*

d. Between which two lines would the graph of $y = \frac{2}{3}x - \frac{12}{7}$ lie? Justify your answer. *c and d*

e. Explain what happens to the graph of $y = \frac{2}{3}x$ when a positive number is added to the right-hand side of the equation. What happens when a negative number is added to the right-hand side of the equation? *When a positive number is added, the graph shifts upward that number of units. When a negative number is added, the graph shifts downward that number of units.*

f. Write an equation for each of lines a through f. $\quad$ a: $y = \frac{2}{3}x + 6\frac{1}{3}$; b: $y = \frac{2}{3}x + 3\frac{2}{3}$; c: $y = \frac{2}{3}x + \frac{2}{3}$; d: $y = \frac{2}{3}x - 2\frac{2}{3}$; e: $y = \frac{2}{3}x - 4$; f: $y = \frac{2}{3}x - 6$

WR-107. Rewrite each of the following equations in slope-intercept form:

a. $2y = 4x + 8 \quad y = 2x + 4$

b. $x = \dfrac{2y - 1}{3} \quad y = \frac{3}{2}x + \frac{1}{2}$

c. $6x + 3y - 5 = 0 \quad y = -2x + \frac{5}{3}$

d. $\dfrac{2}{y} = \dfrac{6}{x + 3} \quad y = \frac{1}{3}x + 1$

e. Use the y-intercept and the slope to graph each equation *without* making a table of values.

WR-108. The equation of any line can be written in slope-intercept form, $y = mx + b$.

a. For what value(s) of m does the graph slope upward? $\quad m > 0$

b. For what value(s) of m is the line horizontal? $\quad m = 0$

c. For what value(s) of m does the line slope downward? $\quad m < 0$

d. To your Tool Kit, add information about the relationship between the *sign of the slope* (positive, negative, or zero) of a line and the *direction of the slant* (upward, downward, or horizontal) of the line from left to right.

WR-109. Without drawing a graph, determine which, if any, of the points given here are on the graph of the equation $y = -2x - 3$.

a. $(-1, -1)$ *yes* $\quad$ b. $(-2, -2)$ *no* $\quad$ c. $(3, -7)$ *no* $\quad$ d. $(4, -11)$ *yes*

WR-110. When asked to complete the expression $x^2 + \_\_x - 15$ so that the resulting polynomial could be factored, the four people in Bernie's group got four different correct answers. Bernie noticed that the coefficient of x could be -2 because

$$x^2 + (-2)x - 15 = x^2 - 2x - 15 = (x - 5)(x + 3).$$

Patricia wrote $x^2 + 14x - 15 = (x + 15)(x - 1)$.

a. What answers did the other two members of Bernie's group find?

b. Complete the expression $x^2 + \_\_x - 10$ in four different ways so that the resulting polynomial can be factored, and write each of your polynomials as a product of its factors. $\quad x^2 - 9x - 10 = (x - 10)(x + 1)$; $x^2 + 9x - 10 = (x + 10)(x - 1)$; $x^2 - 3x - 10 = (x - 5)(x + 2)$; and $x^2 + 3x - 10 = (x - 2)(x + 5)$

In problem WR-111, students examine their solutions to find the domain (possible radii) for the problem.

WR-111. While stuck in commuter traffic, Professor Speedi thought of the following problem:

In the figure below, what is the area of the shaded region in terms of r and π?

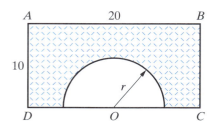

a. Identify and solve two subproblems. Then use your work to solve the problem. *Shaded area = area of rectangle − area of semi-circle* $= 200 - \dfrac{\pi r^2}{2}$ *sq units*

b. If segment DO is the same length as segment OC, for what values of r is the expression you found in part a valid? $0 < r < 10$

WR-112. Evan thinks that $\dfrac{2x + 10}{2}$ equals $x + 10$. Check to see whether this is so by evaluating each side of the equation for $x = 3$.   *It's not true since* $\dfrac{2(3) + 10}{2} = 8$ *and* $3 + 10 = 13$.

WR-113. Solve each of the following equations. Check each answer.

a. $3(4 - x) - (x + 1) = 0$ $\dfrac{11}{4}$

b. $\dfrac{4}{x + 1} - 4 = \dfrac{2}{3}$ $x = -\dfrac{1}{7}$

c. $(x + 17)(2x + 8) = 0$ $-17, -4$

WR-114. A certain rectangle has a diagonal of length 15 centimeters, and the ratio of the width to the length to the diagonal is 3:4:5.

a. Find the width and length of the rectangle.   *9 cm, 12 cm*

b. Find the area of the rectangle.   *108 cm²*

WR-115. Use any method you choose to multiply these binomials:

a. $(7x - 4)(3x + 2)$ $21x^2 + 2x - 8$

b. $(9x + 7)(4x - 3)$ $36x^2 + x - 21$

WR-116. A Difference of Squares An expression such as $x^2 - 4$ is called a *difference of squares*. Other examples of differences of squares are:

$$T^2 - 36,\ 5^2 - 3^2,\ 100 - N^2,\ 81 - 64,\ \text{and}\ 4x^2 - 25.$$

However, none of these expressions is a difference of squares:

$$x^2 - 5,\ T - 6^2,\ N^2 - 20,\ \text{or}\ 2x^2 - 81.$$

a. In a complete sentence, describe what makes an expression a difference of squares. *A difference of squares is an expression of the form "the square of a quantity or expression minus the square of another quantity or expression" or "something squared minus another thing squared."*

b. For each of following differences of squares, draw and complete a Diamond Problem, and then use it to factor the polynomial. The first one has been done as an example.

Difference of Squares	Diamond Problem	Factorization
Example: $x^2 - 4$	top: −4, left: −2, right: 2, bottom: 0	$x^2 - 4 = (x - 2)(x + 2)$
$T^2 - 36$	top: −36, left: −6, right: 6, bottom: 0	$T^2 - 36 = (T - 6)(T + 6)$
$P^2 - 100$	top: −100, left: −10, right: 10, bottom: 0	$P^2 - 100 = (P - 10)(P + 10)$
$N^2 - 1$	top: −1, left: −1, right: 1, bottom: 0	$N^2 - 1 = (N - 1)(N + 1)$

WR-117. Factor each of the following quadratic polynomials, if possible. Treat the set as a puzzle—if the numbers you try don't work for one expression, they may work for another.

a. $x^2 - 10x - 24$
 (x − 12)(x + 2)

b. $x^2 - 23x - 24$
 (x − 24)(x + 1)

c. $x^2 + 8x + 24$
 not factorable

d. $x^2 + 25x + 24$
 (x + 1)(x + 24)

Problem WR-118 is optional. In it students use the tools of factoring and the zero product property to solve quadratic equations. A facility for solving quadratics is not a major issue at this time—the problem serves as a preproblem for a focused development of solving quadratic equations by factoring in Chapter 6.

WR-118. Use factorizations you found in problem WR-117, along with the zero product property, to help you solve these equations:

a. $x^2 - 23x - 24 = 0$ *x = 24, −1*

b. $x^2 + 25x + 24 = 0$ *x = −1, −24*

WR-119. Exponent Practice Write each of the following expressions in a simpler exponential form:

a. $\dfrac{-6x^{-1}y^2}{10x^{-2}y^3}$ *$\dfrac{-3x}{5y}$ or $-0.6xy^{-1}$*   b. $\dfrac{8x^3y^4}{4x^2y}$ *$2xy^3$*

c. $(m^{-2})^{-1}$ *m^2*   d. $\dfrac{-7}{t^{-3}}$ *$-7t^3$*

WR-120. For each statement, find at least one value for n that makes the statement true.

a. $(3^{2n})^4 = 3^{5n+6}$ *2*

b. $\dfrac{x^n}{x^n} = 1$ *all values of n, unless x = 0*

c. $(6n)^2 = 144$ *−2, 2*

d. $n^n = 1$ *1 (Although it's often stated that 0^0 is undefined, there is growing acceptance of the definition $0^0 = 1$.)*

WR-121. The total area of the United States is about 3,623,420 square miles. This includes 79,537 square miles of water in the form of rivers, lakes, and streams. If a satellite fell from its orbit and crashed into the United States, what is the probability that it would land in water? Write the probability as a ratio (fraction), as a decimal, and as a percent.

$\dfrac{79,537}{3,623,420} \approx 0.022 = 2.2\%$

5.6 USING GRAPHS TO WRITE EQUATIONS OF LINES

SECTION MATERIALS
CHAPTER SUMMARY
DRAFT RESOURCE
PAGE (OPTIONAL, FOR WR-141)

After groups do WR-122, have them share the different ways they found to solve the problem.

WR-122. The points (2, 334) and (788, 934) lie on a line. Edward wants to find the coordinates of a third point on the line. Explain to Edward how he could use the idea of slope to help him do this, and then find a third point on the line.   *Some possibilities: (133, 434), (395, 634) or, (1574, 1534). Note that the slope is $\dfrac{100}{131}$.*

Problems WR-123 and WR-124 reinforce the graphical and algebraic relationships for slope, intercepts, and linear equations. The algebraic "check" outlined in WR-123f is just a preview of a more thorough development of writing an equation of a line without graphing points that students will encounter in Section 6.1. We don't expect students to be able to apply it on their own at this early stage.

WR-123. Sierra River Reservoir The Sierra River is one of many that is part of a water management system designed to control seasonal flooding and to provide community water sources and recreation. Sierra River Reservoir is a small lake formed behind a dam built over 50 years ago on the middle portion of the river. Its water level is replenished each spring by the runoff from the melting winter snowpack and is diminished throughout the summer and fall. In particularly wet years, water must be released from the reservoir to make room for water from the snowpack and spring rains. The approximate water level is measured on a large pole (calibrated in feet) near the dam's spillway.

This season, water releases began as the reservoir approached its capacity level. Water was released at a constant rate, and after 10 weeks, the water level measured at the spillway was 123.50 feet. After three more weeks of releasing water at that same rate, the reading was 116.75 feet. If possible, hydrologists would like to model this situation with an equation for use in making decisions.

a. On graph paper, set up and label a pair of coordinate axes, and then plot the water-level data.

b. Using a straightedge, connect the two points you plotted and draw a slope triangle. Then use a slide to find two additional points on the graph.   *Plot points (10, 123.50) and (13, 116.75). Possible slide: down 6.75 and to the right 3. Possible additional points: (7, 130.25), (16, 110.00), and (19, 103.25).*

c. Draw a trend line through the four points from part b. Does this trend line model the water level of the reservoir over time, or should the water-level graph be discrete? Explain why you think so.   *The trend line models the water level over time.*

d. Find the slope, including units, of the line you drew in part c. Use units to interpret the slope ratio as a rate.
$$m = \frac{-6.75 \text{ ft}}{3 \text{ wk}} = \frac{-2.25 \text{ ft}}{1 \text{ wk}} = -2.25 \text{ ft/wk}$$

e. Use the rate you found in part d to find the *y*-intercept of the line. What information does the *y*-intercept represent in this situation? What is the *x*-intercept of the line, and what information does it represent about the situation?   *Starting from data point (10, 123.50), repeat the slide "up 2.25 and to the left 1" until the y-axis is reached. The y-intercept, (0, 146), represents the water level before water was released. The x-intercept, (≈65, 0), represents the number of weeks it would take for the water level to reach 0 ft at the given rate of release.*

f. Here's an algebraic way to check the accuracy of the y-intercept you found from the graph of the water-level data:

First let (0, B) represent the actual y-intercept, and use the slope you found in part d to write the slope-intercept equation of the line. Then, since you know that the two given data points lie on the line, substitute values from either one of them into the equation, and solve for B.

Using the algebraic method, what do you get for the y-intercept of the line? How does the y-intercept you found algebraically compare with the y-intercept you found from the graph? *y = −2.25x + B; B = 123.50 + 2.25(10) = 146, or B = 116.75 + 2.25(13) = 146; so the y-intercept is (0, 146).*

g. Write an equation to model the water level of Sierra River Reservoir after the start of water release. *y = −2.25x + 146, or its equivalent in another form*

WR-124. Temperature Conversion On a thermometer, 20° Celsius is equal to 68° Fahrenheit. Also, 10° Celsius is equal to 50° Fahrenheit. The conversion from degrees Celsius to degrees Fahrenheit is linear, which means that if the conversion data were plotted as points, the points would lie on a line.

a. On graph paper, set up and label a pair of coordinate axes with degrees Celsius on the horizontal axis. Then plot the temperature conversion data. *Plot points (20, 68) and (10, 50).*

b. Use a slide to find two more points on the temperature conversion graph. Connect the points with a line, and find its slope. Be sure to include units. Use units to interpret the slope ratio as a rate.

Possible slide: up 18 and to the right 10; possible additional points: (30, 86) and (40, 104); $m = \dfrac{18°F}{10°C} = \dfrac{9°F}{5°C} = \dfrac{9}{5}°F$ per degree Celsius.

c. Use your graph in part b to estimate the y-intercept of the temperature conversion line.

d. Use information from parts b and c to write an approximate equation for the temperature conversion line. *$F = \dfrac{9}{5}C + 32$, or its equivalent in another form*

e. The linear equation you wrote in part d could be used as a formula for converting degrees Celsius to degrees Fahrenheit. Use your temperature conversion equation to determine the temperature in degrees Fahrenheit for 100° C. Check this with your graph. *212°F*

WR-125. Draw a sketch of two line segments that have the same slope but are not part of the same line.   *Sketch any two parallel line segments.*

WR-126. Is the following situation possible? Justify your response.

Segments *UC*, *CD*, and *UD* have slopes $\frac{1}{2}$, $\frac{1}{2}$, and -2, respectively. Moreover, points *U*, *C*, and *D* all lie on the same line.   *No. If U, C, and D lie on a line, all three segments will have the same slope.*

EXTENSION AND PRACTICE

WR-127. Points $(-23, 345)$ and $(127, 311)$ are two points on a line. Find the coordinates of a third point on the line.   *One possibility is (277, 277). Note that the slope is $\frac{-17}{75}$.*

WR-128. A certain line with slope $\frac{2}{7}$ goes through the point $(10, 3)$. Find two other points on the line.   *Some possible points: (17, 5), (24, 7), and (3, 1)*

WR-129. Suppose a line with slope $\frac{2}{7}$ goes through the point $(0, 4)$. Find an equation of the line.   *$y = \frac{2}{7}x + 4$, or its equivalent in another form*

WR-130. Without drawing a graph, determine which, if any, of the following points are on the graph of $y = -3x - 5$: $(0, -5)$, $(2, -11)$, and $(-3, -14)$.   *(0, −5) and (2, −11)*

WR-131. If the line $y = 2x + K$ passes through the point $(2, 1)$, what must be the value of *K*?   *−3*

WR-132. Graph each of the following lines on the same set of coordinate axes. What do you notice?

   a. $y = 2x + 3$ **b.** $2x - y + 3 = 0$ **c.** $4x - 2y + 6 = 0$

All three lines are the same because all three equations are forms of $y = 2x + 3$.

WR-133. Write an equation for the following problem, and then solve the equation:

Mae started on the bike trail at a rate of 10 miles per hour. Lydia started one hour later and rode at a rate of 15 miles per hour. At these rates, how long would it take for Lydia to catch up with Mae?   *(Hint: If Mae bikes 7 hr, how long would Lydia bike? How far would each travel at these rates?) Possible equations: 15L = 10(L + 1) or 10M = 15(M − 1). Lydia would catch up after biking for 2 hr (Mae would have biked for 3 hr).*

WR-134. Seymour has $2000. He put part of it in an account yielding 6% annually. He put the rest in a restricted savings account yielding 8% annually. After one year the two accounts will have a total of $2136.50.

a. What is the total interest that Seymour's accounts will earn? *$136.50*

b. How much money did Seymour invest at 6%?
0.06x + 0.08(2000 − x) = 136.50; Seymour invested $1175 at 6%.

c. Seymour had been hoping to have a total of $2500 in his accounts by the end of the year so he could buy a new stereo system. What percent interest would Seymour need to earn on $2000 to reach his goal? *$\frac{500}{2000} = 25\%$*

WR-135. Find the area of a rectangle with one side 10,000 millimeters long and the other side 20 meters long. Think carefully about the units. *200 m² or 2·10⁸ mm²*

WR-136. Solve each of the following equations for x:

a. $1234x + 23{,}456 = 987{,}654$ *$x = \frac{964198}{1234} \approx 781.36$*

b. $\frac{10}{x} + \frac{20}{x} = 5$ *x = 6*

c. $x(x - 1)(x - 2) = 0$ *x = 0, 1, or 2*

WR-137. $\triangle RAT$ is similar to $\triangle FNK$ with $\angle R = \angle F$ and $\angle A = \angle N$:

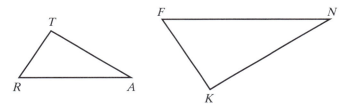

Suppose $|RA| = 10$ units, $|RT| = 5$ units, $|AT| = 9$ units, and $|FN| = 7.5$ units. Copy and label the diagram, and then find $|NK|$.

$\frac{7.5}{10} = \frac{|NK|}{9}$; |NK| = 6.75 units

WR-138. For each of the following differences of squares, draw and label a generic rectangle and factor the polynomial.

Example: $4x^2 - 25 = (2x - 5)(2x + 5)$

	$2x$	$+\ 5$
$2x$	$4x^2$	$10x$
$+\ (-5)$	$-10x$	-25

▲

a. $4x^2 - 9$ *4x² − 9 = (2x − 3)(2x + 3)*

b. $T^2 - 81$ *T² − 81 = (T − 9)(T + 9)*

c. $36N^2 - 1$ *36N² − 1 = (6N − 1)(6N + 1)*

d. Add information about what it means for an expression to be a *difference of squares* to your Tool Kit. Include two examples of factoring a difference of squares.

WR-139. Marci isn't sure whether $5^0 = 0$ or $5^0 = 1$. Use what you know about exponents to show Marci two different ways to figure out what the value of 5^0 should be. *Use a pattern of decreasing exponents: We know $5^3 = 125$, $5^2 = 25$, and $5^1 = 5$. For each decrease of 1 in the exponent, the value is one-fifth the previous value. So for 5^0 we get $\frac{1}{5}$ of 5, which is 1. Use a fractional form of 1: $1 = \frac{5^2}{5^2} = 5^{2-2} = 5^0$.*

WR-140. Matching Exponential Expressions Use what you've learned about rewriting exponential expressions to match each expression in the left-hand column with one or more equivalent expressions in the right-hand column. An expression may be used more than once. Part a has been started for you as an example.

Example: a. $\dfrac{(-16x^4)(4x^5)}{8x^6} = \dfrac{-64x^9}{8x^6} = \dfrac{-64}{8} \cdot x^{9-6} = \underline{}$ ▲

a. $\dfrac{(-16x^4)(4x^5)}{8x^6}$ $-8x^3 = (-2)^3 \cdot x^3$ 2^{-3}

b. $3^2 \cdot x^2$ $(3x)^2$ $\dfrac{1}{x}$

c. $(x^2)^{-3}$ x^{-6} $-8x^3$

d. $\dfrac{(27x^3)(9x)}{(3x)^5}$ $x^{-1} = \dfrac{1}{x}$ $(3x)^2$

e. $\left(\dfrac{x^2}{16}\right)(2x^{-2})$ 2^{-3} x^{-1}

f. $(-2x)^3$ $-8x^3 = (-2)^3 \cdot x^3$ $(-2)^3 \cdot x^3$

g. $x^2 \cdot x^{-3}$ $x^{-1} = \dfrac{1}{x}$ x^{-6}

h. $\dfrac{1}{2^3}$ 2^{-3}

WR-141. Chapter 5 Summary Rough Draft In this chapter you have used your understanding of the concept of ratio to learn techniques for writing equations of lines from data or graphs.

a. What are the main ideas and skills for this chapter?

b. Make photocopies of the Chapter Summary Draft Resource Page (or follow the same format on your own paper). Use a page to complete the following steps for each main idea or skill you listed in part a:

- State the main idea or skill.
- Select and recopy a problem that is a good example of the idea or skill, and which you did well.
- Include a completely worked-out solution to the selected problem.
- Write one or two sentences that describe what you learned about the idea or skill.

As in previous chapters, you will have the opportunity to revise your work, so your focus should be on the content of your summary rather than on its appearance. Be ready to discuss your responses with your group at the next class meeting.

c. How have you used the idea of ratio in this course?

d. Suppose you are given two points on a graph. What subproblems would you solve to write an equation of the line through those points?

e. If you were to choose a favorite problem from the chapter, which would it be? Why?

f. Find a problem that you still cannot solve or that you are unsure you would be able to solve on a test. Write out the question and as much of the solution as you can. Then explain what you think is keeping you from solving the problem. Be clear and precise in describing the hard part.

5.7 SUMMARY AND REVIEW

SECTION MATERIALS

POSTER PAPER OR OVERHEAD TRANSPARENCIES (OPTIONAL, FOR WR-164)

Start with groups discussing their Chapter 5 Summary rough drafts, as you circulate and listen.

The Extension and Practice problems vary in level of difficulty. Assign the problems that best fit your students' needs.

We ask students to pull together the big ideas from the course to date in the Course Summary (WR-164). This will be a challenge for most students. If you have the time, one approach would be to make it a group assignment. After discussing their ideas, each group could record their ideas on poster paper or overhead transparencies, and then report to their classmates. Posters hanging on the classroom wall, or handouts photocopied from the transparencies, are useful reminders of what the students have learned so far.

WR-142. Chapter 5 Summary Group Discussion Take out the rough-draft summary you completed in WR-141. Use some class time to discuss your work; use homework time to revise your summaries as needed.

To start your discussion, have group members take turns reading their lists of main ideas and skills for the chapter. Try to come to a consensus of four to six ideas and/or skills. For each item on the group's list, choose one member of the group to lead a short discussion. When it is your turn to be the discussion leader, you should do the following:

- Explain the problem you chose to illustrate your main idea or skill.
- Explain why you chose that particular problem.
- Explain, as far as possible, the problem from WR-141f that you still cannot solve.

This is also your chance to make sure your summary is complete, to update your Tool Kit, and to work together on problems you may not be able to solve yet.

In problem WR-143, some students may notice right off that the graphs of the lease plans should be discrete. Let their observations stimulate a class discussion about what the graphs represent in part d and the usefulness of graphing trend lines rather than discrete points.

Chapter 6 will introduce students to solving systems of linear equations algebraically. Don't do it now.

WR-143. Leasing a Car Otto had been saving to buy a used car, while dreaming of owning a brand new Saab. With $3574.89 in the bank, he'd just about reached his goal. However, when he visited Sven's Scandanavian Motors showroom just to look around, a sales representative convinced him that he didn't need to settle for a used car when his dream car was right within reach. Yes, Sven had just the perfect car-leasing plan for him! In fact, Otto could choose between two options. He could pay $220 per month plus an initial fee of $600 if he wanted to sign up for the Low-Down Plan and leave some cash in the bank. Or, he could choose the Low-Monthly Plan and pay only $155 per month and an initial fee of $1500. Otto's unsure what to do. He needs some advice about which plan to choose.

a. Each car-leasing plan involves a constant rate of payment. State each rate, and describe what it means in terms of the slope of the graph associated with the plan. Because the payment rate of each plan is constant, what shape would you expect the graph associated with each plan to have? *Low-Down Plan (LPD), $220/mo; m = 220. Low-Monthly Plan (LMP): $155/mo; m = 155. We expect both graphs to be linear.*

b. What does the initial fee represent in terms of the graph associated with each plan? *the y-coordinate of the y-intercept*

c. For each plan, write an equation that represents the cost of leasing a car for a given number of months. *LDP: $y = 600 + 220x$ and LMP: $y = 1500 + 155x$.*

d. Graph each equation you wrote in part a. Label the vertical axis "Total Cost (dollars)" and the horizontal axis "Number of Months." Be sure to label each graph.

e. Describe how your two graphs relate to the solution of the problem. When are the costs for both plans the same? When is the Low-Down Plan a better choice? When is the Low-Monthly Plan a better choice? Which plan would you advise Otto to choose and why? *At $\frac{180}{13} \approx 13.8$ mo, the costs are the same. The LDP costs less if the number of months is less than $\frac{180}{13}$; otherwise the LMP is cheaper.*

f. Sven's Scandanavian Motors does not allow leasing of a car for portions of a month. How does this information change your graph in part d and answers in part e? *The graphs should be discrete, with whole-number input values. Since $\frac{180}{13}$ is not a whole number, for no number of months are the costs the same. The LDP costs less if the number of months is less than 14; otherwise the LMP is cheaper.*

WR-144. Rules for Working with Integer Exponents Since the beginning of this course you have been using the meaning of integer exponents and patterning to solve exponent problems in ways that made sense. In problem LT-131 you summarized how to rewrite exponential expressions of the form $x^a \cdot x^b$, and $(x^a)^b$ for positive integers a and b, by explaining how to rewrite the expressions $x^{47} \cdot x^{21}$ and $(x^{103})^6$ with a single exponent. In EF-77 you saw that an expression of the form x^{-a} means $\frac{1}{x^a}$ and in EF-108, you explained how to simplify fractions such as $\frac{x^5}{x^3}, \frac{x^3}{x^5}$, and $\frac{x^4}{x^4}$, in which the numerator and denominator have the same base but different exponents. You have also applied what you know about integer exponents to expressions of the form $(2x)^3$ and $(xy)^5$.

a. Refer to problems LT-131, EF-77, EF-108, and WR-140, and previous Tool Kit entries to refresh what you have observed about rewriting exponential expressions. Use symbols to write your observations as five general rules.

Rules for Working with Integer Exponents

1. $x^a \cdot x^b =$ x^{a+b}

2. $(x^a)^b =$ x^{ab}

3. $\dfrac{x^a}{x^b} =$ x^{a-b}

4. $x^{-a} =$ $\dfrac{1}{x^a}$

5. $(xy)^a =$ $x^a y^a$

b. Add your group's *Rules for Working with Integer Exponents* to your Tool Kit.

WR-145. Tool Kit Check Your Tool Kit should be quite a substantial resource by now. New additions for this chapter should include at least the following topics:

- definition of the slope of a line   *WR-8, WR-11*
- how to find the slope of a line   *WR-11*
- interpreting the slope of a line   *WR-9*
- rewriting equations in *y*-form   *WR-35*
- the two-point slide method for graphing lines   *WR-34 through WR-38*
- the slope-intercept form of a linear equation.   *WR-61*
- rules for finding the slope and the *y*-intercept of a line in the form $y = mx + b$ *WR-61*
- the point-slope method for graphing lines   *WR-65*
- the relationship between the sign of the slope of a line and the direction of the slant of the line from left to right   *WR-108*
- what it means for an expression to be a difference of squares   *WR-138*
- Rules for Working with Integer Exponents   *WR-144*

You are not limited to topics on this list. If there are other ideas you'd like help remembering, add them to your Tool Kit as well.

a. To be sure that your Tool Kit entries are clear and accurate, exchange Tool Kits with members of your group, and read each other's entries.

b. At home, make any necessary clarification, revisions, or updates to your Tool Kit.

EXTENSION AND PRACTICE

WR-146. In this problem you will practice working with lines that are parallel.

a. Use the point-slope method to graph the line $y = -3x + 6$.

b. What can you say about the slope of any line parallel to the line $y = -3x + 6$? *slope is −3.*

c. Plot the point $(5, -2)$ on the coordinate grid you used in part a. Write an equation for the line that is parallel to the line $y = -3x + 6$, and that passes through the point $(5, -2)$. *$y = -3x + 13$*

WR-147. Rewrite each of the following equations in slope-intercept form. Then use the point-slope method to graph each line. Label each graph with its slope-intercept equation.

a. $5y = 7.5x - 12.5$
   $y = 1.5x - 2.5$

b. $6y + 3x = 12$
   $y = -0.5x + 2$

c. $4 = y - \frac{1}{2}x$
   $y = 0.5x + 4$

d. $4(y - 3x) = 24$
   $y = 3x + 6$

WR-148. Find the x-intercept, the y-intercept, and the slope of the line given by each of the following equations:

a. $3x + 8y = 12$ *x-intercept is (4, 0); y-intercept is (0, 1.5); slope is $-\frac{3}{8}$*

b. $8x - 3y = -12$ *x-intercept is (−1.5, 0); y-int. is (0, 4); slope is $\frac{8}{3}$*

c. $3y = 8$ *no x-intercept; y-intercept is (0, $\frac{8}{3}$); slope is 0*

WR-149. Without drawing a graph, determine which, if any, of the points are on the graph of the given linear equation:

a. $(8, 4), (4, 0), (4, 3), (3, 4)$ for $y = \frac{1}{4}x + 2$ *yes, no, yes, no*

b. $(0, -7), (2, -1), (4, 5), (6, 11)$ for $y = 3x - 7$ *yes, yes, yes, yes*

WR-150. What is the slope of the line that passes through the points $(5, -3)$ and $(95, 2)$? Explain how you got your answer. *$m = \frac{5}{90} = \frac{1}{18}$ Possible explanation: sketch a slope triangle for the two points, and find the lengths of the vertical and horizontal sides. The vertical change (up or down) is the numerator of the slope, and the horizontal change (right or left) is the denominator.*

WR-151. To write an equation for the line that passes though the points (5, 4) and (−10, 1), first graph the two points and draw the line that contains them. Then find the slope of the line and estimate the y-intercept of the line. Finally, use what you know about the slope and the y-intercept of a line to write an equation for the line that passes though the two given points. *$m = \frac{1}{5}$; y-intercept is (0, 3); $y = \frac{1}{5}x + 3$, or its equivalent in another form*

WR-152. Sketch a graph and label the axes for each of the following descriptions.

a. As the temperature increases, the volume of the material increases at a steady rate. *Possible solution:*

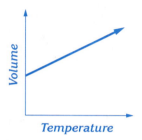

b. As the temperature increases, the number of people at the beach increases until the temperature reaches 110°, then the number levels off. *Possible solution:*

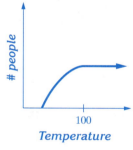

WR-153. When James saw the following diagram in his algebra text, he was reminded of his problem of finding the height of the center support for his skateboard ramp:

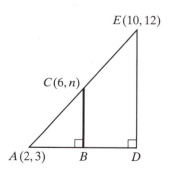

a. Copy the diagram and find the coordinates of points D and B.
   B is (6, 3) and D is (10, 3).

b. Use part a to find $|AB|$, $|AD|$, and $|DE|$, and to write an expression for $|BC|$.   *$|AB| = 4$, $|AD| = 8$, $|DE| = 9$, and $|BC| = n - 3$.*

c. Write an equation and solve it to find n. $\frac{|AB|}{|AD|} = \frac{|BC|}{|DE|}$, so $\frac{4}{8} = \frac{n-3}{9}$; $n = 7.5$ $\frac{|DE|}{|AD|} = \frac{|BC|}{|AB|}$, so $\frac{9}{8} = \frac{n-3}{4}$; or

d. What is the ratio of the lengths of corresponding sides of the two similar triangles shown in the figure? Use this fact to find n without solving an equation. $\frac{|AB|}{|AD|} = \frac{1}{2}$, so $|BC| = \frac{1}{2}|DE| = 4.5$; $n = 3 + 4.5 = 7.5$

WR-154. Exponent Practice Use your group's Rules for Working with Exponents to rewrite each of the following expressions in a simpler exponential form.

a. $\frac{(-6x^{-1})^2}{2x^3}$ $18x^{-5}$

b. $\frac{24x^3y}{16x^{-2}y^4} \cdot \frac{3x^5}{2y^3}$

c. $(5^{-2})(5^{x-1})$ 5^{x-3}

d. $\frac{12^2}{(2m)^3}$ $18m^{-3}$

WR-155. Rabies was recently reported to affect 3 of every 10 skunks in a mountain county. Sampling studies (similar to sampling fish populations) have revealed that there are about 22,400 skunks in the county. About how many of the county's skunks would be expected to carry rabies?   *about 6720*

WR-156. Is $\frac{x-2}{x+3} = \frac{-2}{3}$? Explain why or why not.   *only if $x = 0$*

WR-157. The perimeter of a certain rectangle is 28 centimeters. The length of the rectangle is 4 centimeters more than its width. Write an equation to represent this information, and then find the length and width. Finally, write the ratio of length to width.   *$2x + 2(x + 4) = 28$; 5, 9; $\frac{9}{5}$*

WR-158. Arturo has earned 134 points so far in his algebra class. To get an A he needs to earn 90% of the total possible points. One more test, worth 45 points, is coming up. It will bring the total number of points to 195. How many points must Arturo get on this test in order to earn an A in the class? Write an equation and solve the problem.   *41.5 more points*

WR-159. Use any method you choose to multiply these binomials:

a. $(2x - 5)(3x - 10)$ *$6x^2 - 35x + 50$*

b. $(5x - 4)(3x - 2)$ *$15x^2 - 22x + 8$*

WR-160. Factor each difference of squares:

a. $36 - T^2$ $\quad 36 - T^2 = (6 - T)(6 + T)$

b. $25x^2 - 1$ $\quad 25x^2 - 1 = (5x - 1)(5x + 1)$

c. $9P^2 - 64$ $\quad 9P^2 - 64 = (3P - 8)(3P + 8)$

WR-161. More Exponent Practice Solve each of the following equations for n:

a. $\dfrac{6^{3n}}{6^{n-4}} = 6^{10}$ $\quad n = 3$

b. $a^6 \cdot a^n = 1$ $\quad n = -6$

WR-162. Use the figure to answer the following questions:

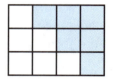

a. What is the area of the *unshaded* region? **6 sq units**

b. If the unshaded region in part a is called a three-step region, what would be the area of a five-step region? Draw a diagram. **15 sq units**

c. What would the area of a 10-step region be? **55 sq units**

d. Describe how you would find the area of a region with 157 steps. What would the area of a 157-step region be? *Possible answers: (1) add $1 + 2 + 3 + \cdots + 157$ to get 12,403 sq units; (2) in a rectangle 157 units wide and 158 units long, the unshaded portion is half the area of the rectangle, so its area $= \dfrac{(157)(158)}{2} = 12{,}403$ sq units.*

e. Describe how you would find the area of a region with N steps.

Possible answers: (1) Add $1 + 2 + 3 + \cdots + N$ sq units. (2) Calculate $\dfrac{N(N+1)}{2}$ sq units.

WR-163. Chapter 5 Summary Revision This is the final summary problem for Chapter 5. Using your rough draft from WR-141 and the ideas your group discussed for WR-142, spend time revising and refining your Chapter 5 Summary. Your presentation should be thorough and organized, and should be done on paper separate from the rest of your assignments.

WR-164. Course Summary It may seem strange that this problem is called "Course Summary" when you haven't even completed the course yet. But the idea of writing a summary of what you've learned so far in the course is really not so strange. Recall that the idea of writing a chapter summary was introduced in Chapter 1 (in problem OP-104) by this statement:

> In this course, as in most courses you study, you will continue to work on many ideas, concepts, and skills. There is a lot of information in this course. It may be difficult to process this information and organize it for yourself, so that you can easily remember *all* of it. It often helps to summarize it in the context of big ideas.
>
> Writing a summary is an important part of the learning process, and you will be expected to write a summary for every chapter in this course.

A summary of the main ideas you've learned in the course so far will help you see how much you've learned and how it all fits together. Use your chapter summaries to identify the important ideas you've learned. List at least five, and no more than eight, really important ideas, and choose one or two problems to represent each idea.

CHAPTER SIX—FOR THE INSTRUCTOR

FITNESS FINANCES
Solving Systems of Equations

MATERIALS
- Graph paper and straight-edges (needed throughout the chapter)
- Dot paper
- Graph for FF-54 prepared ahead on a transparency or flip-chart grid paper
- Algebra tiles
- Transparencies and pens for the overhead projector (optional)
- Resource Pages for FF-48, FF-55, and FF-113–FF-116; also, Summary Draft Template (optional)

By focusing on interpreting graphs of systems of equations, this chapter helps students continue to develop their understanding of the relationship between equations and their graphs. The main focus is on linear systems: their graphs, the points of intersection of the graphs, and the relationship between the intersections and solutions of systems. Students interpret the graphs of pairs of linear equations and use the method of substitution to solve simple linear systems algebraically. Throughout we emphasize the relationship between the graphical and algebraic methods of solution to help students develop an appreciation for the power of the algebraic method.

Most of the graphs associated with this chapter's word problems are linear, but not all are continuous. Because it is easier to draw a trend line than to plot many discrete points, students are directed to graph the line of best fit for the (linear) data. They can then use the lines to help analyze or solve the problem.

Students also explore intersections of linear and nonlinear graphs and systems of linear and quadratic equations. Again, the focus is on the visual representation of a system (by graphing), followed by an algebraic approach (in this case factoring to use the zero product property). Students will see that the system $y = 0$ and $y = ax^2 + bx + c$ can be solved using the single equation $0 = ax^2 + bx + c$.

Before starting the chapter, read through the Investigation in Section 6.7. There are several ways to do the Investigation: as a homework project; or as oral and/or poster presentation to fill one class meeting; or as a two-class project—one meeting for students to develop responses to problems FF-114 and FF-115 (assigned after Section 6.3), and a second for presentations.

If you plan to have students do the Investigation, assign problem FF-21 in Section 6.1 to have them read through the whole project. We have found it works best to decide on due dates for all four parts of the project (FF-113, FF-114, FF-115, and FF-116) as soon as possible, and to introduce the project to students along with Section 6.1. A sample assignment and grading table is included in the "Investigation: Using Algebra to Model and Analyze Linear Data Resource Page" for FF-113 through FF-116.

You'll need to assign FF-113, the project proposal, as you start the chapter to give students time to collect their data. Students could start working on problems FF-114 and FF-115 after Section 6.2.

Throughout this chapter, emphasize cooperative work and neat and careful graphing.

CHAPTER SIX

FITNESS FINANCES
Solving Systems of Equations

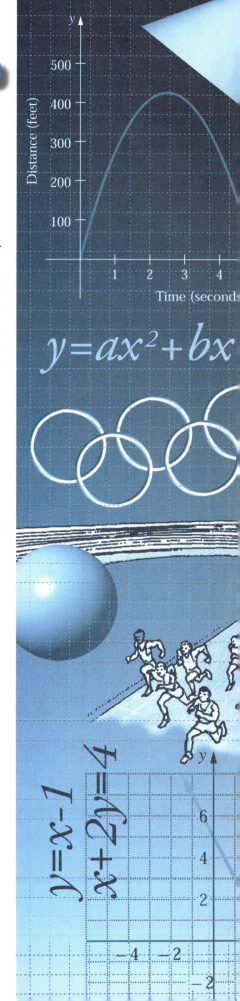

So far in this course you've graphed a lot of equations and focused a lot of attention on solving word problems by writing and solving equations. In this chapter you'll extend these skills while examining problems that can be modeled by pairs of related equations. In solving these "systems of equations" you'll see how the algebraic approach to solving equations is related to the graphical approach. You'll apply these two approaches as you learn algebraic ways to solve problems like the one presented in the Fitness Finances situation:

> Adrian and Sylvester have each decided to join a gym. Adrian wants to join permanently, while Sylvester would prefer to join for only a few months. Two athletic clubs in their neighborhood offer similar services but with different terms. One has a higher initiation fee with lower monthly payments, and the other has relatively high monthly dues and a low initiation fee. What do you think Adrian and Sylvester should do?

Many of the problems in this chapter require graph paper. Be sure you have an adequate supply.

IN THIS CHAPTER, YOU WILL HAVE THE OPPORTUNITY TO:

- interpret graphs of data;
- write an equation of a line given either a point and a slope, or two points;
- use the intersection of two graphs to solve problems involving two sets of information that lead to two equations;
- solve a system of two equations algebraically by the substitution method and the addition method;
- find the intersection of the line $y = 0$ and the parabola $y = ax^2 + bx + c$ by solving the equation $ax^2 + bx + c = 0$ using factoring and the zero product property.

MATERIALS	CHAPTER CONTENTS		PROBLEM SETS
Graph paper and straight-edges (needed throughout the chapter)	6.1	Olympic Records: Writing an Equation of a Line Given Two Points	FF-1–FF-21
	6.2	Using Graphs or Algebra to Solve Systems of Linear Equations	FF-22–FF-37
Dot paper (for FF-48); FF-48 Looking at a Difference of Squares Geometrically Resource Page	6.3	Solving Systems of Linear Equations by the Substitution Method	FF-38–FF-53
Graph for FF-54 on a transparency or flip-chart grid paper (prepared ahead); FF-55 Estimating Rocket Height Resource Page; algebra tiles (for FF-67)	6.4	Using Graphs to Solve Systems of Nonlinear Equations	FF-54–FF-73
	6.5	Solving Quadratic Equations: Graphing and the Zero Product Property	FF-74–FF-91
Summary Draft Template Resource Page (for FF-112)	6.6	Solving Systems of Equations by the Addition Method	FF-92–FF-112
FF-113–FF-116 Investigation Resource Page	6.7	Investigation: Using Algebra to Model and Analyze Linear Data (Optional)	FF-113–FF-116
	6.8	Summary and Review	FF-117–FF-139

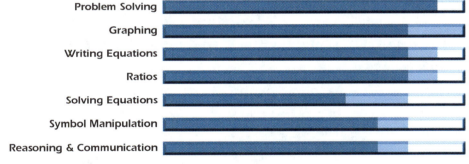

6.1 OLYMPIC RECORDS: WRITING AN EQUATION OF A LINE GIVEN TWO POINTS

SECTION MATERIALS
GRAPH PAPER (NEEDED THROUGHOUT THE CHAPTER)
STRAIGHT EDGES (NEEDED THROUGHOUT THE CHAPTER)

Students may have trouble finding an accurate slope for the Olympic Records data in FF-1. You may want to start the class with a warm-up problem such as "draw a generic slope triangle, and use it to find the slope of the line through the points (2.5, 495) and (7.8, 110)" so they see a "messy" slope, $\frac{-385}{5.3}$.

If you plan to have students do the Investigation in Section 6.7, be sure to assign FF-21.

In Chapter 5, you wrote the equation of a line by first graphing the line, and then used the graph to find the slope and estimate the y-intercept. This method is adequate if the numbers are "nice," or if an approximation suits your needs. But what if you need more accuracy, or if the two points you are given make it difficult to graph the line? In this section, you will use an algebraic method to write the *exact* equation of a line through two points. You will also see how to use the equation you have written to make predictions about real situations.

FF-1. Olympic Records The Olympic motto—*Citius, Altius, Fortius* ("Faster, Higher, Stronger")—is certainly demonstrated by the efforts of the men and women who are track and field athletes. In the modern Olympic Summer Games (which have usually been held every four years since 1896), gold medal winning times for men in the 100-meter dash decreased significantly from 10.8 seconds in 1928 to 9.99 seconds in 1984. For the women's 100-meter dash, the winning time of 10.82 seconds in 1992 was quite a bit faster than the 1956 time of 11.5 seconds. Based on the two data points for each gender, what would you predict the winning times for the men's and women's 100-meter dash in the Olympic Summer Games of the year 2000? *Save your work for this problem; you'll need it to refer back to and use later.*

a. It often helps to visualize data by drawing a coordinate graph. Set up and label coordinate axes with the number of years since the first modern Olympics along the horizontal (x) axis and winning times in seconds along the vertical (y) axis. A convenient way to scale the horizontal axis would be to let every unit represent 4 years. If x represents the number of years *since* 1896, then $x = 0$ would correspond to the year 1896, and $x = 4$ would correspond to the year 1900. Similarly, $x = 32$ would correspond to the year 1928, and so on. Scale the vertical axis in increments of 0.5 seconds, starting with 0 seconds.

b. Use two different colors to plot the given data points—one color for the men's data and another for the women's. The men's data for 1984 corresponds to the point (88, 9.99). Draw a trend line through each pair of data points, and then draw a slope triangle on each line. Using the slope triangles, find the slope of each line and write it as a decimal accurate to the nearest thousandth.
For men's, plot (32, 10.8) and (88, 9.99); $m = \frac{-0.81}{56} \approx -0.014$. For women's, plot (60, 11.5) and (96, 10.82); $m = \frac{-0.68}{36} \approx -0.019$.

c. What are the units of the slopes from part b? Write a sentence interpreting each slope in terms of the Olympic Records problem. *For men's, $m = \frac{-0.81 \text{ sec}}{56 \text{ yr}} \approx -0.014$ sec/yr; so men's winning times are decreasing by 0.014 sec each year. For women's, $\frac{-0.68 \text{ sec}}{36 \text{ yr}} \approx -0.019$ sec/yr; so women's winning times are decreasing by 0.019 sec each year.*

d. Estimate the y-intercept of each trend line. *For men's, $b \approx 11.3$, and for women's, $b \approx 12.6$. Expect lots of variation in students' answers.*

e. Use information from parts b and c to write an equation in slope-intercept form ($y = mx + b$) for each trend line. *men's: $y = -0.014x + 11.3$; women's: $y = -0.019x + 12.6$*

f. How confident are you of the accuracy of your equations to model the 100-meter dash data? Explain. *Confidence is based on accuracy of estimates of y-intercepts and values used for slope, as well as on whether one agrees that the data are actually linear.*

g. According to your graphs of the given data points, are the men's and women's winning times in the 100-meter dash decreasing at the same rate? Explain your answer in complete sentences. *No, although the lines appear parallel, their slopes are not the same—the women's times are decreasing faster than the men's.*

h. What do your equations predict for the men and women who compete in the Olympics in the year 2000? *Answers will vary, but for the equations given in part e, men's is 9.84 sec and women's is 10.62 sec.*

The example in FF-2 guides students to take advantage of substitution. It is not necessary to develop another method for the whole class. Some students may discover other methods on their own. However, you should only encourage those groups who discover another method and want to use it.

FF-2. How to Find an Equation of a Line Without Graphing In FF-1 you found an equation of the line passing through two points by drawing a graph, using the slope triangle, and estimating where the graph crossed the y-axis. But the y-intercept was difficult to find accurately. What if you want a more accurate equation for your line? What if the y-intercept is 396 or $1\frac{3}{7}$? In cases where it is hard to use a graph to find an equation, it's time to use algebraic ideas. In WR-123, you were introduced to this method as a check on the accuracy of the graphing approach. In this problem, you'll develop thoroughly the algebraic method for finding an equation of the line passing through two points. To do this, you'll examine a simple situation.

a. Read the following example carefully, and then answer the questions in parts b and c.

Example: Find an equation of the line passing through the points (1, 5) and (3, 8).

First Subproblem: Write a general equation to model a line. You know you can write an equation for any line like this:

$y = mx + b$, where m is the slope, and b is the y-intercept.

Notice that what you need to find is the slope and the y-intercept.

Second Subproblem: Find the slope. Calculate the slope by drawing a generic slope triangle as usual. An accurate graph is not needed, but a rough sketch is helpful:

So the equation must look like this:

$$y = \frac{3}{2}x + b.$$

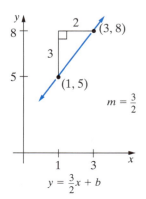

Third Subproblem: Calculate the y-intercept. $y = \frac{3}{2}(x) + b$
The hard part is to find the *y*-intercept, *b*, without drawing an accurate graph. But remember, you know that the point (3, 8) is *on* the line. This means that $x = 3$ and $y = 8$ make the equation true:

$$8 = \frac{3}{2}(3) + b$$

So, the *y*-intercept is 3.5.

$8 = 4.5 + b$
$3.5 = b$

Final Subproblem: Write the specific equation for the given two points. Using the values we found for the slope *m* and the *y*-intercept *b* you can write the equation for the line through (1, 5) and (3, 8):

$$y = \frac{3}{2}x + 3.5 \quad \blacktriangle$$

b. In the third subproblem you used the point $(x, y) = (3, 8)$. Show what happens if you use the point (1, 5) instead. Explain your results in a sentence.   *The y-intercept will still be b = 3.5.*

c. Check your equation by drawing an accurate line on graph paper through the two points (3, 8) and (1, 5) and marking the *y*-intercept.

d. Does the line you graphed show the same slope and *y*-intercept that you found in part a?   *yes*

Whenever students are required to write an equation of a line, encourage them to draw a rough sketch of a generic slope triangle in order to reinforce the geometric meaning of slope.

FF-3. Use the method described in FF-2 to find an equation of the line passing through each pair of points. In each case, write the equation in the slope-intercept form ($y = mx + b$).

a. (4, 6) and (−2, 10) $y = \frac{-2}{3}x + 8\frac{2}{3}$

b. (−3, 2) and (4, 5) $y = \frac{3}{7}x + \frac{23}{7}$

FF-4. Use the method shown in FF-2 to find a more accurate equation for each of your "Olympic Records" lines from FF-1. Write your final equations in slope-intercept form.   *men's: $y \approx -0.014x + 11.263$; women's: $y = -0.019x + 12.633$*

FF-5. If you know the slope of a line and the coordinates of one point on the line, you can write an equation for the line.

a. A certain line with slope $\frac{1}{3}$ goes through the point (5, −3). Find an equation of the line. $y = \frac{1}{3}x - 4\frac{2}{3}$

b. Suppose the line $y = -2x + b$ goes through the point (1, 4). Find an equation of the line. $y = -2x + 6$

FF-6. Here is some information about how long it takes to roast a turkey:

Pounds (x)	Time in Hours (y)
10	4
22	6

a. Sketch a graph of the situation by putting time on the vertical axis. What are the coordinates of the two data points you should plot?   *(10, 4) and (22, 6)*

b. The formula for finding the roasting times for turkeys of various weights is a linear equation (that is, its graph is a line). Use the given information to find an equation relating the weight of a turkey to its roasting time. $y = \frac{1}{6}x + \frac{7}{3}$

c. How long should an 18-pound turkey be roasted?.   *$5\frac{1}{3}$ hr, or 5 hr, 20 min*

d. State the slope (including the units) of your linear turkey-roasting equation, and interpret it in terms of the problem.   *$\frac{1 \text{ hr}}{6 \text{ lb}}$, or $\frac{1}{6}$ hr/lb, or 10 min/lb. Add 10 min for each additional pound of turkey.*

You may need to encourage groups who are having difficulty with FF-7 to sketch a graph.

FF-7. Find an equation for the line that is parallel to the line $y = 2x - 5$ and goes through the point (1, 7). $y = 2x + 5$

After groups have done FF-8, have several volunteers read their explanations to the class.

FF-8. What pieces of information do you need in order to write the equation of a line?   *You need the slope and a point, or two points.*

Section 6.1 Olympic Records: Writing an Equation of a Line Given Two Points **351**

> **EXTENSION AND PRACTICE**

FF-9. The slope of line AB is $\frac{5}{9}$, and the coordinates of point B are (18, 6).

a. What are the coordinates of the y-intercept of line AB? *(0, −4)*

b. What is an equation for the line AB? $y = \frac{5}{9}x - 4$

FF-10. Write an equation for the line passing through each pair of points.

a. $S\,(2, 3)$ and $F\,(5, 1)$ $y = \frac{-2}{3}x + \frac{13}{3}$

b. $L\,(1, 7)$ and $V\,(4, 5)$ $y = \frac{-2}{3}x + \frac{23}{3}$

c. $N\,(-3, 2)$ and $Y\,(1, 6)$ $y = x + 5$

FF-11. The population of Dodge was 18,500 in the year 1960, but by 1970 it had grown to 19,250. The 1980 census showed that Dodge had finally grown to 20,000 people!

a. Plot the census data for Dodge on a coordinate graph with years on the horizontal axis and population on the vertical axis. You should have three data points on your graph. *(1960, 18500), (1970, 19250), and (1980, 20000)*

b. From the given data, do you think a formula for the population of Dodge should be a *linear* equation? Explain your answer. *Yes; the given points lie on a line.*

c. Write the equation of a line that would pass through the three data points you plotted in part a. Be sure to state what your variables represent. *If x represents the year, $y = 75x - 128{,}500$. If x represents the number of years since 1960, $y = 75x + 18{,}500$.*

d. What are the units of the slope of your equation from part c? Write a sentence interpreting the slope in terms of the population of Dodge.

   $m = \frac{75 \text{ people}}{1 \text{ yr}}$. *This means that the population of Dodge is increasing by 75 people each year.*

e. Use your equation to predict the population of Dodge in the year 1998. *21,350*

FF-12. In WR-68, WR-122, and WR-126, you examined the relationship between the slope of a line and the slopes of segments between pairs of points on the line.

a. Suppose points H, S, and U all lie on a line. What must be true about the slopes of segments HS, SU, and HU? *The slopes are all equal.*

b. Using what you know about the slopes of line segments that are on the same line, show that points C, A, and L lie on the same line.

$$C\ (4, 2) \qquad A\ (17, 28) \qquad L\ (7, 8)$$

slope of CA = 2 = slope of AL = slope of CL

FF-13. Solve each of the following equations.

a. $\dfrac{2x - 3}{5} + 1 = \dfrac{2x}{4}$ *4*

b. $31 = 5 - 2(3x + 4) - x$ $\dfrac{-34}{7} \approx -4.86$

FF-14. Factor each difference of squares:

a. $z^2 - 49$ $z^2 - 49 = (z - 7)(z + 7)$

b. $9x^2 - 121$ $9x^2 - 121 = (3x - 11)(3x + 11)$

c. $P^2 - 400Q^2$ $P^2 - 400Q^2 = (P - 20Q)(P + 20Q)$

FF-15. Consider the following problem:

In the figure, the radius of the large circle is 14 cm and the radius of the smaller circle is 9 cm. Find the area of the shaded region.

a. To solve the problem, copy and solve each of the subproblems listed here. Write your answers to the subproblems in complete sentences.

Recall that the area of a circle is πr^2.

Subproblem 1: Find the area of the larger circle. *$196\pi \approx 615.75$ sq cm*

Subproblem 2: Find the area of the smaller circle. *$81\pi \approx 254.47$ sq cm*

b. Explain how to use part a to find the area of the shaded region, and then do it. *Subtract the area of the smaller circle from the area of the larger circle. $115\pi \approx 361.28$ sq cm*

FF-16. When a scientific calculator displays 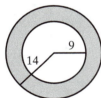, it means $1.35 \cdot 10^{12}$, or 1.35 trillion. Imagine that you are talking on the phone to a friend who just got a new calculator, and she would like some help.

a. Explain to your friend the difference between 1.35 12 on a calculator display and 1.35^{12} written anywhere else. *The calculator display means $1.35 \cdot 10^{12}$. Written anywhere else, 1.35^{12} means "Use 1.35 as a factor 12 times."*

Section 6.1 Olympic Records: Writing an Equation of a Line Given Two Points **353**

b. Explain how to enter $1.35 \cdot 10^{12}$ on your calculator. *Enter 1.35, and then multiply by 10 raised to the power 12. Use the $\boxed{x^y}$ key to do this.*

c. Explain how to calculate 1.35^{12}. *To raise 1.35 to the power 12, enter 1.35, press the $\boxed{x^y}$ key, then enter 12, and press the $\boxed{=}$ key.*

You may have used the following puzzle as a Head Problem in Section 3.7. In problem FF-17, students write an algebraic expression for the same puzzle.

FF-17. Here's a puzzle:

Start with the number of doughnuts in a baker's dozen. Subtract any number of your choice between 5 and 10. Multiply the difference by the largest number of dots on a side of a standard fair six-sided die. Add twice the fifth prime. Divide the sum by 2. Subtract half of 10^2, then divide by your number. What is your result?

a. Write an algebraic expression to represent the puzzle.
$\{[(13 - x)(6) + 2(11)] \div 2 - 0.5(10^2)\} \div x$

b. If the number you choose is 6, what would be your result? What if the number you choose is 10? $-3, -3$

c. Simplify the expression from part a and then explain your results in part b. Write your observations in a sentence.   *The result is always -3.*

FF-18. Rewrite each exponential expression using as few exponents as possible:

a. $(7^3)^3 \cdot 7^{-5}$ **b.** $2^4 \cdot 3^4$ **c.** $(2 \cdot 3^2)^3$ **d.** $\dfrac{2^4 \cdot 3^{10}}{2 \cdot 3^7}$

7^4 6^4 $2^3 \cdot 3^6$ $2^3 \cdot 3^3 = 6^3$

FF-19. Simplifying Fractions Rewrite each of the following fractions using as few digits or variables as possible. Assume that $x \neq 0$ and $y \neq 0$.

a. $\dfrac{4x^3y}{12xy^3}$ $\dfrac{x^2}{3y^2}$ **b.** $\dfrac{(x+3)^6}{(x+3)^8}$ for $x \neq -3$ $\dfrac{1}{(x+3)^2}$

c. $\dfrac{x(y^2)^3 \cdot (3x^4y)^2}{-15x \cdot 6y}$ $-0.1x^8y^7$ **d.** $\dfrac{x^7(x+6)^4}{x^8(x+6)^3}$ for $x \neq -6$ $\dfrac{x+6}{x}$

FF-20. Stewart invests \$4000 in savings bonds that promise 8% interest, and \$7500 in stocks that will pay $10\tfrac{3}{4}\%$ interest. How much will he earn altogether?   *\$1126.25*

Assign FF-21 only if you plan to have students do the Investigation, FF-113 through FF-116.

FF-21. Section 6.7—Investigation: Using Algebra to Model and Analyze Linear Data—consists of four problems that will guide you through the development of a project. In preparation for completing that project, read through the entire Section 6.7 to find out what you will be expected to do.

6.2 USING GRAPHS OR ALGEBRA TO SOLVE SYSTEMS OF LINEAR EQUATIONS

Many situations can be modeled using single equations. Many other problems, such as making airline schedules or allocating club funds, require that we consider two (or more) equations at the same time in order to find the solutions. The **solution to a system of equations** is the point or (points) that make both equations true at once.

You can *estimate* the coordinates of the point that makes two equations true simultaneously (at the same time) by carefully graphing each equation. Sometimes you can even find a *precise* answer using the graph. Most of the time, however, it is very difficult to find an exact answer by graphing; the graphs give, at best, close approximations of the x- and y-values we seek. Algebraic techniques are usually needed to produce exact results.

In this section you will see how a pair of equations can be used to solve a problem, either by graphing or by using algebraic techniques.

Problems FF-22 and FF-23 prepare students for the introduction, in Section 6.3, to solving a system of linear equations by substitution.

To help students see that there are at least two correct forms of the equations in FF-22, part a, you might have students write their equations on the board. Although in mathematics we usually agree to write linear equations in the form $y = mx + b$, students who write their equations in the $y = b + mx$ form might find it interesting to know that other disciplines (such as economics and statistics) would consider $y = b + mx$ to be standard.

FF-22. Changing Populations For the following situation, you will write equations to describe two different populations, and then use a graph to find when the two populations are equal in number.

The population of Oakwood is 3000 this year, and is growing at the rate of 50 people every year. Elm Creek, on the other hand, has a population of 5500 this year, but is declining at the rate of 75 people every year. When will the populations of the two towns be the same?

a. Write two equations, one to model each of the populations x years from now. *For Oakwood: $y = 50x + 3000$; for Elm Creek: $y = -75x + 5500$*

b. State the slope of each equation, including the units. Write a sentence to interpret each slope in terms of the towns of Oakwood and Elm Creek. *For Oakwood, $m = 50$ people/yr, so Oakwood's population is increasing by 50 people each year. For Elm Creek, $m = -75$ people/yr, so Elm Creek's population is decreasing by 75 people each year.*

c. Graph the two equations from part a on the same set of axes. (A convenient scale might be to mark the vertical axis in increments of 200 people, and the horizontal axis in increments of 2 years.)

d. Locate the point where the two lines intersect and label it "*P*."

e. Draw a vertical line through point P to the x-axis, and draw a horizontal line through point P to the y-axis. *vertical line, x = 20; horizontal line, y = 4000*

f. Write the coordinates (x, y) of point P by reading the x-value from where your vertical line crosses the x-axis and the y-value from where your horizontal line crosses the y-axis. *(20, 4000)*

g. Human populations don't usually change at a constant rate. Assuming, however, that the populations of Elm Creek and Oakwood do keep changing at a constant rate, in how many years will they be the same? How many people will live in each town that year? *In 20 yrs each town will have 4000 people.*

If students' population graphs from FF-22 are inaccurate, the point of intersection they found will not make the equations true when, in FF-23, they try to check by substituting. As you circulate around the room, help students fix inaccuracies.

FF-23. Point of Intersection In problem FF-22, the point P is called the **point of intersection** of the graphs for the two equations $y = 50x + 3000$ and $y = -75x + 5500$. Point P lies on *both* lines. This means that the x-value and y-value at point P make *both* equations true; together, these coordinates make a *common solution* to the two equations.

a. To show that P is truly the point of intersection of the two lines from FF-20, you must show that it lies on both lines.

   i. Show that the coordinates of P make the equation $y = 50x + 3000$ true. *4000 = 50(20) + 3000*

   ii. Show that the coordinates of P make the equation $y = -75x + 5500$ true. *4000 = -75(20) + 5500*

b. Add the information about the *point of intersection of two lines* to your Tool Kit.

c. Maggie has graphed the lines $y = 3x - 2$ and $y = 4x + 5$. She thinks that the point of intersection is $(-5, -17)$. What do you think? Use the method of part a to explain your answer. *(−5, −17) is a point on y = 3x − 2 but not on y = 4x + 5.*

FF-24. In FF-22, you found the point of intersection of the graphs of $y = 50x + 3000$ and $y = -75x + 5500$ by graphing the two equations and then making an estimate. To solve the same problem in a different way, you could have used algebraic methods to find the one pair of numbers (x, y) that makes both equations true. The problem is to find the number of years, x, that will produce the same population, y, in each town's equation:

$$y = 50x + 3000$$
$$y = -75x + 5500$$

When will both towns have the same population y? The population of Oakwood is

$$50x + 3000,$$

and the population of Elm Creek is

$$-75x + 5500.$$

If both populations are the same size after x years, then the expressions $50x + 3000$ and $-75x + 5500$ must result in the same y-value. So you can write

$$50x + 3000 = -75x + 5500.$$

You've solved equations like this before, which means you've just taken a new problem and turned it into something familiar.

a. Solve the equation $50x + 3000 = -75x + 5500$ for x.   **$x = 20$**

b. In part a you found the x-coordinate of the point of intersection. You also need to find the y-coordinate. To do this, substitute the solution for x into $y = 50x + 3000$.   **$y = 4000$**

c. Repeat part b, this time substituting the solution for x into $y = -75x + 5500$. What do you notice about the results?   **$y = 4000$; they're the same.**

d. Write the coordinates of the point where the graphs of the two equations intersect.   **(20, 4000)**

e. Write one or two complete sentences to compare the two methods—graphing and algebraic—for finding the point of intersection of two lines. Which seems easier for you? Why?

FF-25. The graph of the equations $y = 5 - 2x$ and $y = \frac{2}{3}x - 4$ is shown here.

a. Use the graph to estimate the coordinates of the point of intersection of the lines

$$y = 5 - 2x \quad \text{and} \quad y = \frac{2}{3}x - 4.$$

≈ (3.5, −1.8)

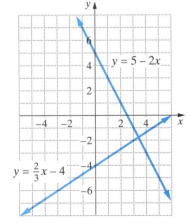

b. To check to see whether you have accurately estimated the point of the intersection, substitute the coordinates into each of the two equations. Does the point you found lie on both lines? Explain why your point may not make both equations exactly true. *The point is close to, but not exactly on, both lines since it's only an estimate.*

c. Use algebra to find the exact coordinates of the point of intersection. *(3.375, −1.75)*

d. Check your answer from part c in both equations,

$$y = 5 - 2x \quad \text{and} \quad y = \tfrac{2}{3}x - 4.$$

$-1.75 = 5 - 2(3.375)$ and $-1.75 = \tfrac{2}{3}(3.375) - 4$

EXTENSION AND PRACTICE

FF-26. Use the algebraic procedure in problem FF-24 to find the coordinates of the point where each of these pairs of lines intersect:

a. $y = -x + 8$ *(5, 3)*
   $y = x - 2$

b. $y = -3x$ *(2, −6)*
   $y = -4x + 2$

c. $y = -x + 3$ *(0, 3)*
   $y = x + 3$

d. $y = -x + 5$ *(2, 3)*
   $y = \tfrac{1}{2}x + 2$

e. Graph the equations in part a to check your answer.

f. To check your answer in part b without graphing, show that the point of intersection makes both equations true.
$-6 = -3(2)$ and $-6 = -4(2) + 2$

FF-27. Rewrite each of the following equations in slope-intercept form ($y = mx + b$), and then graph it.

a. $y - 4x = -3$

b. $3x + 2y = 12$

c. $3y - 3x = 7$

$y = 4x - 3$

$y = \tfrac{-3}{2}x + 6$

$y = x + \tfrac{7}{3}$

FF-28. Solve each of the following equations. Show how you check your answers.

a. $\tfrac{2x}{3} - 5 = -13$ *−12*

b. $4(2x - 1) = -10(x - 5)$ *3*

c. $4(x + 5) - 3(x - 2) = 26$ *0*

d. $3 + \tfrac{2x}{7} = \tfrac{4}{5} + x$ *$\tfrac{77}{25}$*

e. $(x + 3)(2x - 1) = 0$ *−3, 0.5*

FF-29. Justine graphed the lines $y = 3x - 7$ and $y = \frac{2}{3}x + 7$ to see if the point (6, 11) was on either one.

a. *Without drawing a graph,* determine whether the point (6, 11) is on the graph of the line $y = 3x - 7$.   *yes, 11 = 3(6) − 7*

b. *Without drawing a graph,* determine whether the point (6, 11) is on the graph of the line $y = \frac{2}{3}x + 7$.   *Yes, 11 = $\frac{2}{3}$(6) + 7*

c. What is the point of intersection of $y = \frac{2}{3}x + 7$ and $y = 3x - 7$?   *(6, 11)*

FF-30. Temperature Conversion Revisited On a thermometer, 0° Celsius is equal to 32° Fahrenheit. Also, 100°C is equal to 212°F. This means that the points (0°C, 32°F) and (100°C, 212°F) would lie on a temperature conversion graph.

a. Without graphing, use these two data points and the fact that conversion from degrees Celsius to degrees Fahrenheit is linear to find an equation for the line through the given points. Write the equation in the slope-intercept form ($y = mx + b$).   *F = 1.8C + 32*

b. Compare the equation you wrote in part a with the one you found using a graph in part d of the Temperature Conversion problem, WR-124. *The two equations should be equivalent.*

c. On September 13, 1922, the temperature in El Azizia, Libya, reached 58°C, which is still the highest temperature ever recorded in the world. Use your equation to determine the Fahrenheit temperature that corresponds to 58°C.   *136.4°F*

d. Vostok, Antartica, holds the record for the lowest recorded world temperature. On July 21, 1983, Vostok reached −129° Fahrenheit. Use your equation to determine the Celsius temperature that corresponds to −129°F.   *−89.4°C*

FF-31. Use the method described in FF-2 to find an equation of each of the lines described below. In each case write the equation in the slope-intercept form ($y = mx + b$).

a. the line through points (9, 1) and (3, 4) $y = \frac{-1}{2}x + \frac{11}{2}$

b. the line through points (6, 3) and (5, 5) $y = -2x + 15$

c. the line with slope $\frac{1}{3}$ through the point (0, 5) $y = \frac{1}{3}x + 5$

FF-32. Find x and y for each of the following pairs of similar triangles:

a. 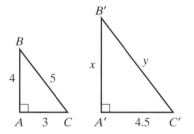   b.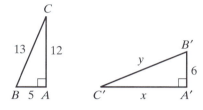

c. 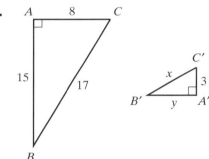   d.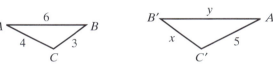

(a) $x = 6$, $y = 7.5$; (b) $x = 12.4$, $y = 15.6$; (c) $x = 6.375$, $y = 5.625$; (d) $x = 3.75$, $y = 7.5$

FF-33. Sonja notices that the lengths of the three sides on a right triangle are consecutive even numbers. The triangle's perimeter is 24 units. Write and solve an equation to find the length of the longest side. *If x = length of shortest side, then $x + (x + 2) + (x + 4) = 24$; $x = 6$, so the longest side is 10 units. If x = length of the longest side, then $x + (x - 2) + (x - 4) = 24$; $x = 10$.*

FF-34. Rewrite each of the following expressions by multiplying the binomials in parts a and b and factoring the polynomials in parts c through f.

a. $2x(5x^2 + 3x - 4)$ $10x^3 + 6x^2 - 8x$

b. $(x + 3)(x^2 - 3x + 9)$ $x^3 + 27$

c. $x^2 - x - 6$ $(x - 3)(x + 2)$

d. $x^2 - 121$ $(x + 11)(x - 11)$

e. $x^2 + 8x - 20$ $(x - 2)(x + 10)$

f. $6x^3 - 30x^2 + 24x$ (Look for common factors.) $6x(x^2 - 5x + 4) = 6x(x - 4)(x - 1)$

FF-35. A penny is 0.00072 meters thick. How tall, in kilometers, would a stack of 1 billion pennies be? Write your answer in both standard form and scientific notation.   *720 km = 7.2 · 10² km*

FF-36. Identify each of the following expressions as a difference of squares or not a difference of squares. If an expression is a difference of squares, then factor it. If an expression is not a difference of squares, then explain why not.

a. $x^2 + 25$ *No; $x^2 + 25$ is a sum of squares, not a difference.*

b. $x^2 - 8$ *No; 8 is not a square of an integer.*

c. $25 - T^2$ *Yes; $25 - T^2 = (5 - T)(5 + T)$*

d. $N^2 - 36N - 25$ *No; there are three terms.*

e. $49x^2 - 100$ *Yes; $49x^2 - 100 = (7x - 10)(7x + 10)$*

FF-37. Write the fraction form and the decimal form of each of the following exponential expressions:

a. 2^{-3} $\frac{1}{8} = 0.125$

b. 10^{-5} $\frac{1}{100{,}000} = 0.00001$

c. 5^0 1

d. 8^{-2} $\frac{1}{64} = 0.015625$

6.3 SOLVING SYSTEMS OF LINEAR EQUATIONS BY THE SUBSTITUTION METHOD

SECTION MATERIALS

DOT PAPER (FOR FF-48)

FF-48 LOOKING AT A DIFFERENCE OF SQUARES GEOMETRICALLY RESOURCE PAGE

Problem FF-38 is written as an investigation. Groups first write and graph equations to solve the problem, and then solve it algebraically. Although the graphs for these problems are not continuous, it is useful to view them as linear trends and draw in the lines.

FF-38. Fitness Finances For their New Year's resolutions this year, Adrian and Sylvester have each decided to join a gym. Adrian wants to join permanently, while Sylvester would prefer to join for only five months, until the weather gets better and he can exercise outdoors. After making several phone calls, Adrian found two athletic clubs in their neighborhood that offer similar services. So she is wondering which club offers the best deal. To join the MidCity Gym, members must pay a $230 initiation fee plus $25 per month. At Downtown Fitness, the initiation fee is only $98, but the monthly fee is $46. Adrian and Sylvester are unsure what to do and would like some advice.

a. Let x represent the number of months and y represent the total in gym fees (in dollars). Write an equation for *each* club option.
MidCity Gym: y = 25x + 230; Downtown Fitness: y = 46x + 98

b. State the slope of each equation, including the units. Write a sentence to interpret what each slope means to anyone considering joining the gyms.
MidCity: $m = \dfrac{\$25}{1\text{ mo}}$, so the total cost increases by \$25 every month. For *Downtown Fitness*, $m = \dfrac{\$46}{1\text{ mo}}$ so the total cost increases by \$46 every month.

c. Explain why the graph of each club option equation is discrete. Then graph the trend line for each option. (You might find it convenient to scale the *x*-axis as 1 month for every 2 units, up to 10 months, and to scale the vertical axis as \$20 per unit up to \$500.)

d. Use the graphs from part b to estimate the number of months that give the same total cost regardless of the club Adrian or Sylvester chooses. Estimate the total cost that corresponds to this number of months.
about 6; about \$380

e. Use your equations from part a to solve for the exact number of months that give the same total cost, and then find the associated cost. Refer to problem FF-22 if you need help. Compare your answers with your estimates in part c. $6\tfrac{2}{7}$ *mo; \$387.14*

f. Now conclude the problem by writing two or three sentences to advise Adrian and Sylvester as to what they should do. *MidCity Gym is better for those joining for more than 6 months. Downtown Fitness is better for those joining for 6 months or less.*

FF-39. The Substitution Method for Solving Pairs of Linear Equations In problems FF-24 and FF-38d, we found a single solution for two equations in *y*-form by making one equation from the two and eliminating a variable. The same thing can be done even if only one equation is in *y*-form (or "*x*-form").

Example: Here is how to use the **substitution method** to find a common solution to a pair of linear equations. Suppose you want to find the point where the graphs of these two lines intersect:

$$y = 10 - x$$
$$2y - x = 2$$

Just as in problem FF-24, there are two equations and two variables.	$y = 10 - x$ $2y - x = 2$
Once again we can *substitute for y* (that is, replace it) with something that is equal to it. The equation $y = 10 - x$ tells us that y is equal to $10 - x$, so we can replace y in the second equation with $10 - x$.	$2y - x = 2$ $2(\mathbf{10 - x}) - x = 2$

The resulting equation looks like many we have solved before.	$20 - 2x - x = 2$ $20 - 3x = 2$ $-3x = -18$ $x = 6$
Remember that we are finding the point where the two lines cross, so we must find the y-value also. We can use either equation to obtain $y = 4$.	$y = 10 - x = 10 - 6 = 4$ or $2y - x = 2$ $2y - 6 = 2$ $2y = 8$, so $y = 4$

So, the common solution for the two equations is:

$$x = 6 \quad \text{and} \quad y = 4 \qquad \blacktriangle$$

a. If you had graphed the lines $y = 10 - x$ and $2y - x = 2$, where would they cross? *(6, 4)*

b. Use the substitution method to solve this pair of equations:

$$y = x + 3$$
$$x + 3y = 5$$

(If you need help, try substituting into the second equation, $x + 3(\_\_\_) = 5$.) *(−1, 2)*

c. Add to your Tool Kit information about using the *substitution method* for solving pairs of linear equations.

FF-40. Use the substitution method to find the point of intersection (x, y) for each of these pairs of linear equations.

a. $y = -3x$
$4x + y = 2$

b. $2x + 3y = -17$
$y - x = -4$

c. $x = 2y + 5$
$3x + 7y = 19$

(This system of equations is a perfect candidate for substitution: one equation is in y-form already.)

(Sometimes, you must first rewrite one of the equations in y-form. Choose the easiest one!)

(If one equation is already in "x-form," you may substitute for x instead of y.)

(2, −6) *(−1, −5)* *$(5\frac{8}{13}, \frac{4}{13})$*

d. Show that your solution to part a is correct by checking it in both equations. *$y = -3x = -3(2) = -6$ and $4x + y = 4(2) + (-6) = 2$*

EXTENSION AND PRACTICE

FF-41. Use the substitution method to find the point of intersection (x, y) for each of the following pairs of linear equations:

a. $y = x + 3$
   $3y - x = 8$
   $(-0.5, 2.5)$

b. $y = 2x - 3$
   $x - y = -4$
   (Hint: $y = 2x - 3$)
   $(7, 11)$

c. $x + y = 4$
   $x = y - 2$
   $(1, 3)$

d. Show, in both equations, that your solution to part a is correct.

FF-42. Is the point $(10, 4)$ the point of intersection of the graphs of $2x - y = 16$ and $x + y = 14$? To show why or why not, evaluate each equation at the point $(10, 4)$.   *Yes, since (10, 4) makes both equations true, the point is on both lines.*

FF-43. Nell is certain that $\dfrac{x - 10}{x} = -10$.

a. Pick a value for x, and use it to show Nell that, in general, $\dfrac{x - 10}{x}$ is not equal to -10.

   For example, pick $x = 15$. Then $\dfrac{x - 10}{x} = \dfrac{15 - 10}{15} = \dfrac{5}{15} = \dfrac{1}{3}$ not -10.

b. Do you think there are any values of x that make the equation $\dfrac{x - 10}{x} = -10$ true? If so, find them (or it). If not, explain why not.
   Yes, $x = \dfrac{10}{11}$

FF-44. Solve each of the following equations for x, if possible:

a. $4 + 2.3x = -5.2$
   -4

b. $2(4x - 7) = 8x + 14$
   impossible

c. $6(x - 2) = 5(x - 11) - 21$
   -64

d. $3(x - 5) = \frac{1}{5}(10x - 25)$
   10

e. $\dfrac{x}{x + 4} = \dfrac{9}{2}$
   $\dfrac{-36}{7} \approx -5.14$

f. $x(2x - 3) = 0$
   $0, \dfrac{3}{2}$

FF-45. Rewrite each of the following expressions by multiplying the factors in parts a through c, and by factoring the polynomials in parts d through f.

a. $2x^2 \cdot 5x$
   $10x^3$

b. $(2x^2 + 4)(5x - 9)$
   $10x^3 - 18x^2 + 20x - 36$

c. $(x + 2)(2x^2 - 3x + 6)$
   $2x^3 + x^2 + 12$

d. $x^2 - 64$
   $(x + 8)(x - 8)$

e. $x^2 - 6x - 72$
   $(x + 6)(x - 12)$

f. $24 - 16x$
   $8(3 - 2x)$

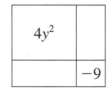

 FF-46. In Chapter 3, you used a guess-and-check table to write an equation that could be used to solve a problem. In the following problem, you will see how to write a system of *two* equations that can be used to solve the problem. First read the problem, and then complete parts a through c to solve it.

At a football game, 2000 tickets were sold. General public tickets sold for $7.50 and student tickets for $5.00. The total revenue was $11,625. How many of each type of ticket were sold?

a. You will need two variables to write the two equations. If x represents the number of student tickets, and y represents the number of general public tickets, how are x and y related? Write an equation about the *number* of tickets sold. **$x + y = 2000$**

b. For your second equation, think about how the *revenues* from the tickets are related. Then write an equation about the money collected at the football game. **$7.50y + 5.00x = 11{,}625$**

c. Finally, solve the pair of equations from parts a and b using the substitution method from problem FF-39. Don't forget to write your final answer as a sentence. **1350 student tickets and 650 general public tickets were sold.**

FF-47. Complete each of the following diagrams, and write an equation to express "the area as a product equals the area as a sum."

a.
$2x^2$	$6x$
$5x$	15

b.
$4y^2$	
	-9

$(2x + 5)(x + 3) = x^2 + 11x + 15$ **$(2y + 3)(2y - 3) = 4y^2 - 9$**

 FF-48. Looking at a Difference of Squares Geometrically Dale and Roy, owners of Turf's Up! Landscaping, have been contracted to install sod for lawns in the backyards of a new condominium complex. The backyards are square, but vary in size from 16.5 feet on a side to 24.75 feet on a side, and each one has a gate in one corner. The plans call for the entire backyard area to be laid with sod, except for a 3-foot by 3-foot square of bare ground that is to be left in the corner by the gate.

Dale and Roy started to figure out many square feet of sod would be needed for each backyard, but they soon realized they could save time by drawing a diagram and letting L represent the length of the side of the backyard square, such that $16.5 \le L \le 24.75$. They labeled four of the lawn's six sides with their lengths, as shown here.

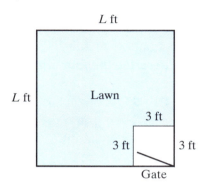

a. Copy the diagram on dot paper. Label the two sides of the lawn (the shaded region) that Dale and Roy forgot to label. *Both are $L - 3$ ft.*

b. Dale remarked, "I see two squares." and quickly scribbled this relationship beside the diagram:

$$\text{Lawn area} = L^2 - 9.$$

Pat, who works for Dale and Roy, then walked in, looked at the paper and asked, "How'd'ja figure that out so fast?" How could Dale explain her thinking to Pat? *Simply subtract the area of the small square from the area of large square.*

c. Meanwhile, Roy stared at the diagram and then subdivided the lawn into two regions by extending a line from one side of the unshaded square through the shaded region. Then he remarked "I see two rectangles," and then jotted down this relationship:

$$\text{Lawn area} = (L + 3)(L - 3).$$

Dale and Pat asked in unison, "How did you get $L + 3$?" and Roy replied, "I rearranged the pieces."

On your FF-48 Resource Page ("Looking at a Difference of Squares Geometrically"), subdivide the lawn in Fig. 2 into two regions, either by placing a straightedge along one side of the small unshaded square and drawing a line through the shaded region (as Roy did), or by folding along an edge of the small square and making a crease through the shaded region. Cut the two regions apart along the new line or crease, rearrange the two regions to form a rectangle, and glue or tape them to your paper. Label the sides of the rectangle with their lengths. Use the rectangle to explain Roy's approach to figuring out the area of the lawn.

FF-49. Find an equation for each of the following lines:

a. the line passing through points $(-1, 4)$ and $(2, 1)$ *$y = -x + 3$*

b. the line passing through points (5, 342) and (30, 2047) $y = 68.2x + 1$

   c. the line with slope $\frac{-2}{3}$ which passes through the point (7, 1)
   $y = \frac{-2}{3}x + 5\frac{2}{3}$

FF-50. Explain to a student who was absent today how you would find an equation of the line through two given points. *Look back at FF-2, and describe the four subproblems.*

FF-51. For each of the following lines, make a table with three input values. Then graph the lines on the same coordinate system. Label each line with its equation.

   a. $y = -3.75$ *horizontal line through (0, −3.75)*
   b. $x = -1.5$ *vertical line through (1.5, 0)*
   c. $y = 0$ *x-axis*
   d. $x = 0$ *y-axis*

FF-52. *Without graphing,* write a sentence to describe each of the following lines so that someone who is absent today could graph the lines from your descriptions.

   a. $y = 42$ *horizontal line passing through (0, 42)*
   b. $x = 1991$ *vertical line passing through (1991, 0)*

FF-53. Suppose the slope of line AB is 5, where A is $(-3, -1)$ and B is $(2, n)$. Draw a diagram, and find the value of n. $n = 24$

6.4   USING GRAPHS TO SOLVE NONLINEAR SYSTEMS OF EQUATIONS

SECTION MATERIALS

GRAPH FOR FF-59 ON TRANSPARENCY OR FLIP CHART

FF-55 ESTIMATING ROCKET HEIGHT RESOURCE PAGE

ALGEBRA TILES (FOR FF-67)

The substitution method is a powerful tool for solving systems of equations because it can be used for nonlinear as well as linear systems. To reinforce the general usefulness of the substitution method, and to show how universal an approach it is, we have students apply it at this point to nonlinear systems. After students have had a chance to become familiar with the substitution method, in Section 6.6 we will introduce the addition method (also known as the elimination method) for solving linear systems.

In Problems FF-54 and FF-55, students solve a nonlinear system of equations by looking for points of intersection. In FF-54 students use an algebraic approach, and in FF-55 they use graphs to estimate solutions.

We highly recommend that you have your own graph for problem FF-54, prepared in advance on an overhead transparency or on flip-chart grid paper. After students have finished FF-54, use it to focus a discussion on the points of intersection.

In this section you will see how the substitution method is a powerful tool that can be used in many different situations.

FF-54. You have found solutions to systems of *linear* equations in two ways: by graphing both equations and finding the point of intersection of the graphs; and by using the substitution method to solve the equations algebraically. You can also find solutions to systems that contain equations that are not linear. A pair of equations is a **nonlinear system** when at least one of the two equations does not produce a straight line when graphed.

Use a full sheet of graph paper to graph the given pair of equations on the same set of coordinate axes. (You will need to use your calculator to make a table of values to help you graph the parabola, but you should use the slope and y-intercept to graph the line more quickly.) Use your graph to estimate the points of intersection.

$$y = x^2 - 4x$$
$$y = 2x - 5, \quad \text{for } -2 \le x \le 7$$
(1, −3) and (5, 5)

Use problem FF-55 to lead a short discussion about the fact that there are two answers in part b. As part of the discussion you might ask students to write a question about the graph for which there is no solution.

FF-55. Estimating Rocket Height A rocket is launched and its distance above sea level is recorded until it lands. The graph shows how the rocket's height (in feet) depends on the number of seconds since the rocket was launched. Use your copy of the FF-55 Estimating Rocket Height Resource Page to complete the following examination of the graph.

a. Estimate the rocket's height above sea level 3.5 seconds after firing by drawing a vertical line on your copy of the graph. Label its point of intersection with the curve. About how far above sea level is the rocket 3.5 seconds after firing?   *(3.5, 360); about 360 ft*

b. Estimate how long after being launched the rocket is 250 feet above sea level by drawing a horizontal line on your copy of the graph. Label its points of intersection with the curve. About how long after being launched is the rocket 250 feet above sea level?
(1, 250) and (4.25, 250); about 1 and 4.25 sec

c. Write equations for the vertical line in part a and the horizontal line in part b.   *x = 3.5, y = 250*

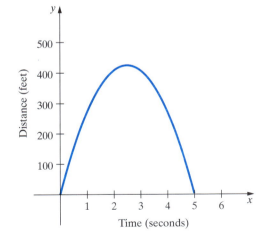

FF-56. This problem concerns the following nonlinear system of equations:

$$y = x^2 \quad \text{and} \quad y = 4.$$

a. Graph both equations on one set of axes. Label each graph with its equation.

b. Give the coordinates of the points where the line intersects the parabola. *(2, 4) and (−2, 4)*

FF-57. Solve the equation $x^2 = 9$ by completing the subproblems in parts a and b.

a. Rewrite the equation $x^2 = 9$ using the meaning of exponents. *x · x = 9*

b. Show that there are two possible values of x that will make the equation true. *(3) · (3) = 9 and (−3) · (−3) = 9*

FF-58. Each of the following equations also has *two solutions*. Find both solutions, and round each one to the nearest 0.01 unit.

a. $x^2 = 4$ *x = ±2.00*
b. $x^2 = 25$ *x = ±5.00*
c. $s^2 = 30$ *s ≈ ±5.48*
d. $z^2 = 148$ *z ≈ ±12.17*
e. $y^2 − 1 = 15$ *y = ±4.00*
f. $t^2 + 4 = 24$ *t ≈ ±4.47*

FF-59. Compare problems FF-56b and FF-58a. Why do you think the x-values in the two problems match? *You could solve y = 4 and y = x² by substituting for y to get x² = 4.*

EXTENSION AND PRACTICE

FF-60. Solve each of the following nonlinear systems of equations by graphing the two equations and estimating the points of intersection to the nearest 0.5 unit.

a. $y = x^2 − 2$
   $y = 4 − x$
   (−3.0, 7.0) and (2.0, 2.0)

b. $y = 5 − x^2$
   $y = 3$
   about (1.5, 3.0) and (−1.5, 3.0)

FF-61. It may surprise you to learn that the solutions to some systems of equations cannot be found by algebraic methods. Devising computer programs to derive approximate solutions to such problems has become a very important problem for modern mathematicians.

The following system of equations is an example of one that cannot be solved algebraically. Although there is no algebraic method of solution, you can estimate its solution by graphing. Graph both equations on the same coordinate axes, and then estimate the point(s) of intersection to the nearest 0.5 unit.

$$y = 2^x \text{ for } x = −2, −1.5, −1, \ldots, 3.5, 4$$
$$y = 3x + 1 \text{ (for } −2 \leq x \leq 4)$$

(0, 1) and (3.5, 11.5)

Section 6.4 Using Graphs to Solve Nonlinear Systems of Equations **369**

FF-62. Solve each system of equations for x and y. (That is, where do the lines intersect?)

a. $y = x - 1$ (2, 1)
   $x + 2y = 4$

b. $y = 3x - 8$ (15, 37)
   $y = 2x + 7$

c. $x = 2y$ (4, 2)
   $x + 3y = 10$

d. $y = 3$ (−5, 3)
   $2x + y = -7$

FF-63. Use what you know about the equations of vertical and horizontal lines to complete this problem.

a. Marie correctly graphed an equation and got a vertical line that contained the point (0, 0). What equation did Marie graph? *x = 0*

b. When Max graphed his equation, he got a horizontal line that contained the point (0, 0). Assuming that the graph was correct, what equation did Max use? *y = 0*

FF-64. Recall from your work in earlier chapters that $-x^2 \neq (-x)^2$. How could you convince a skeptical new student that the two expressions are different? Show the difference with numbers, diagrams, or some other approach. *Some possibilities: evaluating $-x^2$ and $(-x)^2$ for several values of x, showing that the graphs of $y = -x^2$ and $y = (-x)^2$ are different, or explaining how to use the order of operations for each of the two expressions.*

FF-65. Write each of the following exponential expressions in a simpler form:

a. $(2x^5)^2$ *$4x^{10}$*

b. $(2 \cdot 10^5)^2$ *$4 \cdot 10^{10}$*

c. $(5.32 \cdot 10^{15})^2$ *≈ $2.83 \cdot 10^{31}$*

d. $\dfrac{4.2 \cdot 10^{15}}{2.1 \cdot 10^6}$ *$2 \cdot 10^9$*

e. $(2x^5)^{-2}$ *$0.25x^{-10}$*

f. $(2 \cdot 10^5)^{-2}$ *$0.25 \cdot 10^{-10}$*

FF-66. Factor each of the following quadratics. You may want to draw generic rectangles.

a. $x^2 + 4x + 3$ *(x + 3)(x + 1)*

b. $x^2 - 7x + 12$ *(x − 3)(x − 4)*

c. $x^2 + 10x + 16$ *(x + 2)(x + 8)*

d. $x^2 + 5x - 24$ *(x + 8)(x − 3)*

e. $2x^2 - 9x - 18$ *(2x + 3)(x − 6)*

f. $x^2 - 121$ *(x + 11)(x − 11)*

FF-67. In problem FF-48 you saw a way to form a rectangle from a large square with a missing smaller square by following Roy's approach of subdividing and rearranging. Here's a way you can connect that approach to the use of algebra tiles in factoring a difference of squares.

a. On a piece of scratch paper, trace and cut out an x^2 tile (a large square algebra tile). Place a "1" tile (a small square algebra tile) in one corner of the cut-out square, then trace it and cut it out.

b. Label each of the six sides of the remaining portion of the large square with its length.

c. Glue or tape the labeled figure to your paper. Next to the figure, write its area as a difference. Then repeat parts a and b to make and label a second figure.

Solution: 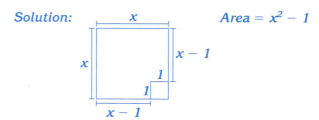 Area = $x^2 - 1$

d. Follow Roy's approach to subdivide the second figure into two pieces that can be rearranged to form a rectangle. Glue or tape the rectangle to your paper, and then label the length of each side. Finally, next to the rectangle, write its area as a product.

Solution: 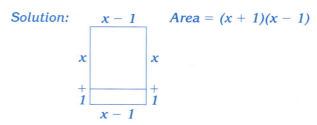 Area = $(x + 1)(x - 1)$

e. Explain how parts a through d can be used to show that $x^2 - 1 = (x + 1)(x - 1)$. *The areas of the two cut figures are the same, so the area of the large square minus the area of the small square is the same as the area of the rectangle.*

FF-68. What is the slope of the line that passes through the points $(5, -3)$ and $(7.5, 2)$? Explain how you get your answer. *Possible answer: Plot the two points and find a slide; m = 2.*

FF-69. John has 380 out of 440 points to date in an algebra class. There are only 60 remaining points possible in the grading period. What is the least number of points John will need to have an average of 80%?
$\dfrac{(380 + P)}{(440 + 60)} = 0.80$; *20 points of the 60 possible*

FF-70. Sliding a Line

a. Graph the line $y = 2x - 5$.

b. Slide the line up 2 units, and then draw the new line. If you have trouble, slide two of the points on $y = 2x - 5$ up 2 units each, and then draw the new line.

c. What is the solution for the system of equations shown by these two lines?   *No solution; the lines are parallel.*

FF-71. Write an equation for each of these lines:

a.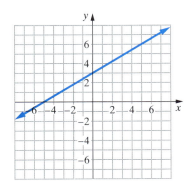

b.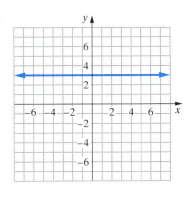

c.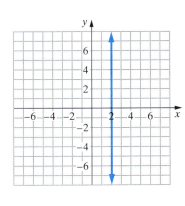

$y = \frac{3}{5}x + 3$

$y = 3$

$x = 2$

FF-72. Write an equation for the x-axis. $y = 0$

FF-73. Consider this system of lines: $y = \frac{2}{3}x + 5$ and $y = 0$.

a. Find the point of intersection of the two lines. $(-7.5, 0)$

b. What is the point you found in part a commonly called?   *x-intercept*

6.5 SOLVING QUADRATIC EQUATIONS: GRAPHING AND THE ZERO PRODUCT PROPERTY

In this section, students get another chance to see how to solve a system of equations either by graphing, or by using substitution to find a related equation to solve.

In FF-74, students examine the solution of a system of a quadratic equation and a linear equation (indeed, a constant) in two ways. By graphing both equations and finding the points of intersection, they solve the system graphically. Then they use substitution to rewrite the system as a quadratic equation, which they solve by factoring and using the zero product property. In FF-75, we reinforce the use of factoring and the zero product property to solve quadratic equations.

After the groups have done FF-74, a class discussion about the results is worthwhile.

In this section you'll examine how solving a certain kind of nonlinear system of equations can be related to solving an associated quadratic equation. As you did with linear systems, you will find solutions by using both algebraic and graphing methods.

First you'll solve systems of nonlinear equations in which one of the equations is quadratic—that is, of the form $y = ax^2 + bx + c$—and the other equation is $y = 0$. As you saw in the previous section, you can solve this type of system by graphing both equations and estimating the points of intersection. Then you'll use the method of substitution to solve the system, just as you did with linear equations. After substituting, you'll use factoring and the zero product property to solve the resulting quadratic equation algebraically. This approach provides an algebraic method for solving the original system.

FF-74. In this problem you will solve the system $y = x^2 + 4x - 5$ and $y = 0$ using two different methods.

a. **Solve the system by graphing:** Graph both equations on the same set of axes and find the points of intersection of the line and the parabola.

$$y = x^2 + 4x - 5$$
$$y = 0 \quad (\text{for } -6 \leq x \leq 5) \quad \textit{(−5, 0) and (1, 0)}$$

b. **Solve the system using algebra:** Use the method of substitution to write a single quadratic equation from the two equations in part a. Since you are trying to find the point of intersection, that means the y-coordinates should be the same, so you may set them equal to each other. (Look back at FF-24 if you need help.) Then solve the resulting quadratic equation by factoring and using the zero product property. *$0 = x^2 + 4x - 5$; $x = -5, 1$*

c. How are the solutions to the quadratic equation related to the points of intersection? *The solutions to the quadratic equation are the x-coordinates of the points of intersection.*

In FF-75, students use the zero product property to solve factorable quadratic equations. Later when they face quadratic equations that are not easily factored, the need for another solution method will arise. At that time we will introduce the quadratic formula, but for now we focus on the graphical interpretation we've been developing and let students apply their factoring skills.

Following FF-75, lead a class discussion to help reinforce two fundamental concepts: (1) When we solve the single equation $0 = ax^2 + bx + c$, we are really using substitution to solve a system of equations of the form $y = 0$ and $y = ax^2 + bx + c$. (2) The zero product property is useful because it gives us a way to solve the quadratic equation $0 = ax^2 + bx + c$ once we have factored the quadratic expression $ax^2 + bx + c$.

FF-75. Using the Zero Product Property to Solve Equations By now you have factored quite a few quadratic expressions similar to $x^2 + 2x - 3$. You have also graphed several parabolas in y-form where a quadratic expression appears on the right side of the equation, such as $y = x^2 + 2x - 3$, for example. Now you'll put these two ideas together by using the zero product property.

a. Factor $x^2 + 2x - 3$. *(x − 1)(x + 3)*

b. Graph $y = x^2 + 2x - 3$ for $-4 \leq x \leq 3$. *vertex (−1, −4); (1, 0), (−3, 0)*

c. Estimate the *x*-intercepts of the graph in part b. *(1, 0), (−3, 0)*

d. What are the values of *y* at the points where the graph crosses the *x*-axis? What is the equation of the *x*-axis? *0; y = 0*

e. Describe how the zero product property helps you solve the equation $(x + 3)(x - 1) = 0$. (Refer to your Tool Kit if you need help.) Then use the zero product property to solve the equation $(x + 3)(x - 1) = 0$ for *x*. *Since the product of x + 3 and x − 1 is zero, at least one of the factors must equal 0; x = −3 or 1*

f. Compare your results in part e to the *x*-values of the intercepts you estimated in part c—what do you notice? Write a sentence or two to describe the relationship between the results of the two methods you used: (1) finding where the graph of the equation $y = x^2 + 2x - 3$ crosses the *x*-axis and (2) solving the equation $0 = x^2 + 2x - 3$ using algebra. *The x-coordinates of the x-intercepts of the graph are the solutions of the equation.*

FF-76. The Zero Product Method In this problem you'll apply what you observed in FF-75f to another quadratic equation.

a. Graph the parabola $y = x^2 - x - 2$ for $-2 \leq x \leq 3$. Mark the *x*-intercepts and label them with their coordinates. *vertex $(\frac{1}{2}, -2\frac{1}{4})$; (2, 0) and (−1, 0)*

b. Substitute $y = 0$ into the equation $y = x^2 - x - 2$ to obtain the single equation. Then find two solutions to the new equation by first factoring the expression $x^2 - x - 2$. *x = 2, −1*

c. Explain how solving the equation $0 = x^2 - x - 2$ gives you enough information to name the *x*-intercepts of the parabola $y = x^2 - x - 2$ without having to draw the graph. *The solutions of the equation are the x-coordinates of the x-intercepts of the graph.*

d. Why might some people call the method you followed in parts b and c the "zero product method"? *We used factoring and the zero product property to solve an equation.*

FF-77. Without graphing, use the zero product method developed in the previous problems to find where each of the following graphs crosses the *x*-axis. (*Hint:* What is the equation of the *x*-axis?)

a. $y = (x - 3)(x - 4)$
(3, 0), (4, 0)

b. $y = x^2 + 5x - 24$
(−8, 0), (3, 0)

c. $y = x^2 + 10x + 16$
(−8, 0), (−2, 0)

374 Chapter 6 Fitness Finances: Solving Systems of Equations

> **EXTENSION AND PRACTICE**
>
> **FF-78.** Solve each of the following quadratic equations:
>
> a. $0 = (x + 3)(x - 6)$ b. $x^2 + 4x - 32 = 0$ c. $0 = 2x^2 + 7x + 6$
> $x = -3, 6$ $x = -8, 4$ $x = -2, \frac{-3}{2}$
>
> d. Substitute the two values that you found for x in part a into the equation to show that both values are correct solutions.
>   *Using $x = -3$: $(-3 + 3)(-3 - 6) = (0)(-9) = 0$.*
>   *Using $x = 6$: $(6 + 3)(6 - 6) = (9)(0) = 0$*
>
> e. Find the coordinates of the x-intercepts of the parabola $y = x^2 + 4x - 32$ without graphing. (*Hint:* Use what you found in part b.) *(−8, 0), (4, 0)*
>
> **FF-79.** Jamilla and Kelley wondered if they could use the zero product method to solve the equation $x^2 = 64$, so they first wrote $x^2 - 64 = 0$, and then tried to factor $x^2 - 64$. Here's how their paper looked:
>
> $$x^2 = 64$$
> $$x^2 - 64 = 0 \qquad \text{Subtracted 64 from each side.}$$
> $$x^2 + 0x - 64 = 0 \qquad \text{Added 0 to the left side.}$$
> $$(x + 8)(x - 8) = 0 \qquad \text{Factored.}$$
> $$x + 8 = 0 \text{ or } x - 8 = 0 \qquad \text{Used the zero-product property}$$
> $$\text{So, } x = -8 \text{ or } x = 8 \qquad \text{Subtracted 8 from each side;}$$
> $$\text{added 8 to each side.}$$
>
> a. Copy Jamilla's and Kelley's work, and describe what they did in each step. (One step has already been filled in.) Then check that both solutions are correct.
>
> b. Now use the zero product method to solve each of the following quadratic equations. Check your solutions.
>
>   i. $x^2 = 169$ $x = \pm 13$ ii. $z^2 - 0.07 = 0.57$ $z = \pm 0.8$
>   iii. $y^2 - 0.1 = 1.11$ $y = \pm 1.1$ iv. $t^2 + 100t = 0$ $t = 0, -100$
>
> **FF-80.** Plot the points $A\,(0, 0)$, $B\,(3, 0)$, and $C\,(3, 2)$. Connect the points to form $\triangle ABC$. Slide $\triangle ABC$ 3 units to the right and up 2 units. Then draw the new triangle and describe the results of the slide. What are the coordinates of the new vertices? *(3, 2), (6, 2), (6, 4)*

FF-81. Reread the introduction in the Olympics Records problem, FF-1, and take out the graphs you made for the men's and women's data for that problem.

The equations you wrote were for lines that pass through the data points given in FF-1. How well do your trend lines model other Olympic data? To answer this question, complete parts a through g.

a. Identify the point you would use from your graph from FF-1 to predict the men's winning time in the first modern Olympic Summer Games. Based on the graph, what prediction would you make?
y-intercept; ≈ 11.4 sec.

b. Thomas Burke won the 100-meter dash in the first modern Olympic Summer Games in 12 seconds. Plot a point on your graph corresponding to this winning time. How close was your prediction? *off by about 0.6 sec*

c. Identify the point you would use from your graph from FF-1 to predict the men's winning time in the 1992 Summer Olympics. Based on the graph, what prediction would you make? *At x = 96 the winning time is about 9.9 sec.*

d. Linford Christie from Britain won the 100-meter dash in 1992 with a time of 9.96 seconds. Plot a point on your graph corresponding to his winning time. How close was your prediction? *off by about 0.1 sec*

e. What would you predict for the women's winning time in 1928, the first year women ran the 100-meter dash in the Summer Games? *≈ 12.0 sec*

f. How close is your prediction to Elizabeth Robinson's 1928 gold medal time of 12.2 seconds? What factors might have affected the accuracy of your prediction? *It's off by about 0.2 sec. The data are not exactly linear; the graph may be drawn inaccurately.*

g. Do you think your equations are good predictors for winning times in the men's and women's 100-meter dash? *Answers will vary.*

FF-82. A **mathematical model** is an equation that can be used to make predictions. Your equations for the men's and women's Olympics data are examples of mathematical models. An equation may be a good model for a particular time interval, but yield inaccurate predictions in other time intervals. In this problem, you will try to estimate a time interval over which your Olympics equations are useful mathematical models.

a. If four points lie on a line, what would be true about the slope ratios between any pair of points? What if the four points lie very close to the line?   *The slope ratios would be equal or almost equal.*

b. What does your model, or equation, for the women's 100-meter dash in FF-4 predict for the winning time in the year 2000? What does it predict for the winning time in the year 2200?   *year 2000: 10.67 sec; year 2200: 6.89 sec.*

c. Suppose the data for the women's winning times in the 100-meter dash were linear—that is, all the data points were on a line. If the Summer Games were continued to be held, what would the winning time eventually be?   *Eventually, the line would intersect the x-axis, and the winning time would then be 0 sec!*

d. What does your model, or equation, for the women's 100-meter dash predict for the winning time in the year 3000?   *−8.22 sec*

e. Which of the following graphs would be a better model than yours? Explain your reasoning.

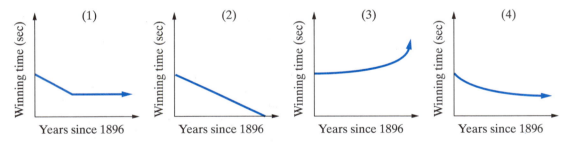

Graph 4 since winning times gradually level off; students could also make a case for graph 1.

f. For which time interval do you think your original equation for the women will be a pretty good model?   *Answers will vary.*

FF-83. Sketch a graph to represent each of the situations or relationships described below. In part b, you will need to set up and label the axes appropriately.

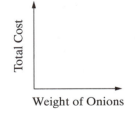

a. The cost of a bag of onions depends on its weight.

b. Albert is raising rabbits. He notices that there are twice as many rabbits every two months! Set up the axes with the number of months on the horizontal axis.

Solutions: **a.** **b.**

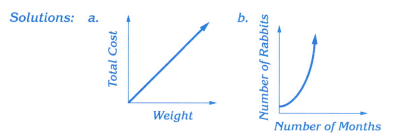

FF-84. When the employees of the accounting firm of Scrooge & Marley were asked how long they'd been with the firm, the responses were recorded on a graph. Each × on the graph below represents one employee.

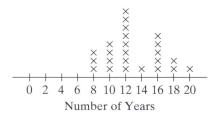

a. Calculate the average number of years an employee has worked for Scrooge & Marley.   *about 12.92 yr*

b. What percent of the employees has been with Scrooge & Marley at least 12 years? $\frac{17}{24} \approx 71\%$

c. If you had to choose one of the employees at random to interview, what would be the probability that the chosen person has worked fewer than 12 years for Scrooge & Marley? $\frac{7}{24} \approx 29\%$

FF-85. Factor each of the following quadratic polynomials. (It may help to draw generic rectangles.)

a. $2x^2 - 6x$ $2x(x - 3)$

b. $a^2 - 4a + 3$ $(a - 3)(a - 1)$

c. $2m^2 - 11m + 5$ $(2m - 1)(m - 5)$

FF-86. Exponent Practice Write each of the following expressions in a simpler exponential form. Refer to your Rules for Working with Exponents in your Tool Kit if you need help.

a. $\dfrac{3x^3y^2}{6xy^5}$ $\dfrac{x^2}{2y^3} = \dfrac{1}{2}x^2y^{-3}$

b. $\dfrac{(4x^2y^3)^{-2}}{2x}$ $\dfrac{x^{-5}y^{-6}}{32}$

c. $\dfrac{2(x^5y)^2}{(2xy)^3}$ $\dfrac{x^7}{4y} = \dfrac{1}{4}x^7y^{-1}$

d. $(6^{-3})(6^{x-1})$ 6^{x-4}

378 Chapter 6 Fitness Finances: Solving Systems of Equations

  e. $\dfrac{(52x^4)(13x)}{169x^7}$ $4x^{-2} = \dfrac{4}{x^2}$ f. $\dfrac{4}{m^{-5}}$ $4m^5$

FF-87. Solve each equation by any method you choose:

  a. $9s - 1 = 4$ $\dfrac{5}{9} \approx 0.56$ b. $32 = \tfrac{1}{7}r + r$ 28
  c. $(x - 7)(2x + 6) = 0$ $7, -3$ d. $x^2 - 4x - 12 = 0$ $6, -2$
  e. $2x(x - 5) = 0$ $0, 5$ f. $1.5(w + 2) = 3 + 2w$ 0

FF-88. Solve the each of the following systems of linear equations.

  a. $x = -2$ b. $y = 3x - 8$ c. $y = 3x - 1$
     $2x + y = -7$ $2x + 3y = 12$ $x + 2y = 5$

  $(-2, -3)$ $\dfrac{36}{11}, \dfrac{20}{11}$, or $\approx (3.27, 1.82)$ $(1, 2)$

FF-89. Write a system of two equations with two variables, and use it to solve the following problem. Refer to FF-46 if you need help getting started.

  A total of $18,000 is deposited into two simple interest accounts. On one account the annual simple interest rate is $4\tfrac{1}{2}$ percent; on the second account the annual simple interest rate is 8 percent. How much should be invested in the 8% account so that both accounts earn the same amount of interest? $x + y = 18{,}000;\ 0.045y = 0.08x;\ \$6480$

FF-90. Find the dimensions of each generic rectangle. Write an equation that expresses the area of each generic rectangle as a product of its factors.

  a.

$3a^2$	
	10

  b.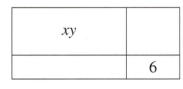

$(3a + 2)(a + 5) = 3a^2 + 17a + 10$, $(x + 2)(y + 3) = xy + 2y + 3x + 6$,
or $(3a + 5)(a + 2) = 3a^2 + 11a + 10$,   or $(x + 3)(y + 2) = xy + 3y + 2x + 6$,
or $(3a + 1)(a + 10) = 3a^2 + 31a + 10$,   or $(x + 1)(y + 6) = xy + y + 6x + 6$,
or $(3a + 10)(a + 1) = 3a^2 + 13a + 10$ or $(x + 6)(y + 1) = xy + 6y + x + 6$

  c. Compare your solutions for parts a and b with those of other members of your group. Write one or two sentences to describe what you notice.

FF-91. Mr. Smith went to the furniture store and bought a sofa for his living room. He paid $496 for the sofa. The sales clerk mentioned that the markup was 55 percent. How much did the furniture store pay the manufacturer for the sofa? Define a variable, and then write and solve an equation to solve the problem. Write your answer in a complete sentence.   *Let $x =$ the cost to the furniture store; $x + 0.55x = 496$; $x = \$320$*

6.6 SOLVING SYSTEMS OF LINEAR EQUATIONS BY THE ADDITION METHOD

SECTION MATERIALS
SUMMARY DRAFT
TEMPLATE RESOURCE
PAGE (FOR FF-112)

In this section, students solve systems of equations by the addition method. The overall strategy is first to eliminate one of the variables from a system of equations by adding the equations, and then to solve for the other variable. Stress the idea, not the mechanics, as groups work through the examples.

FF-92. The Addition Method for Solving a System of Equations

a. Use graphing to solve this system of equations:

$$2x + y = 11 \quad (5, 1)$$
$$x - y = 4$$

b. Now solve the same system of equations by substitution. *(5, 1)*

c. So far in this course, when you've been given a pair of two-variable equations to solve, you've used one of two ways to find a solution: you've either graphed the two equations to find the point(s) of intersection (as you did in part a), or you've used the substitution method (part b). Now you will learn about a third strategy for solving systems of equations—and it uses subproblems! You know how to solve single-variable equations. When there's a pair of two-variable equations, you can *eliminate one of the variables to obtain one single-variable equation.* You can do this by *adding the two equations together.* For this reason, it is called the **addition method** (or sometimes the **elimination method**) for solving systems of equations.

Example: To solve the system of equations given in part a, you can eliminate the *y* terms by adding the two equations together,

$$2x + y = 11$$
$$\underline{x - y = 4}$$
$$3x + 0y = 15$$

and then solve the resulting equation for *x*.

$$3x = 15$$
$$x = 5 \quad \blacktriangle$$

Now that you know the *x*-coordinate of the point of intersection, find the *y*-coordinate in the usual way by substituting into one of the original equations. Then verify that the addition method gives the same point of intersection as the other two methods you used in parts a and b.
2(5) + y = 11, y = 1 and 5 − y = 4, y = 1

d. Add information about the *addition method* for solving systems of linear equations to your Tool Kit.

FF-93. Solve the following systems of equations by using the addition method to eliminate one of the variables.

a. $4x + 2y = 14$ *(3, 1)*
   $x - 2y = 1$

b. $3x + 5y = 25$ *(5, 2)*
   $-3x + 7y = -1$

c. Show two checks to make sure that your answers to part a are correct.
   Substitute (3, 1) into $4x + 2y$ to get $4(3) + 2(1) = 14$. Substitute (3, 1) into $x - 2y$ to get $(3) - 2(1) = 1$.

After groups have done FF-94, have a class discussion to summarize what they've learned about the addition method for solving a system of equations.

FF-94. Revisiting the Addition Method for Solving a System of Equations Here's another example of solving a system of equations by adding the two equations together to eliminate one of the variables.

Example: Suppose you want to solve this system of equations:

$$3x + 2y = 11$$
$$4x + 3y = 14$$

Again, *the subproblem is to eliminate either x or y* when you add the equations together. However, in this case if you just add the equations without changing them, neither of the two variables will be eliminated—You still have *an equation with both x and y*.

$3x + 2y = 11$
$4x + 3y = 14$
───────────
$7x + 5y = 25$

In order to eliminate one of the variables you'll need to do something to both equations before adding them. If you want to eliminate *y*, you could …

multiply both sides of the first equation by 3 to get …

$3(3x + 2y) = 3(11)$
$9x + 6y = 33$

and multiply both sides of the second equation by -2 to get …

$-2(4x + 3y) = -2(14)$
$-8x - 6y = -28$

Why multiply the first equation by 3?

Why multiply the second equation by -2?

What could you do now to eliminate the *y* terms?

Answer: You can eliminate the *y* terms by adding the two new equations.

$9x + 6y = 33$
$-8x - 6y = -28$
───────────
$x + 0y = 5$

Now you know $x = 5$ and can easily find that $y = -2$. (*How?*) Thus, the solution is $(5, -2)$. ▲

a. Discuss the example in your group. What would have happened if we had chosen instead to multiply the first equation by 4 and multiply the second equation by −3? Check by writing out all the steps. *The x terms would be eliminated instead of the y.*

b. Add information to your Tool Kit about using multiplication with the addition method for solving systems of linear equations.

FF-95. Use the addition method to solve each of the following systems of equations. In each case you will have the subproblem of eliminating either x or y from the equations, and will need to do something before adding. What could you do? There is more than one way to solve these, so expect a *variety of methods* in your group, but the *same solutions*, (x, y).

a. $2x + 3y = -1$ *(−2, 1)*   b. $2x + 5y = 27$ *(1, 5)*
   $5x - 2y = -12$ $x + 3y = 16$

c. Write out the two checks that your solution to part b is correct.

EXTENSION AND PRACTICE

FF-96. Use the addition method to solve each of the following systems of equations:

a. $2x + y = 7$ $\left(\dfrac{23}{8}, \dfrac{5}{4}\right)$ b. $x + 3y = 4$ *(1, 1)*
   $2x + 5y = 12$ $3x - y = 2$

c. Write out the two checks that your solution to part b is correct.

FF-97. Which of the three methods for solving the given system of equations—graphing, substitution, or addition—was easiest for you to understand and do? Write one or two sentences to explain your response.

FF-98. Egg Toss Revisited If an egg is tossed straight up into the air at a rate of 80 feet per second, its height y in feet above the ground after x seconds is given by $y = 80x - 16x^2$. In problem SP-92 (in Chapter 2) you sketched the graph of this equation and answered questions about the height of the egg by using your graph. In this problem you will use algebraic techniques to answer questions about the height of the egg.

a. What is the value of y when the egg hits the ground? Use this value of y to write an equation you could solve to find out how many seconds will pass before the egg hits the ground. Then use the zero product method to solve the equation. *The value of y will be zero, so $0 = 80x - 16x^2$; $x = 0$ or 5. The egg will hit ground in 5 sec.*

b. After how many seconds will the egg be 64 feet above the ground? Write an equation to represent this situation, and then use algebra to solve it. *$64 = 80x - 16x^2$; $x = 1$ or 4. The egg will be 64 ft above ground at 1 sec and again at 4 sec.*

c. If you toss a second egg straight up into the air at a rate of 128 feet per second, its height y above the ground (in feet) after x seconds will be $y = 128x - 16x^2$. How long will it take the second egg to hit the ground? When will it be exactly 192 feet above the ground? *The egg will hit the ground after 8 sec. It will be 192 ft above ground at 2 sec and again at 6 sec.*

FF-99. Find an equation of the line described by each of the following statements.

a. A line goes through the points (80, 25.6) and (85, 22.4).
$y = -0.64x + 76.8$

b. The line $y = \frac{1}{2}x + b$ goes through the point (2, 3). $y = \frac{1}{2}x + 2$

*Editorial Board, *University of California at Berkeley Wellness Letter*, Wellness Made Easy, 365 Tips for Better Health (©Health Letter Associates, 1995).

FF-100. According to *Wellness Made Easy**, a book of health tips, food that is kept warm at buffet tables is often not hot enough to prevent food poisoning. To prevent bacteria from growing, hot food should be kept around 145°F to 165°F.

a. If the temperature in degrees Fahrenheit of your potluck lasagna is represented by F, use inequalities to represent the safe temperature range.
$145°F \leq F \leq 165°F$

b. Suppose your potluck dinner is held in England, where temperature is measured in degrees Celsius. If $C = \frac{5}{9}(F - 32)$, use inequalities to algebraically represent the safe temperature range in Celsius.
$62.8°C \leq C \leq 73.9°C$

FF-101. Given the following statement,

"the smaller the pies were, then the more pies he could fit into the oven"

sketch a graph to represent the relationship between the number of pies and the size of the pies that will fit in an oven. You will need to set up and label the axes appropriately.

Solution:

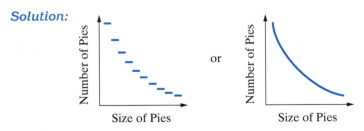

FF-102. Find each of the following sums of fractions:

a. $\frac{1}{2} + \frac{1}{3}$ $\frac{5}{6}$

b. $\frac{1}{2} + \frac{x}{3}$ $\frac{3 + 2x}{6}$

c. $\dfrac{x}{2} + \dfrac{x}{3} \quad \dfrac{5x}{6}$ d. $\dfrac{1}{2} + \dfrac{1}{x} \quad \dfrac{x+2}{2x}$

e. $\dfrac{1}{3} + \dfrac{1}{2x} \quad \dfrac{2x+3}{6x}$

FF-103. Simplifying Fractions Rewrite each of the following fractions using as few digits or variables as possible:

a. $\dfrac{70x^2}{14x^2}$ for $x \neq 0$ b. $\dfrac{8(x-5)^4}{16(x-5)^3}$ for $x \neq 5$

5 $\dfrac{1}{2}(x-5)$

c. $\dfrac{(x+4)(x-4)}{(x-4)}$ for $x \neq 4$ d. $\dfrac{(x+0.3)^4}{x+0.3}$ for $x \neq -0.3$

$x + 4$ $(x+0.3)^3$

FF-104. Write two equations for the following problem. State what each variable represents. Solve the problem and state your answer as a sentence. Refer to FF-46 if you need help getting started.

At the last night of the play, *The Miser,* 141 people were in attendance. Admission for students was $5, but the general public paid $8 to attend. When the receipts were totaled, the box office noticed that only $10 more was spent on general public tickets than on student tickets. How many of each ticket was sold?   *55 general public, 86 student*

FF-105. What percent of the total area of rectangle *ABCD* is shaded? Identify and solve several subproblems you could use to help find an answer. *Find the area of the rectangle; find the area of the unshaded triangle; $66\dfrac{2}{3}\%$*

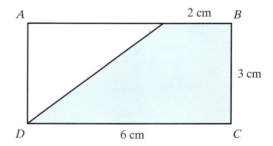

FF-106. If two 10-inch pizzas cost the same as one 15-inch pizza, which is the better buy? What if the crust is your favorite part of a pizza? Would that make a difference? Explain your answers. *The area of two 10-in. pizzas is $50\pi \approx 157.1$ sq in., while the area of one 15-inch pizza is $56.25\pi \approx 176.7$ sq in. If you prefer crust, two 10-in. pizzas have a total circumference of 20π in. compared with 15π in. on the 15-in. pizza.*

FF-107. Solve each of the following equations for the given variable by rewriting the equation so you can factor and use zero product method. Show how you check your solution.

a. $122 - 58 = B^2$ for B ± 8

b. $R^2 + 37 = 181$ for R ± 12

FF-108. Solve each of the following equations:

a. $\dfrac{x}{3} = x + 4$ -6 b. $\dfrac{x+6}{3} = x$ 3

c. $\dfrac{x+6}{x} = x$ $-2, 3$ d. $\dfrac{4x+1}{3x+1} = x + 1$ 0

FF-109. In problem FF-94 you saw an example of solving a system of linear equations by multiplying each equation by a number and then adding to get $0x$ or $0y$. To solve the system

$$3x + 2y = 11$$
$$4x + 3y = 14$$

both sides of the first equation were multiplied by 3, and both sides of the second equation were multiplied by -2. There are many other choices of multipliers that you could use to eliminate either the x terms or the y terms from the system. Choose another pair of numbers that will work as multipliers, and use them to solve the system. Write out all the steps. *For example, to eliminate y, multiply the first equation by 6 and the second by -4; or, to eliminate x, multiply the first by 4 and the second by -3.*

FF-110. Solve each system of equations for x and y:

a. $x + 2y = 5$ $(5, 0)$
   $x + y = 5$

b. $2x + 3y = 5$ $(1, 1)$
   $x + 3y = 4$

c. $x + 2y = 16$ $\left(\dfrac{20}{3}, \dfrac{14}{3}\right)$
   $x - y = 2$

d. $3x + 2y = 11$ $\left(\dfrac{13}{7}, \dfrac{19}{7}\right)$
   $4x - 2y = 2$

e. $3x + 3y = 15$ $\left(\dfrac{11}{2}, \dfrac{-1}{2}\right)$
   $x - y = 6$

f. $2x^2 - y = 5$ $(1, -3)$ or $\left(\dfrac{-5}{4}, \dfrac{-15}{8}\right)$
   $x + 2y = -5$

FF-111. Woodland has a population of 12,200. Its population has been increasing at a rate of 300 people per year. Riverdell has a population of 21,000 and is declining by 250 people per year. Assuming that the rates do not change, in how many years will the populations be equal?

Define two variables, and write two equations to represent the problem. Visualize the graphs and then sketch them. Then use your algebraic skills to write a single equation and solve it. How large will the populations be when they are equal? What are the coordinates of the point where the graphs meet? *Let x = number of years from now; y = population; $y = 12{,}200 + 300x$ and $y = 21{,}000 - 250x$. In 16 years, both populations will be 17,000; (16, 17,000).*

FF-112. Chapter 6 Summary Rough Draft The introduction to the chapter mentioned five main ideas that would be developed in the problems you did. Reread the introduction, and then look back through the chapter to find where each main idea was developed.

Write your answers to the following questions in rough draft form on *pages separate from your other work*, and be ready to discuss them with your group at the next class meeting. Focus on the *content,* not neatness or appearance, as you write your summary draft. You will have the chance to revise your work after discussing the rough draft with your group.

a. In this chapter you've worked solving systems of equations both algebraically and graphically. From your work, select two problems that best show your understanding of

   i. how to solve systems of equations algebraically, and

   ii. how the solution of a system of equations is related to the graphs of the equations.

   Write complete sentences to describe how you did each of the selected problems. Then tell why you chose the problems that you did.

b. Kyran, a classmate, still doesn't understand two of the big ideas of the chapter—how to interpret graphs of data, and how to write an equation of a line given either a point and a slope, or two points. Which problems from the chapter would you suggest he study?

c. Find a problem that you still cannot solve or that you are unsure you would be able to solve on a test. Write out the question and as much of the solution as you can. Then explain what you think is keeping you from solving the problem. Be clear and precise in describing the hard part.

6.7 INVESTIGATION: USING ALGEBRA TO MODEL AND ANALYZE LINEAR DATA (OPTIONAL)

SECTION MATERIALS
FF-113—FF-116
RESOURCE PAGE

If you have time, we highly recommend having students give oral or poster presentations. They give students additional opportunities to share their work and learn from each other.

Here are some suggestions for making the project successful:

- *Set up due dates for each part of the project in advance to help students plan ahead so they can do more thorough work. An assignment and grading table is provided on the FF-113 through FF-116 Investigation Resource Pages.*
- *Some instructors have had success assigning the project to pairs or groups of students. This is a good way to take up less class time with oral presentations; it also gives students another chance to work together in teams.*
- *Projects are often more successful if instructors meet with students outside of class to go over their work in writing the linear equations for problem FF-114.*
- *Students may need help setting up the time variable so that $t = 0$ represents the initial time.*
- *You will probably need to discuss the advantages and disadvantages associated with using a linear equation to model data that are not actually linear.*

You have been learning how equations can be used to model situations. A quantity—weight, cost, distance, or area, for example—that is changing at a constant rate can be modeled using a linear equation. In this chapter, you have also examined problems that could be modeled by pairs of related linear equations. To solve problems involving two sets of information, you have written two linear equations and then used graphing and algebraic techniques to solve the problem.

In this four-part investigation, problems FF-113 through FF-116, you will learn more about how mathematical ideas are used to model real situations. Make note of the due dates and grading information that your instructor provides.

FF-113. Project Proposal The first step is to choose two quantities you'd like to compare that have been either growing or shrinking over time. In this project you will assume that the two quantities are each changing at a *steady* rate; part of the project will be to analyze whether the rates are steady or not, or steady for at least some time interval. As you saw in the Olympic 100-meter dash problem (FF-81), for example, the men's and women's winning times were decreasing at an almost constant rate for a number of years, but it didn't make sense for the rate to decrease steadily forever.

Be creative in choosing the quantities you compare! Choose two quantities you are really interested in comparing. You are free to choose one of the following topics or one of your own.

- cancer rates of two different countries or two states
- world record times over the years in some race or event
- winning times for men and for women in some Olympic event
- crime rates among two different age groups
- car prices
- the number of lawyers in the United States or in your state
- the cost of gasoline
- the populations of two countries (perhaps your family's country of origin and your country of origin).

Your instructor can help you determine an appropriate way to use your topic in this project.

Once you have chosen your topic, you will need to collect the data. Be sure to make clear notes about where you find the data. For each of the two quantities you are comparing, you will need at least three data points. For example, if you are comparing car prices for Mustangs and VW Beetles, you could find the average price of each type of car at three different ages: brand new, five years old, and ten years old.

Next, write a proposal. In your proposal, state your topic and explain why you chose it. Include a table of the data you propose to study, and describe exactly where you found your data. This description must be specific enough that another person could find the original data using the reference information in your proposal.

FF-114. Mathematical Modeling Your job in this part of the project is to construct a *linear mathematical model*—that is, you must write a linear equation—for each of your two data sets, and then use the models to make predictions about the future and to make some comparisons.

a. First, use a full sheet of graph paper to plot one set of data points. Then draw a line that you think approximately models the situation—your line does not have to pass through any of the three points, but should come as close as possible to all three points.

b. Write an equation for the line you drew using algebraic techniques you have learned.

c. Next, plot the second set of data points on the same axes, draw a line to model the second situation, and write a second equation.

d. If the two graphs intersect, estimate the point of intersection, and then use an algebraic method to find the point of intersection more accurately.

Your write-up for this part should be as complete as possible. Show all steps, and explain your mathematical process in a clear way. Clearly label your equations and work.

FF-115. Analysis Your analysis should explain what your equations have to "say" about the two quantities you are comparing. Write several paragraphs to analyze the situation. This question is open-ended in the sense that you should decide how to organize the information, but you should think about the following questions as you do your analysis:

- What are the *x*- and *y*-intercepts of your lines? How would you interpret each of these points in terms of your data sets?
- What are the slopes of your lines?
- What do your equations predict for the future? What are the consequences of these predictions?
- Have the two quantities you are comparing ever been equal, or will they be equal at some time in the future?
- When you drew lines to approximately model each set of data, you were assuming that the data would change at a *steady* rate as time passed. Do you think this was a justifiable assumption?
- Are there any times when your equations don't make sense or are obviously wrong?

Do not limit yourself to these questions—your analysis should be as thorough as possible. The data you chose might have unique elements associated with it. A typewritten paper is easier to read.

FF-116. Presentation Options You instructor may assign one or more of the following ways for you to present your findings:

Oral Presentation: Prepare a five-minute class presentation explaining your project.

Poster Presentation: Make a poster to explain your project. Your poster will be displayed in the classroom, and should be easily understood by someone reading the poster even if you are not there to answer questions. You should include a title, your graphs, and some analysis.

Written Presentation: Your written report should include your original proposal, the work for your mathematical models, and your final analysis. The proposal and the analysis should be typewritten, but your mathematics for finding the equations should be handwritten neatly. Your process for all parts must be clear and organized.

6.8 SUMMARY AND REVIEW

SECTION MATERIALS
TRANSPARENCIES AND PENS (OPTIONAL, FOR FF-122)

Start with groups discussing their Chapter 6 Summary rough drafts, as you circulate and listen. Note that less guidance was provided in FF-112 than in the rough-draft summary problems for previous chapters. The goal is for students to develop the ability to identify and summarize the big ideas of a chapter, and to illustrate them with examples from their work.

Select problems for review and practice as needed by your students.

FF-117. Chapter 6 Summary Group Discussion Take out the rough-draft summary you completed in problem FF-112. Use class time to discuss your work, and use homework time to revise your summary as needed.

Take turns with the members of your group describing the problems each of you chose to illustrate your understanding of how to solve systems of equations, and how the solutions to systems of equations are related to their graphs. When it is your turn to share, you should:

- Explain the problems you chose to illustrate the main ideas.
- Explain why you chose those particular problems.
- Explain, as far as possible, the problems from FF-112c that you still cannot solve.

This is also your chance to make sure your summary is complete, to update your Tool Kits, and to work together on problems you may not be able to solve yet.

FF-118. Tool Kit Check Throughout this chapter you've been told to make entries in your Tool Kit. In this chapter you should have made entries for at least the following topics:

- the point of intersection of two lines *FF-23*
- using the substitution method to solve a system of equations *FF-39*
- using the addition method to solve a system of equations *FF-92, FF-94*

a. Check that your Tool Kit entries are up-to-date. Exchange Tool Kits with members of your group, and read each other's entries to check for clarity and accuracy.

b. At home, make sure that your Tool Kit includes examples and explanations that help you remember and understand the ideas. Make any revisions, clarification, or updates that will reflect your developing understanding of mathematics.

FF-119. Country Road On a late Spring Saturday, Carol and Christina both leave Blue Ridge and travel on the same road. Christina, riding her bike, travels at a steady rate of 6 miles per hour. Carol, who left five hours earlier than Christina, walks at a steady rate of 2 miles per hour. How long will it take for Christina to catch up with Carol? How far outside Blue Ridge will the two meet up?

So far in this course, you've learned several ways to solve such a problem: you could set up and use a guess-and-check table to solve it, you could use a guess-and-check table to write an equation that could be used to solve it, or you could write a system of two equations with two variables and then solve the system graphically or algebraically. Reread the problem, and then complete parts a through d to solve it algebraically.

a. If x represents the number of hours that Christina travels, and y represents the number of miles she travels, how are x and y related? Write an equation. $y = 6x$

b. When Christina catches up with Carol, how does the distance Carol has traveled compare with the distance Christina has traveled?
The distances are equal.

c. Use x and y to write an equation to represent the distance Carol travels.
$y = 2(x + 5)$

d. Finally, solve the pair of equations from parts a and c. Use your solution to answer the questions posed in the problem. Write your answers in one or two complete sentences. $2(x + 5) = 6x$; $x = 2.5$ *hrs; the two meet up after 2.5 hrs, 15 mi outside Blue Ridge.*

In FF-120, groups write two equations, $L = 3W$ and $(L - 4)(W - 4) = LW - 176$, and solve the system. To guide them in writing the second equation, it helps to ask how the area of a rectangle is related to its length and width. If groups are reluctant to substitute 3W for L, reassure them that the resulting equation, $(3W - 4)(W - 4) = (3W)W - 176$, will not be as difficult to solve as it may first seem.

FF-120. Write a system of two equations with two variables, and use it to solve the following problem. (Refer to FF-46 if you need help getting started.)

A rectangle is three times as long as it is wide. If the length and width are each decreased by four units, then the area is decreased by 176 square units. What are the dimensions of the original rectangle?
$L = 3W$ *and* $(L - 4)(W - 4) = LW - 176$; $W = 12$ *units,* $L = 36$ *units*

FF-121. A rocket is launched from a launch pad 98 meters high at noon on Tuesday. The height, h, in meters, of the rocket above the ground after t seconds is given by the following equation from physics:

$$h = -4.9t^2 + 39.2t + 98$$

a. Find the value of h when $t = 0$. What does this value of h represent?
  When $t = 0$, $h = 98$. This represents the height of the rocket when it is sitting on the launch pad before takeoff.

b. How long will it take for the rocket to fall back and hit the ground? *10 sec*

c. At what times will the rocket be exactly 171.5 meters above the ground? *3 sec, 5 sec*

In FF-122, students can apply what they've learned to a "real" situation. Although the numbers are messy, a calculator makes quick work of them. When groups are done, they could present their solutions on transparencies.

FF-122. The Burning Candle Suppose it's your friend's birthday and you want to surprise her by walking into the room carrying a piece of cake with a lighted candle. Could you predict how long before the candle goes out?

In this problem you will use two data points,

(50 seconds, 0.78 grams)
and
(3 minutes 50 seconds, 0.57 grams),

to find the slope-intercept form of the equation of the line, and then use the equation to find when the candle would burn out.

a. Use a generic slope triangle to calculate the slope of the line. Interpret the slope in terms of the burning candle. $\frac{-0.21}{180} \approx -0.0012$ *g/sec, so the candle is burning up at the rate of 0.0012 g each second.*

b. Write an equation of the line in slope-intercept form, messy numbers and all. *$y = -0.0012x + 0.84$*

c. What value will y have when the candle burns out? Use that value to replace y and solve the equation you found in part b for x. How long before the candle goes out? *$y = 0$, so $0 = -0.0012x + 0.84$; $x = 700$ sec ≈ 11.67 min*

Section 6.8 Summary and Review 391

> **EXTENSION AND PRACTICE**

As a follow-up to FF-123 at the next class meeting, you could have students demonstrate to the class, or discuss in their groups, the method they chose to solve the systems.

FF-123. Solve each of the following systems of equations. Identify which method you used: graphing, substitution, or addition.

a. $3x - 2y = 4$ (2, 1)
    $4x + 2y = 10$

b. $p + q = 4$ (−1.5, 5.5)
    $-3p + 2q = 15.5$

c. $y = 3x + 1$ (−1, −2)
    $x + 2y = -5$

d. $x = y + 3$ (0, −3)
    $x + 2y = -6$

e. $y = x^2 - 5x + 4$ (4, 0), (1, 0)
    $y = 0$

f. $y = x^2 - 3x$ (4, 4), (1, −2)
    $y = 2x - 4$

FF-124. Changing Account Balances Susan has $20 and is saving at the rate of $6 per week. Jeanne has $150 and is spending at the rate of $4 per week. After how many weeks will each person have the same amount of money? *13 wk*

Solve this problem in a step-by-step manner by writing an equation for each person and drawing the graphs to estimate the solution. (Scale the *x*-axis at 1 week per tick mark up to 20 weeks. Scale the *y*-axis at $10 per tick mark up to $160.) Then show an algebraic method for finding the solution exactly.

FF-125. Graph the equation $y = -3x + 4$.

FF-126. Without drawing a graph, determine which of the following points are on the graph of the linear equation $y = -2x + 3$.

$$(0, 3), (3, 0), (-3, 3), (1, 1), (5, -7), (5, 7)$$

yes, no, no, yes, yes, no

FF-127. Look back at the Olympic Records problem, FF-4.

a. With the two trends lines you found in FF-4, use an algebraic method to find the point at which the men's and women's winning times will be the same. What is that time? *With $y = -0.014x + 11.26$ (men's) and $y = -0.019x + 12.63$ (women's); $x \approx 274$, $y \approx 7.42$ sec*

b. The scaling suggested for the *x*-axis in FF-1 made the graphing easier by starting at 1896, the year the modern Olympic Games were first held. This way the year represented by $x = 32$ is 1928 since

$$1896 + 32 = 1928.$$

Use this idea to find the year represented by the *x*-coordinate of the point where the winning times will be the same. *in 2170*

c. What does your answer in part b suggest about the practicality of solving the problem by graphing? *Although the equations in part a are messy, it is easier to solve the system algebraically than it is to do so graphically.*

FF-128. Ant Thermometers A certain kind of ant can be used as a thermometer! The ants travel faster as the temperature increases. To calculate the temperature we can use the equation

$$T = 11S + 39,$$

where S represents the ants' speed in inches per second and T represents the temperature in degrees Fahrenheit.

a. Graph this rule with the temperature, T, on the vertical axis and the ants' speed, S, on the horizontal axis. Scale the vertical axis so that each tick mark represents 5°, and mark it up to 100°. Scale the horizontal axis so that 4 units of graph paper represent a speed of 1 inch per second, and mark it up to 6 inches per second.

b. For what temperatures is the temperature equation impractical?
for $T < 39°F$

c. From your graph, estimate the temperature when the ants are traveling at a speed of 2 inches per second by drawing a vertical line at $S = 2$. Record your estimate. *Answers will vary, depending on the accuracy of students' graphs. The algebraic solution, not asked for here, is 61°F.*

d. Draw a line on your graph to represent a temperature of 90°Fahrenheit. What is the ants' speed when the temperature is 90°F? *$\frac{51}{11} \approx 4.64$ in./min*

e. Use your graph from this problem and from the Cricket Thermometers problem, WR-83, to answer the following question:

   If a cricket chirps 108 times per minute, what is the approximate speed of the ants? *about 3 in./min (algebraic solution: 2.27 in./min)*

FF-129. Solve each of the following equations for x. Show all your work.

a. $2x - 3 = 0$ *$\frac{3}{2}$*

b. $0 = (2x - 3)(x + 5)$ *$\frac{3}{2}, -5$*

c. $0 = 2x^2 + 7x - 15$ *$\frac{3}{2}, -5$*

d. $2x^2 + 7x - 15 = -21$ *$\frac{-3}{2}, -2$*

e. $8(x + 6) + 23 = 7$ *-8*

f. $\frac{5 - 2x}{3} = \frac{x}{5}$ *$\frac{25}{13} \approx 1.92$*

g. $3x - 11 = 0$ *$\frac{11}{3} \approx 3.67$*

h. $Dx - C = 0$ *$\frac{C}{D}$*

i. $x^2 - 4x + 4 = 0$ *2*

j. $x^2 + 6 = 6$ *0*

FF-130. Find the x-intercepts for the graph of each of the following equations:

a. $y = x^2 + 10x + 21$ *(−7, 0) and (−3, 0)*

b. $y = 2x - 1$ $\left(\frac{1}{2}, 0\right)$

c. $y = 2x^2 - 5x + 2$ *(2, 0) and $\left(\frac{1}{2}, 0\right)$*

FF-131. Solve this system of nonlinear equations by drawing careful graphs, as you did in problem FF-54:

$$y = \frac{3}{x} \quad \text{and} \quad y = 3^x$$

Use $x = -2, -1.5, -1, \ldots, 2, 2.5$ for both graphs. *(1, 3)*

FF-132. In the figure shown here, the curves are quarter circles cut out of a rectangle. The radius of the smaller circle is one unit.

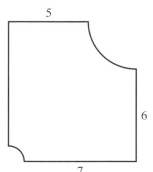

a. Find the dimensions of the original rectangle.
8 by 9

b. Find the area of the figure.
$72 - \frac{\pi \cdot 1^2}{4} - \frac{\pi \cdot 3^2}{4} \approx 64.15$ *sq units*

c. Find the perimeter of the figure.
$7 + 6 + \frac{2\pi(3)}{4} + 5 + 8 + \frac{2\pi(1)}{4} \approx 32.28$ *units*

FF-133. Virginia claims that $\frac{x^2 + 4x}{x^2}$ is equal to $1 + 4x$.

a. Decide whether Virginia's claim is correct by evaluating each expression for $x = 10$. *When $x = 10$, $\frac{x^2 + 4x}{x^2} = 1.4$, but $1 + 4x = 41$, so the expressions are not equal.*

b. Jim thinks that he can simplify the fraction if he factors the numerator first, like this:
$$\frac{x^2 + 4x}{x^2} = \frac{x(x + 4)}{x \cdot x} = \frac{x + 4}{x}$$

Decide whether Jim's method is correct by evaluating all three fractions for $x = 10$. *It's correct, since all three expressions are equal to 1.4 when $x = 10$.*

FF-134. Use the zero product method to solve each of the following quadratic equations. Check your solutions.

a. $x^2 = 169$ *$x = \pm 13$*   b. $z^2 - 0.64z = 0$ *$z = 0$ or 0.64*

c. $y^2 - 0.1 = 1.11$ *$y = \pm 1.1$*   d. $P^2 - 10P = 24$ *$P = 12$ or -2*

FF-135. Solve each of the following equations for the given variable by rewriting the equation so you can factor and use zero product method. Show how you check your solution.

a. $97 - x^2 = 4^2$ for x *± 9*   b. $H^2 = 5^2 + 200$ for H *± 15*

FF-136. In this problem you'll relate what you know about x-intercepts and solutions of quadratic equations.

a. Make a table and draw the graph of $y = x^2 - 16$. Mark and label the x-intercepts.   *vertex $(0, -16)$; x-intercepts $(4, 0)$ and $(-4, 0)$*

b. Find the points of intersection of the line $y = 0$ and the parabola $y = x^2 - 16$ by solving a system of equations. (You can factor the polynomial $x^2 - 16$ by rewriting it as $x^2 + 0x - 16$.)   *Solve $0 = x^2 - 16$ to find the x-coordinates of the points of intersection; $(4, 0)$ and $(-4, 0)$.*

c. In a complete sentence, explain how your solutions in parts a and b are related.   *The solutions to the equation $0 = x^2 - 16$ are the x-coordinates of the x-intercepts of the parabola $y = x^2 - 16$.*

FF-137. Sketch a graph to represent the relationship between the temperature of the oven and the time since it was turned on. You will need to set up and label the axes appropriately.

The temperature in an oven set to 350° is regulated by a thermostat. When the temperature gets a little too hot, the oven automatically turns off. When the temperature gets a little too cold, the oven turns back on again.

Solution:

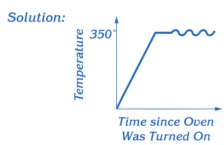

FF-138. The area of this generic rectangle is 60 square units:

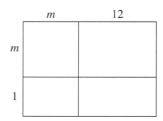

a. Write an equation, clean it up, and then solve to find the value(s) of m.
   $(m + 1)(m + 12) = 60$, so $m^2 + 13m - 48 = (m + 16)(m - 3) = 0$; $m = 3$

b. Can there be more than one value for m? Explain why or why not.
   There is only one value for m in this context. Although the equation $(m + 1)(m + 12) = 60$ has two solutions, $m = 3$ and $m = -16$, the rectangle cannot have sides of lengths -15 and -4 units. The value of m must be 3 in order for there to be a rectangle.

FF-139. Chapter 6 Summary Revision Use the ideas that your group discussed in problem FF-117 to revise your rough draft of Chapter 6 Summary. Your presentation should be thorough and organized, and should be done on paper separate from your other work.

CHAPTER SEVEN—FOR THE INSTRUCTOR

THE BUCKLED RAILROAD TRACK
From Words to Diagrams to Equations

MATERIALS
- Metric rulers
- 3 by 5 cards (one for each student)
- Piece of stiff paper, some string, a straw or stick, and masking tape for each group
- Shoe box, sticks of various lengths, and sticky notes for each group
- Scissors for each group
- Transparency of the RT-53 Resource Page
- Resource Pages for RT-3, RT-25, RT-26/30, RT-53, RT-54, and RT-64; also, Dot Paper and Summary Draft Template (optional)
- Optional: Materials for Pythagorean Puzzle (see Section 7.5) The Theorem of Pythagoras videotape (see Section 7.6)

The problem-solving focus of this chapter is drawing diagrams. Although some problems in this chapter could be solved using other strategies—such as guess-and-check—we focus now on guiding students to create diagrams and write equations based on their diagrams. Students may not appreciate how helpful diagrams can be until they use them to help solve difficult problems, but they need to learn how to set up and label diagrams in the context of simpler problems. We encourage students to practice making diagrams, even when they may be able to solve a problem without the help of a visual representation. You may also want to extend some problems to include a graph of the resulting equation.

We use the Pythagorean relationship in this chapter as a way of generating problems involving quadratic equations. Situations involving the Pythagorean relationship also provide a good way for students to see how useful diagrams can be for organizing information and writing equations. Moreover, such situations give students opportunities to build their number sense with square roots and develop familiarity with square-root arithmetic.

As noted in the materials list, the video *The Theorem of Pythagoras* by Project MATHEMATICS! shows several different proofs of the Pythagorean theorem. If time allows, you might use the Pythagorean puzzle demonstration in Section 7.5, and follow-up by showing the video on a summary day. However, keep in mind that because the Pythagorean relationship is used in this chapter mainly as a vehicle for other ideas, it should not be given rigorous treatment. Our only reason for showing a proof of the theorem at this time is to demonstrate how powerful a tool a picture can be in algebra. As one student remarked, "One picture is worth a thousand equations!"

There may be more problems in the Extension and Practice portion of each section than are reasonable or necessary for you to assign. Select problems that will help your students strengthen new skills and maintain those from previous chapters. Problems that aren't assigned for homework could be used for assessment or as part of course-review assignments in preparation for a final exam.

CHAPTER SEVEN

THE BUCKLED RAILROAD TRACK
From Words to Diagrams to Equations

Mathematical problems that arise outside of a mathematics class rarely come in the form of equations that are ready to be solved. Usually mathematical problems arise in situations and are stated in words, so they resemble "word problems" more than straightforward "solve for x" type problems. When trying to solve a word problem, it's often helpful to draw a picture or a diagram to represent the information in the problem. Many times the diagram can be used to help write an equation that models the problem. Solving the original problem then becomes a matter of solving an equation and checking that the solution(s) makes sense in the problem.

In this chapter you'll be strengthening your problem-solving and algebra skills as you work on problems similar to this one about a mythical buckled railroad track:

> A new railroad line was installed in Placid Valley. In order to reduce derailments along a two-mile stretch, the track was made with straight one-mile-long rails. The rails were laid in the winter, and they expanded in the heat the following summer. Indeed, each mile-long rail expanded 1 foot in length! Ordinarily, because the rails do not bend, they would jut to the side. However, in this strange case, the rails jutted upward where their ends met. How high above the ground were the expanded rails at the joint?

IN THIS CHAPTER, YOU WILL HAVE THE OPPORTUNITY TO:

- develop your skill at drawing diagrams to make it easier to write equations that can be used to solve problems posed originally in words;
- use the Pythagorean relationship to write quadratic equations;
- use factoring to solve quadratic equations;
- develop a sense of approximate sizes of square roots of numbers and build an understanding of some properties of square-root arithmetic.

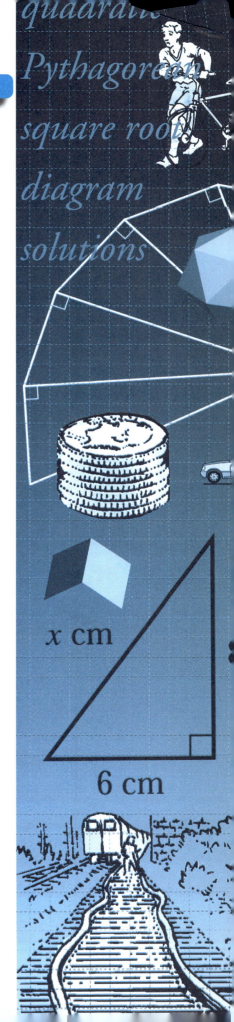

Chapter 7 The Buckled Railroad Track: From Words to Diagrams to Equations

MATERIALS	CHAPTER CONTENTS		PROBLEM SETS
Metric rulers and protractors or 3 by 5 cards (for RT-1); RT-3 Sides of Right Triangles Resource Page	7.1	Right Triangles and the Pythagorean Relationship	RT-1–RT-24
RT-25 Spiral of Right Triangles Resource Page; 3 by 5 cards (one per student, for RT-25); RT-26 Locating Square Roots on a Number Line Resource Page; RT-30 Square-Root Values Resource Page	7.2	Picturing Square-Root Lengths	RT-25–RT-50
Stiff paper, some string, a straw or stick, and masking tape (for each group, for RT-51); a shoe box, sticks of various lengths, and sticky notes (for each group, for RT-52); RT-53 Area and the Pythagorean Relationship Resource Page; scissors (for each group, for RT-53); transparency of RT-53 Resource Page; RT-54 Resource Page; dot paper (for RT-57); RT-62 Are Segment Length and Slope Related? Resource Page	7.3	Models, Diagrams, and Right Triangles	RT-51–RT-75
Dot paper (for RT-82 and RT-85)	7.4	Distances, Times, and Rates: More Diagrams and Equations	RT-76–RT-99
Pythagorean Puzzle pieces (optional); Chapter Summary Rough Draft Resource Page (optional, for RT-123)	7.5	Calculating with Square Roots	RT-100–RT-123
Theorem of Pythagoras videotape (optional)	7.6	Summary and Review	RT-124–RT-156

- Problem Solving
- Graphing
- Writing Equations
- Ratios
- Solving Equations
- Symbol Manipulation
- Reasoning & Communication

7.1 RIGHT TRIANGLES AND THE PYTHAGOREAN RELATIONSHIP

SECTION MATERIALS

METRIC RULERS (FOR RT-1)

PROTRACTORS OR 3 BY 5 CARDS (FOR RT-1)

RT-3 SIDES OF RIGHT TRIANGLES RESOURCE PAGE

Ask a volunteer to read the railroad track problem from the introduction of the chapter and give the class a few moments to think about it before you initiate a discussion about the usefulness of pictures and models for visualizing the problem. Have someone help model the situation using two sticks (or rulers) for two 1-mile-long rails and two slightly longer sticks for the jutting expanded rails. Make a simple sketch, indicating that there are right triangles and that each elevated rail is 1 foot longer than it was when it was laid on the ground:

Ask, "Who thinks the buckled rails would rise 1 foot above the ground? less than 1 foot? more than one foot?" Don't try to solve the railroad track problem now; students will return to it in problem RT-125.

In problem RT-1 students test a common misconception that if the hypotenuse of a right triangle is 1 unit longer than one of the legs, then the other leg will be exactly 1 unit long. Then in RT-3, students use patterning to find a relationship among the legs and hypotenuse of a right triangle. In RT-6 they write the relationship they observed—the "Pythagorean relationship"—in the form of a word equation.

RT-1. Before you work through the following problem, re-read the description of the buckled-railroad-track problem in the introduction of the chapter.

When Casey heard about the buckled-railroad-track problem, he tried to draw a sketch of a similar situation using smaller numbers. He figured that if each rail were originally 4 centimeters long and expanded 1 centimeter to 5 centimeters long, then the jutting ends would be 1 centimeter above the ground, as sketched here:

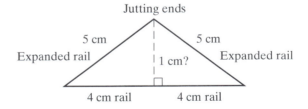

a. To make an accurate drawing to check Casey's conjecture, first make a right angle with sides at least 4 centimeters long. Mark the right angle with a small square. Along one side of the right angle, measure and mark a segment exactly 4 centimeters in length. Your drawing should look like this:

To make the side that is 5 centimeters long, place the 0-centimeter mark of a ruler at the free end of the 4-centimeter segment. Adjust the ruler so that it intersects the other side of the angle at the 5-centimeter mark of the ruler, and then use the ruler to draw the side. Label the side with its length.

b. Measure the height of your right triangle to the nearest 0.1 centimeter, and record the measurement. Is the height of the triangle 1 centimeter? What does this tell you about Casey's conjecture—if the longest side of a right triangle is 1 centimeter longer than another side, is the length of the third side always 1 centimeter? *No, the third side is 3 cm long. The conjecture is not correct.*

RT-2. In a right triangle, the longest side is called the **hypotenuse**. The two shorter sides, which form the right angle, are called **legs**. Add this information, along with a labeled diagram, to your Tool Kit.

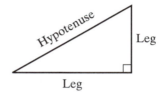

RT-3. Sides of Right Triangles Apparently the relationship among the sides of a right triangle is not as simple as Casey first thought.

a. How complicated is the relationship among the legs and hypotenuse of a right triangle? To help answer this question, complete the table for these five right triangles on the RT-3 Sides of Right Triangles Resource Page:

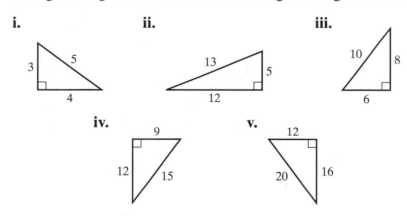

	Length of Leg # 1	Length of Leg # 2	Length of Hypotenuse	(Length of Leg # 1)^2	(Length of Leg # 2)^2	(Length of Hypotenuse)^2
i.						
ii.						
iii.						
iv.						
v.						

b. Look for a pattern in the three columns on the far right of your table in part a; that is, look for a relationship between the squares of the lengths of the legs and the square of the length of the hypotenuse of a right triangle. Describe what you observe by copying and completing this statement:

"In each of the right triangles that we checked, …"

RT-4. In RT-3 you were asked to detect a relationship between the squares of the lengths of the legs and the square of the length of the hypotenuse of a right triangle. The side lengths of the triangles in RT-3 were all integers. Does the relationship hold if the side lengths aren't all integers, as in the following triangles? Check it out by copying each right triangle on your paper and checking to see whether the relationship you saw in RT-3 holds true for it.

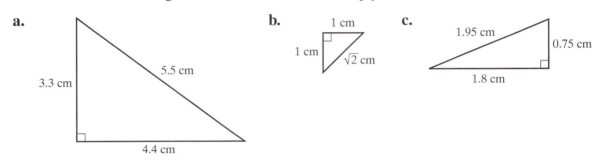

In RT-5, students will write a "word equation" to describe the relationship they saw in their data:

(length of leg #1)² + (length of leg #2)² = (length of hypotenuse)²,

or, recalling from Chapter 4 that the symbol | | denotes the length of a segment, they could write

|leg #1|² + |leg #2|² = |hypotenuse|².

Although some students will already "know" the formula $a^2 + b^2 = c^2$, avoid summarizing the relationship in this form. We want to reinforce the relationship among the legs and hypotenuse of the triangle; the formula by itself is not meaningful unless a, b, and c are defined in context.

After groups have completed RT-5, and before they move on to RT-6, ask representatives to present their word equations and supporting data to the class.

RT-5. Does the pattern you observed in RT-3 between the squares of the lengths of the legs and the square of the length of the hypotenuse of a right triangle seem to be true for *every* right triangle, including those whose side lengths are not integers?

- If so, with your group, write a **word equation** (that is, an equation in words) to describe the relationship you noticed. Be prepared to present your group's description to the class.

- If you think that the pattern does not apply to all right triangles, with your group, draw and label a triangle for which the relationship is not true. Be prepared to present your group's example to the class.

RT-6. The Pythagorean Relationship The relationship you recorded in the previous problem is often referred to as the **Pythagorean relationship** (or the *Pythagorean theorem*). You might have worked with the Pythagorean relationship in a different form in a previous class. In this course it will be helpful to remember it as you wrote it in RT-5, as a relationship between the squares of the lengths of the legs and the square of the length of the hypotenuse of a right triangle. Add your word equation for the Pythagorean relationship to your Tool Kit, starting with "In a right triangle, … ."

Have several students share their descriptions from RT-7, since the responses can be quite creative, and sharing may help those who don't "get it" the standard way.

RT-7. Sketch a right triangle. (Don't forget to draw a small square to indicate the right angle!)

a. Write a sentence or two to explain to a friend how to decide which side is the hypotenuse and which sides are the legs.

b. Now share your response to part a with your group. If all the responses to part a used the same idea, discuss another way to tell the hypotenuse from the legs, and write it down. If more than one approach was used, write down all of them. *The hypotenuse is the longest side; it must cover a distance greater than the longest leg. The legs form the right angle, and the hypotenuse is opposite the right angle.*

RT-8. Use your word equation for the Pygthagorean relationship to write an equation for each of the following right triangles. Then solve your equation using algebra to find the unknown side length.

a.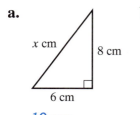

10 cm

b.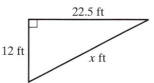

25.5 ft

c.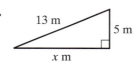

12 m

d.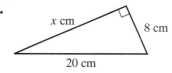

$\sqrt{336} \approx 18.3$ *cm*

e. Each of the quadratic equations you wrote in parts a through d has two solutions. Why do we choose to use only the positive solution in these cases?

As you circulate, make sure that students include units when they label the sides of the triangle in their diagrams for RT-9. By getting into the habit of including units on their diagrams, students will have less difficulty with situations in which they must distinguish distances from speeds in future problems.

RT-9. Dave leaned a 10-foot ladder against a wall so that the foot of the ladder was 3.5 feet away from the wall. He wondered how far up the wall the top of the ladder reached.

a. Draw a diagram and label it with the ladder's length and the distance from the ladder's foot to the wall. Include the units in the label for each side, and label the unknown length "*h* feet."

Possible diagram:

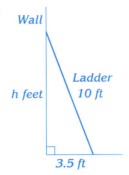

b. Write what your variable represents by copying and completing the following sentence. Be specific.

"Let *h* represent …"

Let h represent the height above the ground of the top of the ladder.

c. How many feet above the ground does the top of the ladder touch the wall? Use your diagram and your word equation for the Pythagorean relationship to write an equation. Solve the equation and then show how you check your answer. *$h^2 + (3.5)^2 = 10^2$. It touches the wall $\sqrt{87.75} \approx 9.4$ ft above the ground.*

In RT-10 and subsequent problems in which students simplify algebraic fractions, we reinforce the idea that we "simplify" algebraic fractions by finding factors to make forms equal to 1. In this chapter students will learn to recognize algebraic fractions of the form $\frac{ax+b}{ax+b}$ (where a and b are not both zero), as being equivalent to 1.

We are careful to avoid any reference to "canceling like factors" because students often think of canceling as subtracting to get zero, as in $x - x = 0$. When we see students doing things such as $\frac{x+6}{x} = 6$, we suggest they pick a value for x and check it.

RT-10. Rewrite each of the following algebraic fractions in simpler form using as few numbers and variables as possible.

a. $\dfrac{12(x-2)^2}{3(x-2)}$ for $x \neq 2$ *$4(x-2)$*

b. $\dfrac{6(m+1)^3}{6(m+1)}$ for $m \neq -1$ *$(m+1)^2$*

c. Explain in one or two sentences how your group got its solution to part a.

EXTENSION AND PRACTICE

RT-11. In each of the following right triangles the lengths of two sides are given. Use your word equation for the Pythagorean relationship to write an equation, and then solve the equation to find the length of the third side. Check each result by substituting the side lengths in your word equation for the Pythagorean relationship.

a.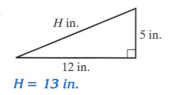

H = **13 in.**

b.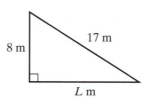

L = **15 m**

c.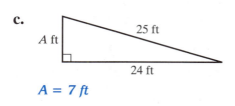

A = **7 ft**

d.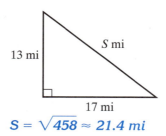

S = $\sqrt{458} \approx$ **21.4 mi**

RT-12. Draw coordinate axes on a piece of graph paper. Mark the point (10, 6) and label it *S*. Slide point *S* 8 units to the left, 7 units down, and then stop. Label the stopping point *P*.

a. What are the coordinates of *P*?   **(2, −1)**

b. Use the Pythagorean relationship to find |*SP*|, the distance between the starting and stopping points. $\sqrt{113} \approx$ **10.6 units**

c. What is the slope of the segment *SP*? $\dfrac{7}{8}$

RT-13. Solve each system of equations for *x* and for *y*:

a. $y = 7 - 2x$ **x = 2, y = 3**
   $3x - y = 3$

b. $3x + 2y = 7$ **x = 2, y = 0.5**
   $2x - 2y = 3$

RT-14. Evaluate each of the following expressions for $x = 3$, and then for $x = -2$:

a. $-x^2$ **−9; −4**

b. $(-x)^2$ **9; 4**

c. $3x^3$ **81; −24**

d. $(3x)^3$ **729; −216**

RT-15. Factor each of the polynomials in parts a through d. Solve each equation in parts e and f.

a. $b^2 - 4b + 4$ **(b − 2)(b − 2)**

b. $2x^2 - 3x + 1$ **(2x − 1)(x − 1)**

c. $b^2 - 5b + 4$ *(b – 4)(b – 1)*   d. $2x^2 - 3xy + y^2$ *(2x – y)(x – y)*

Hint for part d:

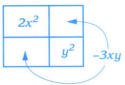

e. $b^2 - 5b + 4 = 0$ *b = 4, 1*   f. $2x^2 - 3x + 1 = 0$ $x = \frac{1}{2}, 1$

RT-16. Solve the equation $8^2 + C^2 = 17^2$ for C. Do this by rewriting the equation so you can factor and use zero product property.   *C = 15 or –15*

RT-17. In this problem you will find the x-intercepts of a parabola first by graphing and then by using algebra, as you did in FF-136.

a. Make a table and draw the graph of $y = x^2 - 9$. Mark and label the x-intercepts.

b. Find the x-intercepts of $y = x^2 - 9$ algebraically.   *(3, 0) and (–3, 0)*

RT-18. Expanding Metals Builders, engineers, architects, and mechanics know that certain solid materials will expand or contract when subjected to changes in temperature. Machinery and buildings that are made from these materials must be designed to allow for the amount of expansion (or contraction) that is expected to occur as the temperature rises (or falls).

To calculate how much an object will grow (or contract) in length, you can use a ratio called the **coefficient of linear thermal expansion** which gives the expansion factor for each degree Celsius that the temperature changes. The initial length of the object, its change in length, and the change in temperature are related to each other by the linear thermal expansion equation

$$\frac{\text{Change in length}}{\text{Initial length}} = \left(\begin{array}{c}\text{Coefficient} \\ \text{of linear} \\ \text{thermal expansion}\end{array}\right) \cdot \left(\begin{array}{c}\text{Change} \\ \text{in temperature} \\ \text{in °C}\end{array}\right)$$

The following table lists the coefficients of linear thermal expansion for several metals:

Metal	Coefficient of Linear Thermal Expansion
aluminum	$\frac{24 \cdot 10^{-6}}{1°C}$
copper	$\frac{17 \cdot 10^{-6}}{1°C}$
lead	$\frac{29 \cdot 10^{-6}}{1°C}$

Example: A huge rectangular stained-glass window is made with a 300-foot-long strip of lead along its outer perimeter. One record-breaking day in June, the temperature increased from 25°C to 45°C, so the change in temperature was +20°C. To find how much the strip of lead expanded, replace parts of the equation with known quantities, and then solve for the change in length.

$$\frac{\text{Change in length}}{\text{Initial length}} = \left(\begin{array}{c}\text{Coefficient}\\\text{of linear}\\\text{thermal expansion}\end{array}\right) \cdot \left(\begin{array}{c}\text{Change}\\\text{in temperature}\\\text{in °C}\end{array}\right)$$

$$\frac{\text{Change in length}}{300 \text{ ft}} = \left(\frac{29 \cdot 10^{-6}}{1°C}\right) \cdot (20°C)$$

Multiplying each side of the equation by 300 gives

$$300 \text{ ft} \cdot \frac{\text{Change in length}}{300 \text{ ft}} = \left(\frac{29 \cdot 10^{-6}}{1°C}\right) \cdot (20°C) \cdot 300 \text{ ft}$$

so

$$\text{Change in length} = (29 \cdot 10^{-6}) \cdot (20) \cdot (300 \text{ ft})$$
$$= 0.174 \text{ ft, or } 2.088 \text{ inches.} \quad \blacktriangle$$

Use the table and the linear thermal expansion equation to answer each of the following questions. Express each answer in scientific notation and in standard decimal form.

a. What is the change in length if a 50-centimeter aluminum pipe is heated from 20°C to 38°C?   *0.0216 = 2.16 · 10⁻² cm*

b. Suppose the temperatures of a 100-centimeter copper wire and a 100-centimeter lead wire are both raised 25°C. How much more would the length of the lead wire increase over that of the copper wire?   *The change in length for the lead wire is 0.0725 = 7.25 · 10⁻² cm and for the copper is 0.0425 = 4.25 · 10⁻² cm, so the difference is 0.03 = 3 · 10⁻² cm.*

RT-19. To find the product $2(x + 1)(2x + 3)$ you could proceed in one of two ways:

   i. You could either first multiply together 2 and $x + 1$, and then multiply the result by $2x + 3$;

   ii. or, you could first multiply $x + 1$ and $2x + 3$, and then multiply the product by 2.

Either way, finding the product $2(x + 1)(2x + 3)$ is a two-step process, as shown in the following examples.

Example 1: First multiply 2 and $(x + 1)$, and then multiply the result by $2x + 3$.

$$2(x + 1)(2x + 3) = (2x + 2)(2x + 3)$$
$$= 4x^2 + 10x + 6 \quad \blacktriangle$$

Example 2: First multiply $(x + 1)(2x + 3)$, and then multiply that result by 2.

$$2(x + 1)(2x + 3) = 2(2x^2 + 5x + 3)$$
$$= 4x^2 + 10x + 6 \quad \blacktriangle$$

Use a two-step process to find each of the following products:

a. $3(x - 1)(5x - 7)$ $15x^2 - 36x + 21$

b. $5(w + 3)(2w - 5)$ $10w^2 + 5w - 75$

c. $xy(x - 3y)(x + 2y)$ $x^3y - x^2y^2 - 6xy^3$

d. $2m^2(m - 7)(m + 7)$ $2m^4 - 98m^2$

RT-20. Consider the following problem:

The florists at Fran's Flower Shop plan to create 12-flower mixed bouquets of roses and daffodils. The roses cost $18 per dozen, and the daffodils cost $9 per dozen. The shop would like to sell the mixed bouquets for $12. For this price, how many roses should be included in each mixed bouquet?

a. If roses cost $18 per dozen, how much does a single rose cost?
$\frac{\$18}{12} = \$1.50$

b. How much does a single daffodil cost? $\frac{\$9}{12} = \$0.75$

c. Write an equation (or a pair of equations) to model the problem, and state what each variable represents. Solve the equation using algebra. Check your answer and then write it in a complete sentence.   *Let R represent the number of roses. Then $(1.5)R + (0.75)(12 - R) = 12$, so $R = 4$. Use 4 roses in each bouquet.*

RT-21. Rewrite each of the following expressions in a simpler form using as few numbers and variables as possible:

a. $\dfrac{x(2x - 1)}{x} \cdot \dfrac{1}{2}$ for $x \neq 0$ $\dfrac{2x - 1}{2}$

b. $\dfrac{12}{x + 1} \cdot \dfrac{x}{6}$ for $x \neq -1$ $\dfrac{2x}{x + 1}$

RT-22. Approximately 4.5% of the lightbulbs produced at the Wignut Light Bulb factory are defective, and about 3.1% of the bulbs last much longer than average. The rest of the bulbs produced are considered "average." Wignut has produced 432 million lightbulbs in the past 10 years. About how many of these bulbs were "average?" Write your answer in scientific notation.   *about $3.99 \cdot 10^8$ bulbs*

RT-23. Find the product $\frac{1}{2} \cdot \frac{2}{3} \cdot \frac{3}{4} \cdot \ldots \cdot \frac{98}{99} \cdot \frac{99}{100}$. Recall that an ellipsis, "...", means "continue the pattern." Look at the product carefully and think before reaching for your calculator. *$\frac{1}{100}$*

RT-24. In Chapter 3 you saw how to use a balanced two-pan scale to represent a situation involving equality. To represent the statement $7 = 4 + 3$, you could draw this:

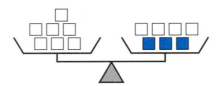

Similarly, you can use an *unbalanced* two-pan scale to represent a situation involving *inequality*, such as $2 < 5$:

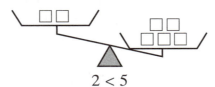

$2 < 5$
Since the side with two tiles is lighter than the side with five tiles, the side with two tiles is higher.

You could also express the relationship $2 < 5$ using "greater than" by writing $5 > 2$ and drawing the scale this way:

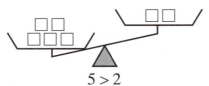

$5 > 2$
Since the side with five tiles is heavier than the side with two tiles, the side with five tiles is lower.

a. Sketch an unbalanced two-pan scale to represent the statement $4 > 1$.

b. Sketch an unbalanced two-pan scale to represent the statement
$$2 + 3 < 1 + 5.$$

c. Rewrite each of the following inequalities using the symbol $>$ for "greater than":

    i. $1 < 6$ **ii.** $-2 < 3$ **iii.** $-7 < -4$
      6 > 1   *3 > -2*   *-4 > -7*

d. Rewrite each of the following inequalities using the symbol $<$ for "less than":

    i. $9.99 > 9.9$ **ii.** $-1.5 > -3.2$ **iii.** $22 > 19.75$
      9.9 < 9.99   *-3.2 < -1.5*   *19.75 < 22*

7.2 PICTURING SQUARE-ROOT LENGTHS

SECTION MATERIALS

- 3 BY 5 CARDS (ONE PER STUDENT, FOR RT-25)
- RT-25 SPIRAL OF RIGHT TRIANGLES RESOURCE PAGE
- RT-26 LOCATING SQUARE ROOTS ON A NUMBER LINE RESOURCE PAGE
- RT-30 SQUARE-ROOT VALUES RESOURCE PAGE

In RT-25, you may need to guide students through the construction phase by having them work along with you as you demonstrate at the board or on the overhead projector. Show how to mark one corner of a 3 by 5 card with a small square to indicate a right angle, and then, using the unit length on the resource page, make a mark 1 unit from the corner along each adjacent edge. Use the right angle and the unit mark to draw a 1–1 right triangle (where each leg is 1 unit long); label its hypotenuse "H_1." Then use the Pythagorean relationship to calculate length H_1.

Repeat the process, drawing a second right triangle with the hypotenuse of the first triangle as one leg, making the length of other leg 1 unit. Then calculate the length of the hypotenuse H_2 of the new triangle. Continue making new triangles by using the hypotenuse of the previous triangle as one leg and making the second leg 1 unit long, until there are seven triangles as shown in the diagram.

RT-25. A Spiral of Right Triangles Karla created a spiral of right triangles that looked like this:

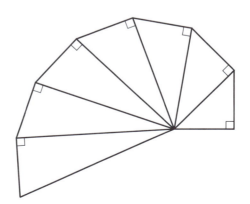

After admiring her drawing, Karla was curious to find out the length of the hypotenuse of the largest triangle. So she labeled her diagram and used what she knew about right triangles and the Pythagorean relationship to figure out the length of the hypotenuse of each triangle in turn, starting with the first triangle she drew. This is what her diagram looked like:

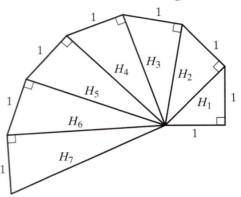

a. Use the following instructions to re-create Karla's spiral of right triangles on the RT-25 Spiral of Right Triangles Resource Page. Be sure to label each side as you draw the triangles one at a time.

Instructions for Drawing Karla's Spiral of Right Triangles: To help make your drawing accurate, mark one corner of a 3 by 5 card with a small square to indicate that it is a right angle. Using the unit length given on the Resource Page, copy and mark off a length of 1 unit along each side adjacent to the right angle you marked.

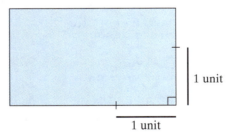

Use your 3 by 5 card to draw, on the Resource Page, a right triangle with each leg 1 unit long. Label the length of the hypotenuse "H_1" as shown:

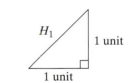

To draw a second right triangle, use the hypotenuse of the first triangle as one leg, and make the other leg 1 unit long, as follows:

Place the 3 by 5 card along the hypotenuse labeled H_1 so that the right angle you marked on the card is at the upper vertex of the first triangle. Use the edge of the card to draw a leg of length 1 unit.

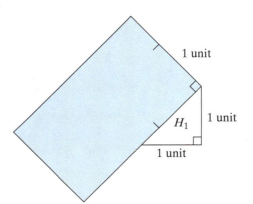

Now draw the hypotenuse of the second right triangle (using the card as a straightedge). Label this hypotenuse "H_2" as shown.

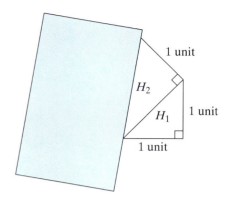

Continue in this manner, using the hypotenuse of the previous triangle as a leg for the subsequent triangle, to make a total of seven right triangles in a spiral.

b. To calculate the length H_1 of the hypotenuse of the first triangle, Karla used the Pythagorean relationship and wrote

$$1^2 + 1^2 = (H_1)^2$$
$$1 + 1 = (H_1)^2$$
$$\text{so } (H_1)^2 = 2.$$

So the length H_1 is a positive number whose square is 2. That number is usually represented by the symbol $\sqrt{2}$, the square root of 2. So

$$H_1 = \sqrt{2} \text{ units.}$$

In general, the symbol $\sqrt{x}$ is read **the square root of x** and represents a positive number whose square is equal to x.

Follow Karla's example and find the lengths H_2, H_3, H_4, H_5, H_6, and H_7. Write each length as a square root. **$\sqrt{3}$, $\sqrt{4}$, $\sqrt{5}$, $\sqrt{6}$, $\sqrt{7}$, and $\sqrt{8}$ units**

c. If Karla had kept drawing right triangles on her spiral, how long would H_{10} be? **$\sqrt{11}$ units**

RT-26. Locating Square Roots on a Number Line Karla realized that the number $\sqrt{2}$ didn't mean much to her, so she decided to make a number line to see where $\sqrt{2}$ fit in. To mark units off on the number line, she used the unit length from the triangles she had drawn, and then marked off integers from -4 to 4. Because she knew that H_1 had length $\sqrt{2}$, she marked a segment the same length as H_1 on a strip of paper. She then put one end of the segment at 0 on the number line and saw that the other end of the segment fell between 1 and 2, as shown:

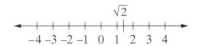

a. Locate each of the numbers $\sqrt{3}, \sqrt{4}, \sqrt{5}, \ldots, \sqrt{8}$ on the number line on your Resource Page for RT-26 ("Locating Square Roots on a Number Line") as was done for $\sqrt{2}$. Do this by using Karla's idea of marking lengths on a strip of paper in order to transfer the length of the hypotenuse of each of your right triangles from RT-25.

b. Locate and mark $-\sqrt{2}$ and $-\sqrt{5}$ on your number line.

c. Suppose you used Karla's process to transfer a segment of length $\sqrt{1}$ to your number line. If you placed one end of the segment at 0, where would the other end of the segment fall? What about a segment of length $\sqrt{9}$? *at 1 or −1; at 3 or −3*

RT-27. In a complete sentence, describe where the number $\sqrt{7}$ lies on the number line. *about two-thirds of the way between 2 and 3.*

RT-28. Approximating Square Roots There are several ways to find an approximate value for the square root of a number. A geometric way is to make a right triangle and then find where the length of the hypotenuse lies on a number line as you did in problems RT-25 and RT-26. In this problem you'll use another, less visual way to approximate square roots by using what you know about *perfect squares* and the relative sizes of numbers. A third, more accurate way is to use a calculator, as you will see in RT-29.

a. Before you use what you know about **perfect squares** and the relative sizes of numbers to approximate the square root of a number, it helps to have a list of perfect squares in mind. The first five perfect squares may come quickly to mind: 1, 4, 9, 16, and 25. List the first 12 perfect squares on your paper. *1, 4, 9, 16, 25, 36, 49, 64, 81, 100, 121, and 144*

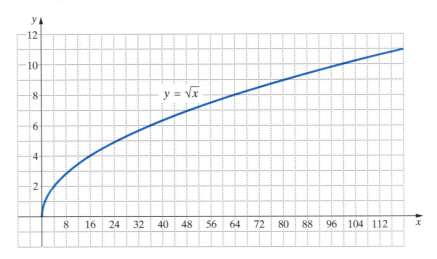

b. **Rough Approximations** To make a rough estimate of the square root of a number, you could use a list of perfect squares to find two consecutive perfect squares that the number lies between, and then write an inequality. You could then use the graph of $y = \sqrt{x}$ to write an inequality statement about the square roots.

Example: Suppose you want to find an approximate value for $\sqrt{53}$. First use the list of perfect squares to find the two consecutive perfect squares that 53 lies between and then write an inequality:

$$49 < 53 < 64.$$

To see how $\sqrt{49}$, $\sqrt{53}$, and $\sqrt{64}$ are related, first locate the numbers 49, 53, and 64 on the *x*-axis. Then use the graph of $y = \sqrt{x}$ to locate the values $\sqrt{49}$, $\sqrt{53}$, and $\sqrt{64}$ on the *y*-axis. By doing this, you see that

$$\sqrt{49} < \sqrt{53} < \sqrt{64},$$

so

$$7 < \sqrt{53} < 8.$$

For a rough approximation, you can write "$\sqrt{53} \approx 7.\text{something}$" since $\sqrt{53}$ is greater than 7 and less than 8 and we don't know exactly what the "something" is. ▲

Use a list of perfect squares and the graph of $y = \sqrt{x}$ to make a rough approximation for each of the following square roots:

   i. $\sqrt{78}$ **ii.** $\sqrt{40}$ **iii.** $\sqrt{136}$ **iv.** $\sqrt{10}$

$\sqrt{78} \approx 8.\text{something}$; $\sqrt{40} \approx 6.\text{something}$; $\sqrt{136} \approx 11.\text{something}$; $\sqrt{10} \approx 3.\text{something}$

  RT-29. More Accurate Approximations In the previous problem, you estimated

$$\sqrt{53} \approx 7.\text{something}$$

To be more specific about the size of the "something," you can use a guess-and-check table and your calculator. Since 53 is closer to 49 than it is to 64, a good first guess might be

$$\sqrt{53} \approx 7.4.$$

Guess: Value for 7.something	Value Squared	*Check:* Value Squared = 53?
7.4	$(7.4)^2 = 54.76$	too big
7.2	$(7.2)^2 = 51.84$	too small
7.3	$(7.3)^2 = 53.29$	too big

a. Copy the guess-and-check table, and make three more guesses to find an approximate value for $\sqrt{53}$.

b. To use a calculator to find an approximate value for the square root of a number, simply use the square root key, $\boxed{\checkmark}$.

Example: To find an approximate value for $\sqrt{53}$, enter 53 and then press $\boxed{\checkmark}$. (On some calculators, you may need to press the square root key first, and then enter 53.) ▲

Use the square root key on your calculator to find an approximation for $\sqrt{53}$ to the nearest hundredth. Then add information about using the *square-root key* to your Calculator Tool Kit.

*The idea in problem RT-30 is for students to examine approximations of square roots, **not** to learn how to simplify radicals. It works well as a whole class activity or discussion. Ask students to suggest values and then use calculators to check. Reinforce the idea that square root notation, $\sqrt{}$, represents an exact value, while a calculator may give only an approximation.*

RT-30. Square Root Values Complete the following table on your RT-30 Square Root Values Resource Page. Write the exact value for each length x and then estimate its value by using the number line from RT-26. Finally, use a calculator to find a decimal approximation for x to the nearest hundredth. *Values estimated from students' number lines will vary.*

Square of the Hypotenuse, x^2	Exact Length of the Hypotenuse, x	Estimated Value of x from Your Number Line	Approximate Value of x from a Calculator (to the nearest 0.01)
1	$\sqrt{1} = 1$	1	1.00
2	$\sqrt{2}$	1.4	1.41
3	$\sqrt{3}$	1.75	1.73
4	$\sqrt{4} = 2$	2	2.00
5	$\sqrt{5}$	2.25	2.24
6	$\sqrt{6}$	2.4	2.45
7	$\sqrt{7}$	2.5	2.65
8	$\sqrt{8}$	2.75	2.83
9	$\sqrt{9} = 3$	3	3.00
10	$\sqrt{10}$	3.25	3.16

Before groups start on RT-31, we suggest having a volunteer read the problem aloud. We usually model the situation by breaking a pencil (the "snap" is dramatic!) and distributing straws to the groups so they can use them as models for cracked telephone poles.

RT-31. Sketch and label a diagram to help you visualize the following problem. Use your diagram to write an equation, and state what your variable represents. Then solve the equation to solve the problem. Check your answer and then write it in a complete sentence.

Telephone poles are anchored by sinking the lowest 5 feet of the pole into the ground. A careless construction worker once drove a forklift into a pole. The pole cracked 7 feet above the ground level, and its top portion fell over as if it were hinged at the crack. The tip of the pole hit the ground 24 feet from its base. The entire pole must be replaced. How long should the replacement pole be?

Possible diagram:

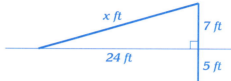

The replacement should be 37 ft long.

EXTENSION AND PRACTICE

RT-32. Sketch and label a diagram to help you visualize the following problem. State what your variable represents. Use your diagram to write an equation, and then solve the equation to solve the problem. Check your answer, and then write it in a complete sentence.

The four baselines of a baseball diamond form a square, and the bases are located 90 feet apart at the corners. How far is it from home plate directly to second base?   *$\sqrt{16{,}200} \approx 127.3$ ft*

RT-33. Without using a calculator, match each exact square-root value in the left-hand column with a description in the right-hand column, as done in the example. (There are choices in the right-hand column that will not be used.)

Example: Since 38 is between 36 and 49, but closer to 36, for an approximation you can write "$\sqrt{38}$ is a little more than 6." ▲

a. $\sqrt{24}$ a little less than 10   *a. $\sqrt{24}$ is a little less than 5.*
b. $\sqrt{17}$ a little more than 4   *b. $\sqrt{17}$ is a little more than 4.*
c. $\sqrt{66}$ exactly 11   *c. $\sqrt{66}$ is a little more than 8.*
d. $\sqrt{98}$ a little more than 8   *d. $\sqrt{98}$ is a little less than 10.*
e. $\sqrt{28}$ a little less than 4   *e. $\sqrt{28}$ is a little more than 5.*

a little more than 5

exactly 16

a little less than 5

RT-34. For each right triangle, use the given information and the Pythagorean relationship to find the length of the missing side. Write the exact length in square-root form ($\sqrt{}$), and give a decimal approximation to the nearest hundredth.

Triangle	Leg #1	Leg #2	Hypotenuse
a.	4	7	$\sqrt{65} \approx 8.06$
b.	13	19	$\sqrt{530} \approx 23.02$
c.	49.77	14	$\sqrt{2673.0529} \approx 51.70$
d.	3	$\sqrt{40} \approx 6.32$	7
e.	6	6	$\sqrt{72} \approx 8.49$
f.	$\sqrt{1104.73} \approx 33.24$	39.6	51.7

RT-35. Rewrite each of the following expressions using a single exponent:

a. $(2x)^3$ b. $(3x)^2$ c. $(-x)^3$ d. $3(x^3)^2$ e. $\dfrac{(3x)^4}{3x^2}$

$8x^3$ $9x^2$ $-x^3$ $3x^6$ $27x^2$

RT-36. Copy the diagram, and then find x to the nearest tenth:

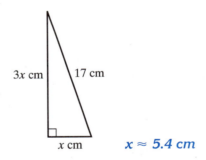

$x \approx 5.4$ cm

RT-37. Draw coordinate axes on a piece of graph paper. Mark the point $(-4, 5)$ and label it A. Slide point A 6 units to the right, 10 units down, and then stop. Label the stopping point B. Find the distance $|AB|$ and the slope of the segment AB.   *B is (2, −5); $|AB| = \sqrt{136} \approx 11.7$ units; slope $= \dfrac{-10}{6} = \dfrac{-5}{3}$*

RT-38. Salina was playing with the $\sqrt{}$ key on her calculator, and she discovered a neat trick. She noticed that for the numbers 25 and 4, if she first multiplied and then took the square root, she got $\sqrt{25 \cdot 4} = \sqrt{100} = 10$. But if she took the square root of each number first and *then* multiplied, she got the same answer: $\sqrt{25} \cdot \sqrt{4} = 5 \cdot 2 = 10$.

a. Salina decided to do some experimenting to see if this pattern would work for operations other than multiplication. Help her by predicting whether each of these three statements is true or false. Then use your calculator to check.

i. $\sqrt{25} + \sqrt{4} = \sqrt{25 + 4}$ F

ii. $\dfrac{\sqrt{25}}{\sqrt{4}} = \sqrt{\dfrac{25}{4}}$ T

iii. $\sqrt{25} - \sqrt{4} = \sqrt{25 - 4}$ F

b. Salina then repeated her experiment using the numbers 49 and 36. Predict whether each statement is true or false. Then use your calculator to check.

i. $\sqrt{49} \cdot \sqrt{36} = \sqrt{49 \cdot 36}$ T

ii. $\sqrt{49} + \sqrt{36} = \sqrt{49 + 36}$ F

iii. $\dfrac{\sqrt{49}}{\sqrt{36}} = \sqrt{\dfrac{49}{36}}$ T

iv. $\sqrt{49} - \sqrt{36} = \sqrt{49 - 36}$ F

RT-39. Solve the equation $14^2 + (2C)^2 = 50^2$ for C. Do this by rewriting the equation so you can factor and use zero product property.   **C = 24 or −24**

RT-40. Fran claims that the equation $x^2 = 225$ will have just one solution because there is just one variable. Ollie is equally certain that there are two values of x that will make the equation true.

a. Settle the argument by rewriting the equation and using what you know about solving difference-of-squares problems.   **$x^2 = 225$ can be rewritten as $(x - 15)(x + 15) = 0$, so there are two solutions: x = 15 or −15.**

b. How is the graph of $y = x^2 - 225$ related to Fran and Ollie's disagreement?   **The x-coordinates of the x-intercepts of the graph are the solutions of the equation.**

c. Settle the argument by describing what the graph of $y = x^2 - 225$ would look like.   **The graph is a parabola that intersects the x-axis twice, so there are two solutions to the equation.**

RT-41. Spare Change Janelle has $40 in her spare-change jar and is adding to it at the rate of $7 per week. Jeanne has $125 hidden away and is putting aside extra cash at a rate of $3 per week. After how many weeks will both women have the same amount of money set aside?

a. Set up coordinate axes, and make a graph to solve the problem.

b. Solve the problem algebraically by writing two equations and solving the system. *For Janelle, y = 7x + 40; for Jeanne, y = 3x + 125. They'll have the same amount of money set aside in 21.25 wk.*

RT-42. Solve each of the following systems of equations for x and y:

a. $y = 4x$ *(−0.2, −0.8)*
   $x + y = -1$

b. $-2x + y = 3$ *(−1.8, −0.6)*
   $x = 3y$

c. $5x + 7y = -16.5$ *(3, −4.5)*
   $2y = 3x - 18$

RT-43. Writing Fractions in Simplest Form You have been simplifying algebraic fractions by finding factors of the numerator and the denominator to make forms that are equivalent to 1, such as $\dfrac{x-1}{x-1}$ for $x \neq 1$, and $\dfrac{x^4}{x^4}$ for $x \neq 0$. When the numerator and denominator of a fraction have been factored, and all possible fractional forms of 1 have been found, we say that the fraction is written in **simplest form**. The trick is to find as many fractional forms of 1 as you can.

Rewrite each of the following algebraic fractions in *simplest form*:

a. $\dfrac{(x+3)^2}{(x+3)(x-2)}$ for $x \neq -3$ or 2 $\dfrac{x+3}{x-2}$

b. $\dfrac{8(2x-5)^3}{4(2x-5)^2(x+4)}$ for $x \neq -4$ or 2.5 $\dfrac{2(2x-5)}{x+4}$

c. $\dfrac{x^2+6x}{x^2+12x+36}$ for $x \neq -6$ You must factor first! $\dfrac{x}{x+6}$

RT-44. Juan has a jar with a lot of coins in it. On Tuesday $\frac{1}{3}$ of the coins were pennies, $\frac{1}{3}$ were nickels, and $\frac{1}{3}$ were dimes. This morning after Juan put 24 more pennies into the jar, the probability of pulling out a penny was $\frac{1}{2}$. How many pennies, nickels and dimes were in the jar on Tuesday?   *24 of each*

RT-45. Find each sum or difference of fractions in parts a, b, and c. Solve the equation in part d.

a. $\dfrac{4}{3} - \dfrac{2}{7}$ $\dfrac{22}{21}$

b. $\dfrac{5}{2x} - \dfrac{6}{x}$ $\dfrac{-7}{2x}$

c. $\dfrac{3}{x} + \dfrac{1}{2} + \dfrac{5}{2x}$ $\dfrac{11+x}{2x}$

d. $\dfrac{3}{x} + \dfrac{1}{2} + \dfrac{5}{2x} = 5$ $x = \dfrac{11}{9}$

RT-46. For the following situation, sketch a graph to represent the relationship between the height of Dorothy's sunflowers and the number of days since they were planted. You will need to set up and label the axes appropriately.

Dorothy was proud of her garden, especially the sunflowers. For a week after she planted them, there was no sign of a sprout. But then the sunflowers grew 2 centimeters every day until they reached their full height.

Solution:

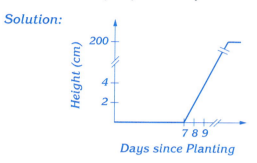

RT-47. Find each of the following products:

a. $-x(4 + x)$ $-4x - x^2$ b. $x(3x + 1)(x + 4)$ $3x^3 + 13x^2 + 4x$

c. $(2x - 7)^2$ $4x^2 - 28x + 49$ d. $(x - 2)(x^2 + 2x + 1)$ $x^3 - 3x - 2$

e. $(2x + 7)(2x - 7)$ $4x^2 - 49$ f. $(3 + 2x)(4 - 3x)$ $12 - x - 6x^2$

RT-48. Factor each of the following polynomials:

a. $y^2 - 3y - 10$ $(y - 5)(y + 2)$ b. $x^2 + 27x + 50$ $(x + 25)(x + 2)$

c. $2x^2 - 7x + 3$ $(2x - 1)(x - 3)$ d. $5x^2 + 6xy + y^2$ $(5x + y)(x + y)$

RT-49. Two neighboring restaurants have competitive lunch specials. At the Pico de Gallo Restaurante, a lunch plate comes with beans, rice, and guacamole for 80¢, and customers choose how many tacos they want at an additional cost of 75¢ per taco. Across the street, the Señorita Taquita charges 90¢ for each taco, but only 35¢ for the beans, rice, and guacamole on its lunch plate.

a. Define two variables and write two equations to represent the situation. Be sure you include the non-optional cost of rice, beans, and guacamole for each restaurant.   *Let t represent the number of tacos and C represent the total cost in cents. At Pico de Gallo, C = 80 + 75t, and at Señorita Taquita, C = 35 + 90t.*

b. For what number of tacos is the cost of a lunch special at both restaurants the same? What is that cost? *Both restaurants charge $3.05 for a three-taco lunch special.*

RT-50. One way to model the inequality $2 < 5$ is to use an unbalanced scale as in RT-24. Another way is to think of locating numbers on a number line. Because 2 is less than 5, the number 2 is located to the left of 5 on the number line:

$2 < 5$

In some of the graphing problems you've done, you've been instructed to pick input values for x that lie between certain numbers, such as $2 \leq x \leq 5$. There are many values of x that fit the condition $2 \leq x \leq 5$, and they are located on a number line between 2 (on the left) and 5 (on the right). To show this, we graph the inequality on a number line by first plotting solid dots at $x = 2$ and $x = 5$ and then connecting the dots with a segment to represent all the values in between:

$2 \leq x \leq 5$

To represent the inequality $-3 \leq x < 4$, we first plot a solid dot at $x = -3$ and a hollow dot at $x = 4$, since x must be less than, and not equal to, 4. Then we draw a segment between the two dots:

$-3 \leq x < 4$

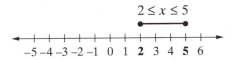

The inequality $x > 1$ states "x is any value greater than 1." To graph the inequality $x > 1$ on a number line, we first plot a hollow dot at $x = 1$ (since $x \neq 1$) and then draw a ray extending to the right of 1:

$x > 1$

Similarly, we graph the inequality $x \leq 3$ by plotting a solid dot at $x = 3$ and drawing a ray extending to the left of 3:

$x \leq 3$

Graph each of the following inequalities on a number line:

a. $-3 \leq x \leq 7$ **b.** $-5 < x \leq 0$

c. $x \geq 4.5$ **d.** $x < -1$

7.3 MODELS, DIAGRAMS, AND RIGHT TRIANGLES

SECTION MATERIALS

STIFF PAPER, STRING, STRAW OR STICK, AND MASKING TAPE FOR EACH GROUP (FOR RT-51)

SHOE BOX, STICKS OF VARIOUS LENGTHS, AND STICKY NOTES FOR EACH GROUP (FOR RT-52)

AREA AND THE PYTHAGOREAN RELATIONSHIP RESOURCE PAGE AND SCISSORS FOR EACH GROUP (FOR RT-53)

TRANSPARENCY OF THE RT-53 AREA AND THE PYTHAGOREAN RELATIONSHIP RESOURCE PAGE

RT-54 RESOURCE PAGE

DOT PAPER (FOR RT-57)

RT-62 ARE SEGMENT LENGTH AND SLOPE RELATED? RESOURCE PAGE

We strongly suggest that you make your own models for RT-51 and RT-52 before class so you'll be familiar with the problems associated with the construction. However, we suggest that you don't take your models to class, since it is more helpful to students to decide for themselves how to make the models rather than to try to copy yours.

We find it worthwhile to follow up RT-51 and RT-52 by having groups share and discuss their solutions.

In solving problems in mathematics, it is often useful to start by drawing a picture or diagram to represent a situation. Labeling a diagram carefully, including appropriate units, can help you to see how to write an equation. However, some problems are difficult to represent on paper because they are three-dimensional. To approach these problems, it is helpful to build a three-dimensional model and then use the model and subproblems to draw separate aspects of the problem on paper.

The first two problems of this section describe situations that involve right triangles. Making models of the situations will help you draw diagrams to use in solving the problems. Just as a diagram is most useful when it includes important information about the problem, a model is most useful when important information is labeled on the model. A model does not need to be made with accurate measurements or even to scale if information from the problem—such as lengths and right angles—is recorded on the model.

For problem RT-51, each group will need some stiff paper, string, a drinking straw or a stick, and some masking tape to make a model of the roof and antenna setup shown in the diagram.

RT-51. Frank's house has a flat roof that is 32 feet wide and 60 feet long. The television antenna rises 30 feet above the center of the roof and is anchored by wires that are attached 5 feet below the top of the antenna. The wires are attached to the house at each corner of the roof and at the midpoint of each edge of the roof. How long must each wire be?

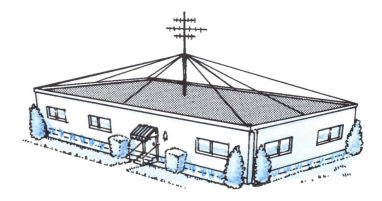

Make a model and use subproblems to draw and label appropriate triangles on your paper. Your answer should state exact lengths and approximations to the nearest tenth. *from corners, $\sqrt{1781} \approx 42.2$ ft; from midpoints of long sides, $\sqrt{881} \approx 29.7$ ft; from midpoints of short sides, $\sqrt{1525} \approx 39.1$ ft*

To make the model asked for in problem RT-52, each group will need a shoe box, sticks of various lengths, and sticky notes (for labeling). We have found that making a model helps students figure out what diagrams to draw.

There are essentially three ways to approach RT-52, depending on which side of the box is used for the base. While students may sketch different right triangles in their diagrams, all three ways lead to the same result.

RT-52. Jay wants to ship a fishing pole to his summer home in Washington. The longest section of the pole is 40 inches. Jay can only find a rectangular box that has dimensions 24 inches by 30 inches by 18 inches. If the 40-inch section of pole is placed corner-to-corner inside the box, will it fit?

Make a model and use subproblems to draw and label appropriate triangles on your paper. Your answer should state exact lengths and approximations to the nearest tenth. *Yes, the longest diagonal is $\sqrt{1800} \approx 42.4$ in.*

We suggest doing RT-53 as a whole-class exercise. We use an overhead transparency to guide students through the problem. Problems RT-57 and RT-82 provide follow-up practice.

Drawing a "tilted square" in part e of RT-53 is challenging for many students. To make sure the quadrilateral they draw truly is a square, you may need to guide students to continue the "over 6, up 8" or "over 8, up 6" pattern started by the two legs around the right angle. (We intend this to be only an informal introduction to the idea of perpendicular lines. Don't get bogged down with unnecessary formalities here.)

RT-53. Area and the Pythagorean Relationship The following problem provides a way to visualize the Pythagorean relationship as a relationship among areas of squares built on the sides of a right triangle.

a. Near the center of the dot paper on your Area and the Pythagorean Relationship Resource Page, draw a right triangle with legs 6 units long and 8 units long, respectively. Mark the right angle, and label the legs with their lengths.

b. Use the Pythagorean relationship to determine the length of the hypotenuse of your triangle.

c. To confirm your calculation in part b, cut off and use a dot-paper measuring strip to measure the hypotenuse in dot-paper units. Label the hypotenuse with its length.

d. Using the shorter leg as one of the sides, draw a square with sides of length 6 units. On the longer leg (8 units long), draw a square with sides of length 8 units. Your drawing should look something like this:

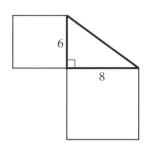

e. Using the hypotenuse as one of the sides, draw a *tilted square*.

f. Inside each of the three squares, write its area.

g. Write a word equation to describe how the areas of the squares on the legs relate to the area of the square on the hypotenuse. *(area of square on short leg) + (area of square on long leg) = area of square on hypotenuse*

h. How is your word equation in part g related to the word equation you wrote in your Tool Kit for the Pythagorean relationship? *We have (short leg)^2 + (long leg)^2 = (hypotenuse)^2. Also, (area of square on short leg) = (short leg)^2, (area of square on long leg) = (long leg)^2, and (area of square on hypotenuse) = (hypotenuse)^2. So the two equations say the same thing in two different ways.*

 RT-54. On your Resource Page for RT-54, draw a right triangle for each graph and determine the length of the given line segment. Give both the exact value and an approximation to the nearest tenth.

a.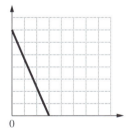

$\sqrt{58} \approx 7.6$ units

b.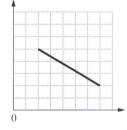

$\sqrt{34} \approx 5.8$ units

c.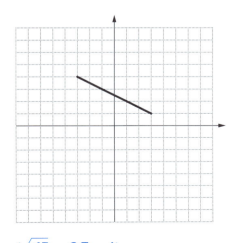

$\sqrt{45} \approx 6.7$ units

d.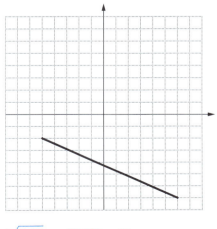

$\sqrt{146} \approx 12.08$ units

EXTENSION AND PRACTICE

RT-55. Graph each pair of points, and find the distance between them. Express each distance in square-root form and as a decimal approximation to the nearest hundredth.

a. $(3, -6)$ and $(-2, 5)$ $\sqrt{146} \approx 12.08$ units

b. $(5, -8)$ and $(-3, 1)$ $\sqrt{145} \approx 12.04$ units

c. $(0, 5)$ and $(5, 0)$ $\sqrt{50} \approx 7.07$ units

RT-56. Sketch and label a diagram to help you visualize the following problem. State what your variable represents. Use your diagram to write an equation, and then solve the equation to solve the problem. Check your answer, and then write it in a complete sentence.

How long must a wire be to reach from the top of a 13-meter telephone pole to a point on the ground 9 meters from the foot of the pole?

Possible diagram:

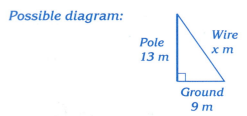

The wire is $\sqrt{250} \approx 15.81$ m.

RT-57. This problem provides another way to visualize the Pythagorean relationship using area.

a. Near the center of a piece of dot paper, draw a right triangle, with legs 3 units long and 5 units long, respectively. Mark the right angle, and label the legs with their lengths.

b. Using the hypotenuse as one of the sides, draw a tilted square on the hypotenuse of your triangle.

c. To calculate the area of the square on the hypotenuse, use the "surround and subtract" method. First draw a large square whose sides are 8 units long; the square should surround both the original right triangle and the tilted square. In your diagram you should see a total of four right triangles the same size as the original triangle. What subproblems need to be solved?

Solution:

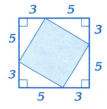

Find the area of the large square; find the area of each right triangle.

d. What is the area of the tilted square?   *34 sq units*

e. Use the Pythagorean relationship to compute the length of the hypotenuse. Do your calculations using the Pythagorean relationship and your calculations from the surround-and-subtract method support or contradict each other? Explain.   *Hypotenuse = $\sqrt{3^2 + 5^2} = \sqrt{34}$ units, so the area of the tilted square is $\sqrt{34} \cdot \sqrt{34} = 34$ sq units. The methods agree!*

RT-58. Find the area of a right triangle with a hypotenuse of length 13 centimeters and one leg of length 5 centimeters. What subproblem did you need to solve?   *30 sq cm; find the length of the other leg.*

 RT-59. When we write the product of a number and a square root, we usually omit the multiplication sign ($\cdot$) just as we do when we write $2x$ for "two times x." For example, $2\sqrt{3}$ means "two times the square root of 3," $17.22\sqrt{10}$ means "17.22 times the square root of 10," and $x\sqrt{7}$ means "x times the square root of 7."

Without using a calculator, match each exact square root value in the left-hand column with a description in the right-hand column. (Some choices in the right-hand column will not be used.)

Example: We know that $5\sqrt{10}$ means "five times the square root of 10" and that $3 < \sqrt{10} < 4$, so
$$5 \cdot 3 < 5 \cdot \sqrt{10} < 5 \cdot 4,$$
or
$$15 < 5\sqrt{10} < 20.$$
Therefore, $5\sqrt{10}$ is between 15 and 20.   ▲

a. $2\sqrt{25}$ *exactly 10*   between 8 and 12

b. $3\sqrt{17}$ *between 12 and 15*   between 3 and 6

c. $4\sqrt{5}$ *between 8 and 12*   exactly 10

d. $3\sqrt{3}$ *between 3 and 6*   between 9 and 11

between 12 and 15

exactly 8

RT-60. Predict whether each statement is true or false. Then use your calculator to check.

a. $\sqrt{100} + \sqrt{16} = \sqrt{116}$ *F*   **b.** $(\sqrt{100})(\sqrt{16}) = \sqrt{1600}$ *T*

c. $\sqrt{100} - \sqrt{16} = \sqrt{84}$ *F*   **d.** $\dfrac{\sqrt{100}}{\sqrt{16}} = \sqrt{6.25}$ *T*

e. Make up two more statements involving square roots: one that you think will be true, and another that you think will be false. Use your calculator to check your statements.   *Answers will vary.*

RT-61. Find each of the following sums or differences:

a. $5x + 2x$
   7x

b. $5m + 2m$
   7m

c. $5\sqrt{6} + 2\sqrt{6}$
   $7\sqrt{6}$

d. $4x - 10x$
   $-6x$

e. $4T - 10T$
   $-6T$

f. $4\sqrt{7} - 10\sqrt{7}$
   $-6\sqrt{7}$

g. $9x + 3x - x$
   11x

h. $9w + 3w - w$
   11w

i. $9\sqrt{15} + 3\sqrt{15} - \sqrt{15}$
   $11\sqrt{15}$

RT-62. Are Segment Length and Slope Related? Use the Resource Page for RT-62 ("Are Segment Length and Slope Related?") to explore the relationship between the slope of a line segment and the length of the segment.

a. On your Resource Page for this problem, draw in the segments AB, CD, and EF.

b. Find the slopes of segments AB, CD, and EF.   *$\frac{3}{6}, \frac{1}{2}, \frac{2}{4}$*

c. How are the slopes of segments AB, CD, and EF related?   *They're equal.*

d. Use the Pythagorean relationship to find the lengths of segments AB, CD, and EF.   *$|AB| = \sqrt{45}, |CD| = \sqrt{5}$, and $|EF| = \sqrt{20}$ units*

e. Does the length of a line segment seem to be related to the slope of the segment? Explain your answer.   *No, the three segments have different lengths but the same slope.*

RT-63. Solve each of the following equations for x:

a. $\frac{x}{2} + \frac{x}{3} = 5$ *6*

b. $\frac{5}{x} - 8 = 12$ *0.25*

c. $2x + y = 29$ *$x = \frac{29 - y}{2}$*

d. $\frac{5 + x}{7} + \frac{12x}{18} = 5$ *$\frac{90}{17} \approx 5.29$*

e. $x^2 + 6x - 7 = 0$ *$-7, 1$*

RT-64. Thermal Expansion of Glass If you know the coefficient of linear thermal expansion for a material, you can apply the linear thermal expansion equation given in problem RT-18 to that material. Here are some additional data:

Material	Coefficient of Linear Thermal Expansion
commercial glass	$\dfrac{11 \cdot 10^{-6}}{1°C}$
Pyrex	$\dfrac{3.3 \cdot 10^{-6}}{1°C}$
fused quartz	$\dfrac{0.5 \cdot 10^{-6}}{1°C}$

Use values from the table and the linear thermal expansion equation to answer each of the following questions:

a. A microscope slide made from commercial glass was heated until its temperature increased 30°C. Before being heated, the slide was 6 centimeters long. How long was the slide when it reached its peak temperature?   *6.00198 cm*

b. How many degrees Celsius would the temperature of a 6 centimeter-long slide made from fused quartz need to rise for the slide to increase in length as much as the 6 cm long commercial glass slide did in part a?   *660°C*

c. The width of a 4-centimeter wide Pyrex microscope slide increased by 0.00012 cm when it was heated. How much did the temperature of the slide increase?   *about 9.09°C*

RT-65. Undoing Square Roots You know that addition undoes subtraction, and division undoes multiplication. What operation will undo the effect of the square-root operation?

a. Here's a mental puzzle: Start with your age in years. Next, imagine finding the square root of your age using your calculator. Now, imagine squaring the result. What would you end up with?   *your age in years*

b. Here's another mental puzzle: This time, start with any positive number you chose. Imagine finding the square root of that number using your calculator. Then imagine squaring the result. What would you end up with this time?   *the original positive number*

c. What operation "undoes" the effect of finding the square root of a positive number?   *squaring the result*

d. Simplify each of the following expressions:

i. $(\sqrt{36})^2$ ii. $(\sqrt{40})^2$ iii. $(\sqrt{x})^2$ for positive numbers x
   36   40 x

RT-66. Find each of the following products:

a. $(2m + 5)(7 - m)$ $-2m^2 + 9m + 35$

b. $(3 + 2x)(4 - 3x)$ $12 - x - 6x^2$

RT-67. Washek claims that the y-intercept of the graph of $2x - y = 5$ is $(0, 5)$. Lida says that it is $(0, -5)$. Which claim is correct? Explain how you know. *(0, −5); y = 2x + (−5)*

RT-68. Solve for T: $\dfrac{4}{T} + \dfrac{3}{5} = \dfrac{7}{10}$.   40

RT-69. Max paid $2379.29 for a computer that was originally priced at $2500.00. What percent of the original price did Max pay? Write an equation and solve it. $\dfrac{\$2379.29}{\$2500} = \dfrac{x}{100}$; 95.17%

RT-70. Write each of the following algebraic fractions in simplest form:

a. $\dfrac{3x(x+4)^7}{12x^2(x+4)^6}$ for $x \neq -4$ or 0 $\dfrac{x+4}{4x}$

b. $\dfrac{12(x+1)^2(x-2)^3}{6(x+1)^3(x-2)^5}$ for $x \neq -1$ or 2 $\dfrac{2}{(x+1)(x-2)^2}$

c. $\dfrac{x^2 + 6x + 9}{2x + 6}$ for $x \neq -3$ $\dfrac{x+3}{2}$

RT-71. Solve each system of equations for x and y:

a. $y = 0$ *(−4, 0), (−1, 0)*
   $y = x^2 + 5x + 4$

b. $y = \dfrac{-2}{3}x + 4$ *(6, 0)*
   $\dfrac{1}{3}x - y = 2$

RT-72. More Practice Factoring Polynomials When factoring a polynomial it is often useful to look first for common factors among the terms. Once the common factors have been identified and factored out, you can then use what you know about generic rectangles to continue factoring, if possible.

Example 1: To factor $8x^2 - 2y^2$, first factor out the common factor 2:

$$\underset{\text{original polynomial}}{8x^2 - 2y^2} = \underset{(\textbf{common factor}) \cdot (\text{remaining polynomial})}{\textbf{2} \cdot (4x^2 - y^2)}$$

In this case, the remaining polynomial $4x^2 - y^2$ is a difference of squares, so we can continue to factor:

$$2(4x^2 - y^2) = 2(2x + y)(2x - y).$$

Since neither $2x + y$ nor $2x - y$ can be factored further, we have factored $8x^2 - 2y^2$ as completely as possible. Thus we have

$$8x^2 - 2y^2 = 2(2x + y)(2x - y). \blacktriangle$$

Example 2: To factor $5x^3 + 15x^2 - 20x$, notice that all three terms have a factor of 5 and a factor of x, so we can factor out the common factor $5x$:

$$\underset{\text{original polynomial}}{5x^3 + 15x^2 - 20x} = \underset{(\textbf{common factor}) \cdot (\text{remaining polynomial})}{\textbf{5x} \cdot (x^2 + 3x - 4)}$$

To continue factoring the remaining polynomial $x^2 + 3x - 4$, we need to find a pair of integers with a product of -4 and a sum of 3. Since -1 and 4 work, we get

$$5x(x^2 + 3x - 4) = 5x(x - 1)(x + 4).$$

Neither $x - 1$ nor $x + 4$ can be factored further, so we have factored $5x^3 + 15x^2 - 20x$ as completely as possible. Therefore,

$$5x^3 + 15x^2 - 20x = 5x(x - 1)(x + 4). \blacktriangle$$

Factor each of the following polynomials as completely as possible:

a. $27x^2 - 3$ *3(3x + 1)(3x − 1)*

b. $4x^3y - xy^3$ *xy(2x + y)(2x − y)*

c. $5x^2y - 30xy + 45y$ *5y(x − 3)(x − 3) or 5y(x − 3)²*

d. $x^4 - 16$ *(x² + 4)(x − 2)(x + 2)*

RT-73. In Chapter 3 we used packets and jelly beans on a two-pan scale to model how to solve simple linear equations. We can also use packets and jelly beans on an *unbalanced* two-pan scale to model how to solve a simple linear inequality such as $4x + 1 < 3x + 10$.

Example: Solve the inequality $4x + 1 < 3x + 10$.

Situation: All packets have the same number of jelly beans. Four packets and one jelly bean are lighter than 3 packets and 10 jelly beans. What can you say about the number jelly beans in each packet?

430 Chapter 7 The Buckled Railroad Track: From Words to Diagrams to Equations

	Description of Process	Diagram (Model with Packets and Beans)	Algebraic Record (Model with Variables)
STEP 1.	Draw a diagram to represent the unbalanced scale corresponding to the inequality $4x + 1 < 3x + 10$.		Let x represent the number of beans in each packet. $4x + 1 < 3x + 10$
STEP 2.	Remove three packets from each side of the scale, and record the algebraic process.		$\begin{array}{r} 4x + 1 < 3x + 10 \\ -3x \qquad -3x \end{array}$ $x + 1 < 10$
STEP 3.	Remove one jelly bean from each side, and record the algebraic process. One packet is lighter than nine jelly beans—so there are fewer than nine jelly beans in each packet: $x < 9$ **Check:** Choose several values of x to check in the original inequality: two values less than 9, two values greater than 9, and 9 itself. Substitute each of your chosen values for x to be sure that only values of x less than 9 make the original inequality true.		$\begin{array}{r} x + 1 < 10 \\ -1 \quad -1 \end{array}$ $x < 9$
STEP 4.	Graph the solution on a number line.	$x < 9$ on number line 0 to 10, open circle at 9	

Use the idea of packets and jelly beans on an unbalanced two-pan scale to solve each of the following inequalities. Graph each solution on a number line.

a. $2x + 7 < 3x + 4$ *3 < x (or x > 3)*

b. $5x + 1 \leq 2x + 7$ *x ≤ 2 (or 2 ≥ x)*

RT-74. You know that if you multiply each side of an equation, say

$$\frac{x}{2} + 1 = 3,$$

by the same number, then the equality is preserved. For example,

$$2 \cdot \left(\frac{x}{2} + 1\right) = 2 \cdot 3.$$

What happens when you multiply an *inequality*, such as $2 < 5$, by a number n? If you multiply both sides of

$$2 < 5$$

by n, it seems natural to assume that you would get

$$2 \cdot n < 5 \cdot n.$$

However, working with inequalities is not so straightforward—you must be careful, as you will see in this problem!

a. Find four values of n that make the inequality $2 \cdot n < 5 \cdot n$ true. *Many possible solutions, such as n = 1, 0.5, 8, or 23*

b. Find four values of n that make the inequality $2 \cdot n < 5 \cdot n$ false. *Many possible solutions; such as n = −1, −0.5, −2, or −17, or 0*

c. For each value of n you chose in part b, could you rewrite the resulting statement to make it a true inequality? If so, show how. If not, explain why not. *If n is negative, reverse the direction of the inequality sign, such as 2(−1) > 5(−1). If n = 0, the resulting statement is an equality.*

d. Is there a value of n that makes the equation $2n = 5n$ true? If so, find one. If not, explain why not. *Yes; if n = 0, then 2(0) = 0 = 5(0).*

RT-75. Consider the inequality $-2 < 5$.

a. Find three values of m that make the inequality $-2m < 5m$ true. *any three numbers > 0*

b. Find three values of m that make the inequality $-2m > 5m$ true. *any three numbers < 0*

c. Find one value of m that makes the equation $-2m = 5m$ true. *m = 0*

7.4 DISTANCES, TIMES, AND RATES: MORE DIAGRAMS AND EQUATIONS

SECTION MATERIALS
DOT PAPER (FOR RT-82 AND RT-85)

As you have seen in previous sections, a diagram, with units included, can help you see how to write an equation for a situation stated in words. In this section many of the problems are about situations that involve distances, times, and rates. To avoid confusion in these cases, it will be especially important to include the units as you label your diagrams for the problems.

RT-76. A Mercedes is traveling east from Elko at 50 miles per hour, and a Yugo is traveling west from Elko at 60 mph. Both cars started at the same time.

a. How far did each car go in 1 hour? *Mercedes: 50 mi; Yugo: 60 mi*

b. How far did each car go in 3 hours? *M: 150 mi; Y: 180 mi*

c. How far did each car go in x hours? *M: 50x mi; Y: 60x mi*

d. Sketch and label a diagram to help you visualize the situation. On your diagram, label the distance each car travels in x hours. Include the units on your diagram. State what the variable x represents.

Possible diagram:

```
        60x miles           50x miles
  Yugo               Elko              Mercedes
```
x represents the number of hours traveled

e. How long will it take before the two cars are 1000 miles apart? Use your diagram to help write an equation, and then solve it. *60x + 50x = 1000; about 9 hr 5 min*

RT-77. Matilda and Nancy are 60 miles apart, bicycling toward each other on the same road. Matilda rides at 11.6 mph and Nancy rides at 8.4 mph.

a. How far has Matilda ridden after 2 hours? How far has Nancy ridden after two hours? *Matilda: 23.2 mi; Nancy: 16.8 mi*

b. How far has Matilda ridden after H hours? How far has Nancy ridden after H hours? *M: 11.6H mi; N: 8.4H mi*

c. In how many hours will Matilda and Nancy meet? Draw and label a diagram, including the units, and state what your variable represents. Then write an equation and solve it. Check your answer and then write it in a complete sentence.

Possible diagram:

```
        11.6H miles      8.4H miles
      Matilda                    Nancy
```
H represents the number of hours traveled.

11.6H + 8.4H = 60. They'll meet after riding 3 hr.

Section 7.4 Distances, Times, and Rates: More Diagrams and Equations

RT-78. Sketch and label a diagram to help you visualize the following problem. State what your variable represents. Use your diagram to write an equation, and then solve the equation to solve the problem. Check your answer, and then write it in a complete sentence.

Cleopatra rode an elephant to the outskirts of Rome at 2 kilometers per hour and then took a chariot back to camp at 10 kilometers per hour. If the total traveling time was 18 hours, how far was it from camp to the outskirts of Rome?

Possible diagram:

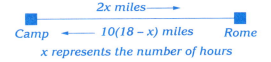

x represents the number of hours

$2x = 10(18 - x)$. *It's 30 km from camp to Rome.*

RT-79. Two trucks leave a rest stop at the same time. One truck heads due east. The other heads due north and travels twice as fast as the first truck.

a. If the eastbound truck travels 100 miles, how far will the northbound truck have gone? $2 \cdot 100 = 200$ *mi*

b. If the eastbound truck travels E miles, how far will the northbound truck have gone? $2 \cdot E$ *mi*

c. Sketch and label a diagram, including the units, to help you visualize the following problem. State what your variable represents. Use your diagram to write an equation, and then solve the equation to solve the problem. Check your answer, and then write it in a complete sentence.

The two trucks lose radio contact when they are 47 miles apart. How far has each truck traveled when they lose contact? (The drivers are obviously using an illegal power amplifier, since FCC regulations limit CBs to 5 watts, with a normal range of about 5 miles.)

Possible diagram:

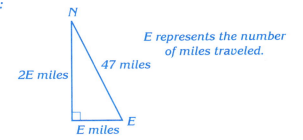

E represents the number of miles traveled.

$E^2 + (2E)^2 = 47^2$. *About 21 mi east and about 42 mi north*

EXTENSION AND PRACTICE

RT-80. A Saab is traveling due west, and a Miata is traveling due north. Both cars started from the same point at the same time going 40 miles per hour.

Sketch and label a diagram (include the units) to help you visualize each of the following situations. State what your variable represents. Use your diagram to write an equation, and then solve the equation to solve the problem in part d.

434 Chapter 7 The Buckled Railroad Track: From Words to Diagrams to Equations

a. How far did each car travel in 1 hour? *Each car traveled 40 mi.*

b. How far did each car travel in 3 hours? *Each car traveled 120 mi.*

c. How far did each car travel in T hours? *Each car traveled 40T mi.*

d. How long will it be until the cars are 500 miles apart? How far will each car have traveled at that point?

Possible diagram: *T represents the number of hours traveled.*

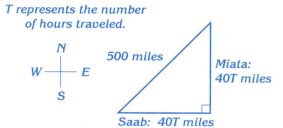

$(40T)^2 + (40T)^2 = 500^2$; They'll meet in 8.84 hr, or about 8 hr and 50 min. Each car will have traveled about 354 mi at that point.

RT-81. Sketch and label a diagram to help you visualize the following problem. Include units on your diagram, and state what your variable represents. Use your diagram to write an equation, and then solve the equation to solve the problem. Check your answer, and then write it in a complete sentence.

Two hikers leave a trail head at the same time. One goes south and one goes west. One is walking at 5.5 kilometers per hour and the other walks at 4.5 kilometers per hour. About how long will it take before the hikers are 15 kilometers apart? How far did each hiker walk in that time?

Possible diagram: *T represents the number of hours traveled.*

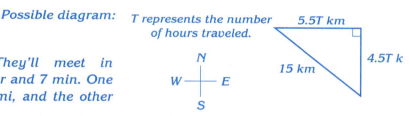

$(5.5T)^2 + (4.5T)^2 = 15^2$. They'll meet in about 2.11 hr, or about 2 hr and 7 min. One hiker walked about 11.6 mi, and the other walked about 9.5 mi.

RT-82. In this problem you will make a visual explanation of the Pythagorean relationship, as you did in RT-57.

a. Near the center of a piece of dot paper, draw a right triangle with legs 1 unit long and 4 units long, respectively. Mark the right angle and label the legs with their lengths.

b. Using the hypotenuse as one of the sides, draw a tilted square on the hypotenuse of your triangle.

c. Use the surround-and-subtract method to calculate the area of the square on the hypotenuse by first drawing a large square surrounding the original right triangle and the tilted square. In your diagram you should see a total of four right triangles the same size as the original triangle. What subproblems need to be solved? *Find the side length and the area of the large square; find the area of each right triangle.*

d. What is the area of the tilted square? What is the length of one side of the tilted square?   *17 sq units; $\sqrt{17}$ units*

e. Use your answer in part d to find the length of the hypotenuse of the original triangle. Then calculate the length of the hypotenuse using the Pythagorean relationship. How do your calculations compare?   *The results should be the same.*

RT-83. Evaluate each of the following expressions. Use a calculator to check your answers in parts a through c.

a. $(\sqrt{121})^2$ *121*
b. $(\sqrt{58.9})^2$ *58.9*
c. $(\sqrt{0.15})^2$ *0.15*
d. $(\sqrt{P})^2$ if P is a positive number   *P*
e. $(\sqrt{x-2})^2$ if $x-2$ is a positive number   *x − 2*

f. What operation undoes the effect of "square rooting" a positive number? Add information about *undoing a square root* to your Tool Kit.   *squaring*

RT-84. Using what you learned in RT-54 and RT-55, find the distance between each pair of points. Write each distance in square-root form and decimal form, rounded to the nearest hundredth.

a. (0, 0) and (4, 4) $\sqrt{32} \approx 5.66$
b. (−2, 4) and (4, 7) $\sqrt{45} \approx 6.71$

RT-85. On dot paper, draw three distinct segments—*AB*, *CD*, and *EF*—so that each segment has a slope of 1 and all three segments have different lengths. Use the Pythagorean relationship to find the length of each segment.   *Answers will vary but every segment length will be of the form $n \cdot \sqrt{2}$, where n is a real number.*

RT-86. Predict whether each statement is true or false. Then use your calculator to check.

a. $\sqrt{18} + \sqrt{3} = \sqrt{21}$ *F*   **b.** $\sqrt{18} \cdot \sqrt{3} = \sqrt{54}$ *T*
c. $\dfrac{\sqrt{18}}{\sqrt{3}} = \sqrt{6}$ *T*   **d.** $\sqrt{18} - \sqrt{3} = \sqrt{15}$ *F*
e. $\sqrt{7} \cdot \sqrt{7} = 7$ *T*   **f.** $\dfrac{\sqrt{25}}{\sqrt{2}} = \sqrt{12.5}$ *T*

RT-87. Rewrite each of the following expressions in a simplified square-root form, if possible, as shown in the example. If it's not possible, explain why not. If you need help, look back at your work in RT-61, or use a calculator to check your work.

Example: $5\sqrt{2} + \sqrt{2} - 2\sqrt{2} = 4\sqrt{2}$ ▲

a. $5\sqrt{3} + 9\sqrt{3}$ *14√3*   b. $19\sqrt{2} - \sqrt{2}$ *18√2*

c. $5\sqrt{6} + 9\sqrt{3}$ *not possible*   d. $\sqrt{5} + 9\sqrt{5} + 6\sqrt{5}$ *16√5*

e. $2\sqrt{7} + 4\sqrt{7} - 10\sqrt{7}$ *−4√7*   f. $12\sqrt{x} + 3\sqrt{x}$ *15√x*

RT-88. Factor each of the following polynomials as completely as possible:

a. $50T^2 - 8$ *2(5T + 2)(5T − 2)*

b. $14m^3 + 77m^2 + 35m$ *7m(2m + 1)(m + 5)*

c. $12x^3y - 3xy^3$ *3xy(2x + y)(2x − y)*

RT-89. Write each of the following algebraic fractions in simplest form:

a. $\dfrac{x^2 - 13x + 40}{10x - 50}$ for $x \neq 5$ or 8 $\dfrac{x-8}{10}$

b. $\dfrac{2x^2 + 7x + 6}{2x^2 + 3x}$ for $x \neq -1.5$ or 0 $\dfrac{x+2}{x}$

RT-90. Solve each of the following quadratic equations:

a. $y^2 + 6y - 16 = 0$ *−8, 2*

b. $x^2 + 8x + 16 = 0$ *−4*

c. $5z^2 - 13z = 0$ $0, \dfrac{13}{5}$

d. $2x^2 + 15x + 7 = 0$ $-7, \dfrac{-1}{2}$

RT-91. To find the x-intercepts of a parabola, you could graph the equation or use algebra.

a. Graph the parabola $y = x^2 + x - 12$, and use the graph to estimate its x-intercepts.

b. Find the x-intercepts of $y = x^2 + x - 12$ algebraically as you did in problems FF-74 and FF-136.   *The x-intercepts are (−4, 0) and (3, 0).*

RT-92. Seth has a 12-foot by 15-foot pen for his goats. He knows they will grow to about four times the size that they are now and will need four times as much room (area) to run around. He decides to build a pen that is four times as large, and would like some help.

a. What should the dimensions of the new pen be?   *24 by 30 ft*

b. What is the area of the original pen?   *180 sq ft*

c. What will the area of the new pen be? Is there four times as much area for the goats?   *720 sq ft; Yes, 4 · 180 = 720 sq ft*

d. Compare the perimeters of the two pens. How much more fencing will Seth need to buy to enlarge the old pen? *$P_{orig} = 54$ ft, $P_{new} = 108$ ft; He'll need 54 feet more.*

RT-93. Find the point of intersection for each of the following pairs of lines:

a. $3x + 4y = -36$ *(8, −15)*
$2x + 8y = -104$

b. $x = 3y - 7$ *(2, 3)*
$3x - y = 3$

RT-94. Sketch and label a diagram (with units) to help you visualize the following problem. State what your variable represents. Use your diagram to write an equation, and then solve the equation to solve the problem. Check your answer, and then write it in a complete sentence.

Luke and Jen both measured the distance between the two large sycamore trees in their backyard. Luke found that the distance is 10 inches short of five broomsticks. According to Jen's measurement, it's 29 inches longer than four broomsticks. Assuming that both measurements are correct, how far apart (in inches) are the two trees?   *If b represents the length of a broomstick, then $5b - 10 = 4b + 29$; $b = 39$ in. The trees are 185 in. apart.*

RT-95. Solve each of the following equations for x:

a. $\dfrac{x-3}{5} = 12(x-1)$ *$\dfrac{57}{59} \approx 0.97$*

b. $\dfrac{10}{2x} + 7 = 10 - \dfrac{4}{x}$ *3*

c. $2x^2 - 6x = 0$ *0, 3*

d. $\dfrac{5+x}{6} = \dfrac{3x}{14}$ *17.5*

RT-96. Copy each of the following pairs of statements and then make each statement true by filling in the box with the appropriate inequality symbol, $<$ or $>$, as shown in the example.

Example: $2 \boxed{<} 3$

   $(-5)2 \boxed{>} (-5)3$ ▲

a. $12 \boxed{>} 7$

   $(-1)(12) \boxed{<} (-1)7$

b. $-3 \boxed{<} 0$

   $(-3)(-3) \boxed{>} (-3)0$

c. $1.75 \boxed{<} 1.80$

   $(5)1.75 \boxed{<} (5)1.80$

d. $-7 \boxed{>} -20$

   $(-2)(-7) \boxed{<} (-2)(-20)$

e. $0 \boxed{<} 4$

   $(0.5)0 \boxed{<} (0.5)4$

f. $-7 \boxed{>} -12$

   $(-1)(-7) \boxed{<} (-1)(-12)$

RT-97. Look back at your results in RT-75 and RT-96. Then copy and complete each of the following statements with the appropriate inequality symbol, $<$ or $>$.

a. If $A < B$ and N is a positive number, then $A \cdot N \boxed{<} B \cdot N$.

b. If $A < B$ and N is a negative number, then $A \cdot N \boxed{>} B \cdot N$.

RT-98. Use the idea of packets and jelly beans on an unbalanced two-pan scale to solve each of the following inequalities. Graph each solution on a number line.

a. $2x + 1 < 3x - 2$ *3 < x (or x > 3)*

b. $\frac{1}{2}(2x + 5) > 2x + 1$ *$\frac{3}{2}$ > x(or x > $\frac{3}{2}$)*

c. $\frac{x}{4} + 1 \geq \frac{x+1}{3}$ *8 ≥ x (or x ≤ 8)*

RT-99. Gene and Roger solved the equation $2x - 8 = 10 - x$ in two different ways.

a. Here's how Gene solved the equation:

$$\begin{array}{r} 2x - 8 = 10 - x \\ +x \qquad\qquad +x \\ \hline 3x - 8 = 10 \\ +8 \quad +8 \\ \hline 3x = 18 \\ x = 6 \end{array}$$

   Added x to each side.

   Added 8 to each side.

   Divided each side by 3.

Copy Gene's work and fill in the blanks to describe what he did in each step. (The last step has been done for you.)

b. Here's how Roger solved the equation:

$$
\begin{array}{rl}
2x - 8 = & 10 - x \\
-2x & -2x \\
\hline
-8 = & 10 - 3x \\
-10 & -10 \\
\hline
-18 = & -3x \\
6 = & x
\end{array}
$$

- *Subtracted 2x from each side.*
- *Subtracted 10 from each side.*
- *Divided each side by −3.*

Recopy Roger's work, and describe what he did in each step.

7.5 CALCULATING WITH SQUARE ROOTS

SECTION MATERIALS

MATERIALS FOR A CLASS SET OF PUZZLES (FOR OPTIONAL PYTHAGOREAN PUZZLE ACTIVITY— SEE INSTRUCTOR'S NOTES)

CHAPTER SUMMARY DRAFT RESOURCE PAGE (OPTIONAL FOR RT-123)

The Pythagorean Puzzle is an optional activity. Based on properties of area, it reinforces what students noticed in RT-53, RT-57, and RT-82 and demonstrates a way for verifying the Pythagorean theorem.

One way to approach the activity is to do it as a student-assisted teacher-guided demonstration, either on the overhead projector with small puzzles made from two colors of transparent plastic, or at the board with large puzzles made from two colors of stiff paper (or plastic) taped to the board.

Another way is to make a set of puzzles out of card stock or stiff art board for each group, and let the groups work through the guided discovery. This takes about 10 to 15 minutes of class time. The puzzle sets require a fair bit of preparation time, but the demonstration will make more sense to the students if they can manipulate the puzzles themselves instead of watching someone else do it. Regardless of what materials you use, do the puzzle for yourself first before deciding whether or not to share it with your students.

Pythagorean Puzzle—A Demonstration of the Pythagorean Theorem—(Optional)

Materials (see Instructor Resource Pages for blackline masters):
Pythagorean Puzzle Triangles reproduced, in one color, on card stock (or stiff art board)— one page (eight triangles) per group

Pythagorean Puzzle Squares reproduced, in a contrasting color, on cardstock (or stiff art board)—one page (three different squares) per group

Pythagorean Puzzle Boards reproduced on white card stock (or stiff art board)—two boards per group

For a group activity, each group will need a set consisting of one each of Puzzle 1 and Puzzle 2, and two printed Pythagorean Puzzle Boards. For the overhead, you'll need a smaller version of each Puzzle made from transparencies in two colors to match the students' game pieces, and two smaller Pythagorean Puzzle Boards marked side by side on a clear transparency.

440 Chapter 7 The Buckled Railroad Track: From Words to Diagrams to Equations

Making the pieces:

For each group, copy one of each of the three blackline master Resource Pages onto card stock or stiff art paper.

For the Pythagorean Puzzle Triangles, use sharp scissors to cut four a by b rectangles for each group from the printed card stock. Cut along a diagonal of each rectangle to make eight right triangles.

For the Pythagorean Puzzle Squares, use sharp scissors to cut three squares—one of side length a, one of side length b, and one square with a side length $\sqrt{a^2 + b^2}$ equal to the diagonal of the rectangle—for each group from the printed card stock.

Organizing the materials:

For Puzzle 1, place the largest square and four right triangles in a resealable plastic bag labeled "Puzzle 1." For Puzzle 2, place the two smaller squares and four of the right triangles in a resealable plastic bag labeled "Puzzle 2." Clip a Pythagorean Puzzle Board to each bag.

Instructions for the Activity:

For this activity, students should work in pairs within their group of four. Distribute one set of puzzles to each group, so one pair works Puzzle 1 using one Pythagorean Puzzle Board, and the other pair shares Puzzle 2 and the second Pythagorean Puzzle Board.

Before starting, ask the pairs within each group to compare their Pythagorean Puzzle Boards. Ask several groups what they noticed. To a response of "They're the same size," ask, "What is the same size?" and other questions to elicit "The squares have the same length sides" and "The puzzle squares have the same area."

Now assign the first task: Tell the pairs that their task is to try to assemble all their (six or five) puzzle pieces to fit inside one of the two Pythagorean Puzzle Boards without any gaps or overlaps (see diagram below for the solution). Walk around observing and encouraging efforts and checking results.

Solution:

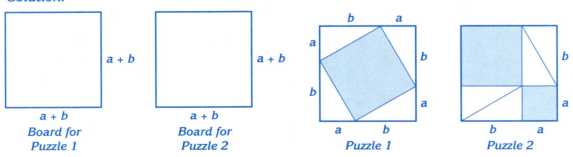

Ask, "What do Puzzle 1 and Puzzle 2 have in common?" and list responses. Be sure to get students to say that both puzzles have the same (total) area and they both contain four right triangles, all eight of which are the same size and shape (congruent).

Tell students to remove the four right triangles from Puzzle 1 and the four triangles from Puzzle 2. Ask, "What must be true of the pieces that remain?"

Section 7.5 Calculating with Square Roots **441**

We're looking for the fact that "the area of the piece remaining in Puzzle 1 must be equal to the area of the pieces remaining in Puzzle 2." In other words, the area of the larger square (from Puzzle 1) is equal to the sum of the areas of the two smaller squares (from Puzzle 2).

Finally, have each group set aside all pieces except one right triangle and the three square pieces. Ask, "Are the square puzzle piece from Puzzle 1 and the two square puzzle pieces from Puzzle 2 related to the right triangle? If so, show how." Walk among groups encouraging and reinforcing their efforts as they put the pieces together with the squares on the matching sides of the right triangle.

Ask for a volunteer to describe or summarize the Pythagorean Puzzle activity. Then paraphrase his or her remarks to conclude the activity. Remember to state the Pythagorean relationship in terms of the "legs" and "hypotenuse" of a right triangle. Discourage students from mindlessly parroting a meaningless "$a^2 + b^2 \ldots$"

In this section you'll examine patterns in square-root arithmetic, that is, in adding, subtracting, multiplying, and dividing numbers in square-root form.

RT-100. Predict whether each of the following statements is true or false. Then use a calculator to check your predictions.

a. $\sqrt{12} = \sqrt{4} \cdot \sqrt{3}$ *T*
b. $\sqrt{48} = 4\sqrt{3}$ *T*
c. $\sqrt{100} - \sqrt{64} = \sqrt{36}$ *F*
d. $\dfrac{\sqrt{10}}{2} = \sqrt{5}$ *F*
e. $\sqrt{12} + \sqrt{13} = \sqrt{25}$ *F*
f. $\sqrt{7} \cdot \sqrt{28} = \sqrt{196}$ *T*
g. $\dfrac{\sqrt{39}}{\sqrt{3}} = \sqrt{13}$ *T*
h. $(\sqrt{5})^2 = 5$ *T*

RT-101. Predict how to rewrite each of following expressions using as few square-root symbols and numbers as possible. Check your answers with a calculator. (Not all of the expressions can be rewritten.)

a. $(\sqrt{3})(\sqrt{7})$
  $\sqrt{21}$

b. $\sqrt{3} + \sqrt{7}$
  $\sqrt{3} + \sqrt{7}$, can't be simplified

c. $\dfrac{\sqrt{20}}{\sqrt{5}}$
  2

d. $\dfrac{\sqrt{144}}{\sqrt{3} \cdot \sqrt{4}}$
  $\sqrt{12}$

e. $\sqrt{7} - \sqrt{3}$
  $\sqrt{7} - \sqrt{3}$, can't be simplified

f. $\dfrac{\sqrt{75}}{\sqrt{3}}$
  5

RT-102. Predict how to rewrite each of following expressions using as few square-root symbols and numbers as possible. Check your answers with a calculator. (Not all of the expressions can be rewritten.)

a. $\sqrt{9} + 5\sqrt{9}$ *$6\sqrt{9} = 18$*
b. $10\sqrt{3} - 2\sqrt{3}$ *$8\sqrt{3}$*
c. $4\sqrt{5} - 3\sqrt{5}$ *$\sqrt{5}$*
d. $4\sqrt{5} + \sqrt{10}$ *can't be simplified*

e. $\sqrt{16} - \sqrt{15}$ **4 − $\sqrt{15}$** f. $2(4\sqrt{13} + \sqrt{13})$ **$10\sqrt{13}$**

After groups have done RT-103, have them present their statements and examples on posters or overhead transparencies. Expect only very general statements. As a follow-up—after they've had more practice—the class could consolidate the groups' posters to create a class poster of statements about square-root arithmetic.

RT-103. Square-Root Arithmetic With your group, look back over the work you have done adding, subtracting, multiplying, and dividing square roots, and look for patterns. Refer to problems RT-38, RT-60, RT-86, RT-87, and RT-100 through RT-102.

a. Complete the following statement to generalize what you have observed about *multiplying* and *dividing unlike* square roots. Also give three examples that support your statement.

"When multiplying or dividing unlike square roots, like $\sqrt{N} \cdot \sqrt{M}$, or $\dfrac{\sqrt{N}}{\sqrt{M}}, \ldots$" **…you can multiply (or divide) first and then take the square root.**

b. Complete the following statement to generalize what you have observed about *adding* and *subtracting unlike* square roots. Also give three examples that support your statement.

"When adding or subtracting unlike square roots, like $\sqrt{N} + \sqrt{M}$, or $\sqrt{N} - \sqrt{M}, \ldots$" **…you cannot add (or subtract) first and then take the square root.**

c. Complete the following statement to generalize what you have observed about *adding* and *subtracting like* square roots. Also give three examples that support your statement.

"When adding or subtracting like square roots, such as $7\sqrt{N} + 3\sqrt{N}$, or $5\sqrt{N} - 2\sqrt{N}, \ldots$" **…you calculate the total number of $\sqrt{N}$ you have.**

d. Add your statements about *square root arithmetic* to your Tool Kit.

In problem RT-104, students continue to work with square roots by examining equivalent ways to express square root values, such as $\sqrt{8} = 2\sqrt{2}$. After groups have completed RT-104 and RT-105, bring the class together to share their responses.

One way to set up RT-104 is to ask students to compare several ratios that all have the same value, such as $\dfrac{2}{3}, \dfrac{4}{6}, \dfrac{10}{15}$, and $\dfrac{200}{300}$. Similarly, 6, $\sqrt{36}$, $(\sqrt{6})^2$, and $\sqrt{6^2}$ all represent the same number. In earlier problems students saw that $\sqrt{M \cdot N} = \sqrt{M} \cdot \sqrt{N}$.

RT-104. Simplifying Square Roots Before calculators were universally available, if people wanted to calculate approximate values for numbers like $\sqrt{45}$, they could use long tables of square-root values or use a guess and check strategy (as you did in RT-29) over and over, depending on how accurate an answer they wanted. They found, however, that if they memorized the square roots of the integers from 1 to 10 and used factoring, they could figure out the square roots of many larger numbers fairly quickly. When factoring a number whose square root they wished to approximate, they found it most useful to look for factors that were perfect squares. Then they used the fact that the square root of a product of two positive numbers is equal to the product of the square roots of the two numbers, as you observed in RT-103.

The process of rewriting square roots by factoring out perfect squares and using facts about square root arithmetic is often called "simplifying" square roots. Here are some examples of how it works:

Example 1: Simplify $\sqrt{45}$.

You want to rewrite $\sqrt{45}$ in an equivalent but "simpler" square-root form. First, factor 45 so that one of the factors is a *perfect square*.
$$\sqrt{45} = \sqrt{\mathbf{9} \cdot 5}$$

Then use the fact that the square root of a product of two positive numbers is equal to the product of the square roots of the two numbers.
$$= \sqrt{\mathbf{9}} \cdot \sqrt{5}$$

Finally, rewrite the square root of the perfect square.
$$= 3 \cdot \sqrt{5}$$

Therefore, $\sqrt{45} = 3\sqrt{5}$.

With your calculator you can verify that $\sqrt{45} \approx 6.71$ and $3\sqrt{5} \approx 3(2.236) \approx 6.71$, so both forms, $\sqrt{45}$ and $3\sqrt{5}$, are equivalent. ▲

Example 2: Simplify $\sqrt{150}$.
$$\sqrt{150} = \sqrt{(\mathbf{25} \cdot 6)} = \sqrt{\mathbf{25}} \cdot \sqrt{6} = \mathbf{5} \cdot \sqrt{6} = 5\sqrt{6} \quad ▲$$

Example 3: Simplify $\sqrt{72}$.
$$\sqrt{72} = \sqrt{(\mathbf{36} \cdot 2)} = \sqrt{\mathbf{36}} \cdot \sqrt{2} = \mathbf{6} \cdot \sqrt{2} = 6\sqrt{2} \quad ▲$$

Notice that we chose to write $\sqrt{72}$ as $\sqrt{\mathbf{36}} \cdot \sqrt{2}$, rather than $\sqrt{\mathbf{4}} \cdot \sqrt{18}$ or $\sqrt{\mathbf{9}} \cdot \sqrt{8}$, because 36 is the largest perfect-square factor of 72. However, since

$$\sqrt{\mathbf{4}} \cdot \sqrt{18} = 2 \cdot \sqrt{18}$$
$$= 2 \cdot \sqrt{\mathbf{9} \cdot 2}$$
$$= 2 \cdot \sqrt{\mathbf{9}} \cdot \sqrt{2}$$
$$= 2 \cdot \mathbf{3}\sqrt{2}$$
$$= 6\sqrt{2},$$

we still get the same answer if we first write $\sqrt{72} = \sqrt{\mathbf{4}} \cdot \sqrt{18}$, and then simplify in several steps. ▲

a. Try simplifying $\sqrt{72}$ by starting with $\sqrt{72} = \sqrt{9} \cdot \sqrt{8}$.

b. Simplify each of the following square roots.

   i. $\sqrt{18}$ $3\sqrt{2}$
   ii. $\sqrt{48}$ $4\sqrt{3}$
   iii. $\sqrt{250}$ $5\sqrt{10}$
   iv. $\sqrt{1000}$ $10\sqrt{10}$

RT-105. Equivalent Square-Root Expressions Without using a calculator, match each square-root expression in the left-hand column with an equivalent simplified square-root expression in the right-hand column, as shown in the example.

Example: $\sqrt{150} = \sqrt{25 \cdot 6} = \sqrt{25} \cdot \sqrt{6} = 5\sqrt{6}$ ▲

a. $\sqrt{12}$ $2\sqrt{3}$ $3\sqrt{5}$
b. $\sqrt{54}$ $3\sqrt{6}$ $3\sqrt{6}$
c. $\sqrt{8}$ $2\sqrt{2}$ $5\sqrt{2}$
d. $\sqrt{50}$ $5\sqrt{2}$ $2\sqrt{3}$
e. $\sqrt{45}$ $3\sqrt{5}$ $2\sqrt{5}$
f. $\sqrt{20}$ $2\sqrt{5}$ $5\sqrt{3}$
g. $\sqrt{75}$ $5\sqrt{3}$ $2\sqrt{2}$

> **EXTENSION AND PRACTICE**

RT-106. Rewrite each of the following expressions in a simplified square-root form, if possible. If not, explain why not.

a. $\sqrt{25} + \sqrt{9}$ 8
b. $(\sqrt{5})(\sqrt{3})$ $\sqrt{15}$
c. $(\sqrt{1.5})(\sqrt{1.5})$ 1.5
d. $\sqrt{9} + \sqrt{8}$ $3 + 2\sqrt{2}$
e. $\sqrt{100} - \sqrt{180}$ $10 - 6\sqrt{5}$
f. $\sqrt{13} - \sqrt{3}$ not possible
g. $\dfrac{\sqrt{90}}{\sqrt{3}}$ $\sqrt{30}$
h. $(\sqrt{3704})^2$ 3704
i. $\sqrt{512^2}$ 512
j. $\sqrt{5^2 \cdot 2}$ $5\sqrt{2}$
k. $\sqrt{63}$ $3\sqrt{7}$
l. $(\sqrt{25})(\sqrt{9})$ 15

m. Do your answers in this problem support your statements in RT-103? If not, describe the conflict. Did you discover anything new?

Students used the graph of $y = \sqrt{x}$ in RT-28, and in SP-109 (Chapter 2). In problem RT-107 they will practice using their calculators and graphing nonlinear rules as they graph $y = \sqrt{x}$ for themselves.

RT-107. You can use a graph of $y = \sqrt{x}$ to approximate square roots if you make it accurately.

a. Graph the equation $y = \sqrt{x}$. Do so by making a table including at least three values of x between 0 and 2, at least three values of x between 2 and 10, and at least three values of x between 10 and 27.

b. Does $x = 0$ work in the rule $y = \sqrt{x}$ Are there any input values that do not work in this rule? Explain.   *Yes, values $x \geq 0$ work. No negative numbers work because when a (real) number is squared, the result is a positive number.*

c. Linda still doesn't understand how to find a rough approximation for the square root of a number by using the square-root graph. Use your graph to explain to Linda why
$$6 < \sqrt{43} < 7.$$
On the x-axis, $36 < 43 < 49$. Since the graph is increasing, we see that $\sqrt{36} < \sqrt{43} < \sqrt{49}$ on the y-axis.

RT-108. Look back at RT-104, and then add information about *simplifying square roots* to your Tool Kit. Then simplify each of the following square roots and use a calculator to approximate each answer to the nearest hundredth.

a. $\sqrt{50}$ $\sqrt{5 \cdot 5 \cdot 2} = \sqrt{5^2} \cdot \sqrt{2} = 5\sqrt{2} \approx 7.07$

b. $\sqrt{24}$ $2\sqrt{6} \approx 4.90$

c. $\sqrt{245}$ $7\sqrt{5} \approx 15.65$

RT-109. Start with the point P at $(900, 400)$. Imagine sliding the point to the right 12 units and up 15 units to the point S.

a. Draw a diagram to illustrate the slide. Your sketch does not need to be made to scale. Label the starting point P and its image S after the slide with their coordinates.   *(900, 400) and (912, 415)*

b. What is the length of the segment that joins points P and S?
   $\sqrt{369} \approx 19.2$ units

c. What is the slope of the segment PS? $\frac{15}{12} = \frac{5}{4} = 1.25$

RT-110. Factor each of the following polynomials as completely as possible:

a. $2x^2 + 2x$ *$2x(x + 1)$*

b. $27x^2 - 3$ *$3(3x + 1)(3x - 1)$*

c. $12x^3y - 3xy^3$
   $3xy(2x - y)(2x + y)$

d. $5x^2y - 30xy + 45y$
   $5y(x - 3)^2$

e. $8 - 64y^2$
   $8(1 - 8y^2)$

f. $x^2 + 36$
   $x^2 + 36$; can't be factored

g. $100x^2 - 49y^2$ *$(10x + 7y)(10x - 7y)$*

RT-111. Solve each of the following equations. Write your answer in fraction form, and show how to check it.

a. $\dfrac{x}{6} + 1 = \dfrac{4}{9}$ $\dfrac{-10}{3}$

b. $\dfrac{6}{x} + 1 = \dfrac{4}{9}$ $\dfrac{-54}{5}$

c. $\dfrac{x}{x+1} = \dfrac{4}{9}$ $\dfrac{4}{5}$

d. $\dfrac{x}{10} + \dfrac{18}{5x} = 2$ *18, 2*

RT-112. Explain why the two triangles in this diagram below are similar. Then use what you know about similar triangles to find x.

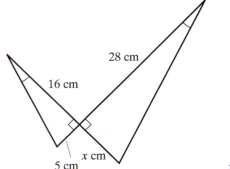

x = 8.75 cm

RT-113. Define two variables and write two equations to model the following problem. Solve the pair of equations to solve the problem.

Patty's Burgers sells three hamburgers and two milkshakes for $3.35. One hamburger and one milkshake cost $1.30. Find the cost of four hamburgers and five milk shakes.   *Let h represent the cost of a hamburger and m the cost of a milkshake in dollars. Then 3h + 2m = 3.35, and h + m = 1.30; h = $0.75 and m = $0.55. The cost for four hamburgers and five milkshakes is 4h + 5m = $5.75.*

RT-114. The lengths of two opposite sides of a rectangle are $2x + 5$ and $3x - 8$ units, respectively. How long is each of the two sides?   *x = 13; each side is 31 units.*

RT-115. More Thermal Expansion The coefficient of linear thermal expansion for iron or steel is approximately $\dfrac{12 \cdot 10^{-6}}{1°C}$. Use the linear thermal expansion equation given in problem RT-18 to solve the following problems.

a. If the temperature of a 10-meter steel railroad track is raised 40°C, about how much would the track expand in length?   *about $4.8 \cdot 10^{-3}$ m, or about $\dfrac{1}{2}$ cm*

b. For a 25-meter steel bridge girder to contract 1 centimeter in length, about how much would its temperature need to be lowered?   *Hint for students: 1 cm = 0.01 m; the temperature would need to decrease about 33.3°C.*

RT-116. Solve each of the following equations for B:

a. $122 - 41 = B^2$ *B = ±9*

b. $2^2 + (2B)^2 = 10^2$ *B = ±√24 = ±2√6 ≈ ±4.9*

c. $A^2 + B^2 = C^2$ *B = ±√(C² − A²)*

RT-117. Write each of the following algebraic fractions in simplest form. Look for a common factor first.

a. $\dfrac{2xy - 4y^2}{xy}$ for $x \neq 0$ and $y \neq 0$ *$\dfrac{2(x - 2y)}{x}$*

b. $\dfrac{3x^2 - 6x + 3}{5x - 5}$ for $x \neq 1$ *$\dfrac{3(x - 1)}{5}$*

RT-118. Solve each of the following systems for x and y:

a. $y = 3x - 5$ *x = 2, y = 1*
   $y = 5x - 9$

b. If $y = 3x - 5$ and $y = 5x - 9$ were graphed on the same set of axes, at what point would the lines intersect?   *(2, 1)*

c. $y = 2x + 5$ *x = 3, y = 11*
   $3x + 2y = 31$

d. $x + 2y = 1$ *x = −1, y = 1*
   $3x - 2y = -5$

RT-119. Copy each of the following pairs of statements, and then make each statement true by filling in the box with the appropriate inequality symbol, $<$ or $>$, as you did in problem RT-96.

a. 4 $\boxed{<}$ 7
   $(4)^2$ $\boxed{<}$ $(7)^2$

b. 5 $\boxed{<}$ 13
   $(5)^2$ $\boxed{<}$ $(13)^2$

c. 0.5 $\boxed{<}$ 8
   $(0.5)^2$ $\boxed{<}$ $(8)^2$

RT-120. Look back at your results in RT-119. Copy and complete the following statement with the appropriate inequality symbol, $<$ or $>$.

If A and B are positive numbers and $A < B$, then A^2 $\boxed{<}$ B^2.

RT-121. In this problem you'll compare two ways to solve the inequality $2x - 8 < 10 - x$.

a. Solve the inequality $2x - 8 < 10 - x$ by following the steps Gene used in RT-99a to solve the equation $2x - 8 = 10 - x$.

b. Solve the inequality $2x - 8 < 10 - x$ by following the steps Roger used in RT-99b to solve the equation $2x - 8 = 10 - x$.

c. Why must you be careful in the last step of part b when following Roger's steps? *When each side of an inequality is multiplied by the same negative number, the direction of the inequality is reversed.*

RT-122. Solve each of the following inequalities. Graph each solution on a number line.

a. $x - 10 < 5x + 2$ *$-3 < x$ (or $x > -3$)*

b. $\dfrac{5x}{2} + 3 \leq \dfrac{5x - 1}{3}$ *$x \leq -4$ (or $-4 \geq x$)*

RT-123. Chapter 7 Summary Rough Draft Some of the problems you did in Chapter 7 focused on developing the four main ideas of the chapter while others reinforced concepts and skills introduced in previous chapters. Reread the chapter's introduction, and then look back through the chapter to find where each main idea was developed.

a. Make photocopies of the Chapter Summary Draft Resource Page (or follow the same format on your own paper). Use a page to complete the following steps for each of the four main ideas of the chapter:

- State the main idea.
- Select and recopy a problem that is a good example of the idea and in which you did well.
- Include a completely worked-out solution to the selected problem.
- Write one or two sentences that describe what you learned about the idea.

As in previous chapters, you will have the opportunity to revise your work, so your focus should be on the content of your summary rather than on its appearance. Be ready to discuss your responses with your group at the next class meeting.

b. In this chapter you've used diagrams to help you write equations that model situations stated in words. From your work, select two problems for which you found making and labeling a diagram particularly useful for solving the problem. Write complete sentences to describe why you chose the problems you did.

c. Cindy drew a right triangle on dot paper. One leg was 9 units long and the other leg was 16 units long. Cindy concluded that the hypotenuse should be 7 units long because $\sqrt{9} + \sqrt{16} = 3 + 4 = 7$. Think of two ways you could convince Cindy that her conclusion is not correct. Write a letter to Cindy in which you clearly describe each way.

d. Find a problem that you still cannot solve or that you are unsure you would be able to solve on a test. Write out the question and as much of the solution as you can. Then explain what you think is keeping you from solving the problem. Be clear and precise in describing the hard part.

7.6 SUMMARY AND REVIEW

SECTION MATERIALS

THE THEOREM OF PYTHAGORAS VIDEO-TAPE (OPTIONAL)

Start with groups discussing their rough drafts of the Chapter 7 Summary, as you circulate and listen. Select problems to provide review and practice as needed by your students.

If you have extra time to spend on review, you might show the video tape entitled The Theorem of Pythagoras. *(©1988 by* Project MATHEMATICS!*) The tape and a workbook can be purchased from California Institute of Technology by mail (Caltech Bookstore, 1-51, California Institute of Technology, Pasadena, CA 91125) or by phone (818-356-6161).*

RT-124. Chapter 7 Summary Group Discussion Take out the rough-draft summary you completed in RT-123. Use this class time to discuss your work; use homework time to revise.

For each of the main ideas of the chapter, choose one member of the group to lead a short discussion. When it is your turn to share, you should do the following:

- Explain the problem you chose to illustrate your main idea.
- Explain why you chose that particular problem.
- Explain, as far as possible, the problem from RT-123d that you still cannot solve.

This is also your chance to make sure your summary is complete and to work together on problems you may not be able to solve yet.

The lengths involved in RT-125 provide an opportunity for students in part b to use the faulty logic that if $a^2 = b^2 + c^2$, then $a = \sqrt{b + c}$ — and still obtain the correct answer. Thus your students may incorrectly write "$H^2 = 5281^2 + 5280^2$" as some of ours have done, and then conclude "so $H = \sqrt{5281 + 5280}$." Watch for this mistake as you circulate among the groups.

Indeed, $H = \sqrt{5281 + 5280}$ in this case, but only because H^2 happens to be a difference of squares whose roots differ by 1:

$$H^2 = 5281^2 - 5280^2 = (5281 + 5280)(5281 - 5280) = (5281 + 5280)(1).$$

RT-125. The Buckled Railroad Track A new railroad line was installed in Placid Valley. In order to reduce derailments along a two-mile stretch, the track was made with straight one-mile long rails. The rails were laid in the winter and they expanded in the heat of the following summer. Indeed, each mile-long rail expanded 1 foot in length! Ordinarily, because the rails do not bend, they would jut to the side. However, in this strange case, the rails jutted upward where their ends met. How high above the ground were the expanded rails at the joint?

Let *H* be the height (in feet) of the tracks above the ground where two rails come together.

a. *Make a guess* of how large you think H might be: big enough for you to stick your arm between the ground and the tracks? big enough to walk through? Could you drive a car under the buckled tracks?

b. Draw a diagram and label it. Use your picture to help calculate H. Note that a mile is 5280 feet long. $H = \sqrt{5281^2 - 5280^2} \approx 102.77 \text{ ft}$

c. How does the value you calculated for H compare with your guess in part a?

RT-126. Sketch and label a diagram to help you visualize the following problem. Include units on your diagram, and state what your variable represents. Use your diagram to write an equation, and then solve the equation to solve the problem. Check your answer, and then write it in a complete sentence.

Two trucks leave the same rest stop at the same time. One truck drives three times as fast as the other truck. The slower truck heads south, and the faster truck goes west. When the two trucks lose CB contact, they are 53 miles apart. How far has each traveled? *The slower truck travels $\sqrt{280.9} \approx 16.8$ mi, and the faster truck travels $3\sqrt{280.9} \approx 50.3$ mi.*

RT-127. List subproblems that will help you solve the following problem. Sketch and label a diagram for each subproblem, and then solve the subproblem. Write your answer to the original problem in a complete sentence.

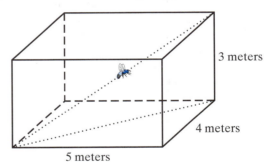

A bee sat on the floor in one corner of a greenhouse. It then flew to the ceiling of the opposite corner of the room. If the bee flew in a straight line, how far was its flight across the greenhouse? (The dimensions of the greenhouse are given in the figure.) *The diagonal of the floor is $\sqrt{41}$ m. The bee flew $\sqrt{41 + 3^2} = 5\sqrt{2} \approx 7.07$ m.*

RT-128. Find at least nine points that are 10 units away from the origin and have integer coordinates. (*Hints:* Sketch a diagram. The point (6, 8) works.) *There are 12 such points: (6, 8), (8, 6), (−6, 8), (−8, 6), (−6, −8), (−8, −6), (6, −8), (8, −6), (0, 10), (10, 0), (0, −10), and (−10, 0).*

RT-129. Sketch and label a diagram (include units) to help you visualize the following problem. State what your variable represents. Use your diagram to write an equation, and then solve the equation to solve the problem. Check your answer, and then write it in a complete sentence.

> Scott rode his bike to Folsom Lake at 20 kilometers per hour. After going for a swim, he found that the front bicycle tire was flat, so he had to walk home. Scott walked at 8 kilometers per hour all the way home. If his round-trip traveling time was 7 hours, how far was it to the lake?
> *It was 40 km to the lake; a long walk.*

RT-130. Sketch and label a diagram to help you visualize the following problem. Use your diagram to write an equation, and then solve the equation to solve the problem. Check your answer, and then write it in a complete sentence.

> Janis is going to fence off a rectangular garden. She will use an existing wall along the back, and she wants the length to be twice as long as the width. The total amount of fencing material she has is 84 meters long. What are the width and length of her garden? *There are two solutions, depending on the orientation of the rectangle: W + 2W + W = 84, where W = 21 m and L = 42 m; or 2W + W + 2W = 84, where W = 16.8 m and L = 33.6 m.*

RT-131. Rules for Working with Square Roots Throughout Chapter 7 you have been predicting ways to rewrite expressions involving square roots. You have also practiced predicting when square-root expressions cannot be rewritten in a simplified form.

To summarize your observations about square-root arithmetic using words rather than symbols, write five complete sentences by completing each statement in the left column with an appropriate choice from the right column. You may use the choices from the right column more than once, or not at all.

a. The square root of a product of two positive numbers ...   ... is equal to the quotient of the square roots of the two numbers.   *c*

b. The square root of a sum of two positive numbers ...   ... is equal to the product of the square roots of the two numbers.   *a*

c. The square root of a quotient of two positive numbers ... ... is equal to the difference of the square roots of the two numbers.

d. The square root of a difference of two positive numbers ... ... is equal to the sum of the square roots of the two numbers.

e. To undo taking the square root of a positive number, ... may not be possible to simplify. **b, d**

... you simply square the result. **e**

RT-132. Tool Kit Check By now your Tool Kits should include new entries for at least the following topics:

Algebra

- the hypotenuse and legs of a right triangle   *RT-2*
- the Pythagorean relationship   *RT-6*
- square-root arithmetic   *RT-103, RT-108, RT-131*

Calculator

- using the square root key   *RT-29*

a. Check that your Tool Kit entries are up to date. Exchange Tool Kits with members of your group, and read each other's entries to check for clarity and accuracy.

b. At home, make sure that your Tool Kit includes examples and explanations that help you remember and understand the ideas. Make any revisions, clarifications, or updates that will reflect your developing understanding of mathematics.

⬡ **EXTENSION AND PRACTICE** ⬡

RT-133. Sketch and label a diagram to help you visualize the following problem. Use your diagram to write an equation, and then solve the equation to solve the problem. Check your answer, and then write it in a complete sentence.

A 10-meter ladder is leaning against a building. The bottom of the ladder is 5 meters from the building. How high is the top of the ladder? $5^2 + h^2 = 10^2$; $5\sqrt{3} \approx 8.66\ m$

RT-134. Sketch and label a diagram, with units, to help you visualize the following problem. State what your variable represents. Use your diagram to write an equation, and then solve the equation to solve the problem. Check your answer, and then write it in a complete sentence.

A train leaves Roseville at 6:00 A.M. heading east at 52 miles per hour. A different train leaves the same station heading west at 7:00 A.M. at 45 mph. What time are the two trains 240 miles apart? *Let x represent the number of hours since 6:00 A.M. Then $52x + 45(x - 1) = 240$; $x \approx 2.94$ hr; at around 8:56 A.M.*

RT-135. Sketch and label a diagram to help you visualize the following problem. Use your diagram to write an equation, and then solve the equation to solve the problem. Check your answer, and then write it in a complete sentence.

Two cars leave St. Louis at the same time going in opposite directions. One travels at 99 kilometers per hour, and the other travels at 81 kilometers per hour. In how many hours will they be 300 kilometers apart?

Let x represent the number of hours; $99x + 81x = 300$; $\frac{5}{3} \approx 1.67$ hr

RT-136. Find the unknown side lengths in each triangle. Write each length in simplified square-root form, and then give an approximation to the nearest hundredth.

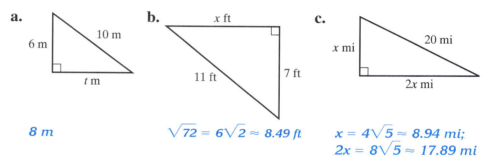

a. 8 m

b. $\sqrt{72} = 6\sqrt{2} \approx 8.49$ ft

c. $x = 4\sqrt{5} \approx 8.94$ mi; $2x = 8\sqrt{5} \approx 17.89$ mi

RT-137. After making and labeling a rough sketch, find the distance between each pair of points. Write the distance in simplified square-root form, if possible, and then find the approximate distance rounded to the nearest hundredth.

a. $(5, -8)$ and $(-3, 1)$ *$\sqrt{145} \approx 12.04$*

b. $(0, -3)$ and $(0, 5)$ *8*

c. $(0, 0)$ and $(25, 25)$ *$\sqrt{1250}$ or $25\sqrt{2} \approx 35.36$*

RT-138. Start with the generic point (x, y). Imagine sliding the point to the right 3 units and up 4 units.

a. Draw a diagram to illustrate the slide. Label the starting point and its "image" after the slide (the stopping point) with their coordinates.
   (x, y) and $(x + 3, y + 4)$

b. Find the slope of the segment that joins the original point and its image after the slide.   *$\frac{4}{3}$*

c. Find the length of the segment that joins the original point and its image after the slide. Does the length of the segment equal the slope of the segment?   *length = 5 units; no*

Note that problem RT-139 requires students to think of similar right triangles while they may have a "Pythagorean relationship" mind-set.

RT-139. Sketch and label a diagram to help you visualize the following problem. Use your diagram to write an equation, and then solve the equation to solve the problem. Check your answer, and then write it in a complete sentence.

A young redwood tree in Muir Woods casts a shadow 12 feet long. At the same time, a 5-foot-tall tourist casts a shadow 2 feet long. How tall is the tree?
Let h represent the height of the tree, then $\frac{h}{12} = \frac{5}{2}$. The tree is 30 ft tall.

RT-140. Match the number in the left-hand column with an equivalent number in the right-hand column. Whenever possible, do this without using a calculator.

a. $\sqrt{96}$ $4\sqrt{6}$ $2\sqrt{15}$
b. $\sqrt{60}$ $2\sqrt{15}$ $3\sqrt{2}$
c. $\sqrt{72}$ $6\sqrt{2}$ $4\sqrt{6}$
d. $\sqrt{24}$ $2\sqrt{6}$ $2\sqrt{6}$
e. $\sqrt{28}$ $2\sqrt{7}$ $6\sqrt{2}$
f. $\sqrt{48}$ $4\sqrt{3}$ $4\sqrt{3}$
g. $\sqrt{18}$ $3\sqrt{2}$ $2\sqrt{7}$

RT-141. Solve each of the following equations for A, if possible.

a. $A^2 = 15^2 - 12^2$ $A = \pm\sqrt{30} \approx \pm 5.48$
b. $A^2 + (26.4)^2 = 29^2$ *no solution*
c. $A^2 + B^2 = C^2$ $A = \pm\sqrt{C^2 - B^2}$

RT-142. Solve each of the following equations for x:

a. $2x^2 - 140 = 4 + x^2$ $-12, 12$
b. $x(x+1)(x+2) = 0$ $0, -1, -2$
c. $1 - \dfrac{5}{6x} = \dfrac{x}{6}$ $1, 5$
d. $2x + 3y = 5$ $\dfrac{5 - 3y}{2}$
e. $\dfrac{3}{x} - 7 = 10$ $\dfrac{3}{17} \approx 0.18$
f. $\dfrac{1}{x} + \dfrac{2}{3x} = 5$ $\dfrac{1}{3} \approx 0.33$

RT-143. Find the point of intersection for each pair of lines:

a. $y = 9 - x$ $(0, 9)$
    $y = 9 + x$

b. $3x - 2y = 20$ $(6, -1)$
    $x = 5 - y$

RT-144. Solve each equation by factoring and using the zero product property:

a. $3x^2 - 4x + 1 = 0$ $1, \frac{1}{3}$

b. $x^2 - 4x = 5$ $5, -1$

RT-145. A certain rectangle has area $3x^2 - xy - 2y^2$ square units. Find its dimensions. *(3x + 2y) by (x − y)*

RT-146. A rectangular garden of area 63 square meters has a width 2 meters less than its length. Find the length and width of the garden. *7 m by 9 m*

RT-147. In the last student-council election at Sierra Nevada College only about 43% of the students voted. A total of 1576 ballots were counted. About how many students attend Sierra Nevada College? *about 3665*

RT-148. Evaluate each of the following expressions with $x = 2$ and $x = -3$:

a. $2x^2$ *8; 18*
b. $(2x)^2$ *16; 36*
c. $(3x)^3$ *216; −729*
d. $-5x^2$ *−20; −45*

e. Write an equivalent expression for $(5x)^2$ without parentheses. *$25x^2$*

RT-149. Follow the instructions in parts a and b to determine the x-intercepts using two different methods.

a. Graph the parabola $y = x^2 - 5x - 6$, and use the graph to estimate its x-intercepts.

b. Find the x-intercepts of $y = x^2 - 5x - 6$ algebraically as you did in FF-74 and FF-136. *(6, 0) and (−1, 0)*

RT-150. If 70 feet of telephone wire weighs 23 pounds, how much would you expect 100 feet of telephone wire to weigh? *$\frac{230}{7} \approx 32.86$ lb*

RT-151. Rewrite each of the following expressions in a simplified square-root form, if possible. If not, explain why not.

a. $\sqrt{3} \cdot \sqrt{12}$ *6*
b. $\sqrt{80} + \sqrt{81}$ *$4\sqrt{5} + 9$*
c. $(\sqrt{7})(\sqrt{10})$ *$\sqrt{70}$*
d. $\sqrt{21} - \sqrt{3}$ *not possible*
e. $\frac{\sqrt{28}}{\sqrt{7}}$ *$\sqrt{4} = 2$*
f. $\frac{\sqrt{35}}{\sqrt{7}}$ *$\sqrt{5}$*
g. $(\sqrt{7})(\sqrt{14})$ *$7\sqrt{2}$*
h. $(\sqrt{134.98})^2$ *134.98*
i. $\sqrt{17^2}$ *17*
j. $\sqrt{162}$ *$9\sqrt{2}$*
k. $\sqrt{192}$ *$8\sqrt{3}$*
l. $\sqrt{18} - \sqrt{2}$ *$3\sqrt{2} - \sqrt{2} = 2\sqrt{2}$*

RT-152. The buckled-railroad situation described in RT-125 is pretty far-fetched. To begin with, rails are not manufactured in mile-long segments. Steel rails do expand when heated, but because they are attached to heavy ties that are anchored down with crushed rock, the expanded rails can't pop up. Instead, in extreme situations they would move from side to side and look wiggly pieces of cooked spaghetti. To allow for some expansion of the rails, construction crews leave small gaps, or joints, between adjacent ends when the rails are laid.

Sacramento's Light Rail System experienced six incidences of buckled rails in its first two years of service. When the tracks were laid, they were heated to 37.8°C (if they weren't already that hot) and secured to the wooden ties before they cooled and contracted. About every mile along the track, a gap of 0.3 to 0.6 centimeters was left between the ends of adjacent rails to allow for expansion. The largest kink occurred where the rails expanded about 16.8 centimeters.

a. In problem RT-18 you were given a word version of the *linear thermal expansion equations,* which describes the relationship among the initial length, the change in length, and the change in temperature for a material:

$$\frac{\text{Change in length}}{\text{Initial length}} = \left(\begin{array}{c}\text{Coefficient}\\ \text{of linear}\\ \text{thermal expansion}\end{array}\right) \cdot \left(\begin{array}{c}\text{Change}\\ \text{in temperature}\\ \text{in °C}\end{array}\right)$$

Useful equations, or formulas, such as this are usually expressed in terms of variables. To represent a change in a quantity, scientists often use the symbol Δ, which is the Greek letter *delta*. The coefficient of linear thermal expansion is usually denoted by the Greek letter α ("alpha").

Rewrite the linear thermal expansion equation in a more compact form. Use ΔL to represent "change in length," L_0 to represent "initial length," α to represent the coefficient of linear thermal expansion, and ΔT to represent "change in temperature." $\frac{\Delta L}{L_o} = \alpha \cdot \Delta T$

b. Use the coefficient of linear thermal expansion for steel that you were given in problem RT-115 and solve the equation you wrote in part a for ΔL. Then solve it for ΔT.

$$\Delta L = (12 \cdot 10^{-6}) \cdot (\Delta T) L_0; \Delta T = \frac{\Delta L}{L_o(12 \cdot 10^{-6})} \text{ or } \Delta T = \frac{\Delta L \cdot 10^6}{L_o \cdot 12}$$

c. To check the accuracy of the report that a rail expanded about 16.8 centimeters, what additional information would be needed? *the initial length and the change in temperature.*

RT-153. Look back at your statements in RT-97.

a. In a complete sentence, describe what happens when each side of an inequality is multiplied by the same positive number. *Multiplying each side of an inequality by the same positive number preserves the direction of the inequality.*

b. In a complete sentence, describe what happens when each side of an inequality is multiplied by the same negative number. *Multiplying each side of an inequality by the same negative number changes the direction of the inequality.*

RT-154. Look back at your statement in RT-120. If both sides of an inequality are positive, describe what happens when each side of the inequality is squared. *If both sides of an inequality are positive, then squaring both sides preserves the direction of the inequality.*

RT-155. Solve each of the following inequalities. Graph each solution on a number line.

a. $x + 6 \leq 4(x - 1)$ $\frac{10}{3} \leq x \text{ (or } x \geq \frac{10}{3})$

b. $4(3x - 1) > 4(3 - x)$ *1 < x (or x > 1)*

RT-156. Chapter 7 Summary Revision Use the ideas your group discussed in problem RT-124 to revise the your rough draft of a Chapter 7 Summary. Your presentation should be thorough and organized, and should be done on paper separate from your other work.

CHAPTER EIGHT—FOR THE INSTRUCTOR

THE GRAZING GOAT
Dealing with Complicated Situations

MATERIALS
- Graph paper and straightedges (needed throughout the chapter)
- A large picture surrounded by a border
- String, 12 in. to 15 in. long (for each group)
- Transparencies and pens for the overhead projector (optional)
- Resource Page for GG-131; also, Summary Draft Template (optional)

In this chapter we look at several ways to deal with complicated problems. First we introduce the quadratic formula as a convenient tool for solving quadratic equations that are not easily solved using the zero product property. After finding exact solutions to quadratic equations, students extend their calculator skills so that they can approximate solutions to problems in practical situations.

Then we apply the strategy of identifying and solving subproblems to simplifying algebraic fractions and to finding areas and/or perimeters of complicated regions. Some of the algebraic fractions arise in the context of finding an average speed over a given distance. The geometric situations require students to use ideas they've learned throughout the course, and we encourage the use of subproblems throughout the chapter.

Finally, to deal with complicated linear data sets, we extend the idea of a trend line to introduce the idea of a line of best fit and, as an extension, the concept of the mean data point. In asking students to write an equation to fit a set of linear data, we reinforce many of the important skills and ideas they've learned in the course— data organization, the graphing of data, and the detection of patterns and trends.

CHAPTER EIGHT

THE GRAZING GOAT
Dealing with Complicated Situations

A goal in this course has been to help you make sense of the important concepts of elementary algebra. In this final chapter you'll tie together many ideas and skills you've learned in the course: you'll focus on solving quadratic equations, using subproblems to deal with complicated situations, and pulling together your graphing and equation-writing skills to write equations to model sets of linear data.

The Grazing Goat Problem is an example of a complicated problem that can be most easily resolved by identifying and solving subproblems:

> On weekends, the Williams tether their goat, Daisy, to a corner of their 15-meter by 25-meter barn. Since the barn sits in the middle of a large, grassy field, Daisy can spend her days grazing. The extent to which she can roam depends on which piece of rope the Williams have used to tie her. Over what area of the field can Daisy graze if the rope is x meters long?

At the beginning of this chapter you will find the areas of complex geometric figures by solving the subproblems of finding areas of simpler, related regions. Later, you will use this same subproblem approach to "uncomplicate" complicated algebraic processes. To use subproblems in an algebraic context, you have to be able to step back from the stated problem and identify useful subproblems before you begin. As was mentioned in Chapter 4, the ability to identify subproblems could be one of the most useful skills you learn in this course.

IN THIS CHAPTER, YOU WILL HAVE THE OPPORTUNITY TO:

- use the quadratic formula to solve quadratic equations that are not easily solvable using factoring and the zero product property;
- extend your calculator skills for use in solving quadratic equations;
- use the strategy of identifying and solving subproblems to solve complicated problems, such as simplifying algebraic fractions and finding the areas of complicated regions; and
- use the idea of the line of best fit to reinforce some important skills and ideas relating graphing data and writing equations of lines.

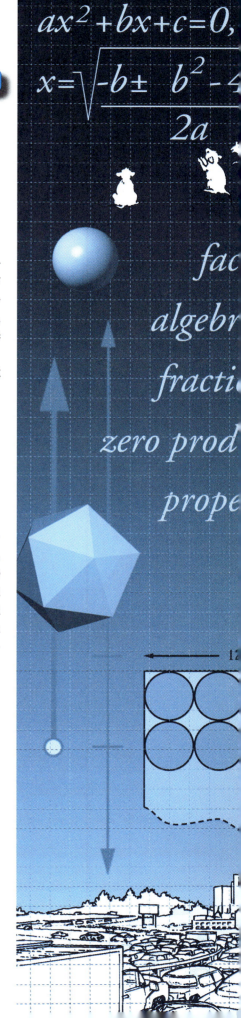

Chapter 8 The Grazing Goat: Dealing with Complicated Situations

MATERIALS	CHAPTER CONTENTS		PROBLEM SETS
Graph paper and straight-edges (needed throughout the chapter)	8.1	Solving Quadratic Equations Revisited: The Quadratic Formula	GG-1–GG-26
	8.2	Complicated Fractions: Using Subproblems to Calculate Average Speed	GG-27–GG-52
Large picture surrounded by a border (for GG-55)	8.3	The Election Poster: Using the Quadratic Formula	GG-53–GG-72
String (12– to 15-in. long, for each group, for GG-73)	8.4	Using Subproblems in Complicated Situations	GG-73–GG-91
Chapter Summary Draft Resource Page (optional, for GG-112)	8.5	Quiz Scores: Data Points and Lines of Best Fit	GG-92–GG-112
	8.6	The Cookie Cutter and the Lunch Bunch: Two Investigations (Optional)	GG-113–GG-122
Large picture surrounded by a border (for GG-124); GG-131 The 100-Meter Freestyle Resource Page	8.7	Summary and Review	GG-123–GG-155

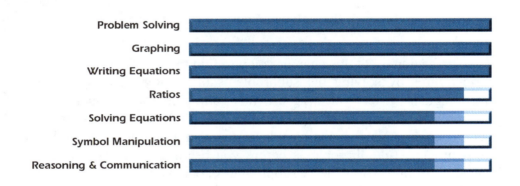

8.1 SOLVING QUADRATIC EQUATIONS REVISITED: THE QUADRATIC FORMULA

SECTION MATERIALS

GRAPH PAPER AND STRAIGHTEDGES (NEEDED THROUGHOUT THE CHAPTER)

Since Chapter 6 you've been solving quadratic equations by factoring and using the zero product property. However, not all quadratic equations are easily factorable—in fact, most quadratic equations that occur in practical situations can't be solved by factoring. In this final chapter, you'll be introduced to another algebraic method for finding solutions to quadratic equations. First, you'll practice using the zero product property on some factorable expressions. Then you'll face the dilemma of finding solutions to a quadratic equation that cannot be factored, and learn an answer to the question, "Now what can we do?"

GG-1. Find as many solutions as you can for each of the following equations:

a. $3x = x$ $x = 0$

b. $x^2 = x$ $x = 1$ or 0

c. $x^3 = x$ $x = -1, 0,$ or 1

d. $3x^2 + 5 = 3x^2 - 12$ none

e. $x^2 + 1 = 2x$ $x = 1$

f. $x(x + 5)(x - 12) = 0$ $x = -5, 0,$ or 12

In GG-2 and GG-3, remind students that sharing the work of calculating points makes the graphing less time-consuming. If students need help with factoring in GG-2, suggest using a generic rectangle to factor geometrically.

GG-2. In this problem you'll examine the relationship between the x-intercept of a parabola and the solutions of a related quadratic equation.

a. Graph the quadratic equation $y = 2x^2 - 7x + 3$ for $-4 \leq x \leq 4$, and use the graph to estimate the x-intercepts of the parabola.
vertex $(\frac{7}{4}, \frac{-25}{8})$; $(\frac{1}{2}, 0), (3, 0)$

b. Solve the equation $2x^2 - 7x + 3 = 0$ for x.   *$\frac{1}{2}, 3$*

c. Write a sentence or two comparing the solutions to parts a and b. How are the x-intercepts of the graph of $y = 2x^2 - 7x + 3$ related to the solutions to the quadratic equation $2x^2 - 7x + 3 = 0$? *The x-coordinates of the x-intercepts in part a are equal to the solutions to the equation in part b.*

GG-3. In GG-2 you saw that you could use the solutions of the quadratic equation $2x^2 - 7x + 3 = 0$ to find the x-intercepts of the parabola $y = 2x^2 - 7x + 3$. In this problem you'll follow a similar process as you examine the x-intercepts of the parabola $y = x^2 + 2x - 6$.

a. Graph the equation $y = x^2 + 2x - 6$, and use the graph to estimate the x-intercepts of the parabola. *vertex $(-1, -7)$; x-intercepts about $(-3.6, 0)$ and $(1.6, 0)$*

b. Substitute the values for x you found in part a into the equation $y = x^2 + 2x - 6$ and then solve for y. *The resulting y-value should be close to zero.*

c. If your estimates in part a were accurate, what value should you have gotten for y in part b? How close to correct were your estimates in part a? *y = 0; accuracy will vary.*

d. Draw generic rectangles and show what happens when you try to factor the polynomial $x^2 + 2x - 6$. Describe what you notice. *It can't be factored. There are four pairs of integers with a product of −6: (1)(−6), (−1)(6), (−3)(2), and (−2)(3), however, their sums are −5, 5, −1, and 1, respectively, not 2.*

e. Does the fact that $x^2 + 2x - 6$ can't be factored mean there are no solutions to the equation $0 = x^2 + 2x - 6$? Explain your answer. *There are no integer solutions because the polynomial can't be factored. There are noninteger solutions because the graph has two x-intercepts.*

In part b of GG-4, check that students correctly identify the coefficient of x^2 as 1.

GG-4. The Standard Form of a Quadratic Equation The equation $0 = x^2 + 2x - 6$ from GG-3 could not be solved by factoring and using the zero product property, and yet it must have two solutions since the graph of $y = x^2 + 2x - 6$ has two x-intercepts. This suggests that we need another method for solving quadratic equations that does not rely on factoring. Fortunately, there is an algebraic tool that accomplishes this task—it's called the *quadratic formula*. (It's introduced in problem GG-6.)

Before you can use the quadratic formula to solve a quadratic equation, the equation must be written in **standard form**. When written in standard form, a quadratic equation reads

$$ax^2 + bx + c = 0 \quad \text{where } a, b, \text{ and } c \text{ are numbers, and } a \neq 0.$$

The coefficient of x^2 is a, the coefficient of x is b, and the constant term is c. The values of a, b, and c are easily recognized when a quadratic equation is written in standard form. You may need to do some work using algebra to rewrite an equation in standard form.

Example 1: The quadratic equation $2x^2 - 13x + 21 = 0$ is already written in standard form. In this case,

$$a = 2, \quad b = -13, \quad \text{and } c = 21. \quad \blacktriangle$$

Example 2: What are a, b, and c for the equation $-x^2 + 25 = 0$? Although the equation $-x^2 + 25 = 0$ may not appear to be written in standard form, you can rewrite it as $-x^2 + 0x + 25 = 0$. You can then see that

$$a = -1, \quad b = 0, \quad \text{and } c = 25. \quad \blacktriangle$$

Each of the following quadratic equations is written in standard form. For each equation, identify the values of the coefficients a and b, and the constant c.

a. $3x^2 - 5x + 4 = 0$ *a = 3; b = −5; c = 4*

b. $x^2 + 9x = 0$ *a = 1; b = 9; c = 0*

c. $-\frac{1}{2}x^2 + 2x - \frac{1}{4} = 0$ $a = \frac{-1}{2}; b = 2; c = \frac{-1}{4}$

d. $0.017x^2 - 0.4x + 20 = 0$ $a = 0.017; b = -0.4; c = 20$

GG-5. If a quadratic equation is not written in standard form, you must convert it to standard form so that you can easily determine a, b, and c. (You will need to know these values in order to apply the quadratic formula.)

Example: The quadratic equation

$$x^2 = 2(7x - 4)$$

is *not* written in the standard form of $ax^2 + bx + c = 0$. To rewrite it in standard form, first use the distributive property to remove the parentheses and get $x^2 = 14x - 8$. Then, subtract $14x$ from each side, and add 8 to each side of the equation. Doing this, you should get

$$x^2 - 14x + 8 = 0, \quad \text{or} \quad x^2 + (-14)x + 8 = 0.$$

Now the equation is written in standard form, with $a = 1$, $b = -14$, and $c = 8$. ▲

Rewrite each of the following quadratic equations in standard form, and then identify the coefficients a and b, and the constant c:

a. $-6x + 5x^2 = 8$ $5x^2 - 6x - 8 = 0; a = 5; b = -6; c = -8$

b. $x(2x + 4) = 7x - 5$ $2x^2 - 3x + 5 = 0; a = 2; b = -3; c = 5$

c. $(x - 3)(x + 4) = 7x$ $x^2 - 6x - 12 = 0; a = 1; b = -6; c = -12$

d. $2x + 1 = \frac{-1}{2}x^2$ $\frac{1}{2}x^2 + 2x + 1 = 0; a = \frac{1}{2}; b = 2; c = 1.$

or $x^2 + 4x + 2 = 0; a = 1; b = 4; c = 2$

GG-6. The Quadratic Formula Once a quadratic equation is written in the form $ax^2 + bx + c = 0$, you can use the values for a, b, and c to calculate the solutions for the equation. That is, you can find those values of x that make the equation true by using the **quadratic formula**:

If $ax^2 + bx + c = 0$ (with $a \neq 0$), then $x = \dfrac{-b \pm \sqrt{b^2 - 4ac}}{2a}$.

This means that:

The exact values of x that make the quadratic equation $ax^2 + bx + c = 0$ true are equal to the opposite of b plus or minus the square root of the quantity, b squared minus the product $4ac$, all divided by the product $2a$.

Notice that, unlike the [+/−] key on your calculator, the $\pm$ symbol in the quadratic formula is *not* a sign change. It is an efficient, shorthand way to write the *two* solutions to the equation at one time. The formula says that the two solutions to the quadratic equation are:

$$x = \frac{-b + \sqrt{b^2 - 4ac}}{2a} \quad \text{or} \quad x = \frac{-b - \sqrt{b^2 - 4ac}}{2a}.$$

This means that, unless the quantity $b^2 - 4ac$ is negative or zero, the graph of $y = ax^2 + bx + c$ crosses the x-axis at two points, namely

$$\left(\frac{-b + \sqrt{b^2 - 4ac}}{2a}, 0 \right) \quad \text{and} \quad \left(\frac{-b - \sqrt{b^2 - 4ac}}{2a}, 0 \right).$$

Copy the *quadratic formula* into your Tool Kit, and then practice saying it aloud. If you are interested in understanding why the quadratic formula gives solutions to a quadratic equation, read Appendix B.

GG-7. Using the Quadratic Formula to Solve a Quadratic Equation

Several steps are involved when you use the quadratic formula to solve a quadratic equation. Read through the following example.

Example: Use the quadratic formula to solve the equation $x^2 - 2x = 4$.

First rewrite the equation in standard form:

$$x^2 - 2x - 4 = 0$$

Then list the numeric values of the coefficients a and b, and the constant c:

$$a = 1$$
$$b = -2$$
$$c = -4$$

Now write the quadratic formula on your paper:

If $ax^2 + bx + c = 0$ (with $a \neq 0$), then $x = \dfrac{-b \pm \sqrt{b^2 - 4ac}}{2a}$.

Substitute the numeric values for a, b, and c into the quadratic formula:

$$x = \frac{-(-2) \pm \sqrt{(-2)^2 - 4(1)(-4)}}{2(1)}$$

Use order of operations and the distributive property to simplify the answer:

$$x = \frac{2 \pm \sqrt{4 + 16}}{2}$$ *Under the square root, first calculated powers, then products.*

$$= \frac{2 \pm \sqrt{20}}{2}$$ *Under the square root, next completed the addition.*

$$= \frac{2 \pm 2\sqrt{5}}{2}$$ *Simplified the square root.*

$$= \frac{2(1 \pm \sqrt{5})}{2}$$ *Factored out the 2.*

$$= 1 \pm \sqrt{5}$$ *Found fractional forms of "1" and reduced.*

Finally, write the two exact solutions:
$$x = 1 + \sqrt{5}, \quad \text{or} \quad x = 1 - \sqrt{5} \quad \blacktriangle$$

a. Copy the steps shown in the example onto your paper. Fill in each blank with a description of what was done in the step. (The first step has been done for you.)

b. Copy the following *procedure for using the quadratic formula* into your Tool Kit:

How to Use the Quadratic Formula to Solve a Quadratic Equation

STEP 1 Rewrite the equation in standard form.

STEP 2 List the numeric values of the coefficients a and b, and the constant c.

STEP 3 Write the quadratic formula on your paper.

STEP 4 Substitute the numeric values for a, b, and c in the quadratic formula.

STEP 5 Use order of operations and the distributive property to simplify the answers.

STEP 6 Write the two exact solutions.

GG-8. In GG-3, you sketched a graph of the parabola $y = x^2 + 2x - 6$ in order to estimate the x-intercepts, but when you tried to use factoring and the zero product property to algebraically find those values, the polynomial $x^2 + 2x - 6$ was not factorable. Now you have the tools to find the exact values of the x-intercepts.

a. Write the standard form of the quadratic equation that you could solve to find the x-coordinates of the x-intercepts of the parabola $y = x^2 + 2x - 6$. $0 = x^2 + 2x - 6$

b. Now follow the six-step procedure "How to Use the Quadratic Formula to Solve a Quadratic Equation," that you copied into your Tool Kit, to find the exact values for the x-coordinates of the x-intercepts.
$x = -1 + \sqrt{7}, -1 - \sqrt{7}$

c. Use your calculator to find an approximation for each of your exact values from part b to the nearest hundredth. $x \approx 1.65, -3.65$

d. In finding the x-coordinates of the x-intercepts of the parabola, how close were your estimates from the graph in GG-3 to the exact values you found using the quadratic formula? *Answers will vary.*

GG-9. Solve each of the following quadratic equations using the quadratic formula. Follow the six-step procedure that you copied into your Tool Kit.

a. $3x^2 - 6x - 45 = 0$ $x = -3, 5$

b. $(x - 3)(x + 4) = 7x$ $x = 3 + \sqrt{21}, 3 - \sqrt{21}$

c. $x^2 + 8x = 2$ $x = -4 + 3\sqrt{2}, -4 - 3\sqrt{2}$

GG-10. In this problem you'll take another look at the equation $3x^2 - 6x - 45 = 0$ from GG-9a.

a. Completely factor the polynomial $3x^2 - 6x - 45$ by first finding a common factor.   *3(x − 5)(x + 3)*

b. Use the your factorization from part a and the zero product property to solve the equation $3x^2 - 6x - 45 = 0$.   *3(x − 5)(x + 3) = 0; x = −3, 5*

c. Compare how you solved the equation $3x^2 - 6x - 45 = 0$ in part b with how you solved the same equation in problem GG-9a using the quadratic formula. Think of some advantages and disadvantages of using the quadratic formula. Similarly, think of some advantages and disadvantages of using factoring and the zero product property. Write down a list of your observations.   *The quadratic formula is a sure process that can be memorized; however, it can be tedious. Factoring can be quick and easy, or may involve a lot of guessing and checking.*

EXTENSION AND PRACTICE

GG-11. In this problem you'll examine how the graph of the equation $y = x^2 - x$ gives information about the solutions to the equation $x^2 - x = 0$.

a. Show that the values $x = 0$ and $x = 1$ satisfy the equation $x^2 - x = 0$.

b. Use a graph of the equation $y = x^2 - x$ to show why no other values of x could also satisfy $x^2 - x = 0$.   *The graph crosses the x-axis only twice—at x = 0 and x = 1.*

GG-12. This problem is similar to GG-3. Here you'll examine the parabola $y = x^2 - 6x + 2$ and the solutions to the equation $x^2 - 6x + 2 = 0$.

a. Graph the equation $y = x^2 - 6x + 2$, and use the graph to estimate the x-intercepts of the parabola.   *vertex (3, −7); (5.6, 0), (0.35, 0)*

b. Using the values for x you found in part a, substitute for x in the equation $y = x^2 - 6x + 2$, and then solve for y.

c. If your estimates in part a were accurate, what value should you have gotten for y in part b? How close to correct were your estimates in part a?   *y = 0; accuracy will vary.*

d. Draw a generic rectangle, and show what happens when you try to factor the polynomial $x^2 - 6x + 2$. Describe what you notice.   *The polynomial can't be factored; the only two negative integers with a product of 2 are −1 and −2, and their sum is −3, not −6.*

e. What do parts a and d imply about the solutions to the equation $0 = x^2 - 6x + 2$? *There are two solutions because the graph has two x-intercepts, and the solutions are not integers because the polynomial can't be factored.*

f. Solve the quadratic equation $0 = x^2 - 6x + 2$ using the quadratic formula. Follow the six-step procedure that you copied in your Tool Kit.
$x = 3 + \sqrt{7}, 3 - \sqrt{7}$

GG-13. Solve each of the following quadratic equations using the quadratic formula. Follow the six-step procedure that you copied into your Tool Kit.

a. $x^2 + 5x - 8 = 0 \quad x = \dfrac{-5 + \sqrt{57}}{2}, \dfrac{-5 - \sqrt{57}}{2}$

b. $3x^2 + 7x = -2 \quad x = \dfrac{-1}{3}, -2$

c. $4x^2 + 36x + 81 = 0 \quad x = -4.5$

d. $-2(2x^2 - 1) = 7x \quad x = \dfrac{1}{4}, -2$

GG-14. Calculator Practice A calculator is a convenient tool for finding the approximate values of complicated expressions containing square roots of positive numbers that are not perfect squares.

a. Work through each of the following examples. Check to see that you can use your calculator to find the correct approximations for the expressions.

Example 1: To calculate $\dfrac{2 - \sqrt{7}}{\sqrt{5}}$ using the proper order of operations, you must calculate the value of the numerator $2 - \sqrt{7}$ before dividing by the denominator $\sqrt{5}$. To do this, you can either use grouping symbols such as [ ], ( ) or { } or press $=$ after entering $2 - \sqrt{7}$.

Write down the sequence of keys you press.

$(\;2\;-\;7\;\checkmark\;)\;\div\;5\;\checkmark\;=$ or $2\;-\;7\;\checkmark\;=\;\div\;5\;\checkmark\;=$

The approximate value, to the nearest hundredth of $\dfrac{2 - \sqrt{7}}{\sqrt{5}}$, is -0.29, so we write $\dfrac{2 - \sqrt{7}}{\sqrt{5}} \approx -0.29$.

Note: If you don't calculate the value of the numerator before dividing, the result will equal $2 - \dfrac{\sqrt{7}}{\sqrt{5}}$. Does $2 - \dfrac{\sqrt{7}}{\sqrt{5}}$ equal $\dfrac{2 - \sqrt{7}}{\sqrt{5}}$? Explain how you know. ▲

Example 2: To calculate $\sqrt{36 - 4(-4)}$, be sure to enclose the quantity $36 - 4(-4)$ in grouping symbols. Record the sequence of keys you press.

$\boxed{(}\ 36\ \boxed{-}\ 4\ \boxed{\times}\ 4\ \boxed{+/-}\ \boxed{)}\ \boxed{\sqrt{\ }}$ or $36\ \boxed{-}\ 4\ \boxed{\times}\ 4\ \boxed{+/-}\ \boxed{=}\ \boxed{\sqrt{\ }}$

Approximating the result to the nearest hundredth, we write $\sqrt{36 - 4(-4)} \approx 7.21.$ ▲

Example 3: To obtain a correct approximation for the expression $\dfrac{\sqrt{5} - 2}{3 - \sqrt{7}}$, use grouping symbols when entering the numerator and the denominator.

Record the sequence of keys that you press, and then use the symbol $\approx$ to write an approximate value to the nearest hundredth for $\dfrac{\sqrt{5} - 2}{3 - \sqrt{7}}$.

$\boxed{(}\ 5\ \boxed{\sqrt{\ }}\ \boxed{-}\ 2\ \boxed{)}\ \boxed{\div}\ \boxed{(}\ 3\ \boxed{-}\ 7\ \boxed{\sqrt{\ }}\ \boxed{)}\ \boxed{=}$

or $5\ \boxed{\sqrt{\ }}\ \boxed{-}\ 2\ \boxed{=}\ \boxed{\div}\ \boxed{(}\ 3\ \boxed{-}\ 7\ \boxed{\sqrt{\ }}\ \boxed{)}\ \boxed{=}\ \approx 0.67$ ▲

b. Use a calculator and what you learned in part a to find an approximate value, to the nearest hundredth, of each of the following expressions:

i. $\dfrac{6 + \sqrt{17}}{\sqrt{55}} \approx 4.53$ ii. $\dfrac{2}{\sqrt{6} - 3} \approx -3.63$

iii. $\sqrt{-13 + \sqrt{179}} \approx 0.62$ iv. $\dfrac{\sqrt{3} + 5}{2} \approx 3.37$

GG-15. In GG-13a, you found two exact solutions to the quadratic equation $x^2 + 5x - 8 = 0$.

a. Use your calculator to find an approximate value, to the nearest thousandth, of each of the two solutions you found. $\approx 1.275, -6.275$

b. Choose one of the two solutions, and substitute into the equation the approximate value for x. Use your calculator to check to see if that value makes the equation true.   *The value is close to, but not exactly, 0.*

c. Now check the other solution by substituting into the equation its approximate value for x and using your calculator to see if that value satisfies the equation.   *The value is close to, but not exactly, 0.*

d. Explain why the values you used in parts b and c did not exactly make the equation $x^2 + 5x - 8 = 0$ true.   *The values were close to the solutions, but they were not exact solutions.*

Problems GG-16 and GG-17 are two of four problems that will help students review work they've done with the area and perimeter of a circle, in preparation for The Grazing Goat problem in Section 8.4.

Heidi's logo design

GG-16. Logo Contest At the Pacific College of Design, seniors compete each year to create a new logo for the graphic designers club. Rules of the contest stipulate that patterns and designs must be composed only of simple geometric shapes—rectangles, squares, right triangles, quarter circles, and semicircles. To make her design, Heidi used a right triangle with legs 4 centimeters long and a semicircle.

a. Copy the design onto your paper, and label the lengths you know.

b. Identify some subproblems that you could use to find the area of Heidi's logo design. Then find the exact area and an approximation to the nearest 0.1 square centimeter. *Find the hypotenuse of the right triangle and the radius of the semicircle; find the area of the triangle and the semicircle. The total area is $8 + 4\pi \approx 20.6$ sq cm.*

c. Identify some subproblems that you could use to find the perimeter of Heidi's design, and then solve them. Find the exact perimeter and an approximation to the nearest 0.1 centimeter. *Find perimeter of the semicircle. The total perimeter is $8 + 2\pi\sqrt{2} \approx 16.9$ cm.*

Miguel's logo design

GG-17. Another Logo Contest Entry Miguel used two sizes of semicircles to create his entry for the logo contest. The smaller semicircle has a radius of 2 centimeters. For the quarter circle, he cut one of the smaller semicircles in half.

a. Copy the design on your paper, and label the lengths you know.

b. Identify some subproblems that you could use to find the area of Miguel's design. Then find the exact area and an approximation to the nearest 0.1 square centimeter. *Find the radius of the large semicircle; find the area of each semicircle and the quarter circle. The total area is 7.5π sq cm ≈ 23.6 sq cm.*

c. Identify some subproblems that you could use to find the perimeter of Miguel's design and then solve them. Find the exact perimeter and an approximation to the nearest 0.1 centimeter. *Find the perimeter of each semicircle and of the quarter circle. The total perimeter is $6\pi + 2 \approx 20.8$ cm.*

Problem GG-18 is preparation for solving equations similar to the one in part e that arise in solving the average-speed problems, which will be introduced in Section 8.2.

GG-18. Solve each of the following equations for x by multiplying each side of the equation by the denominator of the fraction whose numerator is 10.

a. $\dfrac{10}{x} = 15$ $\quad \dfrac{2}{3}$

b. $\dfrac{10}{x+1} = 15$ $\quad \dfrac{-1}{3}$

c. $\dfrac{10}{\frac{x}{3}} = 15$ **2**

d. $\dfrac{10}{\frac{x}{3}+1} = 15$ **−1**

e. $\dfrac{10}{\frac{2}{5}+\frac{8}{x}} = 15$ **30**

GG-19. Two similar right triangles are cut out of the same piece of flat steel plate. The first triangle has a long leg of 17 centimeters, a short leg of 9 centimeters, and weighs 100 ounces. The second triangle has a short leg of 15 centimeters.

  a. Does the weight of each triangle depend on its area, or on its perimeter? **area**

  b. Write an equation using equivalent ratios that you could use to find the weight of the second triangle. How much does the second triangle weigh? $\left(\dfrac{15}{9}\right)^2 = \dfrac{w}{100}$; $w \approx 277.78$ oz

GG-20. Sally has a box full of Ping-Pong balls numbered from 1 through 53. Charlie reached in and pulled out one Ping-Pong ball. Find the probability that the number on the ball he chose was …

  a. even. $\dfrac{26}{53}$

  b. less than or equal to 10. $\dfrac{10}{53}$

  c. a perfect square (1, 4, 9, 16, …). $\dfrac{7}{53}$

  d. a two-digit number. $\dfrac{44}{53}$

GG-21. A rectangle is 2.4 times as high as it is wide and has a diagonal of 52 inches. Find the dimensions of the rectangle. **20 by 48 in.**

GG-22. The two points $(3, -2)$ and $(5, 7)$ determine a line.

  a. Find the slope of the line. $\dfrac{9}{2}$

  b. Find an equation of the line. $y = \dfrac{9}{2}x - 15.5$

  c. Find the distance between the two points. $\sqrt{85} \approx 9.22$ units

GG-23. Find where the graphs of the lines $y = 2x - 7$ and $y = -x + 5$ intersect. **(4, 1)**

GG-24. Rewrite each of the following expressions with a single exponent:

  a. $x^3 \cdot x^{-2}$ x
  b. $n^9 \cdot n^{-4}$ n^5
  c. $w^{-5} \cdot w$ w^{-4}
  d. $A^{89} \cdot A^{-10}$ A^{79}

  e. $\dfrac{x^3}{x^2}$ x
  f. $\dfrac{n^9}{n^4}$ n^5
  g. $\dfrac{w^{-5}}{w^{-1}}$ w^{-4}
  h. $\dfrac{A^{89}}{A^{10}}$ A^{79}

GG-25. Look for similarities and differences among the expressions in GG-24. What do you notice?   *a–d, all products; e–h, all quotients; a and e are equal, as are b and f, c and g, and d and h.*

GG-26. Regions in a Coordinate Plane The graph of any line divides the coordinate plane into three parts: the line itself, and the two regions on either side of the line. In this course you've seen that the relationship between the graph of a line and its equation is that each point of the graph satisfies the equation of the line; that is, when the x- and y-coordinates of a point on the graph are substituted into the equation, the result is a true statement. In this problem you will examine points that lie in either of the two regions made by the line and use patterning to find an algebraic way to express their relationships to the line.

a. On a piece of graph paper, graph the line $y = 3x + 2$.

b. In terms of x, what is the y-coordinate of each point on the line? Locate a generic point on the line, and label it with its coordinates in terms of x. *(x, 3x + 2)*

c. Locate and label each of the following points in the coordinate plane you used in part a:

$$(1, 7), (0, 3), (5, 2), (-2, -4), (-3, -1), (0, 0), (3, 5)$$

d. For each of the points you plotted in part c, describe where the point lies in relationship to the line.

e. Copy and complete the following table. (Two entries have been done as examples.)

Point	x-Coordinate	Corresponding Value of $3x + 2$	y-Coordinate	How Does y Compare with the Value of $3x + 2$?	Location of the Point Compared with the Line $y = 3x + 2$?
(1, 7)	1	3(1) + 2 = 5	7	7 > 3(1) + 2	above the line
(0, 3)	0	*3(0) + 2 = 2*	3	*3 > 3(0) + 2*	*above the line*
(5, 2)	5	*3(5) + 2 = 17*	2	*2 < 3(5) + 2*	*below the line*
(−2, −4)	−2	3(−2) + 2 = −4	−4	−4 = 3(−2) + 2	on the line
(−3, −1)	*−3*	*3(−3) + 2 = −7*	−1	*−1 > 3(−3) + 2*	*above the line*
(0, 0)	*0*	*3(0) + 2 = 2*	0	*0 < 3(0) + 2*	*below the line*
(3, 5)	*3*	*3(3) + 2 = 11*	5	*5 < 3(3) + 2*	*below the line*

f. For each point that lies *above* the line $y = 3x + 2$, how does its y-coordinate compare with the corresponding value of $3x + 2$? For each point that lies *below* the line $y = 3x + 2$, how does its y-coordinate compare to the corresponding value of $3x + 2$? What about for points that lie *on* the line?   *above the line: $y > 3x + 2$; below the line: $y < 3x + 2$; on the line: $y = 3x + 2$*

g. Add information about *regions in a coordinate plane* to your Tool Kit.

8.2 COMPLICATED FRACTIONS: USING SUBPROBLEMS TO CALCULATE AVERAGE SPEED

In simple situations involving motion, an object travels at a constant speed. In such situations, if you know the total distance traveled and the amount of time it took, you can calculate the average speed of the object over the time interval. However, it is more common to have speed vary over time. In some situations, you may need to use two or more speeds over several time intervals to figure out the average speed over the total time of travel. Solving such problems involves working with complicated algebraic fractions.

Since Chapter 4, you have been rewriting algebraic fractions by finding fractional forms of 1 in order to express the fraction in a simpler way. In this section you'll use what you have learned to deal with some complicated algebraic fractions.

GG-27. When Professor Speedi has a weekend off, she likes to make the long drive to Nevada to visit her youngest niece Valerie. When she takes Interstate 80, the drive is exactly 400 miles.

a. Last week, Professor Speedi made the drive in 8 hours and 15 minutes. What was her average speed? $\dfrac{400 \text{ mi}}{8.25 \text{ hr}} = 48.5 \text{ mph}$

b. For the Thanksgiving holiday, Professor Speedi intends to start driving at 8 A.M. If Valerie has asked her to arrive by 2:45 P.M., about what speed will Professor Speedi need to average? $\dfrac{400 \text{ mi}}{6.75 \text{ hr}} \approx 59.3 \text{ mph}$

c. Valerie, who was never in Professor Speedi's class, doesn't understand how to calculate the average speed that Professor Speedi drives. She feels a bit embarrassed to ask her aunt. Write a short note to Valerie explaining how to calculate an average speed.   *average speed = (total distance) ÷ (total time)*

GG-28. One summer, Professor Speedi decided to drive to Los Angeles, 500 miles away, to visit a student who had been in her first math class.

a. The drive to Los Angeles took Professor Speedi only 9 hours and 30 minutes. Describe how to find her average speed in miles per hour, and then find it. *52.6 mph*

b. Unfortunately, when she set out to drive back home, the traffic in the Los Angeles area was thick. It took her 2 hours to drive the first 50 miles! The rest of the trip home took only 7 hours and 30 minutes. Which drive took more time: the drive to Los Angeles, or the drive home? What was her average speed, in miles per hour, on the trip home? *Both drives took the same amount of time; 52.6 mph*

c. What was Professor Speedi's average speed for the first 50 miles of her drive home? What was her average speed for the rest of the trip home? $\frac{50 \text{ mi}}{2 \text{ hr}} = 25 \text{ mph}; \frac{450 \text{ mi}}{7.5 \text{ hr}} = 60 \text{ mph}$

d. Valerie incorrectly calculated her aunt's average speed on the return trip by adding the average speeds for the two segments of her trip, and then dividing by two

$$\frac{25 \text{ mph} + 60 \text{ mph}}{2}$$

to get 42.5 miles per hour. She doesn't understand why this method doesn't give the correct average speed for the return trip. Write a few sentences to help her out. *The travel times were not the same.*

GG-29. Wil's daily commute to work is a 60-mile drive. His average speed on the way to work on Tuesday was 60 miles per hour, but his average speed on the way home was only 30 mp3h because of traffic. In order to find Wil's average speed on his *round-trip* Tuesday, you need to solve several subproblems. To do this, complete parts a through c. *Hint: the answer is not 45 mph.*

a. One subproblem is to find the total distance in Wil's round-trip. Solve it. *The total distance is 2(60) = 120 mi.*

b. Another subproblem is to find how long it took Wil to drive the entire round-trip. Identify and solve two "sub-subproblems" that you need to solve in order find his round-trip time. *How long did it take to drive to work? How long did it take to drive home? The trip to work = $\frac{60 \text{ mi}}{60 \text{ mph}} = 1$ hr, and the return trip = $\frac{60 \text{ mi}}{30 \text{ mph}} = 2$ hr: the round-trip took 3 hr.*

c. What was Wil's average speed on Tuesday? $\frac{120 \text{ mi}}{3 \text{ hr}} = 40 \text{ mph}$

GG-30. Veena is planning to run the Boston Marathon, and has been running 20 miles every weekend to prepare for it. Last weekend, she averaged 6 miles per hour for the first 3 miles, but slowed to 4 mph for the last 17 miles. What was Veena's average speed for the entire practice run?

Complete parts a through d to answer the question.

a. Identify subproblems that will help you find the total time Veena's practice run lasted. *Find the time for the first 3 mi, and the time for the last 17 mi.*

b. Express as a fraction, including units, the time Veena took to run the first 3 miles. $\dfrac{3\ mi}{6\ mph}$

c. Repeat part b for the time Veena took to run the last 17 miles. $\dfrac{17\ mi}{4\ mph}$

d. Express the total time for the practice run as a sum of two fractions. Then use it and your explanation in GG-27c to calculate Veena's average speed on last weekend's practice run. $\dfrac{3\ mi}{6\ mph} + \dfrac{17\ mph}{4\ mph} = \left(\dfrac{3}{6} + \dfrac{17}{4}\right)$ hr; $\dfrac{20\ mi}{\frac{3}{6} + \frac{17}{4}\ hr} \approx 4.2$ mph

GG-31. Race car drivers who want to qualify for the Indy 500 must drive several laps during the qualifying attempts. A participant's fastest lap is used for the qualifying time, but in addition, racers must meet a minimum average speed (posted for each race by the judges) throughout all the qualifying laps.

a. Suppose a particular driver makes two 1-mi laps in the qualifying event. She averages 200 mph on the first lap, but due to mechanical difficulties she averages only 150 mph on the second lap.

Will she qualify for the event if the posted minimum average qualifying speed is 175 mph? Identify and solve subproblems to help answer the question, as you did in problem GG-30. *Find the total distance, the time for each lap, and the total time; no, her average speed is $\dfrac{2\ mi}{\frac{1}{200} + \frac{1}{150}}$ hrs ≈ 171.4 mph.*

b. Another driver ran his first lap at an average speed of 125 miles per hour. What is the least speed he could average on the second lap in order to qualify with a minimum average speed of 175 mph? To answer the question, identify subproblems, define a variable, and then write and solve an equation. *Find the total distance, the time for each lap, and* the total time. Let v be the average speed on the second lap, so

$$\frac{2 \text{ mi}}{\frac{1}{125} + \frac{1}{v}} \text{ hrs} = 175.$$

His average speed for the second lap must be at least 291.7 mph.

Problems GG-32 and its follow-up GG-57 provide more practice with area and subproblems. In each case, the area of the shaded region can be seen as the sum of the areas of two shaded triangles, or as the area of the rectangle minus the area of the unshaded triangle.

GG-32. For each of the following figures, identify subproblems that will help you find the area of the shaded region, and then solve the subproblems to find the area of the shaded regions.

a. 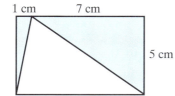   b. 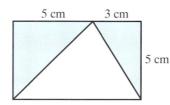   c.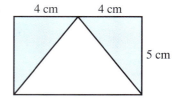

a. $\frac{1}{2} \cdot 1 \cdot 5 + \frac{1}{2} \cdot 7 \cdot 5$, or $8 \cdot 5 - \frac{1}{2} \cdot 8 \cdot 5 = 20$ sq cm;

b. $\frac{1}{2} \cdot 5 \cdot 5 + \frac{1}{2} \cdot 3 \cdot 5$, or $8 \cdot 5 - \frac{1}{2} \cdot 8 \cdot 5 = 20$ sq cm;

c. $\frac{1}{2} \cdot 4 \cdot 5 + \frac{1}{2} \cdot 4 \cdot 5$, or $8 \cdot 5 - \frac{1}{2} \cdot 8 \cdot 5 = 20$ sq cm;

EXTENSION AND PRACTICE

GG-33. Howard hiked from Woods Lake to Round Top Lake via Lake Winnemucca. He covered the first 4 miles of the hike, up to Winnemucca, at a steady pace of 3 miles per hour. He then sped up to cover the final mile to Round Top at a rate of 4 mph. What was Howard's average speed on his hike from Woods Lake to Round Top Lake?

To answer the question, follow the same process of identifying and solving subproblems that you used to find Veena's average speed in problem GG-30. *Find the time for the first 4 mi, and the time for the last 1 mi;*

$$\frac{4 \text{ mi}}{3 \text{ mph}} + \frac{1 \text{ mi}}{4 \text{ mph}} = \left(\frac{4}{3} + \frac{1}{4}\right) \text{ hr}; \quad \frac{5 \text{ mi}}{\frac{4}{3} + \frac{1}{4}} \text{ hr} \approx 3.2 \text{ mph}$$

GG-34. Ivan's Challenge Puzzle Rita's little brother Ivan likes to make up math puzzles that he thinks will torment Rita. This time, the puzzle involves algebraic fractions. Ivan challenges Rita to sketch nine points on the graph of the following equation for $-5 \leq x \leq 5$ by evaluating the expression at different x-values.

$$y = \frac{x^2 + 6x}{(x + 6)^2} \cdot \frac{(x^2 + 7x + 6)}{x^2 - 1}$$

Help Rita out by evaluating the product of algebraic fractions for three x-values, in parts a, b, and c.

a. Find the value of the expression $\frac{x^2 + 6x}{(x + 6)^2} \cdot \frac{(x^2 + 7x + 6)}{x^2 - 1}$ when $x = 5$. What subproblems will you complete? *First evaluate each numerator and denominator; 1.25*

b. Evaluate the expression using $x = -2$. *$\frac{2}{3}$*

c. Evaluate the expression using $x = -1$. What happens in this case? Explain why. *The expression is undefined because you cannot divide by zero.*

GG-35. After doing all the work to evaluate the product of fractions in GG-34 for $x = 5$ and $x = -2$, Rita decided there must be an easier way. She decided to try to rewrite the expression

$$\frac{x^2 + 6x}{(x + 6)^2} \cdot \frac{(x^2 + 7x + 6)}{x^2 - 1}$$

in its simplest form. After looking at it for a moment, she realized she could break up the process into five subproblems by rewriting each of the four polynomials in factored form and then looking for fractional forms of 1. She got:

$$\frac{x(x + 6)}{(x + 6)^2} \cdot \frac{(x + 6)(x + 1)}{(x - 1)(x + 1)}$$

Then, using $(x + 6)^2 = (x + 6)(x + 6)$, Rita rewrote the entire product as this single algebraic fraction:

$$\frac{x \cdot (x + 6) \cdot (x + 6) \cdot (x + 1)}{(x + 6) \cdot (x + 6) \cdot (x - 1) \cdot (x + 1)} *$$

Looking for fractional forms of 1, she rewrote the starred fraction above as a product of fractions with, whenever possible, the same factors in the numerator and denominator:

$$\frac{(x + 6)}{(x + 6)} \cdot \frac{(x + 6)}{(x + 6)} \cdot \frac{x}{x - 1} \cdot \frac{(x + 1)}{(x + 1)} = 1 \cdot 1 \cdot \frac{x}{x - 1} \cdot 1, \text{ or simply } \frac{x}{x - 1}.$$

Section 8.2 Complicated Fractions: Using Subproblems to Calculate Average Speed **477**

Finally, Rita wrote, $\dfrac{x^2 + 6x}{(x+6)^2} \cdot \dfrac{(x^2+7x+6)}{x^2-1} = \dfrac{x}{x-1}.$

In this process Rita learned that *doing each part was easy* and that *doing several small, relatively easy parts allowed her to complete a larger, complicated problem.*

Can the algebraic fraction $\dfrac{x}{x-1}$ be simplified further? Try it and then check by substituting a value for x (such as $x = 3$). **No, it can't be reduced.**

GG-36. In problem GG-35, you found that the expression $\dfrac{x}{x-1}$ had been simplified as much as possible.

a. Evaluate the algebraic fraction $\dfrac{x}{x-1}$ for $x = 5$. **1.25**

b. In GG-34a you evaluated the expression $\dfrac{x^2+6x}{(x+6)^2} \cdot \dfrac{(x^2+7x+6)}{x^2-1}$ for $x = 5$. Compare what you did in GG-34a with part a above. What should be true about the two values you got? **1.25; they should be equal.**

c. Complete the following table of values for
$$y = \dfrac{x^2+6x}{(x+6)^2} \cdot \dfrac{(x^2+7x+6)}{x^2-1}.$$

First fill in the values you already computed in GG-34, and then find a value of y for each remaining value of x. Before you start, think: must you use the expression $\dfrac{x^2+6x}{(x+6)^2} \cdot \dfrac{(x^2+7x+6)}{x^2-1}$ for y, or could you use the expression $\dfrac{x}{x-1}$?

x	-5	-4	-3	-2	-1	0	1	2	3	4	5
y	$\frac{5}{6} \approx 0.83$	$\frac{4}{5} = 0.8$	$\frac{3}{4} = 0.75$	$\frac{2}{3} \approx 0.67$	undefined	$\frac{0}{1} = 0$	undefined	$\frac{2}{1} = 2$	$\frac{3}{2} = 1.5$	$\frac{4}{3} \approx 1.33$	$\frac{5}{4} = 1.25$

d. Now, complete Ivan's Challenge Puzzle from GG-34: use your table from part c above to plot nine points on the graph of
$$y = \dfrac{x^2+6x}{(x+6)^2} \cdot \dfrac{(x^2+7x+6)}{x^2-1}.$$

GG-37. What happened when you tried to substitute $x = 1$ into the expression $\dfrac{x}{x-1}$? Look at the starred fraction in GG-35. What would happen if you tried to use $x = -6$ to evaluate the expression? What would happen if you tried to use $x = -1$ to evaluate the expression? What restrictions should be placed on the values of x that can be used in the expression?

The denominator would equal 0 for $x = -6$ or -1; $x \neq -6$, 1, or -1.

GG-38. Solve each of the following quadratic equations using the quadratic formula. Follow the six-step procedure that you copied into your Tool Kit. Write the exact solutions, and then use a calculator to find approximate values for x to the nearest hundredth.

a. $x^2 + 8x - 5 = 0$ $x = -4 + \sqrt{21} \approx 0.58, -4 - \sqrt{21} \approx -8.58$

b. $2x^2 + 7x - 2 = 0$ $x = \dfrac{-7 + \sqrt{65}}{4} \approx 0.26, \dfrac{-7 - \sqrt{65}}{4} \approx -3.76$

c. $x^2 - 3x - 1 = 0$ $x = \dfrac{3 + \sqrt{13}}{2} \approx 3.30, \dfrac{3 - \sqrt{13}}{2} \approx -0.30$

GG-39. In GG-3 you saw that the equation $0 = x^2 + 2x - 6$ cannot be solved using factoring and the zero-product property.

a. Use the quadratic formula to solve the equation $0 = x^2 + 2x - 6$ from GG-3. $x = \dfrac{-2 \pm \sqrt{28}}{2} = -1 \pm \sqrt{7}$

b. What are the exact coordinates of the x-intercepts of the parabola $y = x^2 + 2x - 6$?   *$(-1 + \sqrt{7}, 0), (-1 - \sqrt{7}, 0)$*

c. Approximately where does the graph cross the x-axis? How close to correct were your original estimates in GG-3?   *$(\approx -3.64, 0)$ and $(\approx 1.64, 0)$; accuracy of estimates will vary*

GG-40. Here's another example of how the quadratic formula can be used to find the x-intercepts of a parabola.

a. Graph the equation $y = 2x^2 - 4x + 1$ and estimate the x-intercepts of the parabola.   *vertex $(1, -1)$; x-intercepts $\approx (1.75, 0)$ and $(0.25, 0)$*

b. Use the quadratic formula to find the exact x-intercepts of the graph of $y = 2x^2 - 4x + 1$.   *$\left(1 + \dfrac{\sqrt{2}}{2}, 0\right)$ and $\left(1 - \dfrac{\sqrt{2}}{2}, 0\right)$*

c. Use a calculator to find approximate values, to the nearest hundredth, for the x-intercepts of the parabola. How close were your estimates?   *$(\approx 1.71, 0)$ and $(\approx 0.29, 0)$*

Problem GG-41 provides more practice with the area and perimeter of a circle in preparation for the Grazing Goat problem in Section 8.4. You may want to assign only part a or part b, as each one involves a fair amount of work.

GG-41. More Logo Contest Designs Copy each of the following logo designs on your paper and label the lengths you know. Then identify and solve subproblems to find the area and the perimeter of each design. Find exact values for the area and perimeter, and then find approximations to the nearest tenth.

a. Marco used a quarter circle, a rectangle, and a right triangle.

b. Elijeo used two right triangles of the same size, and a quarter circle.

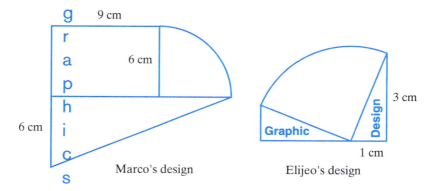

Marco's design

Elijeo's design

(a) For the area: find the second leg of the right triangle; the area is $6 \cdot 9 + \left(\frac{1}{2}\right)(6)(9+6) + \left(\frac{1}{4}\right)(\pi \cdot 36) = 99 + 9\pi \approx 127.3$ sq cm. For the perimeter: find the hypotenuse of the right triangle, and the circumference of a circle with radius 6 cm; the perimeter is $6 + 6 + 9 + \frac{2\pi(6)}{4} + \sqrt{6^2 + 15^2} = 21 + 3\pi + \sqrt{261} \approx 46.6$ cm.

(b) For the area: find the radius of the quarter circle; the area is $2\left(\frac{1}{2}\right)(1)(3) + \left(\frac{1}{4}\right)(\pi(\sqrt{10})^2) = 3 + 2.5\pi \approx 10.9$ sq cm. For the perimeter: find the radius of the quarter circle and the circumference of a circle with that radius; the perimeter is

$1 + 3 + 1 + 3 + \frac{2\pi\sqrt{10}}{4} = 8 + \frac{\pi\sqrt{10}}{2} \approx 13.0$ *cm.*

GG-42. Tony's driver's license was suspended after his last trip to Los Angeles, so he's making the 305-mile trip from Sacramento by plane. The flight takes only 55 minutes (as opposed to more than 5 hours by car).

a. How many seconds does it take the plane to fly 1 mile? *about 10.82 sec*

b. What is the airplane's speed, in terms of miles per hour? *about 332.7 mph*

GG-43. Simplify each of the following expressions:

a. $5x^2 + 3x - x^2 + 2x + 5 - x$ *$4x^2 + 4x + 5$*

b. $5(y^2 - 2y) + 2(y - 3)$ *$5y^2 - 8y - 6$*

c. $4(w + 2) + 3(w + 2) + 2(2 + w)$ *$9w + 18$*

d. $-3(x^2 + 2) + 2(x - 5) - 6(5 - x^2)$ *$3x^2 + 2x - 46$*

GG-44. Solve each of the following equations:

a. $2v + 1 = 7$ **3**

b. $2x^2 + 1 = 7$ **$\pm\sqrt{3} \approx \pm 1.73$**

c. $5(n - 1) + 3(n + 2) + 2[n + 3(n + 1)] = 25$ **$\frac{9}{8}$**

d. $\frac{100}{x^2} = 36$ **$\pm\frac{5}{3}$**

GG-45. Find the x-intercept, the y-intercept, and the slope for each of the following lines:

a. $2x - 3y = 7$ *x-intercept $(\frac{7}{2}, 0)$: y intercept: $(0, -\frac{7}{3})$, $m = \frac{2}{3}$*

b. $y - 4 = 0$ *no x-intercept; y-intercept: (0, 4); m = 0*

c. $2y = 3x$ *x-intercept: (0, 0); y-intercept: (0, 0); $m = \frac{3}{2}$*

d. $x + 44 = 0$ *x-intercept: (−44, 0); no y-intercept; undefined slope*

GG-46. A rectangular box has dimensions of 3 centimeters by 4 centimeters by 5 centimeters. It has eight vertices (corners). Find as many different distances as you can between pairs of vertices. How many different distances did you find?   *six different distances: 3, 4, 5, $\sqrt{34}$, $\sqrt{41}$, $\sqrt{50}$*

GG-47. Plot the points $A\,(5, 2)$ and $B\,(0, -1)$.

a. Describe a slide from point B to point A.   *Move up 3 and to the right 5.*

b. Write the coordinates of two other points on the line that contains A and B.   *(−5, −4), (10, 5), (15, 8), and others*

c. Find the length of the segment AB.   *$\sqrt{3^2 + 5^2} = \sqrt{34} \approx 5.83$*

GG-48. Rewrite each of the following quotients as a product, and then rewrite the resulting expression with a single exponent.

Example: $\frac{x^{-5}}{x^2} = x^{-5} \cdot x^{-2} = x^{-7}$ ▲

a. $\frac{x^{-4}}{x^{-1}}$ b. $\frac{m^{11}}{m^3}$ c. $\frac{n^{-8}}{n^5}$ d. $\frac{w^{-27}}{w^{-33}}$

$x^{-4} \cdot x^1 = x^{-3}$ $m^{11} \cdot m^{-3} = m^8$ $n^{-8} \cdot n^{-5} = n^{-13}$ $w^{-27} \cdot w^{33} = w^6$

GG-49. Rewrite each of the following products as a quotient, and then rewrite the resulting expression with a single exponent.

Example: $x^{-2} \cdot x^7 = \frac{x^7}{x^2} = x^5$ ▲

a. $x^{99} \cdot x^{-6}$ **b.** $b^{-14} \cdot b^{8}$ **c.** $t^{-12} \cdot t^{4}$ **d.** $r^{-7} \cdot r^{-1}$

$\dfrac{x^{99}}{x^{6}} = x^{93}$ $\dfrac{b^{8}}{b^{14}} = b^{-6}$ $\dfrac{t^{4}}{t^{12}} = t^{-8}$ $\dfrac{1}{r^{7} \cdot r^{1}} = r^{-8}$

GG-50. You've been working with integer exponents since the beginning of the course. You used the meaning of integer exponents and patterning to rewrite expressions such as $x^{47} \cdot x^{21}$, $(x^{103})^{6}$, $\dfrac{1}{x^{9}}$, and $\dfrac{x^{15}}{x^{8}}$ with single exponents. In problem WR-144 (in Chapter 5) you summarized the exponent patterns you noticed by writing five general rules for rewriting exponential expressions.

a. Which of the five rules for working with integer exponents did you use in problems GG-48 and GG-49?

b. What do the results in problems GG-48 and GG-49 suggest about the relationship between the rule for rewriting products of the form $x^{a} \cdot x^{b}$ and the rule for rewriting quotients of the form $\dfrac{x^{a}}{x^{b}}$?

GG-51. Graphing a Linear Inequality In Chapter 7 you graphed the solutions to inequalities such as $2x + 3 < 10$ on a number line. You can build on what you observed in problem GG-26 to graph more complicated inequalities that involve two variables, such as $y \geq 3x + 2$.

Example: To graph the inequality $y \geq 3x + 2$, first graph the line $y = 3x + 2$ as you did in GG-26. The line divides the coordinate plane into three parts:

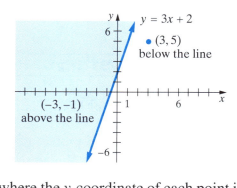

1. The *line itself*, where the y-coordinate of each point is *equal* to the value $3x + 2$,

2. a *region above the line*, where the y-coordinate of each point is *greater than* the value $3x + 2$, and

3. a *region below the line*, where the y-coordinate of each point it *less than* the value $3x + 2$.

In GG-26 you identified three points—(1, 7), (0, 3), and (−3, −1)—that lie above the line $y = 3x + 2$ and saw that each of them satisfies the inequality $y \geq 3x + 2$. To indicate that *all* points that lie on or above the line $y = 3x + 2$ satisfy the inequality $y \geq 3x + 2$, we lightly shade the region above the line. ▲

Graph the inequality $y \leq \frac{1}{2}x - 1$: first graph the line $y \leq \frac{1}{2}x - 1$, then check two points as was done in GG-26, and finally shade the appropriate region. *Shade the region on or below the line $y = \frac{1}{2}x - 1$.*

GG-52. When you graphed the inequality $x < -1$ in Chapter 7, you used a hollow dot to indicate the fact that x is less than, but not equal to, -1. We can use a similar idea to graph inequalities that involve two variables.

Example: Graph $y > -x + 4$:

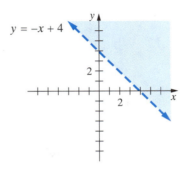

Rather than drawing a solid line for $y = -x + 4$, we can make a *dashed line* to indicate that points whose y-coordinates equal the value of $x - 2$ are *not* part of the solution. ▲

Graph the inequality $y < 1 - 5x$ by first graphing the line $y = 1 - 5x$, then checking two points as done in GG-26, and finally shading the appropriate region. *Shade the region below the line $y = 1 - 5x$.*

8.3 THE ELECTION POSTER: USING THE QUADRATIC FORMULA

Section Materials

A LARGE PICTURE SURROUNDED BY A BORDER (TO REPRESENT AN ELECTION POSTER, FOR GG-55)

Students will need to be patient and careful with their calculator work. They should work in pairs or groups to check each other's calculations.

The central problems of this section are more complicated than those you've seen in the past, and involve some rather "messy" equations. However, strategies you've used in the past—drawing diagrams and labeling dimensions and distances with variables—will help you write equations to model the situations. Then messy equations you get can be solved using the quadratic formula.

Before using a calculator to get an approximate value for a solution, try to estimate the answer. Your estimate will help you determine the reasonableness of your calculator results.

GG-53. Mario drew a right triangle with one leg 2 centimeters longer than the other. The hypotenuse of the triangle was 17 centimeters long.

a. Draw a diagram of a right triangle, and label the length of each side.

   Solution: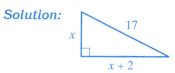

b. Write an equation to find the lengths of the legs. $x^2 + (x + 2)^2 = 17^2$

c. Solve the equation you wrote in part b to find the length of each leg of the triangle Mario drew, and then approximate each length to the nearest hundredth centimeter. $x = \dfrac{-2 \pm \sqrt{574}}{2}$; the legs are about 10.98 and 12.98 cm, respectively.

d. Although there are two solutions to the quadratic equation you wrote in part b, you found the lengths of the legs for only one triangle in part c. Why don't the two solutions lead to two different triangles?   *One of the solutions for x is negative, and a triangle can't have legs of negative lengths.*

GG-54. The Election Poster Malcolm is running for president of the student body association. He orders a poster 7 feet wide and 5 feet high. Irma, Malcolm's campaign manager, decides the poster looks out of proportion and estimates another 75 square feet of area will be needed to make it look just right. If each dimension of the poster is increased by the same amount, what will be the dimensions of the new poster?

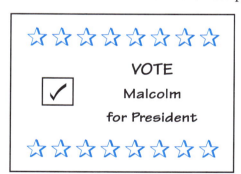

Solution: 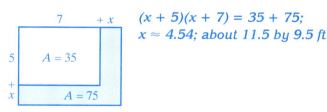 $(x + 5)(x + 7) = 35 + 75$; $x \approx 4.54$; about 11.5 by 9.5 ft

484 Chapter 8 The Grazing Goat: Dealing with Complicated Situations

We've found that it helps students draw an appropriate diagram for GG-55 if we show them a large picture surrounded by a border.

GG-55. Malcolm's opponent, Olivia, also enlarged her original campaign poster. She started with a 4-foot by 5-foot picture of herself. To enlarge the poster she surrounded the picture with a bright border, which had an area of 50 square feet. If adding the border to the picture increased the width of the original poster and the length of the original poster by the same amounts, what are the dimensions of the new poster?

Solution: 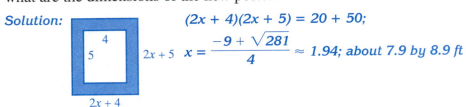 $(2x + 4)(2x + 5) = 20 + 50$; $x = \dfrac{-9 + \sqrt{281}}{4} \approx 1.94$; about 7.9 by 8.9 ft

GG-56. Debra drives due south from the I-5-Capitol Freeway interchange, and Gary, who leaves the same interchange at the same time, drives due east on Highway 50. Debra drives 10 miles per hour faster than Gary, and after 1 hour the two drivers are 108 miles apart (as the crow flies). How fast has each driver been going? Imagine that both highways are actually straight, and use a solution strategy similar to that of problem GG-53.

Solution: 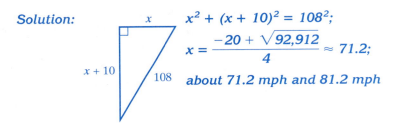 $x^2 + (x + 10)^2 = 108^2$; $x = \dfrac{-20 + \sqrt{92{,}912}}{4} \approx 71.2$; about 71.2 mph and 81.2 mph

In GG-57, students show algebraically that the area of the shaded region does not depend on the placement of the top vertex of the unshaded triangle.

GG-57. Look back at your results in GG-32.

a. What do you notice about the areas of the shaded regions? Does the area of the shaded region seem to depend upon where along the top edge of the rectangle the top vertex of the unshaded triangle is placed? *No, since all the areas are equal.*

b. Find the area of the shaded region in the figure shown here.

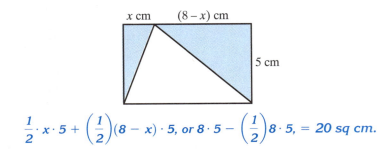

$\dfrac{1}{2} \cdot x \cdot 5 + \left(\dfrac{1}{2}\right)(8 - x) \cdot 5$, or $8 \cdot 5 - \left(\dfrac{1}{2}\right) 8 \cdot 5, = 20$ sq cm.

c. Write an algebraic inequality to show which values of *x* make sense for the figure in part b. *0 ≤ x ≤ 8*

d. What does your work in part b show about the effect of the placement of the top vertex of the unshaded triangle along the top edge of the rectangle? *The placement of the top vertex doesn't affect the area of the shaded region.*

EXTENSION AND PRACTICE

GG-58. A certain rectangle has area 50 square meters, and its length is five more than twice its width. Find the lengths of the sides of the rectangle.

a. Draw a diagram and label the length of each side. Think about whether you want to use *x* for the length or for the width.

Solution:

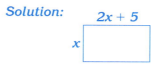

b. Write an equation for the area in terms of the width and length. *x(2x + 5) = 50*

c. How long is each side of the rectangle? *width = x ≈ 3.90 m, length = 2x + 5 ≈ 12.81 m*

Problem GG-59 provides another opportunity for students to work with subproblems in the context of the area and perimeter of a circle as preparation for the Grazing Goat problem in Section 8.4.

GG-59. Doug and Amy are building a circular fountain and tropical-fish pond in their backyard. They want to surround the pond with a 3-foot-wide brick walkway. Doug, who likes making up puzzles, has figured that the outer circumference of the walkway should be 10π feet long.

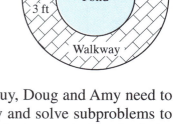

To figure how many bricks they'll need to buy, Doug and Amy need to know the area of the planned walkway. Identify and solve subproblems to find how many square feet of walkway surrounds the pond. Give both the exact value and an approximation to the nearest tenth of a square foot.

Find the radius R of the large circle; find the radius r of the small circle; and find the area of each circle. Circumference = $2\pi R = 10\pi$; R = 5 ft; r = R − 3 ft, so r = 2 ft; area of walkway = $5^2\pi - 2^2\pi = 21\pi \approx 66.0$ sq ft.

GG-60. Identify subproblems that will help you find the area and the perimeter of the following trapezoid. Then, using your subproblems, find the area and the perimeter.

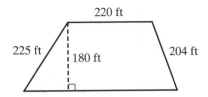

Two subproblems are finding the bases of two right triangles; perimeter = 1100 ft, area = 60,390 sq ft.

GG-61. Rewrite each of the following equations in standard form, and then try to use factoring to solve the equation. If factoring isn't easy, use the quadratic formula to solve the equation. Write the exact solutions. If a solution involves a square root, write its approximate value to the nearest hundredth.

a. $-3x = -x^2 + 14$

   $x = \dfrac{3 \pm \sqrt{65}}{2} \approx 5.53, -2.53$

b. $3x^2 + 4x = 0$

   $x = 0, -\dfrac{4}{3}$

c. $5 = 6x - x^2$

   $x = 1, 5$

d. $-4x^2 + 8x + 3 = 0$

   $x = \dfrac{8 \pm \sqrt{112}}{8} = \dfrac{2 \pm \sqrt{7}}{2} \approx 2.32, -0.32$

e. $0.09x^2 - 0.86x + 2 = 0$

   $x = \dfrac{50}{9}, 4$

f. $25x^2 - 49 = 0$

   $x = \pm\dfrac{7}{5}$

GG-62. To solve the quadratic equation $x^2 - 43x - 3198 = 0$, Larry used the quadratic formula and found that the solutions are $x = 82$ and $x = -39$. Moe says that because the solutions are integers, the polynomial $x^2 - 43x - 3198$ must be factorable.

a. Is Moe's claim true? If so, what are the factors of $x^2 - 43x - 3198$?
   yes; $x^2 - 43x - 3198 = (x - 82)(x + 39)$

b. Write a quadratic equation in factored form that has solutions $x = 4$ and $x = -7$. *$(x - 4)(x + 7) = 0$*

c. Write a quadratic equation in factored form that has solutions $x = 6.5$ and $x = 0$. *$x(x - 6.5) = 0$*

GG-63. Suppose a square of side length s has the same area as a circle of radius r. Write an equation that expresses r in terms of s. *$r = \sqrt{\dfrac{s^2}{\pi}} = \dfrac{s}{\sqrt{\pi}}$*

GG-64. Rewrite the algebraic fraction $\dfrac{x^2(x^2 - x - 2)}{x(x - 2)}$ in a simpler form by finding fractional forms of 1. List any restrictions on the values of x that may be used in the expression.   *x(x + 1); x ≠ 0, 2*

GG-65. Consider the two equations $y = 2x - 4$ and $y = -0.5x + 2$.

a. Use the slope and y-intercept to graph the equations on the same set of axes.

b. Use your graph from part a to estimate the solution to the system:
$$y = 2x - 4$$
$$y = -0.5x + 2. \quad \approx \text{ (2.5, 1)}$$

c. Use algebra to find where the lines $y = 2x - 4$ and $y = -0.5x + 2$ intersect.   *(2.4, 0.8)*

GG-66. Graph the line $y = 4x - 8$.

a. Find the area of the right triangle formed by the line you graphed, the x-axis, and the y-axis.   **8 sq units**

b. Find the perimeter of the triangle described in part a.   **18.25 units**

c. Write the ratio of the length of the long leg of the triangle to the length of the short leg. $\dfrac{8}{2}$ or $\dfrac{4}{1}$

GG-67. Find an equation of the line that ...

a. ... has slope $\tfrac{2}{3}$ and passes through the point $(-6, -1)$. $y = \tfrac{2}{3}x + 3$

b. ... passes through the points $(6, 3)$ and $(5, 5)$. $y = -2x + 15$

GG-68. Find all possible solutions to each of the following equations. Explain how you solved each equation.

a. $x^2 = 9$ **±3**

b. $x^2 = 7$ $\pm\sqrt{7}$

c. $(x - 5)^2 = 36$ **11, −1**

d. $(x + 3)^2 = 49$ **4, −10**

e. $x^2 + 8 = 51$
$\pm\sqrt{43} \approx \pm 6.56$

f. $(x - 4)^2 + 9 = 12$
$4 \pm \sqrt{3} \approx$ **5.73, 2.27**

GG-69. Consider the line segment from the point $(4, 3)$ to the point $(-2, 7)$.

a. How long is the segment? Write your answer in simplified square-root form and as a decimal approximation to the nearest hundredth.
$\sqrt{52} = 2\sqrt{13} \approx$ **7.21 units**

b. What is the slope of the line segment between the points (4, 3) and (−2, 7)? $\frac{-2}{3}$

GG-70. Graph each of the following inequalities on a number line:

a. $4(x - 3) + 6 < 18$ b. $-2(6x + 8) \leq -10$ c. $-3 \leq x - 5 < 9$

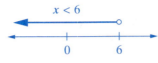
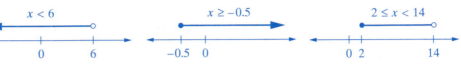

GG-71. The line $y = 4 - \frac{2}{3}x$ divides the coordinate plane into three parts: the line itself, the region described by the inequality $y < 4 - \frac{2}{3}x$, and the region described by the inequality $y > 4 - \frac{2}{3}x$.

a. Without graphing, decide in which part of the plane each of these points lies:

$$(0, 0), (-2, 1), (0, 5), (4, -1), (6, 0), (-3, 6), (-3, 10)$$

Organize your results by copying and completing the following table:

In the Region $y < 4 - \frac{2}{3}x$	On the Line $y = 4 - \frac{2}{3}x$	In the Region $y > 4 - \frac{2}{3}x$
(0, 0)	(6, 0)	(0, 5)
(−2, 1)	(−3, 6)	(−3, 10)
(4, −1)		

b. Which points in part a were the easiest to check? Explain why.   *The easiest points to check are those with an x- or y-coordinate of 0.*

GG-72. Graph the inequality $3x - 2y \leq 8$ by first rewriting the inequality in y-form, and then graphing the appropriate line. Check two points as in GG-51, and then shade the appropriate region.

$y \geq \frac{3}{2}x - 4$; the region is all the points that lie on or above the line
$y = \frac{3}{2}x - 4$.

8.4 USING SUBPROBLEMS IN COMPLICATED SITUATIONS

SECTION MATERIALS

12- TO 15-IN. STRING TO REPRESENT A GOAT TETHER (ONE FOR EACH GROUP, FOR GG-73)

The Grazing Goat Problem (GG-73) is an example of a complicated situation in which several problem-solving strategies are useful: making models, drawing diagrams, and identifying and solving subproblems.

To help students visualize the problem, you might introduce it by "tethering" yourself to a desk with a piece of string. We tape one end of a string to a corner of the desk, wad up the other end in one hand, and wander about the desk. As we lengthen the tether by releasing some of the wadded string, the students can see that the situation becomes more complicated.

One way for students to make their own models of the situation is for each group to use a piece of string as the tether and a textbook as the barn. If they place the string through the text and firmly hold down the cover, the free end of the string can be "walked" around the book to represent the grazing goat. Alternatively, you could have each group use a shoe box as a barn and tape a length of string to one corner. Encourage groups to experiment with various lengths of string (for example, a very short piece, another piece just longer than short side of the barn, and another piece just longer than the long side of the barn). For each length, suggest that the groups notice the farthest places the goat could reach.

Groups should work on parts a through d of GG-73, although part d could be assigned as homework.

GG-73. The Grazing Goat On weekends, the Williams tether their goat, Daisy, to a corner of their 15-meter by 25-meter barn. Since the barn sits in the middle of a large, grassy field, Daisy can spend her days grazing. The extent to which she can roam depends on which piece of rope the Williams have used to tie her.

a. Over what area of the field can Daisy graze if the rope is 10 meters long? Draw a diagram. Give the exact area and an approximation to the nearest hundredth square meter. *grazing area = $75\pi \approx 235.62$ sq m*

b. What if the rope is 20 meters long? Identify subproblems, and solve them to find the exact area of the field over which Daisy can graze. Also give the approximate area to the nearest hundredth square meter.

$300\pi + \frac{25}{4}\pi \approx 962.11$ sq m

c. Repeat part b for a rope that is 30 meters long.

$675\pi + \left(\frac{225}{4}\right)\pi + \left(\frac{25}{4}\right)\pi \approx 2316.92$ sq m

490 Chapter 8 The Grazing Goat: Dealing with Complicated Situations

d. Repeat part b for a rope that is 40 meters long.

$$1200\pi + \left(\frac{625}{4}\right)\pi + \left(\frac{225}{4}\right)\pi \approx 4437.50 \text{ sq m}$$

Diagrams for GG-73 Solutions:

a.
b.
c.

You might guide students in part d of GG-74, by suggesting that they make a table for ropes of lengths 1 m, 2 m, 3 m, and so on.

GG-74. Now suppose that Daisy's rope is x meters long.

a. Write an inequality to express "the rope is shorter than the width of the barn."   *(0 < x < 15) m*

b. In terms of the barn's dimensions, describe what the inequality $15 < x < 25$ represents.   *The rope is longer than the width of the barn and shorter that the length of the barn.*

c. Write an inequality to express "the rope is longer than the length of the barn and shorter than 40 meters."   *(25 < x < 40) m*

d. What is the area of the field over which Daisy can graze if $0 < x < 15$? Express your answer in terms of x and π. $A = \frac{3}{4}\pi x^2$ *sq m*

e. Repeat part d for $15 < x < 25$. $\frac{3}{4}\pi x^2 + \frac{1}{4}\pi(x-15)^2$ *sq m*

f. Repeat part d for $25 < x < 40$. $\frac{3}{4}\pi x^2 + \frac{1}{4}\pi(x-15)^2 + \frac{1}{4}\pi(x-25)^2$ *sq m*

Problem GG-75 may take some time. If groups think there is not enough information to solve the problem, remind them to list all the subproblems and notice the right angles. Some other hints are: break the figure into some rectangles, carefully calculate the width of the lawn (70 yd), and find the area in terms of square feet.

GG-75. The big back lawn at the Wildhorse Country Club is shaped as shown. The rounded pieces are each circular arcs of radius 35 yards. The straight edges meet at right angles. If two pounds of fertilizer are needed for every 100 square feet of grass, about how many pounds of fertilizer are needed for the entire lawn?

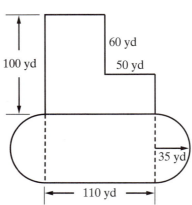

a. Identify a set of subproblems that you could use to solve the problem.

b. Solve each subproblem you listed. Note that a subproblem might have subproblems of its own.

c. Use your work in part b to find approximately how many pounds of fertilizer are needed for the entire lawn. *Area = 15,700 + 35²π ≈ 19,548.45 sq yd ≈ 175,936 sq ft; about 3518.72 lb fertilizer*

GG-76. Cole has only 30 meters of fence sections with which to build a temporary corral for a new horse. Before starting to put the fence sections together to build the corral, he wants to explore the possibilities on paper.

a. On your own, use graph paper to draw four different rectangles, each with whole-unit side lengths and a perimeter of 30 units.

b. Compare your results with what other members of your group drew. Try to come up with as many different rectangles as possible.

c. Record your group's data from parts a and b in a table with four columns: one for the perimeters of the rectangles, one for the widths, another for the lengths, and the fourth column for the areas of the rectangles.

Perimeter	Width	Length	Area
30	1	14	14
30	2	13	26
30	3	12	36
30	4	11	44
30	5	10	50
30	6	9	54
30	7	8	56

d. Graph your data from part c with the possible side lengths on the horizontal axis and area along the vertical axis. *The points lie on a concave-down parabola with vertex (7.5, 56.25) and x-intercepts (0, 0) and (30, 0).*

e. Use your graph from part d to estimate the dimensions of a rectangle with a perimeter of 30 meters and an area of 40 square meters.
about 3.5 by 11.5 m

f. To find the exact length and width of a rectangle whose perimeter is 30 meters and area is 40 square meters, let W represent the width, and then express the length in terms of W. Write an equation to represent the fact that the area is 40, and solve the equation. How close was your estimate in part e? $W(15 - W) = 40$; *width* $= \dfrac{15 - \sqrt{65}}{2} \approx 3.47$ m, *length* $= \dfrac{15 + \sqrt{65}}{2} \approx 11.53$ m.

EXTENSION AND PRACTICE

GG-77. For Katja's party, Lawson, Andrea, Mary, and Ryan are going to hang a piñata from a rope whose ends are attached to the tops of two poles. The poles are each 20 feet high, and they stand 15 feet apart. The piñata, which is 1 foot high, will hang halfway between the two poles. If there is 3 feet of clearance between the bottom of the piñata and the ground, how much rope is needed to hang the piñata? To get started, draw a diagram and identify any subproblems.

Solution:  $2\sqrt{16^2 + 7.5^2} \approx 35.3$ ft of rope

GG-78. Shekhar rides his bicycle to classes at the university each day, a distance of 8 miles. One morning he averaged 5 mph for the first mile. How fast will Shekhar have to ride for the remainder of the trip to average 12 mph over the whole 8 miles?

To answer the question, identify subproblems, define a variable, and then write and solve an equation.

$$\frac{8 \text{ mi}}{\left(\frac{1}{5} + \frac{7}{v}\right) \text{hr}} = 12; v = 15 \text{ mph}$$

GG-79. For each of the following figures, use subproblems to find the area of the shaded region:

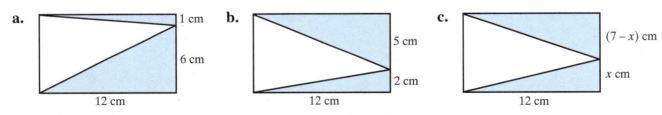

In each case the area is 42 sq cm.

GG-80. Look back at your results in GG-79.

a. What do you notice about the areas of the shaded regions? How is the area affected by the placement of the unshaded triangle's vertex on the right edge of the rectangle? How does your work in part c support your observations? *The areas of the shaded regions are the same; the vertex placement doesn't matter. Part c supports the idea since the vertex is placed x cm from the bottom.*

b. Write an algebraic inequality to show which values of x make sense for the figure in GG-79c. $0 \le x \le 7$

Section 8.4 Using Subproblems in Complicated Situations

GG-81. You can use subproblems to solve equations that involve square roots.

a. Solve the equation $\sqrt{x} = 4.2$ for x. What could you do to both sides of the equation to undo the square root? *square both sides; $x = 17.64$*

b. Solve the equation $\sqrt{3x-1} = 18$ for x. The first subproblem is to undo the square-root process. The second subproblem is to solve the resulting equation. Check your answer to see that it is correct by substituting your value for x in the original equation. *$x = 108\frac{1}{3}$*

GG-82. Rita's Revenge Rita decided to test her little brother Ivan at his own game, and challenged him to sketch six points on the graph of

$$y = \frac{x^2 - 3x}{x^2 - 4} \cdot \frac{(x-2)^2}{x^3 - 9x} \cdot (x+3).$$

a. Find the value of y when $x = 5$. *$y = \frac{720}{1680} \approx 0.429$*

b. Look for at least four subproblems that you could use to rewrite Rita's algebraic expression in a simpler form. List the subproblems, and then solve them to simplify the expression.

c. Finally, rewrite the expression in simpler form. *$y = \frac{x-2}{x+2}$*

d. Evaluate the simplified fraction from part c using $x = 5$. Compare the result to your result in part a—what do you notice? What restrictions should be placed on the values of x that can be used to evaluate the expression $\frac{x^2 - 3x}{x^2 - 4} \cdot \frac{(x-2)^2}{x^3 - 9x} \cdot (x+3)$? *$\frac{3}{7}$; it's equal to $\frac{720}{1680}$; $x \neq 0, -2, 2, -3,$ or 3*

e. Complete the following table of values for

$$y = \frac{x^2 - 3x}{x^2 - 4} \cdot \frac{(x-2)^2}{x^3 - 9x} \cdot (x+3).$$

Which expression for y does it make more sense to use,

$$\frac{x^2 - 3x}{x^2 - 4} \cdot \frac{(x-2)^2}{x^3 - 9x} \cdot (x+3) \text{ or } \frac{x-2}{x+2}?$$

Why did Rita leave $x = -3, 0, 2,$ and 3 out of the table? *Use the simpler expression; the original expression cannot be evaluated for $x = -3, -2, 0, 2,$ or 3, so those values shouldn't be used in the table.*

x	-5	-4	-2	-1	1	4	5
y	$\frac{7}{3} \approx 2.33$	3	error	-3	$-\frac{1}{3}$	$\frac{1}{3}$	$\frac{3}{7} \approx 0.429$

f. Now, complete Rita's Revenge Challenge Puzzle by using your table from part e to plot six points on the graph of

$$y = \frac{x^2 - 3x}{x^2 - 4} \cdot \frac{(x-2)^2}{x^3 - 9x} \cdot (x + 3).$$

GG-83. Write each of the following algebraic fractions as simply as possible. Include any restrictions on the values of the variable that may be used in the expression.

a. $\dfrac{(x-1)(x+2)(x-3)(x+4)}{(x+2)(x+3)(x+4)(x+5)}$ $\dfrac{(x-1)(x-3)}{(x+3)(x+5)}; x \neq -2, -3, -4, -5$

b. $\dfrac{(v^2-1)(v^2-2v-3)}{(v-1)^2(v-2)(v-3)}$ $\dfrac{(v+1)^2}{(v-1)(v-2)}; v \neq 1, 2, 3$

GG-84. In this problem you'll use graphing and algebra to examine the x-intercepts of a parabola.

a. Graph the rule $y = 2(x-1)^2 - 3$.   *vertex (1, −3); (≈ 2.22, 0), (≈ −0.22, 0)*

b. Estimate the points where the graph crosses the x-axis.   *(≈ 2, 0), $\left(\approx -\frac{1}{4}, 0\right)$*

c. Substitute 0 for y, and solve the resulting equation for the actual values of the x-intercepts. $x = \dfrac{2 \pm \sqrt{6}}{2} \approx 2.22, -0.22$

GG-85. Solve each of the following equations for both x and y:

a. $2x + 3y = 12$ $x = 6 - \frac{3}{2}y; y = 4 - \frac{2}{3}x$

b. $x^2 + 6y = 24$ $x = \pm\sqrt{24 - 6y}; y = \dfrac{24 - x^2}{6}$

c. $\dfrac{x}{4} - 2y = 10$ $x = 40 + 8y; y = \dfrac{x}{8} - 5$

d. $x^2 - y^2 = 25$ $x = \pm\sqrt{25 + y^2}; y = \pm\sqrt{x^2 - 25}$

GG-86. A right triangle has an area of 40 square centimeters, and its shortest side has a length of 8 centimeters. Find the length of the hypotenuse.
$\sqrt{164} = 2\sqrt{41} \approx 12.81$ cm

GG-87. The point (3, −7) is on a line with a slope of $\frac{2}{3}$. Find another point on the line.   *… (0, −9), (6, −5), …*

GG-88. On a piece of graph paper, set up a pair of coordinate axes and label the origin as point O. Starting at O, draw a segment with a slope of 3 and a length of at least 5 units. Label the second endpoint A.

Use segment OA to make a diagram of two similar right triangles to find a point P such that segment OP has a slope of 3 and a length of 3 units. What are the coordinates of P?   *If P is (x, y) and A is (2, 6), then $\frac{x}{2} = \frac{3}{2\sqrt{10}}$, so $x = \frac{3}{\sqrt{10}} \approx 0.95$; $P \approx (0.95, 2.85)$.*

GG-89. Use a number line to graph the solution to each of the following linear inequalities:

a. $1 - \frac{5}{9}x < \frac{1}{6}$

b. $-2(4x + 7) - 14 < -10$

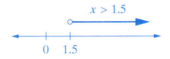

GG-90. Graph each of the following linear inequalities by first graphing the appropriate line, then checking points, and finally shading the appropriate region.

a. $y < -\frac{3}{2}x - 4$

b. $4x - y \leq 7$

GG-91. Graph all the points (x, y) that satisfy each of the following inequalities:

a. $y \geq 3x + 1$ b. $x \geq 0$ c. $y < 2$

8.5 QUIZ SCORES: DATA POINTS AND LINES OF BEST FIT

SECTION MATERIALS
CHAPTER SUMMARY
DRAFT RESOURCE
PAGE (OPTIONAL, FOR GG-112)

Many of the data graphs you've analyzed in this course—such as the graph for the Hawaiian Punch problem in Chapter 2 and the Leasing a Car problem in Chapter 5—have been discrete. To make it easier to use each of those graphs, you drew a trend line through the points and used it to model the situation.

In this section you'll use the idea of a trend line to examine situations where the data are discrete and seem to follow a linear trend, but don't necessarily all lie on a line. In these cases you'll look for a line that follows the trend of the data and that "best fits" the data points on a graph.

Students will use their graphs for GG-92 again in GG-95 and GG-97.

GG-92. Professor Speedi has just finished grading the two most recent quizzes for her algebra class. Each quiz was worth 25 points. At left are the scores for the first 20 students on her class roster.

Student Number	Quiz 11	Quiz 12
1	14	14
2	24	14
3	14	7
4	14	11
5	18	14
6	10	14
7	17	10
8	21	12
9	8	8
10	6	6
11	11	5
12	10	8
13	25	25
14	15	9
15	13	10
16	15	5
17	20	20
18	24	17
19	5	1
20	17	7

a. Examine all the scores, and then describe in a few sentences how the scores on Quiz 12 seem to compare to the scores on Quiz 11. *Quiz 12 scores are generally lower.*

b. Professor Speedi would like to know more specifically how the two quiz scores for each student compare, and in general, how the scores on the two quizzes compare for the class. She could do this by graphing the data. Set up the axes, with scaling and labels, so that the horizontal axis represents scores on Quiz 11 and the vertical axis represents scores on Quiz 12.

Suppose each student had scored exactly the same on the second quiz as on the first quiz. Using a red pen or pencil, plot a few points that represent this situation, and then draw a trend line through them. What is the slope of the trend line? *m = 1*

c. Most of the students did not score the same on both quizzes. For all the students on Professor Speedi's class roster, use a pencil (or blue pen) to plot the Quiz 11 and Quiz 12 scores on the coordinate grid. Describe the general trend or pattern of these scores. If you drew a line showing the trend, would the slope be greater than 1 or less than 1? *less than 1*

d. Suppose the students had generally done better on Quiz 12 than on Quiz 11. How would the graph then look? On a small grid, make a rough sketch. Would the slope of a trend line for this graph have a slope greater than 1, or less than 1? *greater than 1*

e. Save your graph from part c—You'll use it again as you continue your study of trend lines.

GG-93. An important skill needed to work with graphs is the ability to write an equation for a given line. Write an equation in the slope-intercept form for each of following the graphs. You may need to estimate to the nearest 0.5.

a.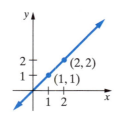

$y = x$

b.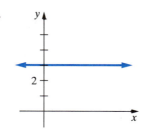

$y = 3$

c.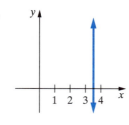

$x = 3.5$

d.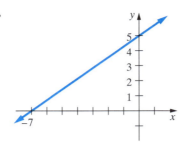

$y = \frac{5}{7}x + 5$

e.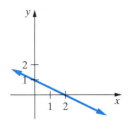

$y = -\frac{1}{2}x + 1$

f.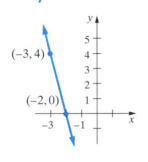

$y = -4x - 8$

g.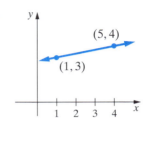

$y = \frac{1}{4}x + 2\frac{3}{4}$

GG-94. During World War II, the United States Navy tried to estimate how many German submarines were sunk each month. After the war, the Navy was able to get the actual numbers.

Month	Estimate	Actual Number of Sunken Subs
1	3	3
2	2	2
3	4	6
4	2	3
5	5	4
6	5	3
7	9	11
8	12	9
9	8	10
10	13	16
11	14	13
12	3	5
13	4	6
14	13	19
15	10	15
16	16	15

a. Make a plot of the estimate x and actual number y of sunken submarines as 16 ordered pairs (x, y). For example, the point associated with month 3 is (4, 6).

b. Suppose the estimated number of sunken submarines was exactly equal to the actual number of sunken subs. Sketch a line to represent this situation, and then write an equation for the line.   **$y = x$**

c. Did the Navy tend to overestimate or underestimate the number of submarines it sank?   **underestimate**

d. Is the line you drew in part b a good representation of the pattern (or trend) of the data?   **yes, fairly good**

GG-95. Look again at Professor Speedi's quiz scores and your graph from problem GG-92.

a. Draw a line on the graph that represents the pattern or trend of the scores. Your line should depict the trend of the data as a whole, but need not pass through any of the actual points. This line is called a **line of best fit**. Add this information to your Tool Kit.

b. Write an equation for the line you drew in part a.   **Answers will vary, but should be close to $y = \frac{4}{5}x - 3$**

c. Check your work with that of other groups.

d. Use your equation from part b to predict a student's score on Quiz 12 if he or she scored 16 on Quiz 11.   **≈ 10**

GG-96. Professor Speedi calculates the *arithmetic mean*—often called the "average"—of her students' scores by adding the scores and then dividing the sum by the number of scores. For example, the mean (average) of 12, 15, and 27 is $\frac{12 + 15 + 27}{3} = 18$.

a. Find the mean (average) of the scores for Professor Speedi's students for Quiz 11 in GG-92.   **15.05**

b. Repeat part a for the scores for Quiz 12.   **10.85**

GG-97. One way to draw a line of best fit is to plot the *mean data point* for a set of data, and then eyeball a line through it.

a. Write the two average quiz scores you found in problem GG-95 for Professor Speedi's class as an ordered pair, (mean for Quiz 11, mean for Quiz 12). This point is the **mean data point** for the set of quiz scores.   **(15.05, 10.85)**

b. On your graph from problem GG-92, plot the mean data point for the two quizzes. Does it fall on your line of best fit? If not, draw a new line of best fit through the mean data point and write an equation. Which of your two lines do you think best represents the data? *Answers will vary.*

EXTENSION AND PRACTICE

GG-98. The following table gives the length and mass of 10 laboratory mice:

Length (centimers)	Mass (grams)
21	43
17	38
22	47
16	32
11	19
13	27
15	26
20	40
16	34
15	30

a. Examine the data, and then set up coordinate axes, with "length" on the horizontal axis. Scale the axes carefully, and then plot the data as 10 ordered pairs (x, y) with x representing length.

b. Draw a line of best fit for the data.

c. Write an equation for your line of best fit. *The linear regression line is $y = 2.4x - 5.7$; sketching by hand and "eyeballing" will yield varying results; equations such as $y = 2.5x - 9$ approximate the data fairly well.*

d. Use your equation from part c to predict the mass of a mouse 25 centimeters long. *53 to 55 g*

GG-99. Julia drives 40 miles per hour faster than Francie. While Francie travels 150 miles, Julia travels 350 miles. Find each person's speed. $\frac{150}{F} = \frac{350}{F + 40}$; *Francie: 30 mph; Julia: 70 mph*

GG-100. A stick 250 centimeters long is cut into five pieces. The two longer pieces are each 14 centimeters longer than the three shorter pieces. Find how long each piece is by defining two variables and writing and solving a pair of equations. *2L + 3S = 250, L = 14 + S; S = 44.4 cm, L = 58.4 cm*

GG-101. Welded wire fencing is sold in rolls of 100 linear feet. Using exactly one roll of wire, Scott plans to build a rectangular rabbit pen with an area of 481 square feet.

Find the dimensions of the pen by drawing a diagram and writing and solving an equation (or a pair of equations). Define the variable(s) you use. *2W + 2L = 100 and W·L = 481, so W(50 − W) = 481; the pen is 13 by 37 ft.*

GG-102. If the 100-foot length of wire in the previous problem weighs 62 pounds, about how much will 240 feet of the same wire weigh? *about 148.8 lb*

GG-103. Consider the equation $\sqrt{9 - 4x} = 30$.

a. State two subproblems you could use to solve the equation for x. *Undo the square-root process, and then solve the resulting equation.*

b. Use your subproblems and solve the equation $\sqrt{9 - 4x} = 30$. Check your answer to see that it is correct by substituting your value for x in the original equation. *x = −222.75*

Problem GG-104 could be assigned as an enrichment problem, but it is not crucial to the development of other ideas of the course. You might assign it to all students, or as extra credit, or only to students who are ready for more problem-solving exercises.

GG-104. Professor Speedi challenged her student to find all solutions to the equation $x^2 = 2^x$.

a. Do you think the equation $x^2 = 2^x$ has any solutions? If so, how many values of x do you think make the equation $x^2 = 2^x$ true? What are some values of x that make the equation true?   *There are three solutions: 4, 2, and ≈ -0.7666647*

b. Check answers in part a by graphing $y = x^2$ and $y = 2^x$ for $-2 \le x \le 5$ and finding any points of intersection of the two graphs.

GG-105. Simplify each of the following algebraic fractions. List any restrictions on the values of x that may be used in the expression.

a. $\dfrac{x^2(x^2 - 3x + 2)}{x(x - 2)}$ *$x(x - 1); x \ne 0, 2$*

b. $\dfrac{3x^2 + 6x + 3}{x^2 + 3x + 2}$ *$\dfrac{3(x + 1)}{x + 2}; x \ne -1, -2$*

GG-106. Consider the polynomial $4x^2 + 12x + 9$.

a. Factor the polynomial by drawing a generic rectangle and labeling its dimensions and area(s).   *$(2x + 3)^2$*

b. The polynomial $4x^2 + 12x + 9$ is called a *perfect-square trinomial*. Explain why this name is appropriate.   *It can be written as the square of an expression.*

GG-107. Sketch each of the following pairs of points, and then use your diagram to find the distance between the points.

a. $(-4, 7)$ and $(29, 76)$ ≈ 76.5 b. $(2, 8)$ and $(2, -20)$ 28

GG-108. Use substitution to find x and y if $y = 2x - 7$ and $x^2 + xy = 100$.   *(7.057, 7.114), (−4.724, −16.447)*

GG-109. Graph each of the following linear inequalities:

a. $4x + 6y \le 12$ b. $\frac{4}{3}x + 2 \ge \frac{2}{5}y$

all points on or below the line $y = -\frac{2}{3}x + 2$ all points on or below the line $y = \frac{10}{3}x + 5$

GG-110. Although it may seem that it would be more complicated to graph the inequality $y \ge x^2 - 1$ than it would be to graph a linear inequality, the process begins the same way: first graph the equation that determines the curve dividing the coordinate plane. Then test points to see which region contains the points that satisfy the inequality.

Graph the inequality $y \ge x^2 - 1$ by first making a table of values and graphing the parabola $y = x^2 - 1$.   *The points lie on or above the parabola.*

GG-111. Graph the inequality $y < 9 - x^2$. Remember to use a dashed line when graphing the parabola $y = 9 - x^2$. *The points lie below the parabola.*

GG-112. Chapter 8 Summary Rough Draft Some of the problems you have done in Chapter 8 focused on developing the main ideas of the chapter and others reinforced concepts and skills introduced in previous chapters. Reread the chapter's introduction, and then look back through the chapter to find where each main idea was developed.

  a. Make photocopies of the Chapter Summary Draft Resource Page (or follow the same format on your own paper). Use a page to complete the following steps for each of the main ideas of the chapter:

  - State the main idea.
  - Select and recopy a problem that is a good example of the idea and in which you did well.
  - Include a completely worked-out solution to the selected problem.
  - Write one or two sentences that describe what you learned about the idea.

  As in previous chapters, you will have the opportunity to revise your work, so your focus should be on the content of your summary rather than on its appearance. Be ready to discuss your responses with your group at the next class meeting.

  b. What did you learn in this chapter that extended or strengthened your understanding of solving quadratic equations? Select one or two problems that illustrate what you now can do. Write complete sentences to describe how you did each of the selected problem(s). Then tell why you chose the problem(s) that you did.

  c. What did you learn in this chapter about solving complicated problems? Write a note to one of your former mathematics teachers in which you describe ways you can get started on a problem that seems difficult at first. As an example, choose a problem in the chapter that seemed very hard but that you found you could solve, and explain what you did to solve it.

  d. Find a problem that you would find difficult to solve on a test. Write out the question and as much of the solution as you can, and then explain what you think is keeping you from solving the problem. Be clear and precise in describing the difficult part.

8.6 THE COOKIE CUTTER AND THE LUNCH BUNCH: TWO INVESTIGATIONS (OPTIONAL)

Because this section is optional, homework problems are not included here. If you wish to assign homework, however, you can choose from the problems in Section 8.7 or assign ones you've skipped in previous sections.

The investigations in this section provide opportunities for you to apply the mathematical skills you've developed in this course to solve problems in two complicated situations, the Cookie Cutter and the Lunch Bunch investigations.

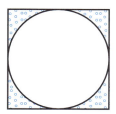

GG-113. Marissa is making a quilt from various sizes of cotton fabric squares. Out of each square she is cutting the largest circle she can.

a. If the square has sides of length 2 centimeters, how much of the square is left over after the circle has been cut out? *(Hint: What is the radius of the circle?) The area of the leftover fabric is $(4 - \pi)$ sq cm.*

b. How much of the square is left over if the length of each side is 10 centimeters? *$(100 - 25\pi)$ sq cm*

c. What *fraction* of each square is wasted in parts a and b?
   (a) $\frac{4 - \pi}{4} = 1 - \frac{\pi}{4}$; (b) $\frac{100 - 25\pi}{100} = 1 - \frac{\pi}{4}$

d. Rewrite each fraction in part c as a percent. *≈21.5%*

GG-114. The Cookie Cutter A cookie baker has a machine that rolls out a sheet of delicate butter cookie dough 12 inches wide and $\frac{3}{8}$ inch thick. His cookie cutter cuts 3-inch-diameter circular cookies, as shown in the figure below.

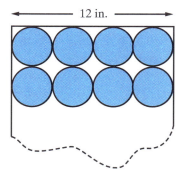

The baker's supervisor has been complaining that too much dough has to be rerolled each time. He wants the baker to use another diameter for the cookie cutter so that a smaller fraction of dough will have to be rerolled.

a. What fraction of the dough would have to be rerolled if the baker used a 4-inch-diameter cookie cutter? *$\approx \frac{3.43365}{16} \approx 21.5\%$*

b. Repeat part a for 6-inch-, 2-inch-, and 1-inch-diameter cookie cutters. *same*

c. What fraction of the dough had to be rerolled when the baker used a 3-inch diameter cookie cutter? *same*

d. What should the baker report to the supervisor? *Any circular cutter whose diameter divides 12 will leave the same fraction of dough to be rerolled.*

GG-115. Imagine that the circle is removed from the shaded square shown here.

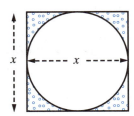

a. In terms of x, what fraction of the shaded square remains after the circle has been removed? $$\dfrac{x^2 - \pi\left(\dfrac{x}{2}\right)^2}{x^2}$$

b. Simplify the expression you wrote for part a. If possible, use the simplified expression to find what percentage of the square remains after the circle has been removed. $1 - \dfrac{\pi}{4} \approx 21.5\%$

GG-116. Write a short paragraph to summarize what you learned in the Cookie Cutter investigation: When a circle *of largest diameter* is removed from a square, what is the ratio of the area that remains to the area of the original square? Does the size of the square affect the ratio? Explain how you know. *As shown in GG-115, the ratio, $1 - \dfrac{\pi}{4}$, does not depend on the size of the square.*

The Lunch Bunch problems are challenging but accessible if students work through them carefully. These problems demonstrate the power of algebra in relation to graphing.

GG-117. The Lunch Bunch Some of Professor Speedi's algebra students get together for lunch every Tuesday and Thursday to study and do homework together. One Tuesday, Michele asked, "Can you find a pair of numbers whose product equals their sum?" Earol replied that it was easy to do since zero times zero equals zero plus zero. So Michele said, "Think of two positive numbers whose product equals their sum." Sam thought of this: $2 \cdot 2 = 2 + 2$. But Michele had another pair of numbers in mind, 1.5 and 3. Then Earol said he'd found two more pairs, 1.25 and 5, and 1.1 and 11.

a. Check to see whether the product of 1.5 and 3 equals their sum. What about the pair 1.25 and 5? the pair 1.1 and 11? *All three pairs check: $1.5(3) = 1.5 + 3$, $1.25(5) = 1.25 + 5$, and $1.1(11) = 1.1 + 11$.*

b. Do you think there's a pattern to the pairs of numbers whose product equals their sum? In this investigation (problems GG-117 through GG-122) you'll use your graphing and equation-solving skills to solve the problem posed by Michele.

Section 8.6 The Cookie Cutter and the Lunch Bunch: Two Investigations (Optional) **505**

First, try your hand at guessing: copy the table onto your paper, and fill in what you know. Then choose three more pairs of numbers, and check to see if their product and sum are equal. Finally, go on to GG-118 through GG-122 to solve the problem without guessing.

One Number	The Other Number	Product of the Two Numbers	Sum of the Two Numbers
0	0	0	0
2	2	4	4
1.5	3	4.5	4.5
1.25	5	6.25	6.25
1.1	11	12.1	12.1

Solutions to Michele's problem lie on the intersection of two generalized families of graphs, $x + y = k$ and $xy = k$, and form a curve, $y = \dfrac{x}{x-1}$. They are positive pairs of the form x and $\dfrac{x}{x-1}$.

GG-118. Another Solution? The Lunch Bunch wants to find out if there are any pairs of positive numbers whose sum is 6 and whose product is 6. When they get tired of guessing and checking, Sam suggests they try graphing.

a. Graph the line $x + y = 6$.

b. On the same coordinate axes, graph $xy = 6$.

c. Estimate where the two graphs meet, and write the approximate coordinates of the point(s).   *From the graph, at about (1.3, 4.5) and (4.5, 1.3)*

d. Explain in one or two sentences how graphing can be used to find two numbers that have a particular sum and product.   *After choosing a target product/sum P, write two equations: $x + y = P$ and $xy = P$. Graph both equations and find the points of intersection.*

GG-119. The Investigation Continues Michele wonders if there are two positive numbers whose sum is her favorite number and whose product is also her favorite number.

a. Choose a variable to represent Michele's favorite number. Then, using that variable, write two equations similar to those in GG-118 a and b to represent the Lunch Bunch problem.   *Let M represent Michele's favorite number. Then the problem is to find x and y such that $xy = M$ and $x + y = M$.*

b. Use substitution and the two equations you found in part a to write a single equation that does not include Michele's favorite number.   *$xy = x + y$ for $x, y > 0$*

c. Rewrite the equation from part (b) in *y*-form. $y = \dfrac{x}{x-1}$

GG-120. If you were to graph all the pairs of numbers that solve the Lunch Bunch problem, do you think the graph would be a straight line? Sam thought so, but Michele and Earol thought they needed more points to check.

When they organized their data into a table, they noticed that they could get additional points for the graph by switching the values of x and y. For example, one pair that works for the problem is $x = 5$ and $y = 1.25$. But the pair $x = 1.25$ and $y = 5$ also works. So once you've found a solution to the problem, you've found two pairs for the data table.

a. Substituting $x = 4$ into the equation from GG-119b, Michele got

$$xy = x + y$$
$$4y = 4 + y.$$

Solve the system for y. $\quad y = \dfrac{4}{3}$ or ≈ 1.33

b. Copy and complete the following tables. To fill them in, use *all* pairs of numbers that the Lunch Bunch has already found plus numbers based on all of their ideas from the discussion above. Also use some ideas of your own!

x	1	1.1	1.125	1.16	1.25	1.33	1.5	2	3	4	5
y	none	11	9	7	5	4	3	2	1.5	1.33	1.25
Fractional Form of y	none	$\dfrac{11}{1}$	$\dfrac{9}{1}$	$\dfrac{7}{1}$	$\dfrac{5}{1}$	$\dfrac{4}{1}$	$\dfrac{3}{1}$	$\dfrac{2}{1}$	$\dfrac{3}{2}$	$\dfrac{4}{3}$	$\dfrac{5}{4}$

x	6	7	9	10	11
y	1.2	1.166	1.125	1.11	1.1
Fractional Form of y	$\dfrac{6}{5}$	$\dfrac{7}{6}$	$\dfrac{9}{8}$	$\dfrac{10}{9}$	$\dfrac{11}{10}$

c. Now use the tables to draw a graph. What happened when you replaced x with the value 1? What would y equal if x were replaced by 1.01? *You get an "error" since $0 \neq 1$; 101*

d. If you were to graph the equation you wrote in GG-119c, how would it compare to the graph in part b above of the data? Explain how you know. *The two graphs would be the same because both arise from the equation $xy = x + y$.*

We suggest assigning problem GG-121 for extra credit.

GG-121. Do you think it's possible to find any negative numbers that would work in the Lunch Bunch problem? Could there be more to the graph in GG-120 than the Lunch Bunch thought? If possible, make a table of values for $x < 1$, and use the values to complete the graph of the situation $xy = x + y$. *Yes, negative values also work to give the other branch of the hyperbola $y = \dfrac{x}{x - 1}$. The graph for $x < 1$ is a reflection about the point $(1, 1)$ of the graph for $x > 1$.*

GG-122. Write a short paragraph to summarize what you learned in the Lunch Bunch investigation about word problems, graphs, and systems of equations.

8.7 SUMMARY AND REVIEW

SECTION MATERIALS

LARGE PICTURE SURROUNDED BY A BORDER (TO REPRESENT AN ELECTION POSTER, FOR GG-124)

GG-131 THE 100-METER FREESTYLE RESOURCE PAGE

There are lots of choices in this section for homework practice. Choose those problems you think will help your students consolidate ideas and/or see how much they've learned in the course.

GG-123. Chapter 8 Summary Group Discussion Take out the rough-draft summary you completed in GG-112. Use class time now to discuss your work, and use homework time to revise it.

For each of the main ideas of the chapter, choose one member of the group to lead a short discussion. When you are the discussion leader, you should:

- Explain the problem you chose to illustrate your main idea.
- Explain why you chose that particular problem.
- Explain, as far as possible, the problem from GG-112d you still cannot solve.

This is also your chance to make sure your summary is complete and to work together on problems you may not be able to solve yet.

Problem GG-124 is challenging. You may need to remind groups that the units (inches versus feet) must correspond. Use a poster to show a picture surrounded by a border. Students might want to compare the poster sizes from problems GG-54 and GG-55.

GG-124. Jody, a final entry into the race for student body president, used an enlargement of a 3-inch by 5-inch photograph for her campaign poster. She surrounded the enlarged photo with a 2-foot border. Including the border, the area of her poster was three times the area of the enlarged photograph.

To find the dimensions of Jody's campaign poster, draw a diagram, identify subproblems, define a variable, and write and solve an equation.

If x is the enlargement ratio, then $(3x + 48)(5x + 48) = 3(3x)(5x)$; $x = \dfrac{32 + 8\sqrt{46}}{5} \approx 17.25$; the poster is about 99.76 by 134.26 in.

VOTE FOR JODY!

In GG-125 students use a scatter plot of data similar to that in the Predicting Shoe Size problem in Chapter 1. The work students do with it here should help them see how much they've learned in the course.

GG-125. Lou's Shoes The goal in the Predicting Shoe Size problem in Chapter 1 was to decide whether or not height was a good predictor of shoe size. You and your classmates graphed your shoe sizes versus your heights on coordinate axes and used the data you gathered to form conclusions.

As Professor Speedi cleared off her desk at the end of the term, she found the data her class had compiled from their Predicting Shoe Size graph:

Student	Height (inches)	Shoe Size*	Student	Height (inches)	Shoe Size*
Rob	70	$11\frac{1}{2}$	Lawson	76	14
Max	66	9	Malcolm	69	12
Jim Bob	68	9	José	65	$8\frac{1}{2}$
Jamilla	62	7	Alicia	58	5
Karen	62	5	Manuel	70	$10\frac{1}{2}$
Christina	68	9	Henry	75	14
Tasha	60	4	Sam	67	10
Arturo	61	6	Marissa	60	$6\frac{1}{2}$
Jen	67	$7\frac{1}{2}$	Earol	62	6
Kelley	65	7	Lim	71	11

*Sizes were adjusted for the differences between women's and men's shoe sizing.

a. Plot the data for Professor Speedi's class. Use height as the independent (x) variable.

b. Use the mean data point to find a line of best fit for the data.

c. Write an equation for the line you drew in part b. $y = \frac{7}{12}x - 30$

d. Lou, who is 6 feet 7 inches tall, was absent the day the data were compiled. Predict his shoe size.   **about 16**

GG-126. Use the mean-data-point approach to write an equation for a line of best fit for the mouse length-and-mass problem, GG-98. Show all your work.   *(16.6, 33.6); the line may be slightly different for some students.*

GG-127. Is it always possible to fit a line to a set of data points? Sketch a graph for which you think it would be impossible to fit a line with any accuracy.   *No, for example, imagine a set of points that lie on or near the edge of a circle, or a parabola.*

GG-128. Is it possible to fit a *curve* to data? Sketch a graph for which you think this would be possible. Give an example of an equation for such a curve.   *It depends on the data. One example is $y = 2^x$ in the dollar bill folding problem in Chapter 1.*

GG-129. Tool Kit Check By now your Tool Kits should include new entries for at least the following topics:

- the quadratic formula   *GG-6*
- using the quadratic formula to solve a quadratic equation   *GG-7*
- regions in a coordinate plane   *GG-26*
- the line of best fit for a set of data points   *GG-95*

a. Check that your Tool Kit entries are up-to-date. Exchange Tool Kits with members of your group, and read each other's entries to check for clarity and accuracy.

b. At home, make sure that your Tool Kit includes examples and explanations that help you remember and understand the ideas. Make any revisions, clarifications, or updates that will reflect your developing understanding of mathematics.

EXTENSION AND PRACTICE

GG-130 Professor Speedi's class measured some can lids, paper tubes, rims of bowls, and other circular items to try to determine a relationship between the circumference of a circle and its diameter. Their data are recorded in the following table:

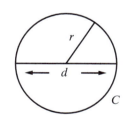

Diameter (in centimeters)	Circumference (in centimeters)
3	10
5	16
10.8	32.8
13	40
10	32.3
6.8	21
4.5	18

a. Plot the data, find the mean data point, fit a line, and write an equation for the line.   *approximately y = 3x*

b. Use what you know about circles, along with the fact that a circle of radius r units has a circumference of length $2\pi r$ units (that is, $C = 2\pi r$), to write an equation for the relationship between the circumference and the diameter of a circle of radius r.   *For diameter d, C = πd.*

c. Compare your equation for the line of best fit from part a to the equation you wrote in part b to relate the circumference and diameter of a circle. How closely does the line of best fit model the actual relationship between the circumference and diameter of a circle?   *The line y = 3x is fairly close to the line y = πx.*

d. Other than measuring more accurately, what else could Professor Speedi's class do to get a more precise description for the circumference-diameter relationship from their data?   *They could use more measurements.*

GG-131. **The 100-Meter Freestyle** The data points on the following graph represent winning times in the Olympic 100-meter free-style swimming race for all Olympic years from 1912 to 1988.

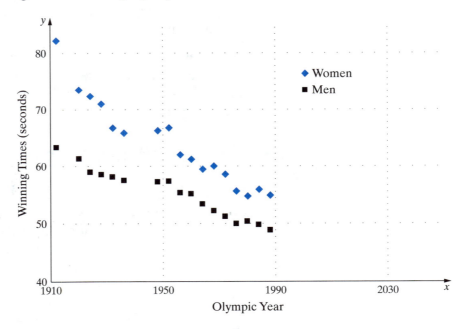

a. Identify two points on the graph and write a sentence explaining what each point represents.

b. On your Resource Page for GG-131 ("The 100-Meter Freestyle"), draw a trend line to show how the winning times for women have been decreasing through the years.

c. Repeat part b for men's winning times.

d. According to the trend lines you drew, which is decreasing faster, the women's or the men's winning times? Explain. Why might one group's winning times decrease faster than the other?   *The women's times are decreasing faster, perhaps as a reflection of improved opportunities and training.*

e. Draw a slope triangle on each trend line to find the average rate of decrease in the winning times for women and for men.

f. If the trends shown on the graph continue, in what year will men and women have the same winning times? Explain in one or two sentences how you determined your answer.   *In 2030—extend the trend lines: the x-coordinate of their point of intersection is the year the times are the same.*

g. What would that winning time be?   *43 sec*

h. Are straight lines always a useful way to estimate a trend? Do you think so in this case? Explain your answer.   *No; assuming data are linear can lead to faulty conclusions.*

GG-132. A rectangular garden has a perimeter of 100 meters and a diagonal of 40 meters. Find its dimensions by first drawing a diagram and then writing and solving an equation (or a pair of equations). Define the variable(s) you use.   *$2W + 2L = 100$ and $W^2 + L^2 = 40^2$, so $W^2 + (50 - W)^2 = 40^2$; about 11.77 by about 38.23 m*

GG-133. The following diagram represents the track around a football field. The total distance around the track for lane 1 is 440 yards. The curved ends of the track are semicircles, each straightaway is 100 yards long, and the lanes have staggered starts.

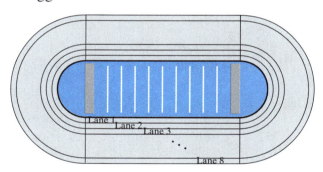

a. If each running lane is 1 yard wide, what distance would a runner on the inside edge of lane 8 cover when going completely around the track, and starting and stopping at the same place?   *$440 + 14\pi \approx 484$ yd*

b. How far ahead of the starting line for lane 1 should the starting line for the runner in lane 8 be so that both runners will run the same distance, 440 yards?   *$14\pi \approx 44$ yd*

GG-134. Solve each of the following four equations:

a. $x^2 - 2x - 4 = 0$ $\dfrac{2 \pm \sqrt{20}}{2} \approx -1.24, 3.24$

b. $x^2 + 6x + 4 = 0$ $x = \dfrac{-6 \pm \sqrt{20}}{2} \approx -5.24, -0.76$

c. $3x^2 + 2x - 7 = 0$ $x = \dfrac{-2 \pm \sqrt{88}}{6} \approx -1.90, 1.23$

d. $2x^2 - 7x - 4 = 0$ $-\dfrac{1}{2}, 4$

GG-135. The two solutions to $x^2 - 6x - 8 = 0$ are $3 + \sqrt{17}$ and $3 - \sqrt{17}$. Using your calculator, substitute each solution into the equation to show that it makes the equation true.

GG-136. Consider the equation $2x^2 - 7x + 3 = 0$.

a. Solve the equation by factoring. $(2x - 1)(x - 3) = 0; x = \dfrac{1}{2}, 3$

b. Solve the equation $2x^2 - 7x + 3 = 0$ by using the quadratic formula.

c. If you had done part b first, you could have used the solution to factor the polynomial $2x^2 - 7x + 3$. Explain in one or two sentences why this is so.   *The solutions found using the quadratic formula make the factored polynomial equal zero.*

GG-137. In the previous problem you discussed how to use the solutions obtained from the quadratic formula to get factors of a polynomial. Use this idea to factor each of the following polynomials:

a. $6x^2 - 7x - 10 = 0$ $(x - 2)(6x + 5)$

b. $9x^2 + 11x + 2 = 0$ $(x + 1)(9x + 2)$

GG-138. Find where the parabolas $y = 2x^2 - 7$ and $y = (x - 4)^2 - 9$ intersect.   *at about $(-9.48, 172.64)$ and $(1.48, -2.64)$*

GG-139. Solve the equation $\sqrt{4x - 3} = x$ for x. The first subproblem is to undo the square-root process. The second subproblem is to solve the resulting equation. Check your answer to see that it is correct. $x = 3, 1$

GG-140. Simplify each of the following algebraic fractions. Include any restrictions on the values of the variable that may be used in the expression.

a. $\dfrac{12x^8(x - 2)^2}{6x(x - 4)^5}$ $\dfrac{2x^7(x - 2)^2}{(x - 4)^5}$ for $x \neq 0, 4$

b. $\dfrac{m^2 - 12m + 27}{m^2 - 9m}$ $\dfrac{m - 3}{m}$ for $m \neq 0, 9$

c. $\dfrac{(4n^3)^2}{2n^4}$ $8n^2$ for $n \neq 0$

d. $\dfrac{2x^2 - 7x - 15}{x^2 - 25}$ $\dfrac{2x + 3}{x + 5}$ for $x \neq 5, -5$

GG-141. Find x: $x = 3$

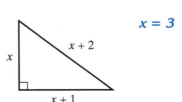

GG-142. One side of a rectangle is 5 centimeters longer than its short side, and its diagonal is 5 centimeters longer than its long side. Find the rectangle's dimensions by drawing a diagram and writing and solving an equation (or a pair of equations). Define the variable(s) you use. *If s represents the length of the short side, then $s^2 + (s + 5)^2 = (s + 10)^2$; 15 by 20 cm*

GG-143. A line parallel to $2x + 3y = 4$ goes through the point $(3, 8)$. Find an equation of the line. *(Hint: Draw a diagram. What is the slope of the two lines?) $y = -\frac{2}{3}x + 10$*

GG-144. Let A be the point $(-5, 2)$ and B be the point $(0, -2)$.

a. Find the distance $|AB|$. *$\sqrt{41} \approx 6.4$*

b. Find the slope of the line through A and B. *$\frac{-4}{5}$*

c. Find the point halfway between A and B (that is, the midpoint of the segment AB). *$(-2.5, 0)$*

d. Write an equation of the line through A and B. *$y = -\frac{4}{5}x - 2$*

GG-145. Solve the following system of equations:

$$4x + 2y = -13 + x + 3y \quad (-3, 4)$$
$$2(x + y) = 5 + x$$

GG-146. Graph the equation $y = \frac{1}{2}x^2 - 6$ for $-5 \leq x \leq 5$. *vertex $(0, -6)$; $(\pm 4, 2)$*

GG-147. Factor each of the following expressions as completely as possible:

a. $x^2 + 4xy + 4y^2$ *$(x + 2y)^2$* b. $x^2 + 2y^2 + 3xy$ *$(x + 2y)(x + y)$*

c. $x^3 + 4x^2 + 4x$ *$x(x + 2)^2$* d. $6x^2 + 12 + 18x$ *$6(x + 2)(x + 1)$*

GG-148. Graph all the points (x, y) with the following restrictions:

a. $x > -2$ b. $y \leq 0$ c. $x - 2y \leq 6$ d. $x < 4y$

(a) *all points to the right of the line $x = -2$;*

(b) *all points on or below the x-axis;*

(c) *all points on or above the line $y = \frac{1}{2}x - 3$;*

(d) *all points above the line $y = \frac{1}{4}x$*

GG-149. Graph each of the following linear inequalities by first graphing the appropriate line, then checking points, and finally shading the appropriate region.

a. $y \leq -\frac{3}{2}x - 4$
b. $2x < y - 7$

(a) all points on or below the line $y \leq -\frac{3}{2}x - 4$;

(b) all points above the line $y = 2x + 7$

GG-150. Is there a point (x, y) that makes the equations $2x - y = -3$ and $4y - 8 = -x$ both true? If so, find it; if not, explain why not. *yes; $(\frac{-4}{9}, \frac{19}{9})$*

GG-151. Find three points (x, y) that make both of these inequalities true:

$$2x - y > -3 \quad \text{and} \quad 4y - 8 \geq -x$$

Some possible points: (0, 2), (4, 1), and (3, 5)

GG-152. Is there a point (x, y) that does not make either of the inequalities $2x - y > -3$ or $4y - 8 \geq -x$ true? If so, find it; if not, explain why not.
Some possible points: (−2, 0), (−3, 1), (−8, 3)

GG-153. Chapter 8 Summary Revision Use the ideas your group discussed in GG-123 to revise your Chapter 8 Summary. Your presentation should be thorough and organized, and should be done on paper separate from your other work.

GG-154. Course Summary Update Congratulations, your hard work has paid off and you've learned a lot in this course! Now is a good time to reflect on all you've learned and try to pull it all together. Read through the Course Summary you wrote at the end of Chapter 5, and then revise it to include important ideas you've learned in the last three chapters. Include copies of homework problems that illustrate those important ideas.

GG-155. Write a note to a student who will be taking this course next term. In your note explain several important mathematical ideas you now understand, or skills you've acquired, through your work in the course. Share some advice on how to be a successful student and make the most of learning about data, equations, and graphs. Also, describe a problem that you liked best or most enjoyed solving in the course, and explain why you chose the problem.

APPENDIX A

DEBITS AND CREDITS
An Introduction to Integer Tiles

The use of tiles and sketches to review integer operations provides justification for previously "learned" rules. It also helps prepare students for using manipulatives to solve equations and factor geometrically. The tiles can be made either from the Integer Tiles Resource Page or from $\frac{3}{8}$-in plastic tile spacers ("+" shaped) with opposing "arms" cut off to form negatives (−), or purchased—small squares from a manufactured algebra tiles set and ceramic tiles both work. The critical requirement is that the tiles have two distinct sides (or "+" and "−" shapes) to represent "positive" and "negative."

Modeling with tiles seems to work best when the students work in **pairs** (groups of four are unwieldy and tend to be unproductive in this activity). Each pair of students needs one set of tiles and an Integer Tiles Balance Sheet Resource Page. When first using manipulative materials, establish your class guidelines for their distribution and use, and allow students some exploration or "play" time. Make sure everyone knows which side of the tiles represents a positive value and which represents a negative one.

We have found that it is very helpful to practice these activities ourselves before trying to demonstrate the use of the tiles to our classes. We introduce the tiles in a familiar context, an account balance sheet, where the negative tiles represent debits and the positive tiles represent credits.

The first concept to consider with your class is that a given integer can be represented with the tiles in many different ways. This is especially important with respect to representing a zero balance with tiles because effective modeling of subtraction of signed numbers requires that 0 be represented with a sufficient number of tiles.

MATERIALS

- Tiles for integer work

Note: Either have students make integer tiles using the Integer Tiles Resource Page or use the small squares from a purchased Algebra Tiles set. Students will use algebra tiles in Chapter 3 to model solving equations and to show the distributive property, and in Chapter 4 to factor quadratics. Algebra tiles are manufactured by Cuisenaire Company of America, Inc., and may be ordered by phoning 1-800-273-0338. Algebra tiles for the overhead projector are also available.

- Integer Tiles Balance Sheet Resource Page
- Integer tiles for the overhead projector
- Calculators

A-1. Representing Numbers with Integer Tiles You've worked with integers before and have probably memorized some "rules" for computing with positive and negative values. Although you may already be able to get the answers, your understanding of the symbolic operations can be reinforced by using pictures or models.

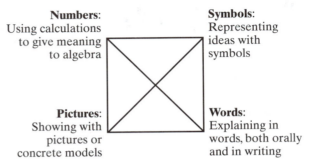

Integer tiles can be used to represent, or *model*, addition, subtraction, and multiplication with integers. Using the tiles may help you increase your understanding of integer operations.

Integer tiles have two distinct sides (or "+" and "−" symbols) to represent "positive" and "negative" respectively. You'll use them in a simple model of an accounting balance sheet where a "+" tile is used to record a credit of $1 and a "−" tile is used to record a debit of $1.

The key idea in this model is that a positive tile and a negative tile "nullify" each other; in other words, a "+" tile combines with a "−" tile to make zero. By accounting for these zeros, you can represent any particular "balance" in many different ways.

Before you can use the tiles to show calculations with integers, you need to know how to represent a single integer. Here are three different ways to represent a net balance of $2:

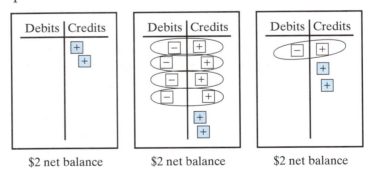

Here are three different ways to represent a net balance of −$1:

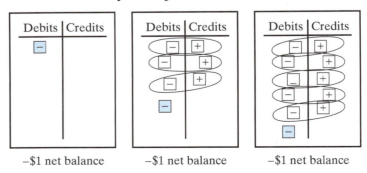

It is important that students understand that 0 can be represented not only by no tiles, but also by an equal number of positive and negative tiles.

It is important to be able to represent a **zero balance** in a number of different ways (the first step in modeling a calculation with integer tiles is to start with a zero balance). To model a net balance of $0, use *equal numbers* of positive and negative tiles:

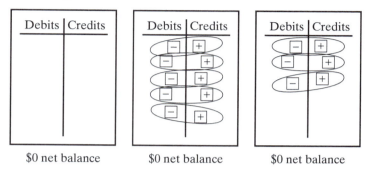

To record your work, make a simple sketch using "+" for a positive tile and "−" for a negative tile. To record the balance sheet on the left, you could make the drawing on the right.

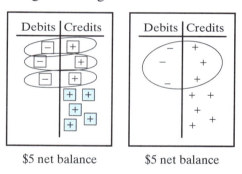

Use integer tiles and an Integer Tiles Balance Sheet Resource Page to represent each of the following integers in three different ways. After building each model with tiles, record your work by sketching the balance sheet using "+s" and "−s."

a. 5   b. 2   c. −7   d. 0

A-2. When representing calculations, it sometimes will be useful to start with a zero balance made of a large number of tiles. Use at least 12 tiles to represent a balance of 0.

A-3. Addition with Integer Tiles You can use integer tiles and a debit/credit balance sheet to model addition with integers.

Example: Model the sum $4 + -3$.

Start with a zero balance (Step 1 below). Then add the appropriate number of tiles—in this example, add 4 "+" tiles, and then 3 "−" tiles—to the zero balance (Steps 2 and 3). Next, match pairs of positive and negative tiles (push them together) to account for zeros (Step 4). Finally, record your results as in Step 5.

Step 1:
Create a zero balance.

Step 2:
Record 4 credits using positive tiles.

Step 3:
Record 3 debits using negative tiles.

Step 4:
Account for zeros.

Step 5:
Record your work and the answer.

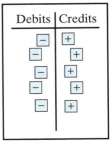

$0 net balance

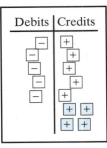

4

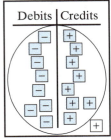

4 + (−3) = ?

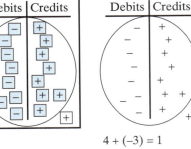

4 + (−3) = 1 ▲

Model each sum using integer tiles. Sketch each model and write the problem with its answer below your drawing, as in the example.

a. −7 + (−2)
b. −1 + 5
c. 3 + (−4)
d. −4 + 0
e. 2 + 5
f. 7 + (−8)

A-4. To model subtraction of integers with tiles, you can physically remove the appropriate number of positive or negative tiles from a debit/credit balance sheet. The tricky part is making sure you've got enough tiles to remove.

For example, suppose you want to model 3 − 7 and you start with no tiles on your debit/credit balance sheet. After recording 3 credits, you would want to *remove* 7 credits. *BUT there aren't enough "+" tiles on the balance sheet for you to remove seven!* So start again …

At the start of the problem, how could you show a zero balance that would allow you to remove 7 credits from a balance of 3 credits? With your group, find a zero balance that works, and then draw and describe the corresponding balance sheet.

A-5. Subtraction with Integer Tiles Here's how to use integer tiles and a debit/credit balance sheet to model subtraction with integers.

Example: Model the difference 2 − (−3).

Start with a $0 net balance (Step 1 below). Represent the first number—in this example the "2"—with tiles (Step 2). To subtract, *physically remove* the appropriate number of positive or negative tiles. In this example, we would remove three negative tiles (Step 3). Finally, match pairs of positive and negative tiles (push them together) to account for zeros (Step 4). Record your work as in Step 5.

Step 1:
Create a zero balance.

Step 2:
Record 2 credits using positive tiles.

Step 3:
Remove 3 debits (negative tiles).

Step 4:
Account for zeros.

Step 5:
Record your work and the answer.

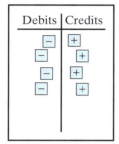
$0 net balance

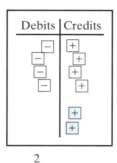
2

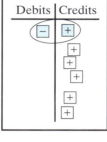
2 − (−3) = ?

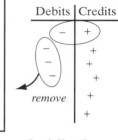
2 − (−3) = 5

Model each difference (subtraction problem) using integer tiles. Sketch each model and write the problem with its answer below your drawing, as in the example.

a. 3 − 5
b. 5 − 3
c. −2 − 1
d. −4 − (−6)
e. 4 − (−4)
f. −4 − 4

A-6. Multiplication with Integer Tiles To model multiplication using integer tiles on a debit/credit balance sheet, you can use the idea that multiplication can be represented by repeated addition. For example, 2(3) = 3 + 3 and 2(−3) = (−3) + (−3). In terms of the integer tiles,

2(3) means add 2 groups of 3 positive tiles;
2(−3) means add 2 groups of 3 negative tiles;
−2(3) means *remove* 2 groups of 3 positive tiles; and
−2(−3) means *remove* 2 groups of 3 negative tiles.

When using the tiles, first create a zero balance **using enough tiles** to allow you to carry out the multiplication. You may not know how many "enough" is at this point, but that's okay—you'll find out in the second step when you try to add or remove **groups** of tiles. If needed, you can go back and add more tiles to the original zero balance.

Example 1: To model $2(-3)$, first create a zero balance (as in Step 1 below) using a sufficient number of tiles. Then add 2 groups of 3 negative tiles (Step 2). Finally, account for zeros as usual (Step 3), and then record your work as in Step 4.

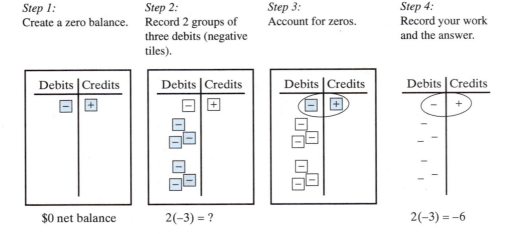

Example 2: To model $-2(3)$, start with a zero balance (as in Step 1 below) made with a sufficient number of tiles. Now remove 2 groups of 3 positive tiles (Step 2). Finally, account for zeros as usual (Step 3), and then record your work as in Step 4.

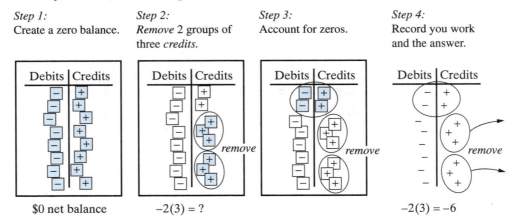

a. Look again at each example. In Step 2, how were the tiles which formed the initial zero balance involved for each situation? When creating a zero balance to begin a calculation, how can you tell what a "sufficient number" of tiles is?

Model each product using integer tiles. Sketch each model and write the problem with its answer below your drawing.

b. $3(-5)$ c. $5(-2)$ d. $-2(5)$

e. $-4(3)$ f. $4(2)$ g. $-4(-2)$

A-7. Copy and complete the given balance sheet for each of the following problems.

a. $-3 + (-1)$

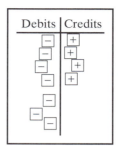

b. $-3 - (-4)$

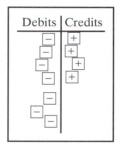

Note: *We haven't found a way to use this model to represent division by negative integers in a way that clarifies the concept. This is an example of the limitations of using physical models to demonstrate mathematical concepts. At this point, students should use the pattern of multiplying negative integers to create a pattern for division. In doing so, they'll use the power of mathematical patterning to extend their knowledge beyond the use of manipulatives.*

The tiles provide a concrete model which justifies the "rules" which the students have seen previously. Throughout this course, we use concrete models to help develop abstract generalizations. Whenever students have to learn abstract rules in order to use a model for a particular situation, the model is no longer useful, for it no longer gives insight into the problem. An important point to emphasize to students is the algebraic power that is gained by moving beyond a model.

Don't require students to use tiles for every integer problem they do. What's important is that they see there are reasons behind the mathematical rules they've been taught.

APPENDIX B

USING SUBPROBLEMS
to Derive the Quadratic Formula

We've included the following derivation of the quadratic formula as an example of how looking for and solving subproblems is useful in mathematics. Stronger students could work through it as a special project.

In this section you'll develop the quadratic formula by solving a series of subproblems. Look for patterns as you solve each problem and discuss the patterns you observe with your group.

In problems B-1 and B-2 students complete a square geometrically using composite rectangles. They'll use this idea in problems B-3 through B-7, as they derive the quadratic formula by solving a series of subproblems.

B-1. Suppose you are given one x^2 tile and 16 small square tiles.

a. It is possible to use some rectangular tiles together with the square tiles you are given to get a composite rectangle that is a square. Use a diagram to show how this can be done.   *Need 8 rectangles: $x^2 + 8x + 16$*

b. Write the area of the composite square you formed in part a as a product and as a sum.   *$(x + 4)^2 = x^2 + 8x + 16$*

B-2. Suppose you are given one x^2 tile and six rectangles.

a. How many small squares will you need to make a composite rectangle that is a square?   *9*

b. Sketch the composite square you formed in part a.

c. Write the area of the composite square as a product and as a sum.   *$(x + 3)^2 = x^2 + 6x + 9$*

d. Now suppose you are given one x^2 tile and 10 rectangles. Repeat parts a, b, and c.   *25; $(x + 5)^2 = x^2 + 10x + 25$*

B-3. Subproblem 1: Solve each of the following equations for *y*. Each equation has *two* solutions.

a. $y^2 = 25$ $y = \pm 5$
b. $y^2 = 19$ $y = \pm\sqrt{19}$
c. $y^2 = 45$ $y = \pm\sqrt{45}$
d. $y^2 = 16b^2$ $y = \pm 4b$
e. $y^2 = 13b^2$ $y = \pm b\sqrt{13}$
f. $y^2 = \frac{36}{49}$ Write the solutions as fractions. $y = \pm\frac{6}{7}$

Stop now to discuss a pattern you see developing in parts a through f.

Use the pattern you see to solve parts g, h, and i.

g. $y^2 = 4a^2$ $y = \pm 2a$
h. $y^2 = \frac{bc}{4a^2}$ $y = \frac{\pm\sqrt{bc}}{2a}$
i. $y^2 = \frac{b^2 - 4ac}{4a^2}$ $y = \frac{\pm\sqrt{b^2 - 4ac}}{2a}$

B-4. Subproblem 2: Completing a Square Here is one way that the expression $x^2 + 6x$ can be represented by tiles:

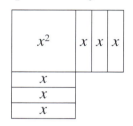

Notice that the lengths of the sides are both the same, namely $x + 3$, so if you complete the diagram you will have a composite **square.** To make the diagram a complete composite square, you would need to add some small squares. Indeed, if you add **9** small square tiles, the composite square formed would have area $x^2 + 6x + 9$.

What number would you need to add to each of the following expressions to make each one represent the area of a composite square?

a. $x^2 + 10x + $ ____ 25
b. $x^2 - 8x + $ ____ 16
c. $x^2 + 7x + $ ____ $\frac{49}{4}$
d. $x^2 - 5x + $ ____ $\frac{25}{4}$
e. $x^2 + bx + $ ____ $\frac{b^2}{4}$
f. $x^2 + \frac{b}{a}x + $ ____ $\frac{b^2}{4a^2}$

B-5. Subproblem 3: Dimensions of Composite Squares From the discussion in problem in B-4, you know $x^2 + 6x + 9 = (x + 3)^2$, so the dimensions of the composite square are $x + 3$ by $x + 3$.

What are the dimensions of each of the completed squares in parts a through f of problem B-4?   a. $(x + 5)^2$; b. $(x - 4)^2$; c. $(x + \frac{7}{2})^2$; d. $(x - \frac{5}{2})^2$; e. $(x + \frac{b}{2})^2$; f. $(x + \frac{b}{2a})^2$

B-6. Subproblem 4: Adding Algebraic Fractions Write each of the following sums as a single algebraic fraction. Remember, before fractions can be added, they need to have a common denominator.

a. $\dfrac{3}{4a} + \dfrac{5}{a}$ $\dfrac{23}{4a}$

b. $\dfrac{b^2}{4a} + \dfrac{2}{a}$ $\dfrac{(b^2 + 8)}{4a}$

c. $\dfrac{7}{4a^2} - \dfrac{c}{a}$ $\dfrac{(7 - 4ac)}{4a^2}$

d. $\dfrac{b^2}{4a^2} - \dfrac{c}{a}$ $\dfrac{(b^2 - 4ac)}{4a^2}$

B-7. Derivation of the Quadratic Formula Now you're ready to derive the quadratic formula. To do this, you will use the last part of each of the subproblems in problems B-3 through B-6 to solve the equation $ax^2 + bx + c = 0$ for x. Your goal is to end up with

$$x = \frac{-b \pm \sqrt{b^2 - 4ac}}{2a}.$$

Fold a piece of lined paper in half vertically, make a crease, then unfold the paper. Copy the algebraic steps shown below onto the left-hand side of your paper. Write your answer to each question to the right of the corresponding algebraic step.

We want to solve the equation $ax^2 + bx + c = 0$.

$x^2 + \dfrac{b}{a}x + \dfrac{c}{a} = 0$ *What was done to get this?*

$x^2 + \dfrac{b}{a}x = -\dfrac{c}{a}$ *What was done to get this?*

$x^2 + \dfrac{b}{a}x + \dfrac{b^2}{4a^2} = \dfrac{b^2}{4a^2} - \dfrac{c}{a}$ *What was done to get this?*

Why do you think $\dfrac{b^2}{4a^2}$ was chosen?

Now make a major replacement for the whole left side of the equation:

$(x + \dfrac{b}{2a})^2 = \dfrac{b^2}{4a^2} - \dfrac{c}{a}$ *What subproblem was used to make this possible?*

Next replace the right-hand side:

$(x + \dfrac{b}{2a})^2 = \dfrac{b^2 - 4ac}{4a^2}$ *What subproblem shows this is correct?*

What operation do you need to do to get the next result?

$x + \dfrac{b}{2a} = \pm\dfrac{\sqrt{b^2 - 4ac}}{2a}$ *What subproblem was used to get this?*

$x = -\dfrac{b}{2a} \pm \dfrac{\sqrt{b^2 - 4ac}}{2a}$ *What was done to get this result?*

Finally, you get to the long-awaited solution:

$x = \dfrac{-b \pm \sqrt{b^2 - 4ac}}{2a}$ *What was done to get this?*

APPENDIX C
EXPLORING QUADRATIC EQUATIONS AND THEIR GRAPHS

To extend students' understanding of quadratic equations, we include an exploration of the graphs of various families of quadratics to discover some general properties of parabolas. If your students have graphing calculators, or if you have access to a set, this is a natural place to use them.

This is an informal treatment of the topic of families of parabolas; we are NOT trying to develop proficiency with the general form $y = a(x - h)^2 + k$. Encourage students to use their discussions and written descriptions to draft clear statements about the equations of parabolas and their graphs. Problem C-6 could be used as a springboard for making a single statement about all aspects of the graph.

Problems C-1 through C-6 are investigations into some general relationships between quadratic equations and their graphs, parabolas. In them, you'll be exploring which part of the equation determines the steepness of the curve and which part determines whether the parabola opens upward or downward. You'll also be moving (translating) parabolas vertically and/or horizontally in a coordinate grid.

C-1. Make a large pair of coordinate axes on a sheet of graph paper by placing the origin at the center of the paper. *Neatly* graph each of the following equations *on this one pair of axes* by making tables and assigning values to x from -4 through 4.

a. $y = x^2$ **b.** $y = 3x^2$ **c.** $y = \frac{1}{3}x^2$

d. $y = -x^2$ **e.** $y = -\frac{1}{3}x^2$

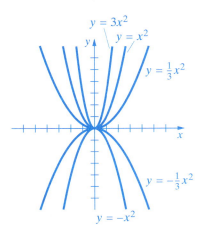

C-2. Look at your graphs for problem C-1. Write a few sentences to describe some patterns you see in the relationship between quadratic equations of the form $y = ax^2$ and their graphs.

C-3. Use your conclusions from the previous problem to decide in your groups how to sketch each of the following equations on the same set of axes you used in C-2. Use a colored pen or pencil to distinguish the graphs.

a. $y = \frac{1}{2}x^2$ *It lies between the graphs of $y = x^2$ and $y = \frac{1}{3}x^2$.*

b. $y = -2x^2$ *It lies beneath the graph of $y = -x^2$.*

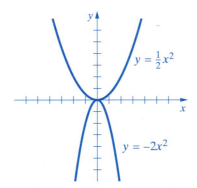

This is a good time to have all the groups put their solutions on a large graph on the board or on overhead transparencies and to have a class discussion of the groups' conclusions.

C-4. Make a large graph on a sheet of graph paper by placing the origin at the center of the paper. *Neatly* graph each of the following equations *on this one graph* by making tables and assigning values for $-4 \leq x \leq 4$.

a. $y = x^2$ **b.** $y = x^2 + 3$ **c.** $y = x^2 - 2$

d. Write a few sentences to describe some patterns you see in the relationship between quadratic equations of the form $y = x^2 + c$ and their graphs.

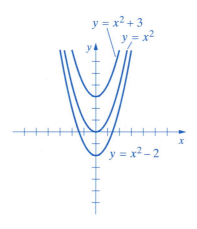

C-5. Make a large graph on a sheet of graph paper by placing the origin at the center of the paper and scaling each axis two squares to each unit. *Neatly* graph each of the following equations *all on this one graph* by making tables and assigning values as suggested.

a. $y = x^2$ for $-4 \leq x \leq 4$

b. $y = (x - 3)^2$ for $-1 \leq x \leq 7$

c. $y = (x + 2)^2$ for $-6 \leq x \leq 2$

d. Write a few sentences to describe some patterns you see in the graphs of these quadratic equations.

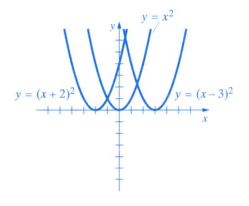

C-6. Summarize your observations from problems C-1 through C-5 by writing a paragraph in which you describe some relationships between quadratic equations and their graphs.

Chapter 1 Instructor's Resource Page

Clue Cards for OP-1 Building Corrals

Three horses boarded in a large rectangular corral at the Equestrian Center have been fighting recently, and need to be separated from each other. Cole subdivided the large rectangular corral to form three new corrals. Use the information on the clue cards to figure how out the new corrals could have been arranged.

BUILDING CORRALS

How are the corrals arranged?

These are your clues to help solve the group's problem. Read them to the group, but do not show them to anyone.

- The large rectangular corral was subdivided to form three smaller corrals.
- Two of the new corrals are squares.

1

BUILDING CORRALS

How are the corrals arranged?

These are your clues to help solve the group's problem. Read them to the group, but do not show them to anyone.

- Cole used fence sections of the same length to build the corrals.
- To build the larger square corral, Cole used twice as many fence sections as he used for the smaller square.

2

BUILDING CORRALS

How are the corrals arranged?

These are your clues to help solve the group's problem. Read them to the group, but do not show them to anyone.

- There are 20 fence sections in all, including those in the original corral.
- The square corrals do not touch each other.

3

BUILDING CORRALS

How are the corrals arranged?

These are your clues to help solve the group's problem. Read them to the group, but do not show them to anyone.

- Cole used six fence sections to subdivide the large rectangle without moving any of the original fence sections.
- The square corrals are not the same size.

4

Distribute the check card **after** the group has solved the problem.

BUILDING CORRALS ✓

Use this clue to check your group's solution to the problem.

- The horse in one corral can roam to all four sides of the large rectangular corral.

Guidelines for Working in Groups

1. You are responsible for your own behavior.

2. You must try to help any group member who asks.

3. You may ask the teacher for help only when all group members have the same question.

4. You must use a "group voice" that only your group can hear.

Chapter 1 Instructor's Resource Page

Clue Cards for OP-4 Cars and Bikes

The Ferrari family and the Schwinn family share a garage. There are four questions your group should answer according to the information given on the clue cards. There are different ways to organize this information, so share your ideas with your group.

CARS AND BIKES

How many cars does the Ferrari family have?

These are your clues to help solve the group's problem. Read them to the group, but do not show them to anyone.

- The Ferrari family and the Schwinn family share a garage.
- The only vehicles in the garage are cars and bikes.

1

CARS AND BIKES

How many bicycles does the Schwinn family have?

These are your clues to help solve the group's problem. Read them to the group, but do not show them to anyone.

- There are three bicycles in the garage.
- The total number of vehicles for each family is the same.

2

CARS AND BIKES

How many cars does the Schwinn family have?

These are your clues to help solve the group's problem. Read them to the group, but do not show them to anyone.

- The Schwinn's vehicles have two more wheels than the Ferrari's.
- Using objects to represent cars and bikes may help you solve the problem.

3

CARS AND BIKES

How many bicycles does the Ferrari family have?

These are your clues to help solve the group's problem. Read them to the group, but do not show them to anyone.

- The Ferrari's vehicles have 12 wheels altogether.
- There are more cars than bicycles.

4

Distribute the check card **after** the group has solved the problem.

CARS AND BIKES ✓

Use this clue to check your group's solution to the problem.

- The Ferraris have the same number of cars as bikes, but the Schwinns do not.

I-RP-5

Chapter 2 Instructor's Resource Page

Colored Cards for use in Algebra Walk (SP-1)

Make five copies of this sheet, one each of blue, green, yellow, orange, and red.

input output	input output
input −4 □	input −3 □
input −2 □	input −1 □
input 4 □	input 3 □
input 2 □	input 1 □
input 0 □	

I-RP-7

The Birthday Problem

Start with the number of the month in which you were born.

Add 4.

Multiply the sum by 10.

Subtract 15.

Multiply by 5.

Add the day of the month on which you were born.

Subtract the number of days in April.

Double the result.

Subtract the day of the month on which you were born.

Add 15.

Tell me the result.

Transparency Master Chapter 2 Instructor's Resource Page

SP-152

"Keisha sold 300 tickets to the Rolling Stones Reunion concert. General admission tickets sold for $35 each, and student tickets were $25 each. Total sales amounted to $8,280. How many students bought tickets to the concert?"

Guess:					Check:
10	$350	290	$7,250	$350 + 7,250 =	$7,600 (too low)
45	$1,575	255	$6,375	$1,575 + 6,375 =	$7,950 (too low)
105					

Pythagorean Puzzle Triangles

On colored cardstock, copy one page of puzzle triangles for each group. Cut four *a* by *b* rectangles, and then cut along a diagonal of each rectangle to make eight right triangles.

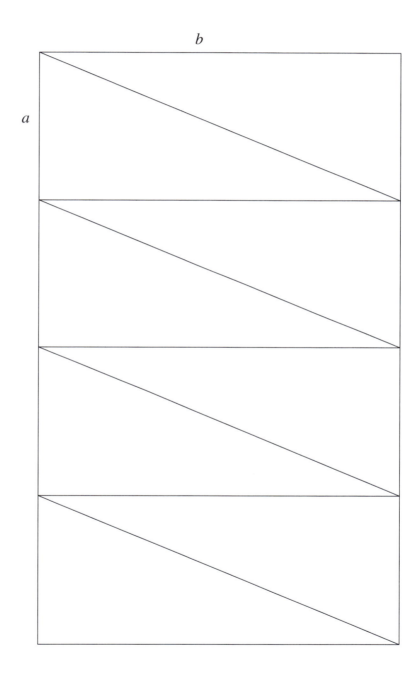

Chapter 7 Instructor's Resource Page

Pythagorean Puzzle Squares

On a second color of cardstock, copy one page of puzzle squares for each group. Carefully cut out each set of squares.

Pythagorean Puzzle Board

(Use this sheet to make additional copies for your use.)

Resource Page Template

DOT PAPER

(Use this sheet to make additional copies for your use.)

Resource Page Template

ALGEBRA TOOL KIT

The "What to do when you don't remember what to do" kit.

RP-3

(Use this sheet to make additional copies for your use.)

Resource Page Template

CALCULATOR TOOL KIT

For instructions on using your calculator

RP-5

(Use this sheet to make additional copies for your use.) Resource Page Template

CHAPTER SUMMARY DRAFT: CHAPTER ___

Use one page per main idea by making copies or following the outline on your own paper.

a. State a main idea of the chapter.

b. Find and write out a problem that is a good example of the idea. Include the problem number.

c. Show a complete solution to the problem.

d. In one or two sentences, describe what you have learned about the idea.

RP-7

Use this sheet to make additional copies for your use.

Resource Page Template

Grid Sheet

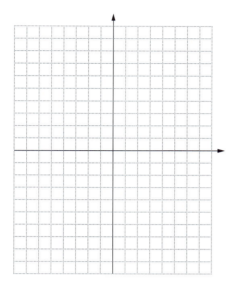

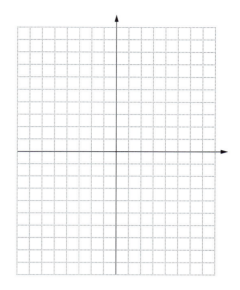

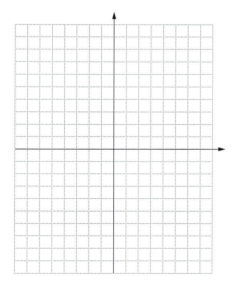

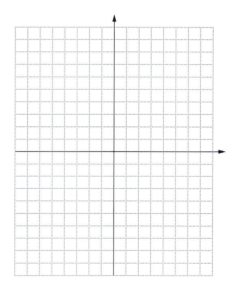

RP-9

OP-7. More Corrals

a. Is it possible to build a rectangle corral using exactly 12 fence sections? Show as many ways as you can. (The dots represent fence posts. A fence section will fit only horizontally or vertically, not diagonally.)

b. Repeat part a using exactly 15 fence sections.

c. Is it possible to use exactly 10 fence sections to make a non-rectangular corral? Show as many ways as you can.

OP-8. Perimeter

Copy each of the figures from OP-8. Label each side with its length and then find the perimeter of the figure.

Chapter 1 Resource Page

ALGEBRA TOOL KIT

The "What to do when you don't remember what to do" kit.

Perimeter

Order of Operations

Area

Integer Arithmetic

Base, Exponent, Exponential Form

RP-15

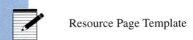

Resource Page Template

ALGEBRA TOOL KIT

The "What to do when you don't remember what to do" kit.

Scientific Notation

Standard Decimal Form

Probability

Chapter 1 Resource Page

OP-15. The Five-Digit Problem

Write expressions for the integers from 0 to 20 using the class digit (the one your class picked) five times along with any "legal" mathematical operations (for example: +, −, ·, ÷, …).

0	
1	
2	
3	
4	
5	
6	
7	
8	
9	

RP-17

OP-15. The Five-Digit Problem continued

10
11
12
13
14
15
16
17
18
19
20

Chapter 1 Resource Page

CALCULATOR TOOL KIT

For instructions on using your calculator

| +/− or (−) | Changing the sign of a number |

| y^x or ^ | Raising a number to a power |

| EE or EXP | Entering a number in scientific notation |

RP-19

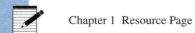

Chapter 1 Resource Page   (Use this sheet to make additional copies for your use.)

CALCULATOR TOOL KIT

For instructions on using your calculator

Chapter 1 Resource Page

CENTIMETER GRID PAPER

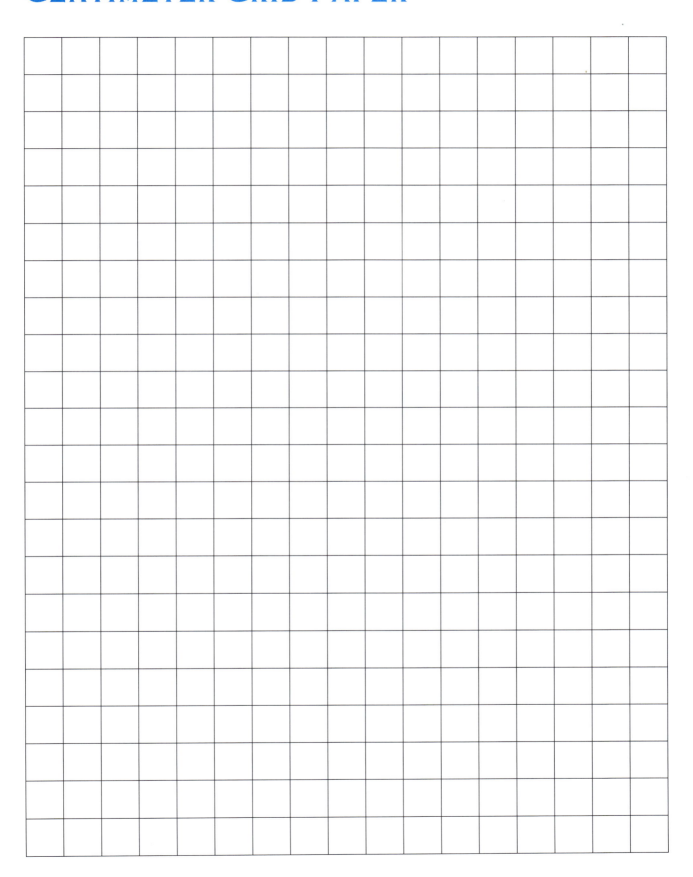

RP-21

OP-29. Area and Perimeter

This is allowed.   This is not allowed.   Overlaps are not allowed.   This is not allowed either!

Chapter 1 Resource Page

RP-23

SP-1. Algebra Walk Observations

Sketch what you see: Describe what you see, and write the rule in words:

Blue		Description: The rule in words:
Green		Description: The rule in words:
Yellow		Description: The rule in words:
Orange		Description: The rule in words:
Red		Description: The rule in words:

Chapter 2 Resource Page

SP-20. Area as a Product and Area as a Sum

a. Label the sides of the following rectangle with their lengths.

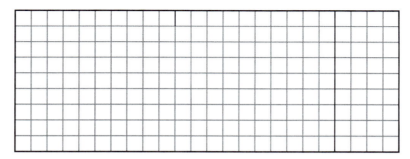

b. Label each of the two large sections with its area.

c. To express the area of the rectangle as a *product*, multiply the length by the width:

$$9 \cdot (20 + 4).$$

To express the area of the rectangle as a *sum*, add the areas of the parts:

$$9(20) + 9(4).$$

Check to see that these two expressions are equal by evaluating each of them using the proper order of operations.

Area as a Product = Area as a Sum

$$9(20 + 4) = 9(20) + 9(4)$$

$$9(24) = 180 + 36$$

$$216 = 216$$

RP-27

SP-21.

For each of the following rectangles, write an equation showing that the area (in square units) can be computed two ways—as a *product* of its lengths and widths (as Derek did in SP-20) and as a *sum* of its parts (as Deanne did). For each rectangle, label the side lengths and the area of each part, and then write an area equation below each rectangle.

a.
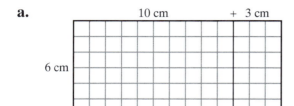

b.
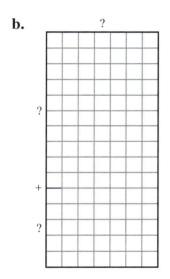

c.
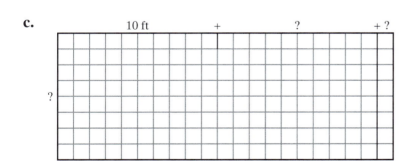

Chapter 2 Resource Page

SP-22.

In each of the following equations, the area of a rectangle is represented both as a product and as a sum. On the grid, sketch the rectangle represented by each equation, and label each section of the rectangle with its area:

a. 4(10 + 5) = 4(10) + 4(5) = 60

b. 7(20 + 1) = 7(20) + 7(1) = 147

c. (10 + 2)(9) = 10(9) + 2(9) = 108

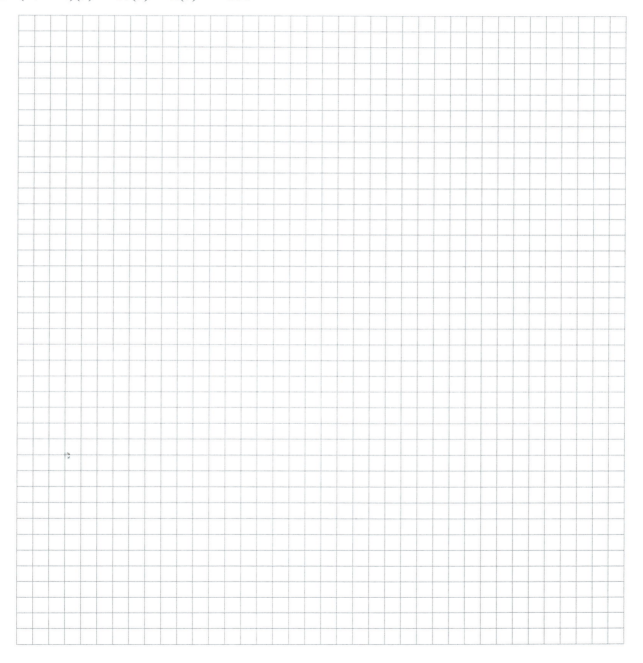

Algebra Tiles

Glue this page to a piece of lightweight cardboard or another piece of paper. Carefully cut out the individual rectangles and the small and large squares. Keep them in a re-sealable plastic bag or envelope. Store the bag of tiles in your notebook.

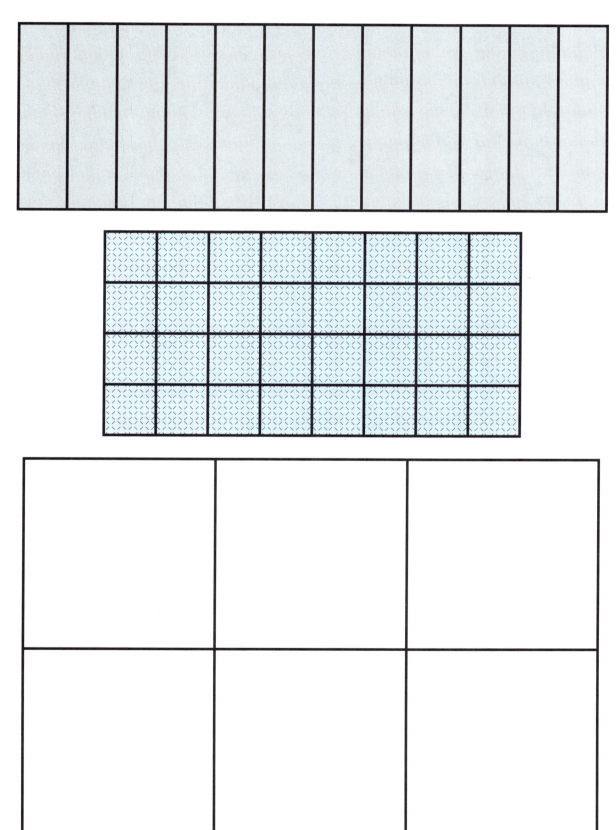

Chapter 2 Resource Page

SP-73. Slicing a Pizza

Suppose you have a perfectly round (circular) pizza. Around the edge of the crust are some olives. (There are no other olives on the pizza.) If you slice the pizza by making cuts that connect the olives, so that every pair of olives is connected by a cut, how many pieces of pizza will you make?

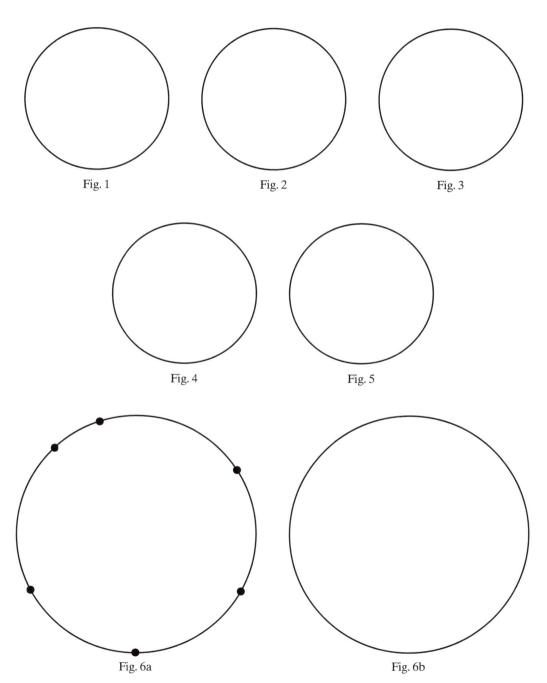

Fig. 1

Fig. 2

Fig. 3

Fig. 4

Fig. 5

Fig. 6a

Fig. 6b

Place 6 dots and complete the figure.

Chapter 2 Resource Page

SP-76. Composite Rectangles and Area

a. Determine if it is possible to build a composite rectangle with each set of tiles indicated in the chart below. If so, on this resource page sketch a composite rectangle you could build; if not, answer "No."

b. For each composite rectangle, label each tile with its area as shown, and then label the length and width of the rectangle.

Use what you have done to write the area of each set of tiles as a sum of the areas of the pieces, and then as a product of the length and the width, if possible.

Number of $1x$ Tiles	Number of 1 Tiles	Is a Composite Rectangle Possible?	Sketch	Algebraic Expression for the Area as a Sum	Algebraic Expression for the Area as a Product
2	10	yes	$x + 5$ 2 [1x | 1 1 1 1 1 / 1x | 1 1 1 1 1]	$2x + 10$	$= (2)(x + 5)$
3	12				
3	5				
1	3				
4	3				
4	4				

RP-37

LT-20. Composite Rectangles and Area

Number of x^2 Tiles	Number of $1x$ Tiles	Number of 1 Tiles	Is a Composite Rectangle Possible?	Sketch	Algebraic Expression for the Area as a Sum	Algebraic Expression for the Area as a Product
1	3	2	yes	(sketch of composite rectangle with x^2, $1x$, $1x$, $1x$, 1, 1; dimensions $x+2$ by $x+1$)	$x^2 + 3x + 2$	$= (x+1)(x+2)$
1	4	4				
2	3	1				
1	6	4				
0	3	6				
1	4	0				
2	7	6				
1	7	10				

LT-26.

Write an algebraic expression for the area of each of the following composite rectangles in two different ways—first as a product, and then as a sum. Use "x^2" to represent each large square, "x" to represent each rectangular tile, and "1" to represent each small square.

a.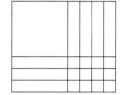

b.

c.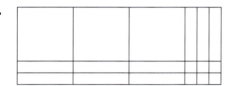

d.

e.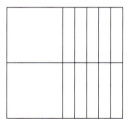

LT-27.

Find the area of each pool in two ways:

1. by adding the areas of the parts, and
2. by multiplying the length of the whole rectangle by the width of the whole rectangle.

a.
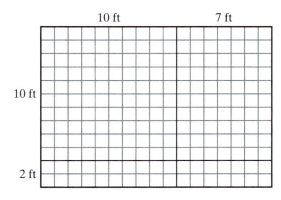

b.
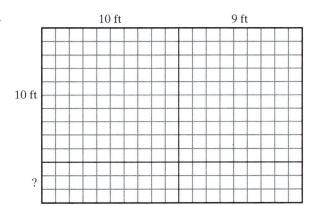

c.
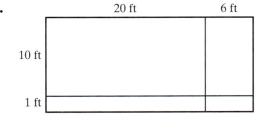

Two-Pan Balance Scale

Use this page to model solving equations with packets and jelly beans.

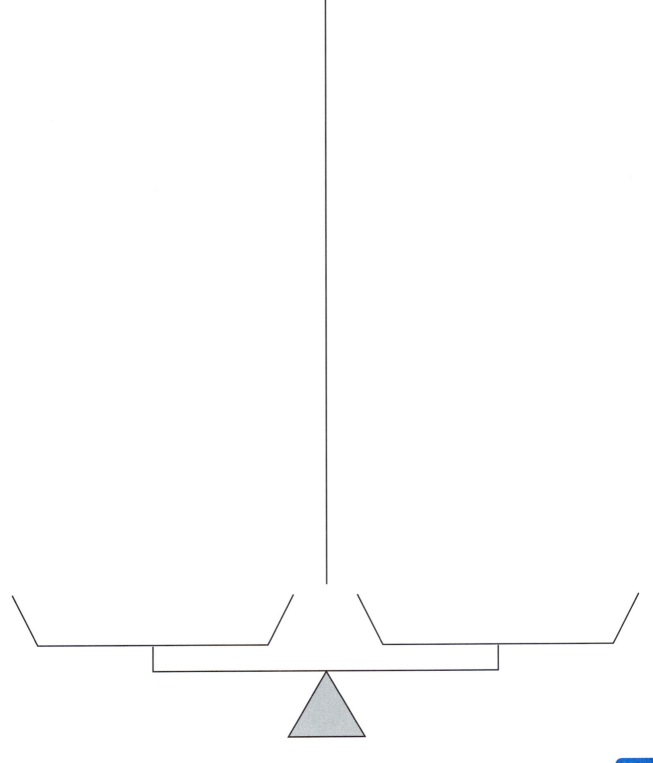

LT-36. Using Packets and Jelly Beans to Solve Linear Equations

Model the following situation on the Two-Pan Balance Scale Resource Page. Record each step of the process by drawing a diagram and by showing the corresponding algebraic steps.

Situation: All packets have the same number of jelly beans. Two jelly beans and 3 packets balance 1 packet and 14 jelly beans. How many jelly beans are in a packet?

Description of Process	Diagram—Model with Packets and Beans	Algebraic Record—Model with Variables
Step 1. Draw a diagram to represent the balanced scale. Then write an equation to model the situation.	(3 packets and 2 beans | 1 packet and 14 beans)	Let P represent the number of beans in each packet.
Step 2. Ask yourself, "What could I do to both sides to eliminate some packets or some jelly beans while keeping the balance?" Record the algebraic process.		
Step 3. Continue. Ask yourself, "What could I now do to both sides to eliminate some packets or some jelly beans while keeping the balance?" Record the algebraic process.		
Step 4. Arrange the diagram so you can see how many beans would balance one packet. Record the algebraic process.		
Step 5. Answer the question, "How many jelly beans are in a packet?" Record the algebraic process.		
Step 6. Check your answer in the original diagram and equation.		

LT-47. Representing Tiles with Polynomials

Represent each of the tile collections in parts a through c with an algebraic expression as shown in the example. Represent the tile collection in part d with an equation.

Tile Collection	Polynomial
Example: (one x^2 square, two x rectangles, three unit squares)	$x^2 + 2x + 3$
a. (one x^2 square, one small square, four small squares, two rectangles, one rectangle)	
b. (three rectangles, five small squares)	
c. 38 small squares, 20 rectangles, 5 large squares	
d. (tile collection equation) Polynomial:	

EF-2. Englargement Ratios

EF-3. Subproblems

Subdivide: Identify two more subproblems you could use to subdivide the shaded region to finish computing the total area. Solve each of your subproblems, and then check to see that the total shaded area is 53 sq units.

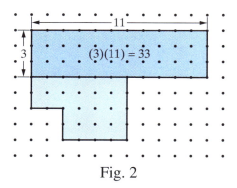

Fig. 2

EF-4.

Use the idea of subproblems to find the area of the shaded region. Show all of your work by stating each subproblem you use.

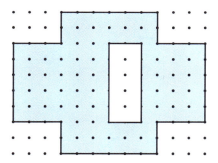

EF-6. Reducing Figures

a.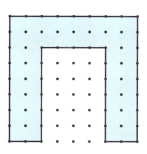

b.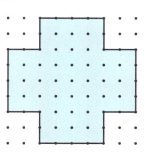

c.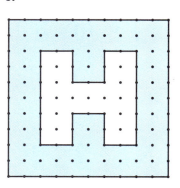

Chapter 4 Resource Page

EF-21. Enlargement and Reduction Ratios Summary

a. Organize your data and look for a pattern. Record your group's data from problems EF-2, EF-6, and EF-7. Look for patterns in your chart. When you enlarge (or reduce) a figure on dot paper, what relationships are there among the ratios for the figure?

	Figure	Enlargement or Reduction Ratio	$\dfrac{S_{new}}{S_{orig}}$	$\dfrac{P_{new}}{P_{orig}}$	$\dfrac{A_{new}}{A_{orig}}$
EF-2a.					
EF-2b.					
EF-2c.					
EF-2d.					
EF-6a.					
EF-6b.					
EF-6c.					
EF-7.	5 ft × 4 ft				

RP-57

Chapter 4 Resource Page

EF-21. Enlargement and Reduction Ratios Summary (continued)

b. Make some conjectures: What generally happens when you enlarge (or reduce) a figure on dot paper? Use this chart to record your predictions for parts b and c of problem EF-21.

	Make Conjectures	Enlargement or Reduction Ratio	$\dfrac{S_{new}}{S_{orig}}$	$\dfrac{P_{new}}{P_{orig}}$	$\dfrac{A_{new}}{A_{orig}}$
EF-21b. (i)	What if the ratio $\dfrac{S_{new}}{S_{orig}}$ is $\dfrac{2}{1}$?				
EF-21b. (ii)	What if the new figure has corresponding side lengths three times the original?				
EF-21b. (iii)	What if the new figure has corresponding side lengths 10 times the original?				
EF-21c.	What if the ratio $\dfrac{S_{new}}{S_{orig}}$ is $\dfrac{N}{1}$?				

Chapter 4 Resource Page

EF-22 AND EF-23. Your Enlargement and Reduction Ratios Conjecture

Record your group's data from EF-22 and EF-23 in the following table. Do the ratios fit your conjecture from part c of EF-21?

	Figure	Enlargement or Reduction Ratio	$\dfrac{S_{new}}{S_{orig}}$	$\dfrac{P_{new}}{P_{orig}}$	$\dfrac{A_{new}}{A_{orig}}$
EF-22b. Test 1	8 ft × 4 ft rectangle				
EF-22b. Test 2	8 ft × 4 ft rectangle				
EF-23b. Test 1	L-shaped figure on dot grid				
EF-23b. Test 2	L-shaped figure on dot grid				

RP-59

EF-63. Similar Triangles

Use a protractor to measure the angles in each of the following triangles. Label each angle as shown in the example, and then state which triangles are similar to each other. Be sure to name the similar triangles by stating the vertices in the same the order as corresponding angles.

Example:

△ZIP is similar to △GUM.

Chapter 4 Resource Page

EF-64. Similar Triangles and Ratios of Side Lengths

Each of the two maps below shows the direct air route—from Sacramento to Los Angeles, then to Reno and back to Sacramento—for ExpressAir, a small air freight company in California. Use this page to record the measure of each angle and the lengths of the sides as you complete EF-64.

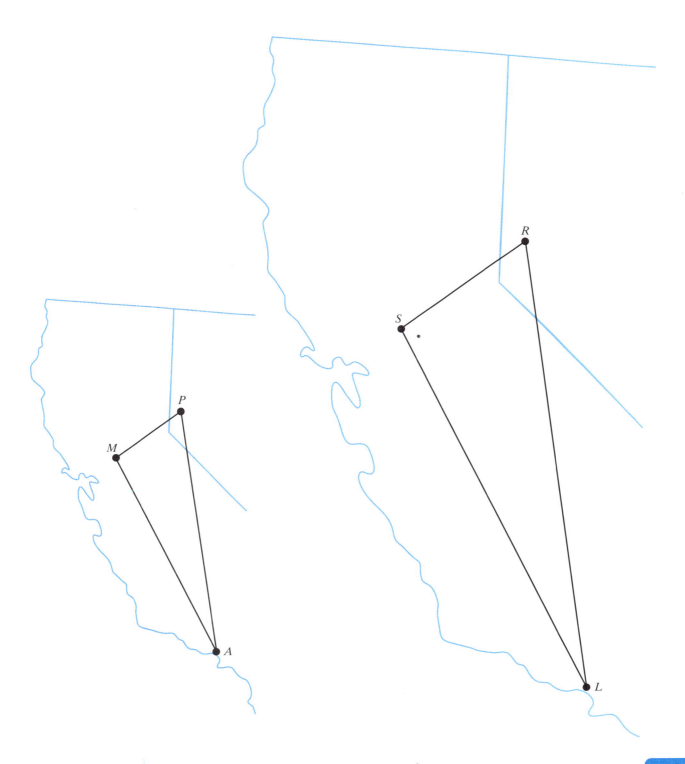

EF-96. Sums and Products: Part I

a. Record integer pairs whose sums and products are indicated in the table. For each sum-product pair, find two integers whose sum is the number in the left-hand column and whose product is the number in the right-hand column.

Example: Find two integers whose sum is 5 and whose product is 6.

Solution: The integers 2 and 3 work, since $2 + 3 = 5$ and $2 \cdot 3 = 6$. ▲

Note: There may be some sum-product pairs for which no pair of integers works.

Sum	Product	Integer Pair	Sum	Product	Integer Pair
5	6	2, 3 (or 3, 2)	−5	6	
7	6		−7	6	
13	12		−11	30	
10	16		−8	16	
6	12		−10	16	
8	16		7	12	
10	0		−10	24	

b. You can use a line in the Sums and Products table to write a quadratic polynomial and its factored form. For example, $x^2 + 5x + 6 = (x + 2)(x + 3)$.

Which line in the Sums and Products table above is related to each of the quadratic polynomials in problem EF-95? Look back at your work, circle the related lines in your Sums and Products table, and then complete the table below.

	Sum	Product	Integer Pair	Polynomial Area as a Sum	=	Factored Form Area as a Product
Example:	5	6	2, 3	$x^2 + 5x + 6$	=	$(x + 2)(x + 3)$
				$x^2 + 10x + 16 =$		
				$x^2 + 8x + 16 =$		
				$m^2 + 5m + 6 =$		

EF-124. Sums and Products: Part II

a. Record integer pairs whose sums and products are indicated in the table. For each sum-product pair, find two integers whose sum is the number in the left-hand column and whose product is the number in the right-hand column, as you did in EF-96.

Sum	Product	Integer Pair
15	56	
−3	−28	
5	−6	
2	−35	
1	−72	
0	−16	
3	−10	

Sum	Product	Integer Pair
5	0	
−9	18	
−5	−6	
−2	−35	
11	30	
−1	−12	
0	−25	

c. You can use a line in the Sums and Products table to write a quadratic polynomial and its factors, such as $x^2 - 5x - 6 = (x - 6)(x + 1)$. Find which line in the Sums and Products table above is related to each of the quadratic polynomials in problem EF-123. Circle the related lines in your Sums and Products table, and complete the table below. Then add two additional examples to your table.

	Sum	Product	Integer Pair	Polynomial Area as a Sum	=	Factored Form Area as a Product
Example:	−5	−6	−6, 1	$x^2 + 5x - 6$	=	$(x - 6)(x + 1)$

EF-130. Estimating Fish Populations

Initial guess of the total number of fish in the lake = _____

| **Initial Outing to Tag Some Fish:** ($1000, required) Number of *tagged* fish in the entire lake = _____ Variable to represent *total* number of fish in the lake = _____ | **Sample #1:** ($800, required) Number of *tagged* fish in the sample = _____ *Total* number of fish in the sample = _____ Ratio Equation: Record your estimate: Total number of fish in the lake ≈ _____ | **Sample #2:** ($800, required) Number of *tagged* fish in the sample = _____ *Total* number of fish in the sample = _____ Ratio Equation: Record your estimate: Total number of fish in the lake ≈ _____ Combined estimate using Sample #1 and Sample #2: |

EF-130. Estimating Fish Populations continued

Sample #3: ($800, optional) Number of *tagged* fish in the sample = _____ *Total* number of fish in the sample = _____ *Ratio Equation:* Record your estimate: Total number of fish in the lake ≈ _____ New combined estimate using Samples #1, #2, and #3:	Sample #4: ($800, optional) Number of *tagged* fish in the sample = _____ *Total* number of fish in the sample = _____ *Ratio Equation:* Final Estimate: _____ Actual Population: _____ Error: _____ Percent Error: _____ Record your estimate: Total number of fish in the lake ≈ _____ New combined estimate using Samples #1, #2, #3, and #4:

EF-134. Sums and Products: Part III

a. Record integer pairs whose sums and products are indicated in the table. For each sum-product pair, find two integers whose sum is the number in the left-hand column and whose product is the number in the right-hand column, as you did in EF-96.

Sum	Product	Integer Pair
−11	28	
14	0	
0	−100	
0	−32	
23	−24	
10	−24	
20	99	

Sum	Product	Integer Pair
2	−120	
18	−63	
−10	−25	
−7	−30	
−1	−42	
−12	−13	
13	−30	

Chapter 5 Resource Page

WR-27 AND WR-28. Graph Comparisons

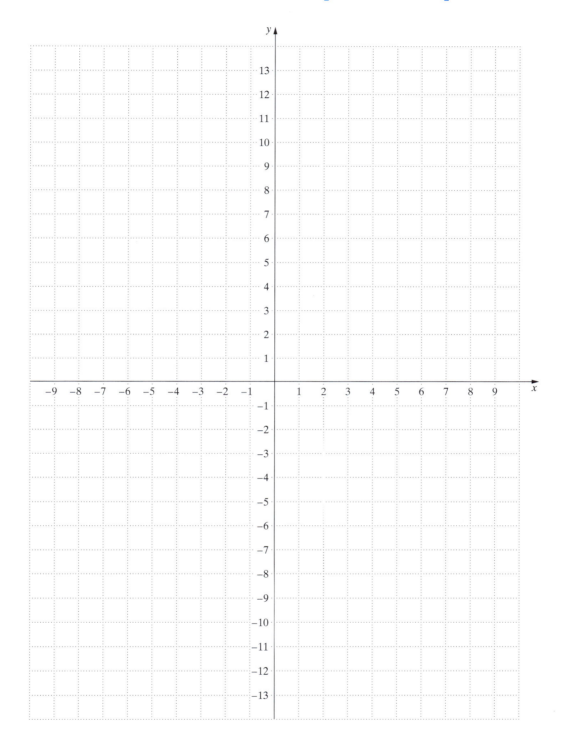

RP-73

WR-45. The Bike Race

Cindy and Dean are engaged in a friendly 60-mile bike race that started at noon. The graph below represents their race. Notice that Dean, who doesn't feel a great need to hurry, stopped to rest for a while.

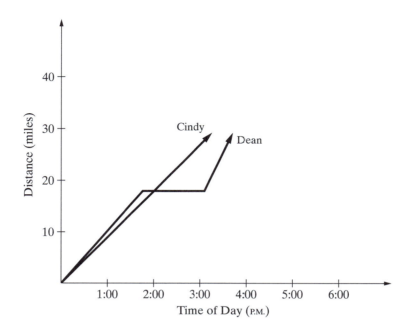

After reading through the questions in parts a through f, make marks on the graph to show how to use it to answer each question. Write your responses to the questions in complete sentences.

a. At what time (approximately) does Cindy pass Dean?

b. About how far has Cindy traveled when she passes Dean?

c. You can see that Dean will catch up to Cindy again if they both continue at their same rates. At approximately what time will this happen? About how far will they each have traveled when they reach this point?

d. Does Dean travel faster or slower after his rest? Why do you think so?

e. About how long did Dean rest? On the graph, show how you know.

f. If Cindy continues at a steady pace, about how long will it take her to ride the entire race?

WR-57. Slopes of Segments

Draw each of the following segments: *DF*, *GH*, *CF*, *EF*, and *EG*. Next to each segment, write its slope.

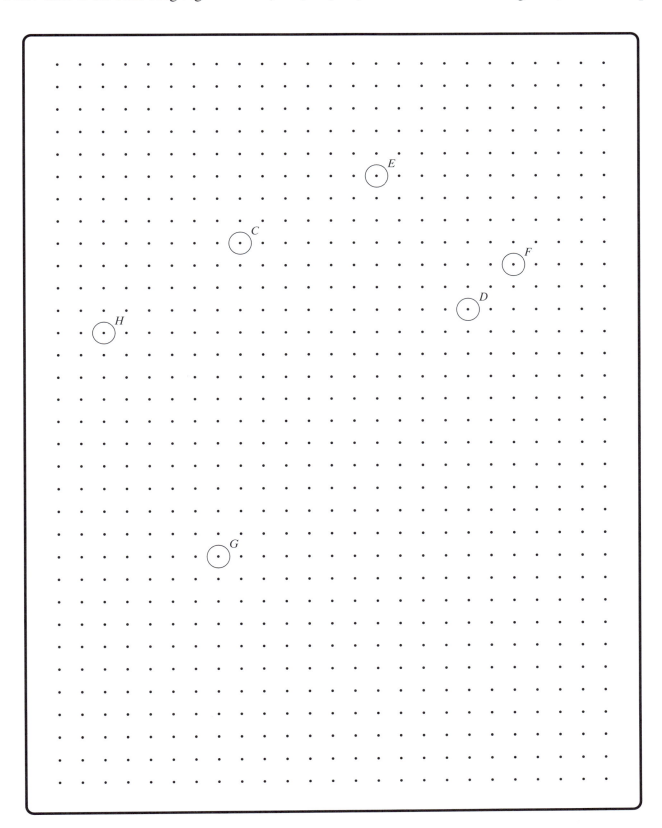

WR-103. Patterns with Lines and Their Equations

Sketch the graphs and look for similarities and differences among the lines and their equations.

a. $y = x$

$y = 2x$

$y = 4.9x$

What happens to the graph of $y = x$ when the coefficient of x is greater than 1—that is, when x is multiplied by a number greater than 1?

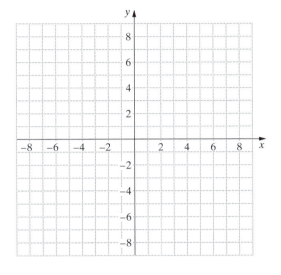

b. $y = x$

$y = \frac{1}{2}x$

$y = \frac{1}{4}x$

$y = \frac{1}{6}x$

What happens to the graph of $y = x$ when the coefficient of x is between 0 and 1?

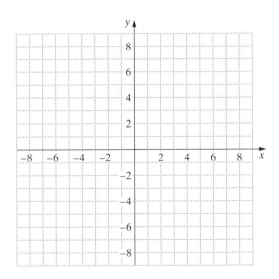

(continued on next page)

Chapter 5 Resource Page

WR-103. Patterns with Lines and Their Equations continued

c. $y = x$

$y = \frac{-1}{4}x$

$y = \frac{-1}{2}x$

$y = -4x$

$y = -2x$

What happens to the graph of $y = x$ when the coefficient of x is negative?

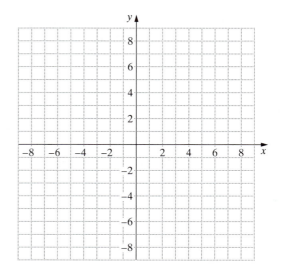

d. $y = x$

$y = x + 2$

$y = x - 2$

$y = x + 4$

$y = x - 4$

What happens to the graph of $y = x$ when a constant (number) is added to x?

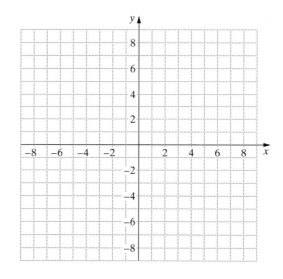

Now complete parts e and f of the problem.

RP-80

WR-104. Patterns with Lines and Their Equations

Look for similarities and differences among the graphs and equations in each set.

a. Find the slope and y-intercept of each of the lines a, b, and c. Near each line, write its equation in slope-intercept form.

b. Look at the lines and their equations in part a. What happens to the graph of $y = x$ when the coefficient of x is greater than 1—that is, when x is multiplied by a number greater than 1?

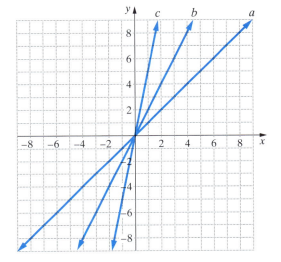

c. Now repeat part a for the lines a, d, and e in the following graph:

d. Look again at the lines in part c. What happens to the graph of $y = x$ when the coefficient of x is between 0 and 1?

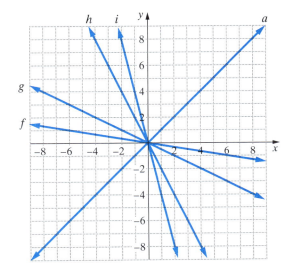

Chapter 5 Resource Page

WR-104. Patterns with Lines and Their Equations continued

e. Repeat part a for the lines $a, f, g, h,$ and i in the following graph:

f. What happens to the graph of $y = x$ when the coefficient of x is negative?

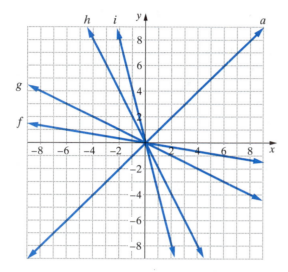

g. Repeat part a for the lines $a, j, k, l,$ and m in the following graph:

h. Look at the lines in part g. What happens to the graph of $y = x$ when a constant (number) is added to x?

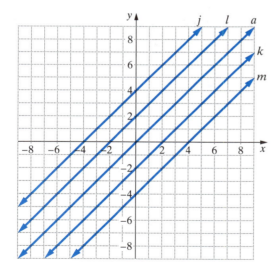

Chapter 6 Resource Page

FF-48. Looking at a Difference of Squares Geometrically

Dale and Roy labeled a diagram of a square backyard covered with lawn except for a 3-foot by 3-foot square in the corner by the gate. Here's how it looked:

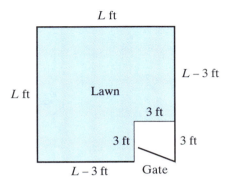

To use Roy's approach for calculating the area of the large square that is shaded, first cut along the edges of the shaded region of the diagram below. Then subdivide the lawn into two regions. Do this either by placing a straight edge along one side of the small square and drawing a line through the shaded region, or by folding along an edge of the small square and making a crease through the shaded region.

Cut the two regions apart, rearrange them to form a rectangle, and glue or tape them to your paper with your other work for problem FF-48. Label the sides of the rectangle with their lengths. Use the rectangle to explain Roy's approach to figuring out the area of the lawn.

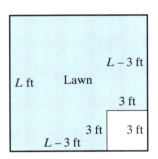

FF-55. Estimating Rocket Height

A rocket is launched and its distance above sea level is recorded until it lands. A graph of the data is shown below.

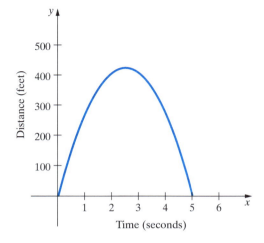

a. Estimate the rocket's height above sea level 3.5 seconds after firing by drawing a vertical line on your copy of the graph. Label its point of intersection with the curve. About how far above sea level is the rocket 3.5 seconds after firing?

b. Estimate how long after being launched the rocket is 250 feet above sea level by drawing a horizontal line on your copy of the graph. Label its points of intersection with the curve. About how long after being launched is the rocket 250 feet above sea level?

c. Write the equation of the vertical line in part a and the equation of the horizontal line in part b.

Chapter 6 Resource Page

FF-113–FF-116.
Investigation: Using Algebra to Model and Analyze Linear Data

Grading and Due Dates

Assignment	Due Dates	Percent of Grade
Proposal	*Assign with Section 6.1; allow about a week*	
Mathematical Models	*Assign with Section 6.3; allow about a week*	
Analysis		
Presentation		
Format Issues: neatness, typewritten, etc.		

RT-3. Sides of Right Triangles

a. How complicated is the relationship among the legs and hypotenuse of a right triangle? To help answer this question, complete the table for the following five right triangles.

i.

ii.

iii.

iv.

v.

	Length of Leg # 1	Length of Leg # 2	Length of Hypotenuse	(Length of Leg # 1)^2	(Length of Leg # 2)^2	(Length of Hypotenuse)^2
i.						
ii.						
iii.						
iv.						
v.						

RT-25. Spiral of Right Triangles

To re-create Karla's spiral of right triangles step-by-step, follow the instructions in your text for part a. Be sure to label each side as you draw.

Use this as the 1 unit length: |———————|
                                  1 unit

                                                                    1 unit
                                                           |————————|
                                                              1 unit

RT-26. Locating Square Roots on a Number Line

Locate each of the numbers $\sqrt{3}$, $\sqrt{4}$, $\sqrt{5}$, ..., $\sqrt{8}$ on the given number line as was done for $\sqrt{2}$. Do this by using Karla's idea of marking lengths on a strip of paper to transfer the length of the hypotenuse of each of your right triangles from RT-25. Also locate $-\sqrt{2}$ and $-\sqrt{5}$.

RT-30. Square Root Values

Complete the following table. Write the exact value for each of each length x and then estimate its value by using the number line from RT-26. Finally, use a calculator to find a decimal approximation for x to the nearest hundredth.

Square of the hypotenuse, x^2	Exact length of the hypotenuse, x	Estimated value of x from a number line	Approximate value of x from a calculator
1			
2			
3			
4			
5			
6			
7			
8			
9			
10			

RT-53. Area and the Pythagorean Relationship

Dot paper measuring strip

RT-54.

For each graph, draw a right triangle and use it to determine the length of the given line segment. Give both the exact value and an approximation to the nearest tenth.

a.

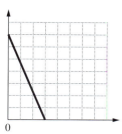

b.

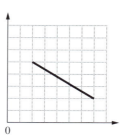

c.

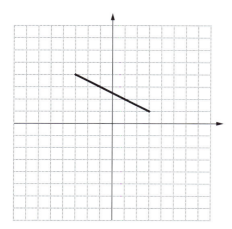

d.

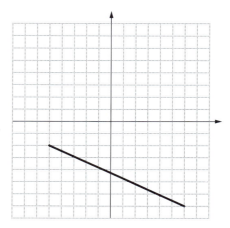

Chapter 7 Resource Page

RT-62. Are Segment Length and Slope Related?

Use this dot paper to explore the relationship between the slope of a line segment and the length of the segment as instructed in the problem.

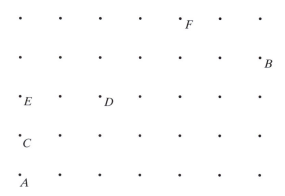

a. Draw in the segments *AB*, *CD*, and *EF*.

b. Find the slopes of segments *AB*, *CD*, and *EF*. $\frac{3}{6}, \frac{1}{2},$ and $\frac{2}{4}$

c. How are the slopes of segments *AB*, *CD*, and *EF* related?   They're equal.

d. Use the Pythagorean relationship to find the lengths of segments *AB*, *CD*, and *EF*.
   $|AB| = \sqrt{45}, |CD| = \sqrt{5},$ and $|EF| = \sqrt{20}$ = units

e. Does the length of a line segment seem to be related to the slope of the segment? Explain your answer.

GG-131. The 100-meter Freestyle

The data points on the graph below represent winning times in the Olympic 100-meter free-style swimming race for all Olympic years from 1912 to 1988.

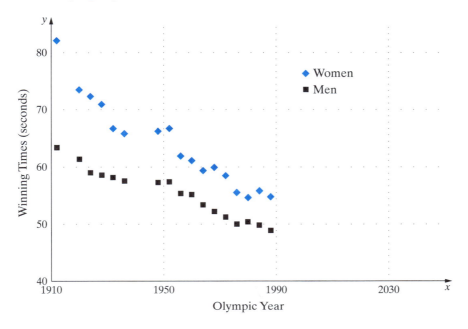

Appendix A Resource Page

Integer Tiles

Glue the entire grid of integer tiles to a piece of lightweight cardboard (an empty cereal box works well) or another piece of paper. Carefully cut out the individual tiles and keep them in a resealable envelope or plastic bag. Store the bag of tiles in your notebook.

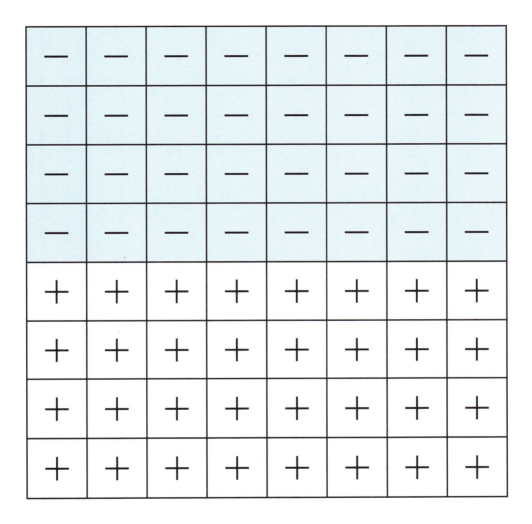

RP-103

Appendix A Resource Page

Integer Tiles Balance Sheet

| Debits | Credits |

SOME ANSWERS AND SOME WAYS TO GET STARTED

CHAPTER 1

OP-5. a. $N = 12, S = 7$ b. $N = 66, S = 17$
  c. 3 and 4   d. $N = 3, W = 1$
  e. $N = 6, S = -5$ f. $N = -4, S = 3$
  i. $N = \frac{1}{4}, S = 1$
  k. $N = x^2, S = x + x$, or $2 \cdot x$

OP-7. a. Since it takes two lengths and two widths to make a rectangle, one length and one width must add up to 6. How many ways are there to get 6?
  b. Is this possible? Why not? Your explanation should include the fact that there are two of everything, two lengths and two widths.
  c. There are four possibilities. Think of shapes like T's, L's, and Z's.

OP-8. a. Perimeter = 14 cm   c. P = 32 units

OP-12. a. $N = -21, S = 4$
  b. $N = \frac{14}{9}, S = 3$
  c. -5 and 1
  f. $W = 0, S = \frac{4}{7}$
  g. $E = 0.5, S = -0.25$

OP-14. a. The first way gives 56; the other way, 380
  b. First, simplify each expression within the **parentheses**. Second, compute the **powers**, if there are any. Third, do the **multiplication** and **division** in order from **left** to **right**. Finally, do the **addition** and **subtraction** in order from **left** to **right**.

OP-17. a. About 0.0022 lb   b. About 2.2 million lb

OP-18. a. $[6] \cdot 6 + 6 = 42$
  b. It belongs on the "7 line."
  c. Try putting parentheses around one of the 6 + 6 groups.

OP-20. b. 29   c. $\frac{-2}{5}$ f. -13
  h. -1 i. 13

OP-25. a. 1000 days, or about 2 years and 9 months

OP-26. a. $N = \frac{-5}{16}, S = \frac{1}{8}$ b. $-2, 3$
  d. $N = \frac{1}{8}, S = \frac{11}{12}$

OP-30. c. 9 sq cm

OP-31. On the tenth day he will receive $2 \cdot 2 \cdot 2 \cdot 2 \cdot 2 \cdot 2 \cdot 2 \cdot 2 \cdot 2 = 512$ pennies, and on the 30th day, 536,870,912 pennies.

OP-32. b. There are several possibilities; $8 \cdot 8 = 8^2$ may be the first one you find, but try using 4's. How many 4's can be multiplied to get 64? Since each $4 = 2 \cdot 2$, you can replace each 4 with $2 \cdot 2$. So how many 2's would you need to multiply together?

OP-33. a. Missing lengths: 11 ft and 4 ft
  Baseboard: 46 ft; Area: 133 sq ft

OP-34. a. $8 \cdot 8 \cdot 8 = 512$
  b. $1 \cdot 1 \cdot 1 \cdot 1 \cdot 1 \cdot 1 \cdot 1 \cdot 1 = 1$
  d. $3 \cdot 3 \cdot 3 \cdot 3 \cdot 3 = 243$
  f. Compute the value of 2^6 and compare it to what you get for 6^2.

OP-36. b. $(5^2)(5^2)(5^2) = (5 \cdot 5)(5 \cdot 5)(5 \cdot 5) = 5^6$
  d. $(5^2)^3 = (5^2)(5^2)(5^2)$ which is the same as part b.

OP-37. Start by multiplying the input number by three. Then what?

OP-38. a. Example: input -5 gives an output -4.
  c. If x represents the input, the output is $3x - 1$

OP-39. b. $\frac{1}{2}bh$ means multiply $\frac{1}{2}$ times b, then multiply the resulting product times h, or, if you recognize this formula for the area of a triangle: $\frac{1}{2}$ times base times height.

OP-40. b. $2a$ c. $22n$

OP-41. a. 43   b. 208   c. -3

OP-42. d. $N = \frac{1}{6}, S = \frac{5}{6}$ e. $E = 8, S = 8\frac{1}{2}$
  f. $N = 2a, S = a + 2$

OP-45. What if you started with a guess of 10 cm for the length? The width would have to be $(\frac{2}{3})(10) \approx 6.67$ cm. What perimeter will that give? Next, try a larger number like 15 or 30.

OP-47. Sequences like 2, 4, 6, 8 and 56, 58, 60 contain **consecutive even** numbers. Two consecutive even numbers could be 92 and 94. On the other hand, 87 and 89 would be **consecutive odd** numbers. Start by guessing an even number such as 56, then the other number will have to be 58. Since their sum is only 114, the guess, 56, is too small.

OP-48. You could start with 40 admissions at $3.95. Then there would have to be 66 more, or 106 at $8.95, but that will turn out to be too low. Try a larger number.

OP-51. Try starting with 100 bouquets of daffodils. Then there are 127 bouquets of freesias. But a total of 227 is too small. Guess a larger number.

OP-52. b. $3^6 = 729$

OP-53. Evaluate the expression $\frac{1}{5}S$ for $S = 18$.

OP-54. a. 2^3
  c. There are two possibilities: 9^2 or 3^4
  d. What small number will divide 243? How many of these would you need to multiply together to get 243?

OP-55. a. $(3^4)^2 = (3^4)(3^4) = (3 \cdot 3 \cdot 3 \cdot 3)(3 \cdot 3 \cdot 3 \cdot 3) = 3^8$
  c. Be sure you show three groups of three 2's each.
  d. Your final result should be either $2^3x^3y^3$ or $8x^3y^3$.

A-1

OP-56. a. F, 34 b. T c. T
d. F, 2 e. T f. F, −7
g. F, 2 h. T

OP-68. Try starting with 100 for the number of Democrats. Then the number of Republicans is thirty more than half, or 50 + 30. The guess 100 is too small, since 100 + 80 doesn't add up to 327.

OP-69. "Four consecutive integers" here just means any four integers in a row. For example, you could guess 28, 29, 30, and 31. In this case the product of the two largest is $30 \cdot 31 = 930$, and the product of the two smallest is $28 \cdot 29 = 812$. Since $930 - 812 \neq 74$, try a different set of four consecutive integers.

OP-70. Evaluate the expression $P - P(0.01r)$ for $P = 350$ and $r = 15$. Then, $S = \$297.50$.

OP-71. a. 0 b. $3\frac{3}{5}$ d. 16 e. 12

OP-73. a. $(6xy)^2 = (6 \cdot x \cdot y)(6 \cdot x \cdot y) = 36x^2y^2$
b. $(6x^2)^4 = (6 \cdot x \cdot x)(6 \cdot x \cdot x)(6 \cdot x \cdot x)(6 \cdot x \cdot x)$
$= 1296x^8$
c. $6(x^2)^4 = 6(x \cdot x)(x \cdot x)(x \cdot x)(x \cdot x) = 6x^8$
e. $(x^2)(6x)^2 = (x \cdot x)(6 \cdot x)(6 \cdot x) = 36x^4$

OP-76. b. A few of the possibilities: $49.2 \cdot 10^{11}$, $492 \cdot 10^{10}$, and $4,920,000 \cdot 10^6$.

OP-77. c. 5,340, $5.34 \cdot 10^3$

OP-78. It would be more than enough to pay off the debt. The contributions would total about $5.25 trillion, so there would be almost $330 billion in overpayments.

OP-79. a. About $18,707
b. One share would weigh about 41.2 pounds.

OP-80. About 231 calculations per second.

OP-81. a. One way to get started would be to figure out how many bunches of 454 bills are in a stack of one billion bills.
b. If each bill is 15.6 cm long, what is the length of a billion of them? Now to get the answer in kilometers you need to know how many centimeters there are in one km. The length is about 156,000 or $1.56 \cdot 10^5$ km. (That amounts to about 97,000 mi.)
c. This problem is similar to part b. The result is about 66,300 or $6.63 \cdot 10^4$ km. That would be about 41,100 mi.

OP-82. First figure out how many dollars you could count in one year at the rate of one per second.

OP-84. About 269 miles by 361 miles.

OP-85. a. $64x^6$ b. $4x^6$ d. $16x^5$

OP-86. c. $[-13 + (-12)](-4) = 100$
d. $(-8)(-8) - (-8) = 72$
g. $(1\frac{1}{2})(6)(0) = 0$ h. $\frac{3}{4}(100) = 75$

OP-88. a. 200.96 b. 631.0144 d. 1250

OP-89. b. What happens when E and W are opposites, such as 3 and −3, or when one of them is 0?

OP-91. c. 10 d. Top

OP-93. a. 120 miles d. $\frac{2}{3}$ hour, 38 miles
e. If you know your speed in miles per hour and you know how many hours you traveled, what mathematical operation do you do to figure out the number of miles you went?

OP-94. c. Part b gives you one guess—here's another: If he drove 3.5 hours at 64 mph, then he drove only 2 hours at 80 mph, but $3.5(64) + 2(80)$ is not enough miles.

OP-95. b. Make sure you have a systematic way of making the list. One way is to make the list based on the possible number of heads: 3, 2, 1, or 0. For three heads there is just one possibility, HHH. For two heads there are three ways: HHT, and two more. For one head there are three ways. And for no heads there is only TTT.

OP-96. a. $\frac{5}{6}$ b. $\frac{3}{6} = \frac{1}{2}$
c. $\frac{26}{52} = \frac{1}{2}$. If you are unfamiliar with a regular deck of playing cards see the Glossary for a more complete description. The deck contains four suits of 13 cards each. Two of the suits are red and two are black.
d. $\frac{4}{52} = \frac{1}{13}$

OP-97. a. Car 1 b. Car 2 c. Car 1
d. It's brand new and still on the dealer's lot.

OP-98. a. True. b. True. c. False. d. True.

OP-99. About 35,000,000 or $3.5 \cdot 10^7$ oz

OP-100. The first number is 9.

OP-101. b. If E and W are both negative what happens to their product?

OP-102. $(6x^5)^2 = (6x^5)(6x^5) = 36x^{10}$
But $6(x^5)^2 = 6(x^5)(x^5) = 6x^{10}$

OP-103. b. −60 d. −175 f. 3
h. ≈8.07 j. 0

OP-106. Now is a good time to just guess 10 and see what happens. In ten years Woody will be 38 and Susie will be 18. Pretty close. Continue guessing and checking.

OP-107. a. This is another problem for which using a guess-and-check table is a good strategy.
b. $\frac{(\text{number of five's})}{8}$

OP-108. a. You'll need to figure out how many seconds there are in one year. About $5.87 \cdot 10^{12}$ miles
b. About $2.48 \cdot 10^{13}$ miles
c. If you knew your friend's house was 150 miles away, and that you could average 60 miles per hour driving there, what operation would you use to figure out how long it would take?

OP-110. b. -4 d. 8
f. -4 h. 0

OP-111. a. On the Cruising Speed graph point A should be to the right of and lower than point B, and on the Range graph, point A is to the right of and higher than point B.
b. It's your choice, but keep in mind that Car A is older, larger, slower, and has a greater range relative to Car B.

OP-112. a. 31 c. 42

OP-113. a. You'll need parentheses to show that the adding should be done first.
b. $\dfrac{N^2}{4}$

OP-114. The point for student A should be farthest to the right, about medium height. The point for student B should be close to the vertical axis and higher than the point for student A. The point for student C should be **on** the vertical axis and lower than the point for student A.

OP-115. a. About 720,000 or $7.2 \cdot 10^5$ cm.
b. 109 or $1.09 \cdot 10^2$ km, (which is about 68 mi)

OP-116. a. $36x^6$ b. $16x^4y^4$
c. 5^8, or 390,625 d. $7x^3y^6$
e. x^9 f. $9x^4y^2$

OP-117. a. 0 d. -18

OP-119. a. 48 b. $-1{,}000$
c. $1{,}000{,}000$ d. -13

OP-120. Try guessing 40. Then the other piece is 58, but $40 + 58$ is too long. So try 30.

OP-121. Try guessing the cost of the lamp first, then figure out the cost of the desk and then the chair. For example, if the lamp costs $10.00, the desk costs $40.00, and the chair costs $17.00.

OP-122. a. $N = -24$, $S = 2$
b. $N = 45$, $S = 14$
c. $N = -2a$, $S = a - 2$
d. E and W are -6 and -2.
e. $N = \frac{-3}{16}$, $S = \frac{1}{2}$
f. E and W are both $\frac{1}{2}$.
g. $N = -52$, $E = 4$

OP-123. The perimeter is 44 ft and the area is 79 sq ft.

OP-125. b. The position of the dot is related to the figure number. If a figure number is a multiple of 4, then that square has a dot in the lower left-hand corner. Use the fact that 24 is a multiple of 4 to determine figure number 27.

CHAPTER 2

SP-3. a. There are about 15 cans at 10:30 A.M. and about 90 cans at 11:00 A.M.
b. It was refilled.

SP-4. e. A(2, 5), B(-3, 4), C(-5, -2), D(0, -3), E(2, 0), F(4, -3), G(5, 2)

SP-5. The best set of axes is iv. To see why the others are not, focus on what is unacceptable about the vertical scale in iii, and look at what is wrong with the horizontal scale in i. Both scales in ii have a problem.

SP-6. When substituting a value for x in the formula $y = -x + 2$, you can think of the $-x$ term several ways. For example, $-x$ means "the opposite of x," or it can also be thought of as "-1 times x".

Input (x)	-4	-3	-2	-1	0	1	2	3	4
Output ($y = -x + 2$)	6	5	4	3	2	1	0	-1	-2

SP-7. An input of -4 gives an output of -11.

SP-9. An input of 2 gives an output of 6, and when 10 is the output the input is 6. The best clue is that when the input is 1000 the output is 1004.

SP-10. a. i. no; ii. yes; iii. yes; iv. no.
b. They are reciprocals because $(1\frac{1}{3}) \cdot \frac{3}{4} = 1$.
c. As long as x is not zero, its reciprocal is $\dfrac{1}{x}$, because $x \cdot \dfrac{1}{x} = 1$.
d. By Max's method the reciprocal of $\frac{3}{4}$ is $\frac{4}{3}$. This is true because the product of $\frac{3}{4}$ and $\frac{4}{3}$ is 1.

SP-11. Zero does not have a reciprocal, and the calculator will give an ERROR message for $\frac{1}{0}$. The problem is that 0 times any number is equal to 0, so 0 times $\frac{1}{0}$ could never be equal to 1.

SP-12. a. The rectangle should be 8 units long and 6 units wide.
b. The area is 48 sq units.

SP-13. b. 4^6 d. 4^5 e. $(4^3)^2$ is different.

SP-14. c. There were 13 nickels and 26 dimes in the slot machine.

SP-16. a. blue: (0, 1); red: (0, -4).
b. blue: $(\frac{-1}{2}, 0)$; red: (2, 0) and (-2, 0).

SP-17. b. The perimeter of Figure 9 is 56 units.

SP-20. b. The area of the larger section is $9 \cdot 20 = 180$ sq feet.
c. Area as a Product = Area as a Sum
$9 \cdot (20 + 4) = 9(20) + 9(4)$
$9(24) = 180 + 36$
$216 = 216$

SP-21. a. In square centimeters,
$6(10 + 3) = 60 + 18$.
b. In square units,
$(10 + 5) \cdot 7 = 70 + 35$
c. In square centimeters,
$8(20 + 1) = 160 + 8$

SP-23. Output values (in order):
5, 4, 3, 2, 1, 0, -1, -2, -3.

SP-24. When the input is −4, the output is 14.
When the input is −2, the output is 2.
When the input is 3, the output is 7.
When the input is 1, the output is −1.

SP-25. a. The rectangle is 18 units by 6 units.
b. The area of one section is 6(10) sq units.
The area of the other section is 6(8) sq units.

SP-26. Some possibilities:
$2^6 = 2^2 \cdot 2^4 = 2^5 \cdot 2^1 = 2^3 \cdot 2^3$

SP-27. a. 10^3 b. 10^7 c. 10^6

SP-28. $(10^2)^3$ is different.

SP-29. a. Think of the questions being numbered: 1, 2, and 3. Then think of the possible answers. All true would be TTT. The first one false and numbers 2 and 3 true would be F, T, T or FTT. You made a similar list for the problem about three coins, OP-95, but maybe it already occurred to you that taking some true-false tests is like flipping coins.

SP-30. "Compute half of your input number and add three." Or, "Divide your input number by 2 and add 3."

SP-31. b. East and West are 6 and −2.
c. East is −16, South is $-15\frac{1}{2}$.
d. North is $\frac{1}{10}$, South is $\frac{7}{10}$.

SP-32. a. −500.3
b. If you can subtract 12 from 92 mentally, then you should be able to do this problem mentally too.
c. $\frac{-1}{2}$
d. $4\frac{1}{5}$ is the same as $\frac{21}{5}$. This improper fraction is easier to multiply by 5 mentally.
e. To do this problem mentally, first compute the division. Then you are adding three positive numbers. Notice that it is easier to add 18 and 72 first, since 8 + 2 is 10.
f. $-2^4 = -2 \cdot 2 \cdot 2 \cdot 2 = -16$, so the final answer is −8.

SP-33. See the answer for OP-47 for a description of consecutive even and odd integers, or look in the Glossary. Suppose you guess 9. Then the four integers would be 9, 11, 13, and 15. To check this guess, take the second one, 11, and add it to twice the fourth, 2 · 15. The result is 11 + 2 · 15 = 41, too small. Try again.

SP-34. When the input is 100 the output is −200. If the input were −43, the output would be 86.

SP-35. b. △ c. △
This problem is based on a pattern similar to the one in OP-125, only the multiple is different because the figure is a triangle rather than a square.

SP-36. a. If she is driving 58 miles per hour, how many miles does she drive in one hour?
b. Look in your Tool Kit for a reminder, or look back at OP-93.

SP-37. c. 1 hour and 15 minutes

SP-38. a. Use the fact that the problem states that it takes Professor Speedi half an hour when she drives 60 mph.
c. For example, if traffic forces Professor Speedi to drive to her college at 25 mph, it will take her $1\frac{1}{5}$ hours, or 72 minutes.

SP-40. Use the graph by starting at the time (15 minutes) on the vertical axis, going across to the graph (or curve), and then down to the speed (about 120 mph) on the horizontal axis. For a more accurate answer, divide the distance to the college by the amount of time given, to get the (constant) rate at which she must drive: $r = \frac{d}{t} = \frac{30 \text{ mi}}{\frac{1}{4} \text{ hr}} = 120$ mph.

SP-41. The method is similar to that in SP-40.

SP-42. b. Notice that $(-4)(-4) = 16$ but the output was 17, but $(3)(3) = 9$ and the output was 10, and, $(-1)(-1) = 1$ but the output was 2. How about an input of 2—does the relationship hold? Check the other inputs and outputs.

SP-43. a. Missing outputs: −5, −14
c. One way to write the rule is $y = \frac{x}{2}$, but there are several others that are also correct.

SP-44. b. The outputs for the table are (in order):
$-0.2, -0.\overline{3}, -0.5, -0.6, -1, -3.03, -5$, Not defined, 5, 3.03, 1, 0.6, 0.5, $0.\overline{3}$, 0.2

SP-45. a. The graph should be in two parts, both shaped a bit like the graph for Professor Speedi's Commute (SP-38) that was done in class. The part on the lower left should look like a diagonal reflection of the piece on the upper right. There should be no points on the x- or y-axes.
b. As x gets larger, y gets smaller.
c. As x decreases from 1 to 0, y gets larger. As x increases from −1 to 0, y gets smaller. (Remember numbers further down the y-axis are *smaller* negative numbers.)
d. Nothing. There are no y values for x = 0.
e. i. x = 0.5; iii. x = −0.5; v. x = −4

SP-46. a. 2^7 b. 2^{12}
c. 8 + 16 = 24   f. x^6

SP-47. a. 10(10 + 3) = 10(10) + 10(3)
10(13) = 100 + 30
b. 2(10 + 3) = 2(10) + 2(3)
2(13) = 20 + 6

SP-48. **a.** 1, 2, 3, 4, 5, or 6
b. (1, 1), (1, 2), (1, 3), (1, 4), (1, 5), or (1, 6)
c. (1, 1), (1, 2), (1, 3), (1, 4), (1, 5), (1, 6), (2, 1), (2, 2), (2, 3), (2, 4), (2, 5), or (2, 6)
d. There are 36 of them. Twelve of them are listed in part c. You need 24 more.
e. $\frac{4}{36} = \frac{1}{9}$

SP-49. Try 50 centimeters for the shorter piece, so the other one is 74 cm. Then the sum is only 124 cm. Try a bigger guess for the shorter piece.

SP-50. **b.** **i.** 36; **ii.** 128; **iii.** 28; **iv.** 8; **v.** 120; **vi.** 136; **vii.** 16; **viii.** 1024
c. **ii.** $2^5 \cdot 2^2 = 2^7$; **iv.** $2^5 \div 2^2 = 2^3$; **vii.** $2^7 \div 2^3 = 2^4$; **viii.** $2^7 \cdot 2^3 = 2^{10}$. What do these expressions have in common?
d. The rest of them. What do the rest have in common?

SP-52. **b.** $4^8, 4^{10}, 4^{12}$ **c.** 6
d. The high school student's answer should be 4, but how do you know that?

SP-56.

x	-4	-3	-2	-1	-0.5	0	0.5	1	2	3	4
$y = x^2$	16	9	4	1	0.25	0	0.25	1	4	9	16

SP-57. **a.** $q + 3m$ **d.** $3y + 11$

SP-58.

x	-1	9	$2\frac{1}{3}$	$\frac{1}{2}$	x
$y = $ ___	6	-4	$2\frac{2}{3}$	$4\frac{1}{2}$	$5 - x$

SP-60.

x	-5	-10	100	x
$y = $ ___	-13	-28	302	$3x + 2$

SP-61. Be sure you mention that all the graphs are straight lines, and that one of the graphs is much steeper than the others. Also, the graph for $y = 5 - x$ goes down from left to right while the other two go up. Because their graphs are straight lines these three equations are called linear equations.

SP-62. **a.** $(5, 0), (-2, 0), (\frac{-2}{3}, 0)$
b. The y-coordinate is zero.

SP-63. **a.** Look at SP-61 parts b and c.

SP-64. **a.** The parabola: $(-1, 0)$ and $(5, 0)$
b. The straight line: $(4, 0)$
c. The "up-down-up" graph (It's called a cubic or polynomial graph.): $(-5, 0), (-2, 0), (3, 0)$
d. The reciprocal graph (part of a hyperbola): no x-intercepts

SP-65. **a.** One way would be to refer back to OP-95 when you made the list for three coins, a penny, a nickel, and a dime. You already have 8 results for flipping the three coins—now think of flipping the quarter. For any of the 8 results the quarter could be heads, or for any of the 8 it could be tails. That's just one way of many to organize your list.
b. $\frac{4}{16}$ or $\frac{1}{4}$ **c.** $\frac{11}{16}$

SP-66. **a.** The area of a rectangle can be found by finding the product of its dimensions; that is, by multiplying its width times its length.
b. The area of the smaller part is 9(5), or 45 square units. Find the area of the larger part, and then add the two together to find the total area.

SP-67. **a.** "To get the output, square the input number, and then multiply the result by 3."
b. "To get the output, multiply the input number by 3, and then square the result."

SP-68. Once you write out each expression using the meaning of exponents, only the expression in part a is a product of 2's. That's why only the expression in part a can be rewritten as a power of 2.

SP-69. **a.** $2,420,000,000 or $2.42 \cdot 10^9$
b. If you make up a similar problem with easier numbers, you can sometimes figure a method to use on the more difficult problem. Try this similar problem with smaller numbers: If 8 bikers spent a total of $840, how much did each spend? What mathematical operation did you use to decide?
c. ≈ 16.7 miles single file; ≈ 8.3 miles in pairs

SP-70. **a.** $y^3 \cdot y^2$ means $(y \cdot y \cdot y)(y \cdot y)$.
b. y^5
c. $y^4 \cdot y$ means $(y \cdot y \cdot y \cdot y)(y) = y^5$

SP-71. What happens when you just start with the standard guess of 10 for the first number?

SP-78. $7(2) + 7(x) = 7(2 + x)$
or $7(2) + 7(x) = 7(x + 2)$

1	1		$1x$
1	1		$1x$
1	1		$1x$
1	1		$1x$
1	1		$1x$
1	1		$1x$
1	1		$1x$

$5(x + 3) = 5(x) + 5(3)$
or $5(x + 3) = 5x + 15$

$1x$	1	1	1
$1x$	1	1	1
$1x$	1	1	1
$1x$	1	1	1
$1x$	1	1	1

SP-79. a. $3(x + 2) = 3(x) + 3(2)$
b. $(x + 5)2 = (x)2 + (5)2$
c. $5(x + 1) = 5(x) + 5(1)$

SP-80. a. D b. F c. none
d. B e. none f. A

SP-81. b. Double the number and add three; or multiply by 2 and add 3.
c. $y = 2x + 3$
f. No, since only finitely many points would be plotted.
g. No.

SP-82. b. It would not make sense to connect the points since the figure numbers are always whole numbers, never fractions or decimals. The graph would be discrete.
c. Because you can calculate a time for any speed greater than 0, the graph would be continuous. So, you can connect the points.

SP-83. a. 22, 25, 28 for the first sequence, and 13, 21, 34 for the second.

SP-84. a. For example: (1, 14), (3, 12), (12.5, 2.5), (10.7, 4.3), (16, −1), (15, 0).
b. $5\frac{2}{3}$
d. Yes—*any* two numbers may be chosen as long as their sum is 15.

SP-85. Suppose Dorothy drove 100 mph, and Frank drove 95 mph. How far did Dorothy travel after 4 hours? How far did Frank go in that same amount of time? Add their distances to find how far apart: $100(4) + 95(4) = 780$ miles.

SP-86. a. $30 - x$ ft
d. Whole amount − one part = other part

SP-87. a. i. 3^7; v. 252; vi. 288

SP-88. b. Choose numbers that are bigger than 8. For example, some integers are 10, 230, and 65; some non-integers are 8.01, $9\frac{1}{2}$, and 27.99.
c. -2 is one choice that works.

SP-89. b. Up to the ninth line, $123456789 \cdot 9 + 10 = 1{,}111{,}111{,}111$

SP-94. a. Your graph should pass through the origin because if you buy 0 gallons of gas, it will cost $0.

SP-95. a. Some possibilities are: (0, 80), (1, 64), and (3, 32).
b. All the points lie on a line.
c. The x-intercept of a graph is always of the form (?, 0).

SP-96. a. $-27, -3$ b. $-3, -3$ d. 8, 0

SP-98. a. $2x + 7$

SP-99. a. $4(x + 9) = 4x + 36$
b. $6x + 3$

x		1

c. $3(x) + 3(x) = 3x + 3x = 6x$
d. $6(x + 1)$

SP-100. a. D b. B c. H
d. J e. I

SP-101. No. Try using the order of operations to calculate each side of Deanne's equation. The values are different, so the distributive property doesn't work this way.

SP-102. a. Look at the x-coordinates in part c.
c. The parabola: $(0, -5)$; the straight line: $(0, 2)$; the "up-down-up" graph (called a cubic polynomial graph): $(0, -30)$; the reciprocal graph (part of a hyperbola): no y-intercept

SP-103. a. 36. Refer to SP-48.
b. 5 c. $\frac{5}{36}$

SP-104. b. 5, −3 c. $E = 4a, S = 5a$

SP-105. You could guess 10: $10 + 5(11) = 65$. What next?

SP-106. The graph should be discrete.

SP-107. 0.011 centimeters

SP-112. a. It's about $(-1.5, 0)$.
b. $x = -1.5$
c. The value for x is the same.

SP-113. c. $2x + 1$

SP-115. b. $x^2 - 3x$ c. $-7c - 56$
d. $4(x + 2)$ f. $3(x + 3)$

SP-116. b. $\approx 2.22 \cdot 10^9$

SP-117. a. 3^7 b. 252

SP-118. a. Your answer should be less than 50 miles, but more than 25 miles, since Ms. Escargot drove for less than an hour, but more than half an hour.
b. 3 hours and 12 minutes.

SP-120. The missing sides are $12\frac{1}{4} + 2\frac{3}{4} = 15$ inches for the top, and $4\frac{3}{8} + 3\frac{5}{8} = 8$ inches for the vertical side. So the perimeter is $12\frac{1}{4} + 15 + 8 + 3\frac{5}{8} + 2\frac{3}{4} + 4\frac{3}{8} = 46$ in.

SP-121. a. Think about combining the numbers in the shaded squares in each figure.

SP-129. Your answer for SP-128 should help you. If you are still having trouble, look back at the graphs and their equations from SP-2, 6, 7, 23, 24, 42, 58, 59, 60, 92, 111. Try to put these into two groups according to how their graphs look. What do the equations from each group have in common? To which group does the graph of $y = 2x - 3$ belong? How about $y = x^2 - 3$?

SP-130. The graph of $y = x^3$ is a curve that passes through the points $(-2, -8)$, $(-1, -1)$, $(0, 0)$, $(1, 1)$, and $(2, 8)$.

SP-131. The graph of $y = x$ is a line through the origin and the point $(3, 3)$.

SP-132. b. Divide by ten, or multiply by one-tenth.
c. $y = 0.1x$ or $y = \frac{x}{10}$ or $y = (\frac{1}{10})x$
d. The graph is a line that is not very steep; it goes through the origin and $(10, 1)$.

SP-133. The graph looks like the graph of $y = x$ (SP-131) moved up 2.5 units, so its y-intercept is $(0, 2.5)$ instead of $(0,0)$.

SP-134. b. $x^2 \cdot x^2 \cdot x^2 = x^6$ c. $16x^8$
e. The simplified form is $x^6 \cdot y^9$.

SP-136. a. $2(3x + 2) = 6x + 4$
b. $2x + 8$

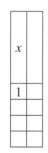

SP-137. a. F c. none d. E
SP-138. c. $8(3 + y)$ d. $-6y + 15$
SP-139. d. The number of dimes is $x + 2$; the value of quarters is $\$0.25(x - 1)$.
SP-140. $3\frac{3}{4}$ qt raspberry; $7\frac{3}{4}$ qt lemon-lime soda
SP-141. a. 6.5 b. 18
SP-142.

Greatest Value	Least Value
$(2^2 \cdot 3^2 - 1)^2 = 1225$	$2^2 \cdot (3^2 - 1^2)^2 = 32$
$(5 + 7 \cdot 3)^2 - 4 = 672$	$5 + 7 \cdot (3^2 - 4) = 40$

SP-143. b. $2x + 6$ c. $2x + 6$

SP-144. a. $2(x + 4) = 2x + 8$
SP-145. b. $y = 2x - 1$
SP-146. It's a parabola with vertex $(-1, 0)$ and y-intercept $(0, 1)$.
SP-147. a. Two possibilities: -3 and 0
b. Two possibilities: -597 and -12.1
c. Two possibilities: 0.1 and 1
d. Two possibilities: 0.508 and 913
SP-148. a. i d. iii
SP-151. Your list should have 12 combinations.
SP-152. c. There were 222 student tickets sold.

CHAPTER 3

LT-7. One possibility: Let A = the number of adults' tickets. $2.50(A) + 1.10(105 - A) = 221.90$; 29 children's and 76 adults' tickets

LT-8. Let the variable represent the number of miles Maude drove. Then Chloe drove that number plus 80. So Maude's miles plus the expression for Chloe's miles must total 520 miles.

LT-9. If W is the width, then how could you represent the length, which is 92 miles longer? The expression for the length times the width equals 97,109. A good guess for W is 250; then the length is 342, but the product is 85,500, which is too small.

LT-10. a. $6x + 12$ d. $6y + 4y$ or $10y$
LT-11. a. (a) 24 (b) 16 b. (c) 54 (d) -30
LT-12. a. 35 b. -15
c. Your explanation should mention the distributive property.
LT-13. a. $(y \cdot y)(y \cdot y)(y \cdot y)$
b. y^6
c. $(x \cdot x \cdot x)(x \cdot x \cdot x) = x^6$
d. $(m \cdot m)(m \cdot m)(m \cdot m)(m \cdot m) = m^8$
e. x^{217}
LT-14. a. $x^{31} \cdot x^7 = x^{38}$, but $(x^{31})^7 = x^{217}$
LT-15. Remember the calculator leaves out the base number 10, so what it means is $1.3 \cdot 10^3 = 1,300$.
LT-16. $(2.1 \cdot 10^5)(3.25 \cdot 10^5) = 6.825 \cdot 10^{10} = 68,250,000,000$
$(2.1 \cdot 10^5) + (3.25 \cdot 10^5) = 5.35 \cdot 10^5 = 535,000$
LT-17. a. It's a fairly steep straight line through the points $(0, -15)$, $(3, -6)$, $(6, 3)$, etc.
b. The x-intercept is $(5, 0)$, the y-value is 0.
c. $x = 5$.
LT-18. d. $\frac{1}{52} \approx 1.9\%$; $\frac{1}{13} \approx 7.7\%$; $\frac{1}{17} \approx 5.9\%$
LT-21. d. sum e. product g. sum

LT-23. Let n = the number of nickels Mr. Jordan has; $0.05(n) + 0.10(n + 3) + 0.25(n - 2) = 5.40$; 14 nickels, 17 dimes, and 12 quarters

LT-24. Let w = the width of the garden. Then the length is $2w + 3$ and you need two of each to total the perimeter. Write the equation. If you are guessing, 10 for the width is not a bad place to start.

LT-25. Let w = the width; $w(2w) = 722$. Notice that 10 cm is way too small. Try 20 cm; it's closer.

LT-26.
- **a.** $(x + 3)(x + 4) = x^2 + 7x + 12$
- **c.** $(x + 2)(3x + 3) = 3x^2 + 9x + 6$
- **d.** $x(x + 3) = x^2 + 3x$
- **e.** $2x(x + 5) = 2x^2 + 10x$

LT-27.
- **a.** $100 + 70 + 20 + 14 = 204$ sq ft
  $(10 + 2)(10 + 7) = 204$ sq ft
- **c.** $200 + 60 + 20 + 6 = 286$ sq ft
  $(10 + 1)(20 + 6) = 286$ sq ft

LT-29. **c.** $x^2 - 2x$ **d.** $3(x + 2)$ **f.** $m(m + 5)$

LT-30. **b.** x^7 **c.** 2^{12}

LT-31.
- **a.** $64x^6$ **d.** $49x^{10}$ **e.** $9x^2$
- **g.** Each pair of expressions has the same result. For example, $4^3(x^2)^3 = (4x^2)^3$
- **h.** Here's an example: $(8x)^3 = 8^3(x)^3$. You should give a different one.

LT-32. The graph of $y = x$ is a straight line through $(0, 0)$. It makes a 45 degree angle with the x-axis. The graph of $y = x + 2$ is a straight line parallel to $y = x$ and two units higher.

LT-33.
- **a.** About 68 tortillas per hour.
- **b.** They'd have to make about 13 tortillas every 53 seconds.

LT-34. **b.** When $x = 2$ the result is -10; when $x = -3$, the result is -30.

LT-35.
- **a.** One set of parentheses will do the job. One of the multiplication problems needs parentheses.
- **b.** You'll need three sets of parentheses. Think first about grouping a multiplication problem and then a subtraction problem. Then one more well-placed set of parentheses will do it.

LT-40. If p is the number of pies Andy ate on the first day, then the second day he ate $p + 2$, and the third day $p + 4$, etc. Add all those up to get 133. Be sure you have an expression to represent the number of pies he ate on each of the seven days. He ate 13 pies the first day.

LT-41. Variable: Let x represent the first number. Equation: $x(100 - x) = 2176$

LT-42. **a.** 11 **c.** 7.75 **d.** $\frac{2}{3}$

LT-43. Each sample packet contains 26 jelly beans.

LT-44. If x represents the first piece, the others are $2x$, $4x$, and $8x$. Remember to use ounces, so all *four* pieces add up to 80 ounces.

LT-45. Let x = the favorite number. Equation: $3x = 2x + 12$.

LT-46. $3x + 23 = x + 33$

LT-47.
- **a.** $2x^2 + 3x + 4$
- **d.** $(x^2 + 2x + 4) + (x^2 + 3x) = 2x^2 + 5x + 4$

LT-48. $(2x^2 + 3x + 1) + (x^2 + 5x + 8) = 3x^2 + 8x + 9$

LT-50.
- **a.** 6^{59} **b.** 6^{850} **c.** $1296x^8$
- **d.** x^3y^3 **e.** $6x^{11}$

LT-51.
- **a.** The result is $32x^{15}$. Be sure you show clearly how to get it.
- **b.** $362{,}797{,}056x^{44}$

LT-52.
- **a.** $(2x + 3)(x + 4) = 2x^2 + 11x + 12$
- **b.** $(2x + 5)(3x + 2) = 6x^2 + 19x + 10$

LT-53. **a.** $4y - 28$ **b.** $7(y + 8)$ **c.** $6z^2 - 12z$ **d.** $m(m + 1)$

LT-54.
- **a.** $x = -1, 2$
- **b.** If $x = -4$, the product is $(0)(-9) = 0$. What other number works?

LT-55. **a.**

x	-3	-2	-1	0	0.5	1	2	3
y	-6	-4	-2	0	1	2	4	6

b. The graph is a fairly steep straight line through the point $(0, 0)$.

c.

x	-3	-2	-1	0	0.5	1	2	3
y	9	4	1	0	0.25	1	4	9

d. The graph is a parabola with its vertex at $(0, 0)$.

LT-56.
- **a.** $N = -27, S = -1.5$
- **b.** One number is 6.
- **c.** One number is 8.
- **d.** One number is 2.

LT-60. **b.** $\frac{5}{24} \approx 0.21$ **c.** $19\frac{1}{2}$ **d.** 8 **e.** $-13\frac{5}{7} \approx -13.71$

LT-61. **c.** $(x + 2)(3x + 5) = 3x^2 + 11x + 10$

LT-62.
- **a.** Undo list: Multiply by 2, add 5, divide by 7, subtract 3.
- **b.** Result: $x = 87$

LT-63. **a.** $\frac{12}{5}$ **c.** $\frac{47}{12} \approx 3.92$ **d.** 7.125

LT-64.
- **a.** $3(x + 4) = 3x + 12$
- **c.** $(x + 2)(2x + 3) = 2x^2 + 7x + 7$

LT-68. **a.** In LT-66, you used the distributive property in the first step. In LT-67, you divided first. Both ways led to the same solution.

LT-68.
b. Some people prefer not to work with fractions such as $\frac{7}{3} \cdot x = \frac{5}{3}$
c. If you didn't have a calculator the multiplication would be pretty ugly, and even if you do have a calculator you could save some steps by dividing first. $x \approx 2.02$

LT-69. a. $x = 188$

LT-70. A guess-and-check table is a good way to start to figure out how to write the equation. The number is -3.

LT-71. Let $q =$ the number of quarters, so twice as many dimes is $2q$, and… If you're not sure how to get the equation try a guess-and-check table. When you figure out the number of each kind of coin, be sure to re-read the problem so you can give the answer to the question that was asked. There are 62 coins.

LT-72. a. $x = 2.5, -3$

LT-73. a. $(10 + 4)(10 + 3) = 100 + 30 + 40 + 12$

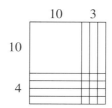

b. $(10 + 1)(20 + 5) = 200 + 50 + 20 + 5$. Draw a composite rectangle.

LT-74. a. Be sure to show the meaning of the expression. The result is x^{10}.
b. $16x^{12}y^8$

LT-75. b. $100x^4$ c. $49x^2$ d. $20x^3$

LT-77. Could Alberto take measurements at any point along the track? How does the speed change?

LT-78. a. $\frac{8}{1440}$ or 1:180
b. Remember there are 2 pints in one quart and 4 quarts in one gallon.

LT-85. a. $x = 4$ b. $x = \frac{12}{b}$

LT-86. b. $x = \frac{43 - c}{4}$
c. $c = 43 - x$

LT-87. a. At the time of purchase $t = 0$. Substitute 0 for t and calculate V. After 10 years, the refrigerator will be worth $245.
b. Substitute 150 for V and then solve the equation to find t. It will be worth nothing in about 18 years.
c. $t = \frac{V - 550}{-30.5}$

LT-88. b. $5(x + 6) = 5x + 30$
c. $(x + 4)(2x + 13) = 2x^2 + 21x + 52$

LT-89. a. $N = 14, S = -9$
c. $N = 24, W = 8$

LT-91. Let m represent the number of men's tickets, so the money taken in for men's tickets is $7.50m$. Equation: $7.50m - 5(750 - m) = 175$. Answer: At the party, 314 men's tickets were sold.

LT-92. There are two ways to think of this: You could start with the original price, p, and subtract $0.27p$ to represent the amount taken off; or you could just say that if they took off 27%, then they only charged 73% of the original price. So $0.73p$ would equal the new price. The original price was $62.

LT-93. a. $\frac{40}{3} \approx 13.33$ b. $\frac{13}{18} \approx 0.72$
c. Remember that $4 - 8(3x - 5)$ means $4 + (-8)(3x - 5)$, and the order of operations implies the multiplication is done first. $x = -2$

LT-94. a. For six hours of sleep, solve $6 = \frac{34 - A}{2}$. According to this formula, 6 hours is enough sleep for someone who is 22 years old.
b. The first step would be to multiply by 2. After that subtract 34. Then what? You could multiply both sides by -1.

LT-95. Here's a possible problem: Amand thought of a number, doubled it, and added 1. Then he added his original number and the new number. The answer was 22. What was the original number?

LT-96. a. Surprised? They're all zero.
b. $x = 0$
c. At least one of the numbers must be 0.

LT-97. a. One of the solutions is $x = -30$ because $(-30 + 30)(-30 - 17) = (0)(-47) = 0$. What is the other solution?

LT-98. a. The graph is a line through the point $(0, -6)$ on the y-axis.

x	0	2	4	6	8	10
y	-6	-4.5	-3	-1.5	0	1.5

b. The x-intercept is $(8, 0)$; the y-value is 0.
d. They should be the same.

LT-99. a. $80x^{12}y^8$ Be sure you showed the meaning of the expression that led you to this result.
b. $8y^{11}$

LT-100. b. The claim is correct. In which section does 3^2 lie? What about 4^{-1}?

LT-101. a. $\frac{1}{6}$ b. $\frac{1}{3}$ c. $\frac{1}{2}$

LT-107. Let s be the length of a side of the original square. How will you represent the length of a side of the new square? Put your expression in parentheses and multiply by 4 and the result should equal 50. Now solve your equation.

LT-108. If x represents the number of canaries, and there are 45 heads all together, how can you represent the number of cats? Multiply the correct number of legs times the appropriate representation of heads, and then show the addition of the results to write the equation. There are 38 canaries.

LT-109. Let x = the number of cubic yards of sand. You could get about 20.89 cubic yards.

LT-110. $(x^2 + 7x + 6) - (x^2 + 5) = 7x + 1$

LT-111. $(2x^2 + 4x + 3) - (4x + 2) = 2x^2 + 1$

LT-112. b. $\frac{14}{3} \approx 4.67$ d. $\frac{4}{3} \approx 1.33$

LT-113. c. Yes; $(-3 - 2)(-3 + 3) = (-5)(0) = 0$

LT-114. a. $(2x - 5)(2x + 5) = 4x^2 - 25$
c. $15x(2x - 1) = 30x^2 - 15x$
d. $(12 - x)(3x + 4) = -3x^2 + 32x + 48$

LT-116. b. 324   d. $100{,}000x^{15}y^{10}$
e. $25x^6$ f. $40x^4$

LT-117. b. $\frac{20}{3} \approx 6.67$ d. 7
e. 76   f. $12, -1$

LT-120. a. $(2, -1)$
b. Move 7 units to the right and 3 units down.

LT-121. a. $\frac{45}{120}$ or 3:8   b. $\frac{16}{12}$ or 4:3

LT-122. a.

x	-1	0	1	1.5	2	3	4
y	6	2	0	-0.25	0	2	6

b. The graph is a line sloping downward from left to right, with y-intercept $(0, 8)$ and x-intercept $(8, 0)$.

LT-131. a. i. x^7 ii. x^{10} iii. $x^5 + x^2$ c. x^{68} d. x^{618}

LT-133. Let s = the length of the side of the original square. How can you represent the dimensions of the new rectangle? How about $s + 12$ and $s - 3$? Now the problem of writing the equation should be a familiar one, because you know you need two lengths and two widths to get the perimeter.

LT-134. b. $C = \frac{5F - 160}{9}$

LT-135. c. 15   d. ≈ 12.02 g. $\frac{-3}{2}$ h. $\frac{-2}{3}$

LT-136. a. $\frac{b}{3}$ b. $\frac{2A}{b}$ c. $2(2r - 5)$ or $4r - 10$
d. $\frac{3}{2}(4s + 6)$ or $6s + 9$

LT-137. Let x = the number of cookies. The number of cookies times the price per cookie plus the cost of the box should equal the total.

LT-138. If x represents the length of one piece, then the piece that is three times as long must be…and together they must equal….

LT-139. So one side is x, another is $2x$, and the third is $2x + 8$. And the total perimeter is 33.

LT-140. Let s = the length of a short piece. How can you represent the length of a long one? Then there have to be four shorts and two longs, and the total length is 152.

LT-141. Let the balance at the start of the year be x. Then add to that $0.08x$ to get the total.

LT-142. c. $0, -16$ d. $4, -4$

LT-143. d. The 3 in the graph of $y = x - 3$ made the whole graph three units lower than the graph of $y = x$. So the line for $y = x - 3$ is parallel to and three units lower than $y = x$.
e. The 3 in the graph of $y = 3x$ made its graph much steeper than the graph of $y = x$, but both still pass through the origin.

LT-144. a. There's a steady increase in speed.
b. There's a quick increase in speed, and then the speed levels off.
c. There is increasing acceleration—the speed gets faster and faster.

LT-145. $3 + 3x = 9 + 2x, x = 6$

LT-146. a. 24   b. 18   c. 4
d. $\frac{10 - c}{3}$ e. $\frac{-3}{2}$ f. $\frac{-3}{2}, -5$
g. $\frac{-(1 + 3c)}{6}$ h. $\frac{y + 5}{c}$

LT-147. b. x^6y^{10} c. $10{,}000x^{24}y^{12}$
d. $x^2 + x^3$ no change, or $x^2(1 + x)$
e. $90x^{12}$ f. $50x^3$

LT-149. Try $\frac{2(2x + 6)}{4} - 3 = \frac{4x + 12}{4} - 3$ and keep going.

LT-150. b. $x(3 + 4) = 3x + 4x = 7x$
d. $(y - 2)(y + 7) = y^2 + 5y - 14$

LT-151. $(3x^2 + 5x + 10) - (2x^2 + 2x + 4) = x^2 + 3x + 6$

LT-152. a. Here's a start:

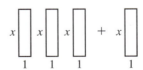

b. Two x-tiles minus one x-tile would leave one x-tile.

CHAPTER 4

EF-2. b. $\frac{S_{\text{new}}}{S_{\text{original}}} = \frac{2}{1}; \frac{P_{\text{new}}}{P_{\text{original}}} = \frac{52}{26} = \frac{2}{1}$; and $\frac{A_{\text{new}}}{A_{\text{original}}} = \frac{88}{22} = \frac{4}{1}$

EF-7. b. The new perimeter is 72 feet, and the new area is 320 square feet.

EF-8. Your explanation should include something about the relationship between 80 and 100.

EF-9. c. Try multiplying $\frac{6}{10}$ by a fraction that is equal to 1 to get an equivalent fraction with a denominator of 100.
d. $x = 6$

EF-10. a. Remember to convert 3.5 hours into minutes before writing the ratio.

EF-11. a. One way would be to use three rectangles, two small—3 by 4 and 2 by 3, and one large—5 by 13.

EF-12. a. To see how to write the equation, remember guess-and-check tables. Try guessing the weight of the bucket, say 2 kilograms.

EF-13. b. $\frac{3}{14} \approx 0.21$
c. To combine the two polynomials on the left hand side, imagine using algebra tiles. Why does $-(2x + 7) = -2x - 7$?

EF-14. a. If you exercise for 2 hours, you'll sweat 8 cups.
b. The graph would be steeper, since the line would go up at a rate of 16 to 1. The equation is $y = 16x$.

EF-15. b. One x-intercept is approximately $(-1.7, 0)$. There is another one.

EF-16. d. $\frac{100}{60} = \frac{5 \cdot 20}{3 \cdot 20} = \frac{5}{3} \cdot \frac{20}{20} = \frac{5}{3} \cdot 1 = \frac{5}{3}$

EF-18. a. $2.744x^6$
b. To check your answer, try substituting a number for x into both the original expression and your simplified expression. If you have simplified correctly, you should get the same result for each. For example, try $x = 10$. Then $(1.4x^2)(2.5x^3) = (1.4 \cdot 10^2)(2.5 \cdot 10^3) = 3{,}500{,}000$. What happens when you substitute $x = 10$ into your simplified expression?
d. $\frac{1}{5}$

EF-19. b. East $= 5z$, South $= 7z$

EF-20. One way to factor $6x^2 + 12x$ is $3x(2x + 4)$. To find other possibilities, try using each of the following for the width (the vertical side) of the rectangle: width $= x$, or $2x$, or $6x$, or 2, or 3, or 6.

EF-26. a. 22.5 sq units c. 2.5 sq units
d. Try using the base of the triangle as one side of a rectangle that surrounds the triangle.

EF-27. a. A ratio equation you could use to solve the problem is $\frac{42}{1} = \frac{x}{6}$.
b. If the side ratio is $\frac{1}{3}$, what is the ratio of areas?

EF-28. a. What should you do to both sides of this equation to undo the division by 3?
b. What should you do to both sides of this equation to undo the division by 5?

d. $\frac{8}{5} = 1.6$
e. In this equation, there are two divisions to undo. $w = \frac{-8}{3} \approx -2.67$

EF-29. If the first number is x, how could you represent the second number? The ratio of the first number to the second is equal to $\frac{1}{5}$.

EF-30. a. Four CDs cost about $22.
f. All the reduced ratios should be the same. What is this reduced ratio?

EF-32. c. 68%
d. Try multiplying by $\frac{20}{20}$. The answer should be more than 100%.

EF-33. b. North $= 8$, West $= -8$

EF-34. b. $|AC| = x + 3$ c. $\frac{1}{4}$

EF-35. b. x^3

EF-37. The answers to parts a and b should be the same.
c. $7^{12} \cdot 10^8$ f. $\frac{-7}{y}$

EF-38. a. Mattel made about $970,000,000 more in 1995, or $9.7 \cdot 10^8$ in scientific notation.
b. Mattel made about 3.26 times as much in 1995.

EF-39. As height increases, weight increases.

EF-40. d. There are two points on the parabola with a y-coordinate of 3, so you should state the x-coordinate of each of them.
e. There are two x-intercepts. One of them is approximately $(0.8, 0)$.

EF-48. a. One of the factors is $x + 4$.
b. The possible factor pairs for 63 are: $1 \cdot 63$, $3 \cdot 21$, and $7 \cdot 9$.

EF-50. a. Try using x as one side of the generic rectangle.
b. Three different rectangles are possible.
c. Try using 3 as one side of the generic rectangle.

EF-51. a. This generic rectangle should have 4 parts. The total area is $2x^2 + 5xy - 3y^2$.
b. This generic rectangle should have only 2 parts. The area of one part will be $15x^3$.
d. This generic rectangle should have 6 parts. The area of one part will be x^3 and another will have area 12.

EF-53. a. about 125.7 in. b. about 12.7 in.

EF-54. b. The diameter of the circle is 8 in., so the radius is 4 in., and the area is $16\pi \approx 50.3$ sq inches.
d. The side of the square and the radius of the circle are each 8 inches. Find one-fourth the area of the whole circle, then subtract it from the area of the square.

EF-55. a. $\frac{4}{50} = 0.08 = 8\%$

EF-57. a. Slide down 20, and to the right 3.
c. 200 lb
f. ≈ 6.7 lb per day. Realistically, the ratio would have to decrease.

EF-58. b. Both equal 0.125.
d. Calculator result: 0.03125

EF-59. c. x^{-7}

EF-60. b. $\dfrac{y^4}{x^2}$

EF-61. Check your Tool Kit to remind yourself of how the distance, speed, and time are related for a moving object. Or, try this simpler problem to help you figure the relationship between distance, speed and time—If you drove 200 miles in 4 hours, how fast on average were you driving?

EF-62. Try a guess-and-check table. If Cassandra deposits $1000 into the 6% account, she'll earn $60 in interest. This is less than the interest she'd earn on $5000 at 4.5%, so she better deposit more than $1000.

EF-63. Note that $\angle O$ and $\angle E$ are each about 123°.

EF-64. b. The corresponding pairs of angles should be equal, even though your measurements with a protractor may not show them to be exactly the same.
d. Because your measurements are not exact, these ratios may not be *exactly* the same, but they should come out the same to the nearest tenth.

EF-68. b. $m\angle G = m\angle N \approx 29°$; $m\angle L = \angle U \approx 130°$
d. $|AL| \approx 63.6$ cm; $|UN| \approx 96.9$ cm

EF-69. c. $\dfrac{\text{Short side}}{\text{Longest side}} = 0.6$; $\dfrac{\text{Short side}}{\text{Middle-length side}} = 0.75$;
$\dfrac{\text{Longest side}}{\text{Middle-length side}} = 1.25$.
d. $\dfrac{\text{Short side}}{\text{Longest side}} = 0.6$; $\dfrac{\text{Short side}}{\text{Middle-length side}} = 0.75$;
$\dfrac{\text{Longest side}}{\text{Middle-length side}} = 1.25$.

EF-70. a. $\dfrac{|DE|}{|AB|} = \dfrac{|EF|}{|BC|} = \dfrac{|DF|}{|AC|} = \dfrac{3}{1}$
b. One pair of equal ratios is $\dfrac{|BC|}{|AB|} = \dfrac{|EF|}{|DE|}$ which equal $\dfrac{9}{17}$. There are two more pairs; one pair is equal to $\dfrac{9}{19}$, and the other is equal to $\dfrac{19}{17}$.

EF-72. This problem is similar to EF-29 where you set up the equation $\dfrac{x}{27-x} = \dfrac{1}{5}$.

EF-73. a. One of the dimensions is $x + 5$.

EF-74. a. The two factors are the same.
c. There are only two possible generic rectangles. Sketch them, and then try them both.

EF-75. b. Area ≈ 8.3 sq cm, Circumference ≈ 11.83 cm

EF-76. Try an easy guess of 2 hours to get your bearings. In 2 hours Maya rides 40 miles and Mike drives 60 miles. That totals 100 miles. Make your next guess a smaller number. Try 1 hour; or 1.5 hours.

EF-77. a.

Exponential Form	Fraction Form	Expanded Form	Decimal (Standard) Form
2^{-4}	$\dfrac{1}{2^4}$	$\dfrac{1}{2\cdot 2\cdot 2\cdot 2}$	0.0625
4^{-1}	$\dfrac{1}{4}$	$\dfrac{1}{2\cdot 2}$	0.25

EF-78. d. $5^{-2} = \dfrac{1}{5^2} = 0.04$

EF-79. b. $3x^7$ d. $4y^2$

EF-80. b. The answer is too large for most calculators.
c. Since you can multiply in any order, $(1.4 \cdot 10^{98}) \cdot (2.3 \cdot 10^5)$ equals $(1.4 \cdot 2.3) \cdot (10^{98} \cdot 10^5)$.

EF-81. a. $\dfrac{1-c}{3}$ b. $1 - 3c$ c. $\dfrac{m+14}{4}$

EF-82. a. Figures 1 and 3 are the only composite rectangles.
b. Figure 1: $2x^2 + 10x = 2x(x+5)$

EF-83. a. Were you surprised to get the same result, 768 tiles, as you did in EF-41?

EF-92. a. The least common denominator is z. Multiply both sides by z to get a new equation: $4 + 1z = 9z$.
b. The least common denominator is 6.
c. The least common denominator is 15. After you multiply both sides by 15, the resulting equation is $15x + 3x + 5x = 1035$.

EF-93. To get started, find a ratio which is equivalent to $\dfrac{|BY|}{1.5}$. Then write and solve an equation.

EF-94. One with an 11.28-in. diameter or in pizza terms, a 12-in. pizza

EF-95. a. One of the factors is $x + 2$.
b. The factor pairs for 16 are $1 \cdot 16$, $2 \cdot 8$, and $4 \cdot 4$.

EF-96. a.

Sum	Product	Integer Pair	Sum	Product	Integer Pair
5	6	2, 3	−5	6	−3, −2
13	12	1, 12	−11	30	−6, −5
6	12	none	−10	16	−8, −2
10	0	10, 0	−10	24	−6, −4

EF-97. a. 1, 12   c. −2, −3

EF-98. If you are making a guess-and-check table remember that to choose consecutive odd numbers, you start with an odd number and increase by two. Try 11, 13, and 15, but remember you only need to use the first one and the third one to check.

EF-99. c. Rewrite each fraction with a denominator of $3x$ first: multiply the first fraction by $\frac{3}{3}$, and multiply the second one by $\frac{x}{x}$. The numerator of your answer will be $15 + 2x$. What is the denominator?

EF-100. a. $\frac{43}{35}$ b. $\frac{3+x}{7}$
c. This one is just like part a, except that the denominator of the first fraction is z instead of 7. Try using the same process as you did in part a. The denominator of your answer should be $5z$.
d. Rewrite each fraction with a denominator of 21 first: multiply the first fraction by $\frac{7}{7}$, and multiply the second one by $\frac{3}{3}$. The denominator of your answer will be 21.

EF-101. One way to be certain is to graph the two points and trace from (3, 0) to (5, 2). Then just describe how far up or down and how far left or right, you moved.

EF-102. Here are the answers for Saturday:
a. On Saturday Lora started with $34, lost steadily, and ran out of money at 2:00 P.M.
b. On Saturday she *lost* at a rate of $\frac{\$34}{8\text{ hrs}}$, so −$4.25 per hour.

EF-103. $1.98 \cdot 10^{30}$ kg

EF-104. d. The two problems are essentially the same.

EF-106. a. $\frac{1}{4^3} = \frac{1}{64} = 0.015625$ b. 1

EF-107. c. $\frac{1}{3^2}$, or 3^{-2}

EF-116. a. You could solve the equation $\frac{100}{1.89} = \frac{350}{x}$ to find the answer.
b. $\frac{\$1.89}{100} \approx 1.9$¢ per vitamin

EF-117. about 573 times

EF-118. a. $\frac{28}{3472}$
b. Percent means the number in relation to 100 that will give you the same result as the fraction in part a. What equation can you write and solve?
c. Refer back to the original problem to notice that 3 million people live in the city, and then you can use either the fraction from part a or your answer to part b to write an equation of equal ratios.

EF-119. b. $\frac{7}{2}$

EF-120. Make a new picture that separates the drawing of the ramp into two triangles. One triangle will be 4 ft on its short side and 18 ft "along the ground." The larger triangle will be 7.5 ft on its short side and $18 + x$ "along the ground." Now write an equation and solve it.

EF-121. Slide 2 units up and 3 units to the right. All the points you find should lie on a straight line.

EF-122. a. ≈13.73 sq units b. ≈28.54 sq units

EF-123. a. One of the factors is $x - 6$.
c. In the generic rectangle, you have −35 in the lower right corner. The possible dimensions of that corner are 1 by −35, −1 by 35, 5 by −7, or −5 by 7. Which one works?

EF-124. a. Some of the integer pairs are given below. If you are having trouble finding a particular integer pair, try listing all the possible ways to makes the product first. If none of those pairs makes the correct sum, then there is not an integer pair that works. In this table, however, you should be able to find an integer pair for every sum-product pair.

Sum	Product	Integer Pair	Sum	Product	Integer Pair
15	56	7, 8	5	0	5, 0
−3	−28	−7, 4	−9	18	−6, −3
5	−6	−1, 6	−5	−6	−6, 1
3	−10	−2, 5	0	−25	−5, 5

EF-125. b. East and West must be −5 and 7, so the polynomial is
$$x^2 + 2x - 35 = (x - 5)(x + 7).$$

EF-126. a. A fraction buster for this equation is 21, but you could also get x by itself simply by multiplying both sides by 3.
b. A fraction buster for this equation is $y(y + 3)$. When you multiply both sides by it, the resulting equation is $5(y + 3) = 2y$.
c. You can find a fraction buster for this equation in a similar way to the one in part b. The solution is $x = \frac{1}{4}$.

EF-127. Check you Tool Kit for a reminder about the relationship between distance, speed, and time.

EF-128. b. $\frac{1}{x^2 y^2}$

EF-129. b. The wider the cylinder, the slower the water will rise.

EF-131. Yes; you'll buy your ticket in about 19 minutes, or at 8:10 P.M. Another way to look at it is that you have enough time—24 minutes—for 185 people to buy tickets ahead of you.

EF-132. a. 0 lb c. ≈ 16.43 lb d. ≈45.65 ft

EF-133. When in doubt set up a guess-and-check table to help organize the information so you can write an equation. You can start this problem either by guessing the number sold at the door first, or by starting with the number sold in advance. How about 100 for an opening guess?

EF-134. a. Some of the integer pairs are given below. If you are having trouble finding a particular integer pair, try listing all the possible ways to make the product first. If none of those pairs makes the correct sum, then there is not an integer pair that works.

Sum	Product	Integer Pair	Sum	Product	Integer Pair
0	-100	$10, -10$	-10	-25	none
23	-24	$-1, 24$	-1	-42	$-7, 6$
20	99	11, 9	13	-30	$15, -2$

EF-135. b. One of the factors is $x + 10$.
c. There are only two possibilities:

[box diagrams: m and $+1$ on sides, $3m$ and $+1$ on top; or m and $+5$ on sides, $3m$ and $+1$ on top]

Test each of them by multiplying.
e. There are four integer pairs whose product is -24. Which of them have a sum of 10?

EF-136. a. $\frac{3}{4}$ b. $\frac{15}{4}$
d. There are two solutions, one positive and one negative.
f. A fraction buster is 12. After multiplying both sides of the equation by 12, the resulting equation is $4(2x - 5) + 12(1) = 3(x) + 4(x + 2)$.

EF-137. Try using a guess-and-check table to help organize the information so you can write an equation. You can start this problem either by guessing the number of student tickets first, or by starting with the number of general public tickets. How about 100 for an opening guess? Careful, this time you don't know the total sales. How can you compare the proceeds from the student tickets with the proceeds from the adult tickets?

EF-138. a. There are 16 ounces in a pound.
b. Write an equation using two equal ratios of baking powder to flour.

EF-139. e. $2x^5y$ f. $8x^9$

EF-140. a. There is enough to go another 275 miles.
c. $\frac{1}{25} = 0.04$ gallons per mile

EF-141. a. The 12-in. whole pizza is bigger.
b. The difference is 4π, or about 12.6 sq in.
c. The 12-in. pizza is about $0.079 or $0.08 per square inch, and the 16-in. half pizza is about $0.084 or about $0.08.

EF-145. To get started, draw a picture of both triangles and label the sides. What subproblems will you need to solve before you can find the perimeter?

EF-150. Remember, decimals and fractions are not integers. What slide could use to go from (3, 1) to (6, 2)? A point on this line with integer coordinates is (18, 6).

EF-151. c. $\frac{\text{length of Side of new figure}}{\text{length of Side of original figure}} = \frac{2}{1}$; $\frac{\text{Perimeter of new figure}}{\text{Perimeter of original figure}} = \frac{2}{1}$; and $\frac{\text{Area of new figure}}{\text{Area of original figure}} = \frac{4}{1}$.

EF-152. Redraw the three triangles so they all face the same way before trying to set up the ratio equations. $|DE| = \frac{25}{9} \approx 2.78$, and $|DF| = \frac{20}{3} \approx 6.67$

EF-153. c. The two factors are the same.
d. There are four possible set ups:

[four rectangle diagrams showing: x by 2 with top labels $2x$ and 5; or x by 5 with top labels $2x$ and 2; x by 1 with top labels $2x$ and 10; or x by 10 with top labels $2x$ and 1]

Test each of them by multiplying, and then make the signs work.

EF-154. For a hint, look back at EF-133 and EF-137.

EF-155. a. ≈ 1244 sq in.   b. ≈ 138 in.

EF-158. a. 6
b. A fraction buster is $x(x + 5)$.
d. A fraction buster is 15. The resulting equation is $5(2x + 5) + 15(3) = 3x$.

EF-159. a. The answer is smaller than 100%.
b. The answer should be slightly less than half of 60, since 45% is slightly less than 50%.
c. 75

EF-160. b. $-12x^5$ c. x^{12}
d. $8x^6$ e. $3xy^2$

EF-161.

Exponential Form	Fraction Form	Expanded Form	Decimal (Standard) Form
3^{-1}	$\frac{1}{3^1}$	$\frac{1}{3}$	≈ 0.333
5^{-3}	$\frac{1}{5^3}$	$\frac{1}{5 \cdot 5 \cdot 5}$	0.008

EF-162. a. Check your graph with the table. Did you graph enough points?

x	-2	-1	0	1	2	3	4
y	0.25	0.5	1	2	4	8	16

b. $x \approx 2.3$ c. $2^{2.3} \approx 4.92$

EF-163. The shaded area is equal to (area of smaller rectangle + area of larger rectangle) − (twice the area of the overlap)

EF-164. How many smaller square tiles, 3 in. on a side, will it take to cover each square tile that is 9 in. on each side?

CHAPTER 5

WR-10. b. The points are (1, 3.6) and (4, 8.4).
  c. The slope is $\frac{4.8}{3} = 1.6$.
  d. (10, 18) and (16, 27.6); slope = 1.6

WR-12. c. Slope = $\frac{\$15.79}{100 \text{ vitamins}} \approx \frac{\$0.16}{1 \text{ vitamin}}$
  d. The price of Mega-vitamins is about 16¢ per vitamin.

WR-13. The storage increased 35,044 acre-feet for an average rate of about 11,681.3 acre-feet per day.

WR-14. b. $|BC| = \frac{9}{5}$
  c. $C = (3, \frac{9}{5})$, $F = (5, 3)$; slope $= \frac{3}{5}$

WR-15. b. $\frac{10 + n}{5n}$ **d.** $\frac{11}{2z}$
  e. $z = 33$

WR-16. a. $x = 5.75$ **b.** $n \approx 1.71 \cdot 10^{21}$ **c.** $R = 2$ or 6

WR-17. a. Solve the equation $3C + 2 = 14$.
  b. Substitute the answer from part a into the expression $5C + 1$.

WR-18. b. 3   **d.** $4(x + 2)$

WR-19. a. $\frac{21}{5}$ **c.** $\frac{126}{37}$

WR-20. b. Percy needs a total of 257.6 points, so he needs 21 points more.
  d. He needs 67 points. Did you use the equation $\frac{237 + x}{380} = \frac{80}{100}$, or did you use a different equation?

WR-21. a. Area of the circle $= 20.25\pi \approx 63.6$ sq in.
  b. Probability of hitting the circle $= \frac{20.25\pi \text{ sq in.}}{81 \text{ sq in.}} \approx 0.785$

WR-22. b. x^{14} **d.** n^7

WR-23. b. $x^2 - 121$ **d.** $2(3 - x^2)$
  f. $(3x + 1)(x + 5)$

WR-24. a. $x^2 - 4x = x(x - 4)$
  b. $x^2 - 4 = (x - 2)(x + 2)$
  e. $x^2 - 10x = x(x - 10)$
  f. no solution with integers

WR-25. c. True; substituting 1 for x, we get $y = 3(1) + 2 = 5$.
  d. (2, 6) False, (−2, −4) True, (0, 1) False, (−1, −1) True

WR-33. b. You don't need to do any computation, but if you must substitute, use $x = 0$.

WR-36. b. Be sure you notice that it is $-y$.

WR-39. a. $y = -4x + 1$ **b.** $y = \frac{1}{2}x - \frac{5}{4}$

WR-40. Using one of the points you found, substitute the x-coordinate for x and the y-coordinate for y in the equation $3x - 6y = -24$. Show that the computations work out correctly.

WR-41. a. Remember: an x-intercept of a graph is a point that has 0 for its y-coordinate. Replace the y with 0 and solve the equation for x.
  b. Remember: a y-intercept happens when $x = 0$. Go back to the original equation, substitute, and solve the equation for y.
  c. This time $x = 10$, so replace x with 10 and solve the equation for y.

WR-42. The two-point slide method works for a and b but not for c which is not a linear equation. What kind of equation is it?

WR-43. b. No   **d.** Yes

WR-44. b. Go from 6 on the H-axis up to the graph and then left to the C-axis. When $H = 6$, $C = \$14.00$.
  c. Go from 20 on the C-axis over to the graph and then down to the H-axis. When $C = \$20$, $H = 10$.

WR-45. c. After his rest Dean will catch up to Cindy at about 4:00 P.M.
  f. If she continues at her steady pace, Cindy will take about 6 hours and 40 minutes to complete the race.

WR-46. Use the fact that Car A goes 120 miles in 2 hours to get both the average rate and the slope of the line for Car A.

WR-47. b. $x = 2$ **c.** $x = \frac{52}{3}$ **d.** $y = \frac{-3}{4}$
  e. There are three answers; $n = -5$ is one of them.

WR-48. b. $2x^2 + 9x + 9$ **c.** $(x + 3)(2x + 3)$

WR-49. a. $3x^2 + 5x + 13$ **c.** $5x$

WR-50. b. $-3x^2$ **c.** $4x^{-3}$ **d.** $27x^3y^3$

WR-51. a. 7   **b.** -4 **c.** 16

WR-52. a. $\frac{1}{3^2} = \frac{3^0}{3^2} = 3^{0-2} = 3^{-2}$

WR-53. b. $x^2 + 5xy + 6y^2$
  c. $(3x + 4)(3x - 4)$
  d. $3x(x - 7)$

WR-54. a. If the number of blue cubes is 6, the total number of cubes is 30, so the probability of randomly selecting a blue cube is $\frac{6}{30}$ or $\frac{1}{5}$.
  c. If the number of blue cubes is 10, the total number of cubes is 34. Then the probability of randomly selecting a blue cube is $\frac{10}{34}$, which is not $\frac{1}{3}$, but it's close. If the number of blue cubes is 16, the total number of cubes is 40, and the probability is $\frac{16}{40} = \frac{2}{5}$ which is still not right. If now the number of blue cubes is B, what is the total number of cubes? Write an equation using the fact that B divided by the total number of cubes has to equal $\frac{1}{3}$.

WR-55. a. $N = -35$, $S = 11.5$
  c. $E = -15$, $S = -14\frac{2}{3}$

WR-66. b. $y = \frac{-3}{2}x + 4$ **d.** $y = 2x + 6$

WR-67. a. Use the fact that speed is the ratio of distance to time.

WR-69. To find other points that lie on the line, notice that the slide "up 3, right 5" goes from $(-3, 2)$ to $(2, 5)$, and from $(2, 5)$ to $(7, 8)$, and keep going. No, $(12, 12)$ is on not the line; the point $(12, 11)$ is.

WR-70. One possible equation is $\frac{14}{8} = \frac{y}{11}$.

WR-72. a. $(x - 6)(x + 4)$ **d.** $(x + 8)(x - 3)$

WR-73. a. Substitute $T = -1002$ for both T's and show that the product equals zero. What happens if you do the same thing with $T = 471$?
b. According to the zero product property, at least one factor must equal 0.

WR-74. a. $6, -4$

WR-75. If w represents the width, the perimeter is $w + w + 5w + 5w$. Write and solve an equation using the value given for the perimeter.

WR-76. a. Write an equation comparing 25 miles in 35 minutes to 90 miles in M minutes.
b. Write an equation as in part a to find how many miles are traveled in 60 minutes.

WR-77. If they travel for 1 hour, the Edsel travels 70 km and the Studebaker travels 57 km, so the total distance is not enough. In 2 hours the Edsel travels 140 km and the Studebaker travels 114 km, for a total of 254 km, still not enough. If t represents the number of hours traveled, then the Edsel travels $70t$ km and the Studebaker travels $57t$. You want the total distance to be 635 km.

WR-78. The intercepts are $(0, 3)$ and $(\frac{-3}{2}, 0)$, and the slope is 2.

WR-79. a. x^5 **d.** $-8n^{17}$ **e.** 99^{x+2}

WR-80. a. You can use 42 as a fraction buster. $n = \frac{-1}{2}$
b. You can use 36 as a fraction buster to get $12(4 - w) + 15 = 2(2w + 1)$. $w = \frac{61}{16}$

WR-81. a. $6x^2 + 22x + 20$
b. $5x^2 + 13x - 6$
c. i. $6x^2 - 17x + 5$ **ii.** $5x^2 - 6x - 8$

WR-82. a. Area of the circle $= 36\pi$ sq in.
b. Area of the shaded part $= 144 - 36\pi$ sq in.
c. Probability of hitting the shaded part is $\frac{144 - 36\pi \text{ sq in.}}{144 \text{ sq in.}} \approx 0.2146$

WR-86. a. How much would it cost to rent a stove for 0 days?
b. It costs $14 for 8 days, $5.25 for 3 days, and $24.50 for 14 days.
c. The slope is the rental rate of $1.75 per day.

WR-87.   **c.** $m = 0, b = 4$ **d.** $m = 1, b = \frac{-3}{2}$

WR-88. b. $y = -\frac{3}{2}x + 6$ **c.** $y = x + \frac{7}{3}$

WR-89. a. $x = -1$ **b.** $y = 3$
c. $z = 0$ **d.** $x = \frac{-28}{25} = -1.12$

WR-90. The water storage was increasing at the rate of 9219 cubic feet per second. To find the amount of change over 12 hours, first figure out how many seconds are in 12 hours.

WR-91. This problem is something like WR-77 about the Edsel and the Studebaker. You can let t represent the number of hours, and use it to represent how far north the CRV went in t hours and how far south the Land Cruiser went in the same t hours. Add the two expressions to get the total distance of 225 miles. Once you have an equation, the solution is easy.

WR-92. First solve two subproblems to find the total amount of gorp and the total cost of the gorp. You can use the results to figure out that the price of the mixture is $4.57 per pound.

WR-93. b. $x = 36, y = 39$ **c.** $x = 6, y \approx 5.2$
d. $x = 7.5, y = 15$

WR-94. $42.00

WR-95. b. $-3x^2 + 2x + 40$
c. $(3x - 1)(x + 2)$

WR-96. b. $m^2 - 0m - 36 = (m - 6)(m + 6)$

WR-97. b. $w^2 - 0w - 1 = (w - 1)(w + 1)$

WR-98. b. The y-intercept is $(0, 0)$ and the x-intercepts are $(0, 0)$ and $(3, 0)$.
c. $-2, 5$

WR-99. $x^0 = 1$

WR-100. a. $n = -11$ **b.** $n = 4$ **d.** $n = -7$

WR-101. b. -2 **c.** $5 - x$

WR-102. No; according to the data, it would take about 6.67 min, or about 25 sec longer than claimed.

WR-106. f. line a: $y = \frac{2}{3}x + 6\frac{1}{3}$
line b: $y = \frac{2}{3}x + 3\frac{2}{3}$
line c: $y = \frac{2}{3}x + \frac{2}{3}$
line d: $y = \frac{2}{3}x - 2\frac{2}{3}$
line e: $y = \frac{2}{3}x - 4$
line f: $y = \frac{2}{3}x - 6$

WR-107. b. $y = \frac{3}{2}x + \frac{1}{2}$ **d.** $y = \frac{1}{3}x + 1$

WR-108. A line slopes upward for $m > 0$, is horizontal for $m = 0$, and slopes downward for $m < 0$.

WR-109. The points $(-1, -1)$ and $(4, -11)$ are on the graph.

WR-110. a. The other two answers are $x^2 + 2x - 15$ and $x^2 + (-14)x - 15$.
b. You could use 9, -9, 3, or -3 for the coefficient of x; then factor the resulting polynomial.

WR-111. a. Find the area of the rectangle and the area of the semi-circle, and then subtract. The area of the rectangle minus the area of the semi-circle equals $200 - \frac{\pi r^2}{2}$ sq units.
b. What is the largest the radius could be?

WR-112. By substituting 3 for x we see that $\frac{2x+10}{2} \neq x + 10$. Could $\frac{2x+10}{2}$ equal $x + 5$?

WR-113. a. $\frac{11}{4}$ b. $\frac{-1}{7}$ c. $-17, -4$

WR-114. b. The area is 108 sq cm.

WR-115. b. $36x^2 + x - 21$

WR-116. b. $P^2 - 100 = (P - 10)(P + 10)$

WR-117. b. $(x - 24)(x + 1)$ c. not factorable

WR-118. a. $-1, 24$ b. $-24, -1$

WR-119. a. $-0.6xy^{-1}$ c. m^2 d. $-7t^3$

WR-120. a. 2 b. all values of n, for $x \neq 0$
c. $-2, 2$

WR-121. about 2.2%

WR-127. One of the possible points is $(277, 277)$.

WR-128. Some possible points are $(17, 5)$, $(24, 7)$, and $(3, 1)$.

WR-129. One possible equation is $y = \frac{2}{7}x + 4$.

WR-130. The points $(0, -5)$ and $(2, -11)$ are on the graph.

WR-131. Substitute 2 for x, and 1 for y, in the equation, to get $1 = 2(2) + K$. Then figure out what K equals.

WR-132. All three graphs are the same line because all three equations are equivalent to $y = 2x + 3$.

WR-133. If Mae bikes 7 hr, how long would Lydia bike? If Mae bikes for M hr, how long will Lydia bike? How far will Mae travel in M hr? How far will Lydia go in $M - 1$ hr? Since Lydia catches up with Mae, the distances they travel will be the equal.

WR-134. b. If you guessed Seymour invested $1400 at 6%, how would you figure out what he had invested at 8%? If he invested x dollars at 6%, represent the amount he invested at 8%. You need an equation that shows that 6% of x and 8% of the rest add up to the total interest he earned, which you figured out in part a. It turns out that he invested $1175 at 6%.
c. Seymour is a dreamer. He'd have to get 25% interest.

WR-135. If you change the millimeter measurement to meters, the area is 200 sq m.

WR-136. a. $\frac{964198}{1234} \approx 781.36$ b. 6 c. 0, 1, 2

WR-137. 6.75 units

WR-138. a. $4x^2 - 9 = (2x - 3)(2x + 3)$
c. $36N^2 - 1 = (6N - 1)(6N + 1)$

WR-140. a. $-8x^3 = (-2)^3 \cdot x^3$ c. x^{-6}
e. 2^{-3} f. $-8x^3 = (-2)^3 \cdot x^3$
g. $x^{-1} = \frac{1}{x}$

WR-146. c. $y = -3x + 13$

WR-147. a. $y = 1.5x - 2.5$ b. $y = -0.5x + 2$
c. $y = 0.5x + 4$ d. $y = 3x + 6$

WR-148. a. x-intercept is $(4, 0)$; y-intercept is $(0, 1.5)$; slope is $\frac{-3}{8}$
b. x-intercept is $(-1.5, 0)$; y-intercept is $(0, 4)$; slope is $\frac{8}{3}$
c. no x-intercept; y-intercept is $(0, \frac{8}{3})$; slope is 0

WR-149. a. Points $(8, 4)$ and $(4, 3)$ are on the graph.
b. All four points are on the graph.

WR-150. $m = \frac{5}{90} = \frac{1}{18}$

WR-151. $y = \frac{1}{5}x + 3$

WR-152. a. The graph should start above the origin and increase steadily left to right—it will be a straight line with a positive slope.
b. The graph should start to the right of the origin, then increase for a while, and then level off gradually until it is horizontal.

WR-153. a. B is $(6, 3)$ and D is $(10, 3)$
b. $|AB| = 4$, $|AD| = 8$, $|DE| = 9$, and $|BC| = n - 3$
c. $n = 7.5$
d. $\frac{|AB|}{|AD|} = \frac{1}{2}$, so $|BC| = \frac{1}{2} \cdot |DE| = 4.5$; $n = 3 + 4.5 = 7.5$

WR-154. a. $18x^{-5}$ b. $\frac{3x^5}{2y^3}$
c. 5^{x-3} d. $18m^{-3}$

WR-155. About 6,720 of the skunks would be expected to carry rabies.

WR-156. Only if $x = 0$.

WR-157. $2x + 2(x + 4) = 28$; 5, 9, $\frac{9}{5}$

WR-158. $\frac{134 + x}{195} = \frac{90}{100}$, $x = 41.5$. Arturo must get at least 41.5 points on the test.

WR-159. a. $6x^2 - 35x + 50$
b. $15x^2 - 22x + 8$

WR-160. b. $25x^2 - 1 = (5x - 1)(5x + 1)$
c. $9P^2 - 64 = (3P - 8)(3P + 8)$

WR-161. a. $n = 3$ b. $n = -6$

WR-162. d. One possibility would be to add all the steps: $1 + 2 + 3 + 4 + 5 + \ldots + 157$. Another way that might be less work would be to notice that the 3-step region is half of a 3 by 4 rectangle, the 5-step region is half of a 5 by 6 rectangle, and the 10-step region is half of a 10 by 11 rectangle. Continuing the pattern, the 157-step region would be half of a rectangle of what dimensions?

e. Again, we could add all the numbers, $1 + 2 + 3 + 4 + \ldots + N$, or we could use the second idea in part d and figure that the area of an N-step region must be half of an N by $(N + 1)$ rectangle.

CHAPTER 6

FF-3. If you draw a generic slope triangle, you can determine the slope.

FF-9. a. $(0, -4)$ b. $y = \frac{5}{9}x - 4$

FF-10. a. $y = \frac{-2}{3}x + \frac{13}{3}$ b. $y = \frac{-2}{3}x + \frac{23}{3}$
c. $y = x + 5$

FF-11. b. Yes; the given points lie on a line.
c. If x represents the year, then $y = 75x - 128{,}500$.
d. The slope is $\frac{75 \text{ people}}{1 \text{ yr}}$, which means that the population of Dodge is increasing by 75 people each year.
e. Substitute 1998 for x in your equation.

FF-12. a. The slopes are all equal.
b. The slope of $CA = 2 =$ the slope of $AL =$ the slope of CL.

FF-13. a. Use fraction busters; as a first step, multiply both sides of the equation by 20. Don't forget to multiply the *entire* expression on each side by 20.
b. Using the correct order of operations to rewrite the right-hand side of the equation, you cannot subtract before multiplying $2(3x + 4)$.

FF-14. b. $9x^2 - 121 = (3x - 11)(3x + 11)$
c. $P^2 - 400Q^2 = (P - 20Q)(P + 20Q)$

FF-15. a. 1. $196\pi \approx 615.75$ sq cm
2. $81\pi \approx 254.47$ sq cm
b. $115\pi \approx 361.28$ sq cm

FF-16. a. The calculator display $\boxed{1.35\ ^{12}}$ means $1.35 \cdot 10^{12}$. Written anywhere else, 1.35^{12} means "Use 1.35 as a factor 12 times."
b. Enter 1.35, and then multiply by 10 raised to the power 12. Use the $\boxed{x^y}$ key to do this.
c. To raise 1.35 to the power 12, enter 1.35, press the $\boxed{x^y}$ key, enter 12, and then press the $\boxed{=}$ key.

FF-17. a. $\{[(13 - x)(6) + 2(11)] \div 2 - 0.5(10^2)\} \div x$, or, in fraction form,
$$\frac{\frac{(13 - x)(6) + 2(11)}{2} - 0.5(10^2)}{x}$$
b. If you chose 6, the result would be -3. If you chose 10, the result would also be -3.
c. The result is always -3.

FF-18. a. 7^4 b. 6^4
c. $2^3 \cdot 3^6$ d. $2^3 \cdot 3^3 = 6^3$

FF-19. b. Write out the meaning: $\frac{(x + 3)^6}{(x + 3)^8} =$
$\frac{(x + 3)(x + 3)(x + 3)(x + 3)(x + 3)(x + 3)}{(x + 3)(x + 3)(x + 3)(x + 3)(x + 3)(x + 3)(x + 3)(x + 3)}$
c. $-0.1x^8y^7$
d. $\frac{x + 6}{x}$

FF-20. $1126.25

FF-23. a. If your population graphs from FF-21 are inaccurate, the point of intersection you found will not make the equations true when you try to check by substituting. If this happens, you can either go back and make a more accurate graph, or go on to FF-24 to find a more accurate point of intersection using algebra.
c. To be a point of intersection, the point $(-5, -17)$ must make both equations true, not just one of them.

FF-26. c. $(0, 3)$ d. $(2, 3)$

FF-27. b. $y = \frac{-3}{2}x + 6$ c. $y = x + \frac{7}{3}$

FF-28. a. $x = -12$ d. $x = \frac{77}{25}$ e. $x = -3$ or 0.5

FF-29. a. Yes, since $11 = 3(6) - 7$.
b. Yes, since $11 = \frac{2}{3}(6) + 7$.

FF-30. a. $F = 1.8C + 32$
b. The two equations should be equivalent.
c. $136.4°F$ d. $-89.4°C$

FF-31. a. $y = \frac{-1}{2}x + \frac{11}{2}$ c. $y = \frac{1}{3}x + 5$

FF-32. a. $x = 6y, y = 7.5$
c. $x = 6.375, y = 5.625$

FF-33. If x is the length of shortest side, then $x + (x + 2) + (x + 4) = 24$.

FF-34. a. $10x^3 + 6x^2 - 8x$ b. $x^3 + 27$
c. One of the factors is $(x - 3)$.
f. $6x(x^2 - 5x + 4) = 6x(x - 4)(x - 1)$

FF-35. A billion is $1{,}000{,}000{,}000$; 720 km $= 7.2 \cdot 10^2$ km

FF-36. a. No, $x^2 + 25$ is a sum of squares, not a difference of squares.
c. Yes; $25 - T^2 = (5 - T)(5 + T)$
d. No, there are three terms.

FF-37. b. $10^{-5} = \frac{1}{100{,}000} = 0.00001$
c. $5^0 = 1$
d. $8^{-2} = \frac{1}{64} = 0.015625$

FF-41. b. Replace y in the second equation to get $x - (2x - 3) = -4$, and then solve for x.
c. $(1, 3)$

FF-42. Yes; $(10, 4)$ makes both equations true, so the point is on both lines.

FF-43. a. For example, pick $x = 15$.
Then $\frac{x-10}{x} = \frac{15-10}{15} = \frac{5}{15} = \frac{1}{3}$, not -10.
b. Yes, $x = \frac{10}{11}$.

FF-44. b. This equation is impossible to solve; no value of x could make it true.
e. $x = \frac{-36}{7} \approx -5.14$ f. $x = 0$ or $\frac{3}{2}$

FF-45. b. $10x^3 - 18x^2 + 20x - 36$
c. $2x^3 + x^2 + 12$
d. Factor the difference of squares.
f. Factor out an 8.

FF-46. a. $x + y = 2000$
b. $7.50y + 5.00x = 11,625$
c. There were 1350 student tickets and 650 general public tickets sold.

FF-47. a. $(2x + 5)(x + 3) = x^2 + 11x + 15$
b. $(2y + 3)(2y - 3) = 4y^2 - 9$

FF-48. a. Both unlabeled sides are $L - 3$ ft.
b. Simply subtract the area of the small square from the area of large square.

FF-49. a. $y = -x + 3$ b. $y = 68.2x + 1$
c. $y = \frac{-2}{3}x + 5\frac{2}{3}$

FF-50. Look back at problem FF-2, and describe the four subproblems.

FF-51. a.

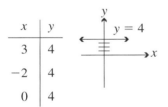

b.

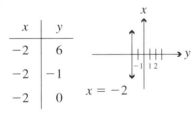

c. the x-axis   d. the y-axis

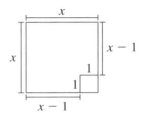

FF-52. a. It's a horizontal line passing through $(0, 42)$.
b. It's a vertical line passing through $(1991, 0)$.

FF-53. $n = 24$

FF-60. a. $(-3.0, 7.0)$ and $(2.0, 2.0)$
b. about $(1.4, 3.0)$ and $(-1.4, 3.0)$

FF-61. $(0, 1)$ and about $(3.5, 11.5)$

FF-62. a. $(2, 1)$ b. $(15, 37)$
c. $(4, 2)$ d. $(-5, 3)$

FF-63. a. $x = 0$ b. $y = 0$

FF-64. Some possibilities include evaluating $-x^2$ and $(-x)^2$ for several values of x, showing that the graphs of $y = -x^2$ and $y = (-x)^2$ are different, or explaining how to use the order of operations for each of the two expressions.

FF-65. a. $4x^{10}$ b. $4 \cdot 10^{10}$
c. $\approx 2.83 \cdot 10^{31}$ d. $2 \cdot 10^9$
e. $0.25x^{-10}$ f. $0.25 \cdot 10^{-10}$

FF-66. b. One of the factors is $x - 3$.
e. One of the factors is $x - 6$.
f. This one is a difference of squares and one of the factors is $x + 11$.

FF-67. c. area $= x^2 - 1$

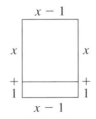

d. area $= (x + 1)(x - 1)$

e. The areas of the two cut figures are the same, so the area of the large square minus the area of small square is the same as the area of the rectangle.

FF-68. A possible answer is "Plot the two points and find a slide."

FF-69. Try a guess-and-check table: if you guess that John earns 10 out of the remaining 60 points, then he will have earned a total of $380 + 10 = 390$ points out of a possible 500. Since $\frac{390}{500} = 0.78$, this is only a 78% average, so the guess 10 was too small.

FF-70. c. No solution; the lines are parallel so they don't intersect.

FF-71. a. $y = \frac{3}{5}x + 3$ b. $y = 3$ c. $x = 2$

FF-72. $y = 0$

FF-73. a. $(-7.5, 0)$ b. It's the x-intercept.

FF-78. a. $x = -3, 6$ b. $x = -8, 4$
c. $x = -2, \frac{-3}{2}$ e. $(-8, 0), (4, 0)$

FF-79. b. i. There are two different numbers you can multiply by themselves to get 169.
ii. $z = 0.8$ or -0.8
iii. First step: add 0.1 to both sides of the equation. There are two solutions.
iv. Careful—the expression $t^2 + 100t$ is not a difference of two squares.

FF-80. (3, 2), (6, 2), and (6, 4)

FF-81. a. You would use the y-intercept and get a winning time of about 11.4 sec.
c. At $x = 96$, the winning time is about 9.9 sec.
e. About 12.0 sec.
f. It's off by about 0.2 sec; the data is not exactly linear and the graph may be drawn inaccurately.

FF-82. a. The slope ratios would be equal or almost equal.
b. To find what your model predicts for the year 2000, you should substitute 104 in place of x in your equation, since $2000 - 1896 = 104$. What should you substitute for x to make predictions for the year 2200?
c. Since the line is sloping downward, it will eventually cross the x-axis. What are the coordinates of the x-intercept? What would this point represent for the women's data?
d. Substitute $x = 1104$ in your equation to make predictions for the year 3000.
e. Graph 4 would be best since winning times gradually level off. You could also make a case for graph 1, but what would the sharp corner in the graph represent?

FF-83. a.

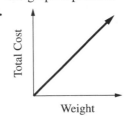

b.

FF-84. a. To calculate the average number of years, you must find the total number of years, and divide by the total number of employees.
b. $\frac{17}{24} \approx 71\%$ c. $\frac{7}{24} \approx 29\%$

FF-85. a. The generic rectangle for $2x^2 - 6x$ has only two regions.
b. $a^2 - 4a + 3 = (a - 3)(a - 1)$
c. $2m^2 - 11m + 5 = (2m - 1)(m - 5)$

FF-86. a. $\frac{x^2}{2y^3}$ or $\frac{1}{2}x^2y^{-3}$ b. $\frac{x^{-5}y^{-6}}{32}$ or $\frac{1}{32}x^{-5}y^{-6}$
c. $\frac{x^7}{4y}$ or $\frac{1}{4}x^7y^{-1}$ d. 6^{x-4}
e. $\frac{4}{x^2}$ or $4x^{-2}$ f. $4m^5$

FF-87. a. $s = \frac{5}{9} \approx 0.56$ b. $r = 28$
c. $x = 7$ or -3 d. $x = 6$ or -2

FF-88. a. This system would be a good candidate for the method of substitution.
b. $(\frac{32}{11}, \frac{20}{11})$, or about (3.27, 1.82)
c. (1, 2)

FF-89. First, define x as the amount invested at 4.5%, and y as the amount invested at 8%. The two investments must total $8000, so $x + y = 8000$. The interest from each account is the same, so $0.045y = 0.08x$. The answer is to invest $6480 in the 8% account.

FF-90. a. $(3a + 2)(a + 5) = 3a^2 + 17a + 10$, or
$(3a + 5)(a + 2) = 3a^2 + 11a + 10$, or
$(3a + 1)(a + 10) = 3a^2 + 31a + 10$, or
$(3a + 10)(a + 1) = 3a^2 + 13a + 10$.
b. $(x + 2)(y + 3) = xy + 2y + 3x + 6$, or
$(x + 3)(y + 2) = xy + 3y + 2x + 6$, or
$(x + 1)(y + 6) = xy + y + 6x + 6$, or
$(x + 6)(y + 1) = xy + 6y + x + 6$.

FF-91. Let x be the cost to the furniture store. Then the equation is $x + 0.55x = 496$, so $x = \$320$.

FF-93. c. Substitute (3, 1) in $4x + 2y$ to get $4(3) + 2(1) = 14$. Substitute (3, 1) into $x - 2y$ to get $(3) - 2(1) = 1$.

FF-94. a. The x terms would be eliminated instead of the y. Don't forget to write out the steps to show that you would get the same answer!

FF-96. You can try to eliminate either x or y when using the addition method.
a. $(\frac{23}{8}, \frac{5}{4})$

FF-98. a. When the egg hits the ground, y will be zero. Substituting 0 for y in the equation gives $0 = 80x - 16x^2$. Solve the equation to get $x = 0$ or $x = 5$.
b. The equation $64 = 80x - 16x^2$ has solutions $x = 1$ or $x = 4$. The egg will be 64 ft above ground at 1 sec and again at 4 sec.
c. The egg will hit the ground after 8 sec. It will be 192 ft above ground twice—once at 2 sec, and then again at 6 sec on its way back down.

FF-99. a. $y = -0.64x + 76.8$ b. $y = \frac{1}{2}x + 2$

FF-100. a. $145°F \leq F \leq 165°F$
b. $62.8°C \leq C \leq 73.9°C$

FF-101.

FF-102.   b. $\frac{3+2x}{6}$ c. $\frac{5x}{6}$
          d. $\frac{x+2}{2x}$ e. $\frac{2x+3}{6x}$

FF-103.   Look for fractional forms of 1. For example, in part c, $\frac{(x-4)}{(x-4)} = 1$ for $x \neq 4$.
          b. $\frac{1}{2}(x-5)$ d. $(x+0.3)^3$

FF-104.   If G represents the number of general public tickets sold, and D represents the number of student tickets sold, then the equations are: $G + D = 141$ and $5D + 10 = 8G$.

FF-105.   Some subproblems: first find the area of the rectangle. You could find the area of the shaded part by subtracting the area of the triangle, but to find the area of the triangle, you must first find $|AD|$ and the missing length. The shaded area is $66\frac{2}{3}\%$ of the area of the rectangle.

FF-106.   The area of two 10-in. pizzas is $50\pi \approx 157.1$ sq in., while the area of one 15-in. pizza is $56.25\pi \approx 176.7$ sq in. If you prefer crust, two 10-in pizzas have a total circumference of 20π in. compared to 15π in. on the 15-in. pizza.

FF-107.   a. $B = 8$ or -8 b. $R = 12$ or -12

FF-108.   Fraction busters may help you solve these.
          a. $x = -6$ b. $x = 3$
          c. $x = -2$ or 3 d. $x = 0$

FF-109.   One way to eliminate y is to multiply the first equation by 6 and the second by -4, and then add the two equations together. There are many other possibilities!

FF-110.   b. $(1, 1)$ c. $\left(\frac{20}{3}, \frac{14}{3}\right)$ e. $\left(\frac{11}{2}, \frac{-1}{2}\right)$
          f. $(1, -3)$ or $\left(\frac{-5}{4}, \frac{-15}{8}\right)$

FF-111.   In 16 years both populations will be 17,000.

FF-119.   b. The distances are equal.
          d. They meet after 2.5 hr, fifteen mi outside of town.

FF-122.   a. The slope is $\frac{-0.21}{180}$, or about -0.0012 g/sec. The candle is burning up at a rate of about 0.0012 g each second.
          b. $y = -0.0012x + 0.84$ g
          c. When the candle burns out, $y = 0$, so $0 = -0.0012x + 0.84$. Solve the equation to find when the candle goes out.

FF-123.   a. $(2, 1)$ b. $(-1.5, 5.5)$ c. $(-1, -2)$
          d. $(0, -3)$ f. $(4, 4)$ and $(1, -2)$

FF-124.   13 wk

FF-125.   The slope is -3 and the y-intercept is $(0, 4)$.

FF-126.   yes, no, no, yes, yes, and no

FF-127.   a. With $y = -0.014x + 11.26$ for the men's time and $y = -0.019x + 12.63$ for the women's time, the winning times are the same when $x = 272$.
          b. in the year 2170
          c. Although the equations in part a are messy, it is easier to solve the system algebraically than it is to do so graphically.

FF-128.   b. The equation is impractical for $T < 39°F$.
          d. When the temperature is $90°F$, the ants' speed is $\frac{51}{11} \approx 4.64$ in./min.
          e. When a cricket chirps 108 times per minute, the ants' speed would be about 3 in./min.

FF-129.   c. $x = \frac{3}{2}$ or -5 d. $x = \frac{-3}{2}$ or -2
          h. $x = \frac{C}{D}$ j. $x = 0$

FF-130.   b. $(\frac{1}{2}, 0)$ c. $(2, 0)$ and $(\frac{1}{2}, 0)$

FF-131. $(1, 3)$

FF-132.   a. 8 by 9
          b. The area is $72 - \frac{\pi \cdot 1^2}{4} - \frac{\pi \cdot 3^2}{4} \approx 64.15$ sq units.
          c. The perimeter is $7 + 6 + \frac{2\pi(3)}{4} + 5 + 8 + \frac{2\pi(1)}{4} \approx 32.28$ units.

FF-133.   a. When $x = 100$, $\frac{x^2 + 4x}{x^2} = 1.4$, but $1 + 4x = 41$, so the expressions are not equal.
          b. It's correct since all three expressions are equal to 1.4 when $x = 10$.

FF-134.   a. $x = 13$ or -13 b. $z = 0$ or 0.8
          c. $y = 1.1$ or -1.1 d. $P = 12$ or -2

FF-135.   a. $x = 9$ or -9 b. $H = 15$ or -15

FF-136.   a. The x-intercepts are $(4, 0)$ and $(-4, 0)$.
          b. The points of intersection are $(4, 0)$ and $(-4, 0)$.
          c. The solutions to the equation $0 = x^2 - 16$ are the x-coordinates of the x-intercepts of the parabola $y = x^2 - 16$.

FF-137.

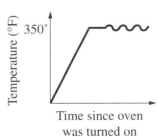

FF-138.
a. The area is $(m+1)(m+12) = 60$, so $m^2 + 13m - 48 = 0$. Factor $m^2 + 13m - 48$ and solve for m.
b. There is only one value for m in this situation. Although $(m+1)(m+12) = 60$ has two solutions, namely $m=3$ and $m=-16$, the rectangle cannot have sides of lengths -15 and -4 units. The value of m must be 3 in order for there to be a rectangle.

CHAPTER 7

RT-11. a. $H = 13$ in. b. Use $8^2 + L^2 = 17^2$.
c. $A = 7$ ft d. $S = \sqrt{458}$ mi

RT-12. a. $(2, -1)$ b. Use $8^2 + 7^2 = |SP|^2$
c. The slope of segment SP is $\frac{7}{8}$.

RT-13. a. Try substituting for y in the second equation.
b. Use the addition method.

RT-14. a. $-9, -4$ b. $9, 4$
c. $81, -24$ d. $729, -216$

RT-15. b. $(2x-1)(x-1)$
d. Hint:

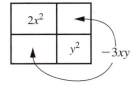

f. $x = \frac{1}{2}$, or 1

RT-16. There are two solutions; one of them is $C = -15$.

RT-17. b. The x-intercepts are $(3, 0)$ and $(-3, 0)$.

RT-18. a. The change in length is $0.0216 = 2.16 \cdot 10^{-2}$ cm.
b. The change in length for the lead wire is $0.0725 = 7.25 \cdot 10^{-2}$ cm, and for the copper wire is $0.0425 = 4.25 \cdot 10^{-2}$ cm. Now calculate the difference.

RT-19. a. $15x^2 - 36x + 21$ b. $10w^2 + 5w - 75$
c. $x^3y - x^2y^2 - 6xy^3$ d. $2m^4 - 98m^2$

RT-20. a. $\frac{\$18 \text{ per dozen roses}}{12 \text{ roses per dozen}} = \$1.50$ per rose
b. Use $\frac{\$9 \text{ per dozen daffodils}}{12 \text{ daffodils per dozen}}$.
c. If R represents the number of roses, then $1.5R + 0.75(12 - R) = 12$. Now solve for R.

RT-21. a. $\frac{2x-1}{2}$ b. $\frac{2x}{x+1}$

RT-22. About $3.99 \cdot 10^8$ bulbs have an average lifespan.

RT-23. Notice that $\frac{1}{2} \cdot \frac{2}{3} = \frac{1}{3}$, so $\frac{1}{2} \cdot \frac{2}{3} \cdot \frac{3}{4} = \frac{1}{3} \cdot \frac{3}{4} = \frac{1}{4}$. Keep going!

RT-24. c. i. $6 > 1$, ii. $3 > -2$,
iii. $-4 > -7$
d. i. $9.9 + 9.99$, ii. $-3.2 < -1.5$,
iii. $19.75 + 22$

RT-28. b. i. $\sqrt{78} \approx 8.$ something,
iii. $\sqrt{136} \approx 11.$ something

RT-32. If D is the distance from home directly to second base, then $90^2 + 90^2 = D^2$; $D = \sqrt{16{,}200}$ ft.

RT-33. a. $\sqrt{24}$ is a little less than 5.
c. $\sqrt{66}$ is a little more than 8.
e. $\sqrt{28}$ is a little more than 5.

RT-34. a. Hypotenuse $= \sqrt{65} \approx 8.06$
b. Hypotenuse $= \sqrt{530} \approx 23.02$
d. Leg #2 $= \sqrt{40} \approx 6.32$
f. Leg #1 $= 33.24$

RT-35. c. $-x^3$ e. $27x^2$

RT-36. Use $(3x)^2 + x^2 = 17^2$.

RT-37. Find the length of AB by using $|AB|^2 = 6^2 + 10^2$. The slope of AB is $\frac{-10}{6}$.

RT-38. a. i. Notice that $\sqrt{25} + \sqrt{4} = 5 + 2 = 7$, while $\sqrt{25 + 4} = \sqrt{29}$,
ii. True iii. False
b. i. True ii. Notice that $\sqrt{49} + \sqrt{36} = 7 + 6 = 13$, while $\sqrt{49 + 36} =$
iii. True iv. False

RT-39. Be sure to square $2C$. There are two answers; one of them is -24.

RT-40. a. Since $x^2 = 225$, $x^2 - 225 = 0$. Now factor the difference of squares to find two solutions.
b. Look at the x-coordinates of the x-intercepts and the solutions to the equation.
c. Use the fact that the graph is a parabola that intersects the x-axis twice.

RT-41. b. An equation for Janelle's situation is $y = 7x + 40$; and for Jeanne's, $y = 3x + 125$. They'll have the same amount of money set aside in 21.25 weeks.

RT-42. a. Substitute $4x$ for y in the second equation; $x = -0.2$, $y = -0.8$.
b. Replace x in the first equation to get $-2(3y) + y = 3$, and then solve for y; $x = -1.8$, $y = -0.6$.
c. Here's one possible approach: rewrite the second equation in the form $3x - 2y = 18$, and then use the addition method by multiplying both sides of the first equation by 3, and both sides of $3x - 2y = 18$ by -5.

RT-43. a. $\frac{x+3}{x-2}$ b. $\frac{2(2x-5)}{x+4}$
c. $\frac{x^2 + 6x}{x^2 + 12x + 36} = \frac{x(x+6)}{(x+6)(x+6)} = \frac{x}{(x+6)}$

RT-44. Let P be the number of pennies and C be the number of coins. Then $\frac{P}{C} = \frac{1}{3}$ and $\frac{P + 24}{C + 24} = \frac{1}{2}$. Rewrite both equations and solve the system. There were 24 of each type of coin on Tuesday.

RT-45. b. $\frac{-7}{2x}$ c. $\frac{11 + x}{2x}$

RT-47. **b.** $3x^3 + 13x^2 + 4x$ **c.** $4x^2 - 28x + 49$
d. $x^3 - 3x - 2$
e. You'll get a difference of squares.
f. $12 - x - 6x^2$

RT-48. **c.** $(2x - 1)(x - 3)$ **d.** $(5x + y)(x + y)$

RT-49. **a.** Let t represent the number of tacos and C represent the total cost in cents. At Pico de Gallo, $C = 80 + 75t$, and at Señorita Taquita, $C = 35 + 90t$.
b. Both restaurants charge $3.05 for a three-taco lunch special.

RT-51. You will need to consider five right triangles. The hypotenuse of two of the triangles is the segment from a corner of the roof to the base of the antenna.

RT-55. **a.** $\sqrt{146} \approx 12.08$ units
b. $\sqrt{145} \approx 12.04$ units
c. $\sqrt{50} \approx 7.07$ units

RT-56. Let w represent the length of the wire. Then $9^2 + 13^2 = w^2$; $w = \sqrt{250} \approx 15.81$ m.

RT-57. **c.** Find the area of the large square; find the area of each right triangle.

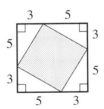

d. 34 sq units
e. The length of the hypotenuse is $\sqrt{3^2 + 5^2} = \sqrt{34}$ units, so the area of the tilted square is $\sqrt{34} \cdot \sqrt{34} = 34$ sq units. The methods agree!

RT-58. Let L represent the length of the second leg. According to the Pythagorean relationship, $L^2 + 5^2 = 13^2$ cm. Now solve for L. The area is 30 sq cm.

RT-59. **a.** $2\sqrt{25}$ is exactly 10
b. $3\sqrt{17}$ is between 12 and 15
c. Use the fact that $4 < 5 < 9$
d. Use the fact that $1 < 3 < 4$

RT-61. **c.** $7\sqrt{6}$ **f.** $-6\sqrt{7}$

RT-62. **b.** The slope of AB is $\frac{3}{6}$; the slope of EF is $\frac{2}{4}$.
d. $|AB| = \sqrt{45}$ units, $|CD| = \sqrt{5}$ units, and $|EF| = \sqrt{20}$ units

RT-63. **c.** $x = \frac{29 - y}{2}$ **d.** $\frac{90}{17}$ **e.** $-7, 1$

RT-64. **a.** It was about 6.00198 cm long.
b. It would need to rise 660° C.
c. The temperature was increased about 9.09° C.

RT-65. **a.** You would get your age in years.
b. You would get your original positive number.
c. Squaring "undoes" square-rooting.
d. i. 36   **iii.** x

RT-66. **a.** $-2m^2 + 9m + 35$

RT-67. Notice that the equation $2x - y = 5$ can be rewritten as $y = 2x + (-5)$.

RT-68. $T = 40$

RT-69. Solve the equation $\frac{\$2379.29}{\$2500} = \frac{x}{100}$.

RT-70. **a.** $\frac{x + 4}{4x}$
b. $\frac{2}{(x + 1)(x - 2)^2}$
c. $\frac{x + 3}{2}$

RT-71. **a.** There are two solutions; you'll need to factor $x^2 + 5x + 4$.
b. Try using a fraction buster first, and then use the substitution method to eliminate y; (6, 0).

RT-72. **a.** One factor is 3.   **b.** One factor is xy.
c. One factor is $5y$.
d. Use the fact that $x^4 - 16 = (x^2)^2 - 4^2$, and factor the difference of squares.

RT-73. **a.** $3 < x$, or $x > 3$ **b.** $x \leq 2$, or $2 \geq x$

RT-75. **a.** One possible value is $m = 4$.
b. One possible value is $m = -10$.

RT-79. **b.** The northbound truck would have gone $2E$ miles.
c. If E represents the number of hours traveled, then $E^2 + (2E)^2 = 47^2$.

RT-80. **b.** Each car traveled 120 miles.
c. Each car traveled $40T$ miles.
d. Use a diagram to write an equation, and then solve the equation:

T represents the number of hours traveled.

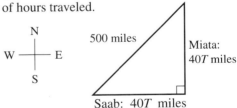

RT-81. Use a diagram to write an equation, and then solve the equation:

T represents the number of hours traveled.

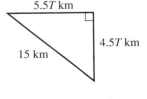

They'll meet in about 2.11 hours, or about 2 hr and 7 min. One hiker walked about 11.6 mi, and the other walked about 9.5 mi.

RT-82. c. Find the side length and the area of the large square; find the area of each right triangle.
d. The area is 17 sq units and each side is $\sqrt{17}$ units long.
e. The result should be the same length.

RT-83. d. P   e. $x - 2$

RT-84. a. $\sqrt{32} \approx 5.66$ units   b. $\sqrt{45} \approx 6.71$ units

RT-86. a. False   d. False
f. True

RT-87. b. $18\sqrt{2}$ c. not possible
d. $16\sqrt{5}$ f. $15\sqrt{x}$

RT-88. a. First factor out 2.
b. First factor out $7m$ and then use a generic rectangle to factor further.
c. One factor is $3xy$.

RT-89. a. To start, factor the numerator and the denominator: $\frac{x^2 - 13x + 40}{10x - 50} = \frac{(x - 8)(x - 5)}{10(x - 5)}$.
Keep going!
b. $\frac{x + 2}{x}$

RT-90. a. $-8, 2$ b. -4 c. $0, \frac{13}{5}$ d. $-7, -\frac{1}{2}$

RT-91. b. The x-intercepts are $(-4, 0)$ and $(3, 0)$.

RT-93. a. You can try to eliminate either x or y when using the addition method; $(8, -15)$.
b. $(2, 3)$

RT-94. If b represents the length of a broomstick, then $5b - 10 = 4b + 29$. Now solve for b and use it to find the distance between the trees.

RT-95. a. $\frac{57}{59} \approx 0.97$ b. 3   c. 0, 3   d. 17.5

RT-96. a. $12 > 7$ and $(-1)(12) < (-1)(7)$
b. $-3 < 0$ and $(-3)(-3) > (-3)(0)$
e. $0 < 4$ and $(-0.5)(0) > (-0.5)(4)$

RT-97. a. If $A < B$ and N is a positive number, then $A \cdot N < B \cdot N$.
b. If $A < B$ and N is a negative number, then $A \cdot N > B \cdot N$.

RT-98. a. $3 < x$, or $x > 3$
b. $\frac{3}{2} > x$, or $x < \frac{3}{2}$
c. $8 \geq x$, or $x \leq 8$

RT-99. a. Added x to each side; added 8 to each side.
b. Subtracted $2x$ from each side; subtracted 10 from each side; divided each side by -3.

RT-104. b. iii. Try starting with $\sqrt{250} = \sqrt{25 \cdot 10}$, and keep going.

RT-106. d. $3 + 2\sqrt{2}$ e. $10 - 6\sqrt{5}$
f. not possible   j. $5\sqrt{2}$

RT-107. a. Your graph should look like the one given in RT-28.
b. Yes, values $x \geq 0$ work. No negative numbers work because when a (real) number is squared, the result is a positive number.
c. Start on the x-axis with $36 < 43 < 49$; go up to the graph of $y = \sqrt{x}$, and then left to the y-axis.

RT-108. c. $7\sqrt{5} \approx 15.65$

RT-109. a. Point S is (912, 415).
b. $|PS| = \sqrt{369} \approx 19.2$ units
c. The slope of PS is $\frac{5}{4}$.

RT-110. a. One factor is $2x$.   b. One factor is 3.
c. First factor out $3xy$.   d. First factor out $5y$.
e. One of the two factors is 8.
f. can't be factored
g. First rewrite it as a difference of squares: $100x^2 - 49y^2 = (10x)^2 - (7y)^2$

RT-111. a. $\frac{-10}{3}$ b. $\frac{-54}{5}$
c. $\frac{4}{5}$ d. 18, 2

RT-112. 8.75 cm

RT-113. If h represents the cost of a hamburger and m represents the cost of a milkshake in dollars, then $3h + 2m = 3.35$. Write another equation, and then solve the system.

RT-114. Start with the fact that $2x + 5 = 3x - 8$, and then solve for x.

RT-115. a. It would lengthen about $4.8 \cdot 10^{-3}$ meters, or about $\frac{1}{2}$ a centimeter.
b. Use the fact that 1 cm is 0.01 m. The temperature would need to decrease about 33.3° C.

RT-116. a. $B = \pm 9$
b. $B = \pm\sqrt{24} = \pm\sqrt{4 \cdot 6}$; keep going!
c. $B = \pm\sqrt{C^2 - A^2}$

RT-117. a. Factor the numerator first; $\frac{2(x-2y)}{x}$
b. Start with $\frac{3x^2 - 6x + 3}{5x - 5} = \frac{3(x^2 - 2x + 1)}{5(x - 1)}$ and keep going.

RT-118. a. $x = 2, y = 1$ c. $x = 3, y = 11$
d. $x = -1, y = 1$

RT-119. a. $4 < 7$ and $(4)^2 < (7)^2$
b. $5 < 13$ and $(5)^2 < (13)^2$
c. $0.5 < 8$ and $(0.5)^2 < (8)^2$

RT-120. If A and B are positive numbers and $A < B$, then $A^2 < B^2$.

RT-121. c. When each side of an inequality is multiplied by the same negative number, the direction of the inequality is reversed.

RT-122. a. $-3 < x$, or $x > -3$
b. $x \leq -4$, or $-4 \geq x$

RT-133. $5\sqrt{3} \approx 8.66$ m

RT-134. If x represents the number of hours since 6:00 A.M., then $52x + 45(x - 1) = 240$. The two trains are 240 miles apart around 8:56 A.M.

RT-135. The two trains are 300 km apart after $\frac{5}{3} \approx 1.67$ hours.

RT-136.   b. $\sqrt{72} = 6\sqrt{2} \approx 8.49$ ft
        c. $x = 4\sqrt{5} \approx 8.94$ mi;
          $2x = 8\sqrt{5} \approx 17.89$ mi

RT-137.   a. $\sqrt{145} \approx 12.04$
        c. $\sqrt{1250}$ or $25\sqrt{2} \approx 35.36$

RT-138.   a. Start the slide at (x, y) and stop at $(x + 3, y + 4)$.
        b. The slope is $\frac{4}{3}$.
        c. The length of the segment is 5 units.

RT-139.   In your diagram you should have two similar triangles. Let h represent the height of the tree, and then use ratios to write an equation. The tree is 30 ft tall.

RT-140.   a. Use $\sqrt{96} = \sqrt{16 \cdot 6}$.
        c. $\sqrt{72} = 6\sqrt{2}$
        f. Use $\sqrt{48} = \sqrt{16 \cdot 3}$.

RT-141.   a. $A = \pm\sqrt{30} \approx \pm 5.48$
        b. Since $A^2 + 404.96$ is positive for all (real) numbers A, there is no solution.
        c. $A = \pm\sqrt{C^2 - B^2}$

RT-142.   a. $-12, 12$
        b. Use the zero product property; there are three solutions.
        c. $1, 5$ d. $\frac{5 - 3y}{2}$
        e. $\frac{3}{17} \approx 0.18$ f. $\frac{1}{3} \approx 0.33$

RT-143.   a. $(0, 9)$ b. $(6, -1)$

RT-144.   a. $1, \frac{1}{3}$ b. $5, -1$

RT-145.   One dimension is $x - y$.

RT-146.   The garden is 7 m by 9 m.

RT-147.   There are about 3665 students in the college.

RT-148.   a. $8; 18$ b. $16; 36$ d. $-20; -45$

RT-149.   b. The x-intercepts are $(6, 0)$ and $(-1, 0)$.

RT-150.   Use ratios to write an equation. The 100 feet of wire would weigh about 32.86 pounds.

RT-151.   a. 6 b. $4\sqrt{5} + 9$
        d. not possible   e. $\sqrt{4} = 2$
        g. $7\sqrt{2}$ h. 134.98
        j. Use $\sqrt{162} = \sqrt{81 \cdot 2}$.
        k. Use $\sqrt{192} = \sqrt{64 \cdot 3}$.
        l. $3\sqrt{2} - \sqrt{2} = 2\sqrt{2}$

RT-152.   a. $\frac{\Delta L}{L_0} = \alpha \cdot (\Delta T)$
        b. $\Delta L = (12 \cdot 10^{-6}) \cdot (\Delta T) L_0$;
          $\Delta T = \dfrac{\Delta L}{L_0(12 \cdot 10^{-6})}$ or $\dfrac{\Delta L \cdot 10^6}{L_0 \cdot 12}$
        c. You would need to know the initial length and the change in temperature.

RT-155.   a. $\frac{10}{3} \leq x$, or $x \geq \frac{10}{3}$ b. $1 < x$, or $x > 1$

CHAPTER 8

GG-11.   a. Substitute 0 for x in the equation, and then do it again for $x = 1$.
        b. How many times does the graph intersect the x-axis?

GG-12.   a. The x-intercepts are about $(5.5, 0)$ and $(0.4, 0)$.
        c. If the estimates were accurate, the y-value would be 0.
        d. Why can't the polynomial be factored?
        e. How do you know that there are two solutions to the equation? Could the solutions be integers?
        f. $x = 3 + \sqrt{7}$ and $x = 3 - \sqrt{7}$

GG-13.   a. $x = \dfrac{-5 \pm \sqrt{57}}{2}$ b. $x = \frac{-1}{3}, -2$
        c. $x = -4.5$ d. $x = \frac{1}{4}, -2$

GG-14.   b. i. ≈ 4.53 ii. ≈ -3.63
          iii. ≈ 0.62 iv. ≈ 3.37

GG-15.   a. $\approx 1.275, -6.275$
        b. The value should be close to, but not exactly, 0.
        d. The values were close to the solutions, but were not exact solutions.

GG-16.   a. Find the hypotenuse of the triangle and the radius of the semicircle; find the areas of the triangle and the semicircle; the area is about 20.6 sq cm.
        b. Find perimeter of the semicircle. The total perimeter is about 16.9 cm.

GG-17.   b. Find the radius of large semicircle: find the area of each semicircle and of the quarter circle. The total area is 7.5π sq cm.
        c. Find the perimeter of each semicircle and of the quarter circle. The total perimeter is $6\pi + 2$ cm ≈ 20.8 cm.

GG-18.   b. $\frac{-1}{3}$
        c. To start, multiply each side by $\frac{x}{3}$ to get $10 = 15(\frac{x}{3})$.
        d. -1 e. 30

GG-19.   a. Look back at OP-29.   b. Solve $(\frac{15}{9})^2 = \frac{w}{100}$.

GG-20.   a. $\frac{26}{53}$ c. $\frac{7}{53}$
        d. The only numbers from 1 through 53 that don't have two digits are 1 through 9.

GG-21.   20 in. by 48 in.

GG-22.   a. Sketch a diagram and think of a slide from $(3, -2)$ to $(5, 7)$.
        b. Start with $y = \frac{9}{2}x + b$ and substitute for x and y to find b.
        c. $\sqrt{85}$ units

GG-23.   You'll get $x = 4$ if you solve the equation $2x - 7 = y = -x + 5$.

GG-24.   b. and f. n^5 c. and g. w^{-4}

GG-25.   A difference is that parts a through d are all products while parts e through h are not.

GG-26. **b.** $(x, 3x + 2)$
f. For points above the line, $y > 3x + 2$; for points below the line, $y < 3x + 2$; and for points on the line, $y = 3x + 2$.

GG-27. **c.** The average speed equals the total distance divided by the total time.

GG-28. **d.** The travel times were not the same.

GG-30. **c.** $\frac{17 \text{ miles}}{4 \text{ miles per hour}}$

GG-31. **b.** Find the total distance, the time for each lap, and the total time.

GG-33. Find the time for the first 4 mi, and the time for the last 1 mi. His average speed was 3.2 miles per hour.

GG-34. **a.** First evaluate each numerator and denominator; 1.25.
b. $\frac{2}{3}$
c. What is the value of the denominator if $x = -1$? Can you divide by that value?

GG-35. The fraction $\frac{x}{x-1}$ cannot be simplified.

GG-36. **a.** 1.25
b. The two results should be equal.
c.

x	-5	-4	-3	0
y	$\frac{5}{6} \approx 0.83$	$\frac{4}{5} = 0.8$	$\frac{3}{4} = 0.75$	$\frac{0}{1} = 0$

x	1	2	3	4
y	undef.	$\frac{2}{1} = 2$	$\frac{3}{2} = 1.5$	$\frac{4}{3} \approx 1.33$

GG-37. The denominator would equal 0, so x should not equal -6, 1, or -1.

GG-38. **a.** $x = -4 + \sqrt{21} \approx 0.58$;
$x = -4 - \sqrt{21} \approx -8.58$
b. $x = \frac{-7 + \sqrt{65}}{4} \approx 0.26$;
$x = \frac{-7 - \sqrt{65}}{4} \approx -3.76$
c. $x = \frac{3 + \sqrt{13}}{2} \approx 3.30$;
$x = \frac{3 - \sqrt{13}}{2} \approx -0.30$

GG-39. **a.** $x = \frac{-2 \pm \sqrt{28}}{2} = -1 \pm \sqrt{7}$;
c. $(\approx -3.64, 0)$ and $(\approx 1.64, 0)$

GG-40. **a.** The x-intercepts are about $(1.75, 0)$ and $(0.25, 0)$.
b. $(1 + \frac{\sqrt{2}}{2}, 0)$ and $(1 - \frac{\sqrt{2}}{2}, 0)$
c. $(\approx 1.71, 0)$ and $(\approx 0.29, 0)$

GG-41. **a.** For the area: find the second leg of the right triangle; the area is $6 \cdot 9 + \frac{1}{2}(6)(9+6) + \frac{1}{4}(\pi \cdot 36) = 99 + 9\pi \approx 127.3$ sq cm. For the perimeter: find the hypotenuse of the right triangle, and the circumference of a circle with radius 6 cm; the perimeter is $6 + 6 + 9 + \frac{2\pi(6)}{4} + \sqrt{6^2 + 15^2} = 21 + 3\pi + \sqrt{261} \approx 46.6$ cm.
b. For the area: find the radius of the quarter circle; the area is $2\frac{1}{2}(1)(3) + \frac{1}{4}(\pi \cdot 10) = 3 + \frac{5\pi}{2} \approx 10.9$ sq cm. For the perimeter: find the radius of the quarter circle and the circumference of a circle with that radius; the perimeter is $1 + 3 + 1 + 3 + \frac{2\pi\sqrt{10}}{4} = 8 + \frac{\pi\sqrt{10}}{2} \approx 13.0$ cm.

GG-42. **a.** It takes about 10.82 sec.
b. The airplane's speed is about 332.7 mph.

GG-43. **a.** $4x^2 + 4x + 5$ **b.** $5y^2 - 8y - 6$
c. $9w + 18$ **d.** $3x^2 + 2x - 46$

GG-44. **b.** $\pm\sqrt{3}$ **c.** $\frac{9}{8}$ **d.** $\pm\frac{5}{3}$

GG-45. **a.** $(\frac{7}{2}, 0); (0, \frac{-7}{3})$; slope $= \frac{2}{3}$
b. no x-intercept; $(0, 4)$; slope $= 0$
d. $(-44, 0)$; no y-intercept; undefined slope

GG-46. There are six different distances; the shortest is 3 units and the longest is $\sqrt{50}$ units.

GG-47. **a.** To help get started, sketch and label a diagram.
b. One other point is $(10, 5)$.
c. Start with $3^2 + 5^2 = |AB|^2$.

GG-48. **a.** $\frac{x^{-4}}{x^{-1}} = x^{-4} \cdot x^1 = x^{-3}$
c. $\frac{n^{-8}}{n^5} = n^{-8} \cdot n^{-5} = n^{-13}$

GG-49. **b.** $b^{-14} \cdot b^8 = \frac{b^8}{b^{14}} = b^{-6}$
d. $r^{-7} \cdot r^{-1} = \frac{1}{r^7 \cdot r^1} = \frac{1}{r^8}$, or r^{-8}

GG-51. Shade the region on or below the line $y = \frac{1}{2}x - 1$.

GG-52. Shade the region below the line $y = 1 - 5x$.

GG-58. If x represents the width of the rectangle, then the area is $x(2x + 5)$ sq m. Solve the equation $x(2x + 5) = 50$.

GG-59. Find radius R of the large circle; find radius r of the small circle, and find the area of each circle. To get started, use the fact that the circumference is $2\pi R$ and solve $2\pi R = 10\pi$. The area of the walkway is $5^2\pi - 2^2\pi = 21\pi \approx 66.0$ sq ft.

GG-60. Two subproblems are finding the bases of two right triangles. The perimeter is 1100 ft and the area is 60,390 sq ft.

GG-61. **a.** $x^2 - 3x - 14 = 0$;
$x = \frac{3 \pm \sqrt{65}}{2} \approx 5.53, -2.53$
c. $x^2 - 6x - 5 = 0$; $x = 1.00, 5.00$
d. $x = \frac{8 \pm \sqrt{112}}{8} = \frac{2 \pm \sqrt{7}}{2}x \approx 2.32, -0.32$

GG-62. a. Yes, one factor is $x + 39$.
b. One factor is $x - 4$.

GG-63. The area of the circle is πr^2 and the area of the sq is s^2. Write an equation to show that the areas are equal and then solve it for r.

GG-64. $x(x + 1)$ for $x \neq 0$ or 2.

GG-65. b. about (2.5, 1)
c. Start by solving the equation $2x - 4 = -0.5x + 2$ to find x.

GG-66. a. The area is 8 sq units.
b. The perimeter is 18.25 units.
c. $\frac{|\text{long leg}|}{|\text{short leg}|} = \frac{8}{2}$

GG-67. a. $y = \frac{2}{3}x + 3$
b. First show that the slope is -2, and then substitute for x and y in the equation $y = -2x + b$. Solve for b.

GG-68. c. Notice that $36 = 6^2$, so $(x - 5)^2 - 36$ is a difference of squares:
$(x - 5)^2 - 6^2 = (x - 5 + 6)(x - 5 - 6)$; $x = -1, 11$
d. If $x^2 + 8 = 51$, then $x^2 - (\sqrt{43})^2 = 0$. Factor the difference of squares and then use the zero product property.
f. If $(x - 4)^2 + 9 = 12$, then $(x - 4)^2 - (\sqrt{3})^2 = 0$. Factor the difference of squares and then use the zero product property; $x = 4 \pm \sqrt{3}$.

GG-69. a. $\sqrt{52} = 2\sqrt{13} \approx 7.21$ units

GG-70. b.
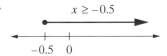
$x \geq -0.5$
$-0.5 \quad 0$

GG-71. a.

In the Region	On the Line	In the Region
$y < 4 - \frac{2}{3}x$	$y = 4 - \frac{2}{3}x$	$y > 4 - \frac{2}{3}x$
(0, 0)	(6, 0)	(0, 5)
(−2, 1)	(−3, 6)	(−3, 10)
(4, −1)		

GG-72. $y \geq \frac{3}{2}x - 4$; the region is all points that lie on or above the line $y = \frac{3}{2}x - 4$.

GG-77. Label the diagram:

$2\sqrt{16^2 + 7.5^2} \approx 35.3$ ft of rope

GG-78. To find the total time, you need to find the time for the first mile and the time for the last 7 miles; then solve $\dfrac{8 \text{ mi}}{\frac{1}{5} + \frac{7}{y} \text{ hr}} = 12$.

GG-79. c. 42 sq cm

GG-80. a. The areas of the shaded regions are the same; the vertex placement doesn't matter.
b. How small could x be? If $x = 7$, what kind of triangle is formed?

GG-81. a. Square both sides. $x = 17.64$
b. If you square both sides of $\sqrt{3x-1} = 18$, you get $3x - 1 = 18^2$. Keep going!

GG-82. a. $y = \frac{720}{1680}$ c. $y = \frac{x-2}{x+2}$
d. $y = \frac{3}{7}$ which equals $\frac{720}{1680}$; $x \neq -2, 0, 2, -3,$ or 3.
e. Use the simpler expression. The original expression cannot be evaluated for $x = -3, -2, 0, 2,$ or 3, so those values shouldn't be used in the table.

GG-83. b. $\dfrac{(v+1)^2}{(v-1)(v-2)}$; $v \neq 1, 2, 3$

GG-84. b. about (2, 0) and $(\frac{-1}{4}, 0)$
c. $x = \frac{2 \pm \sqrt{6}}{2} \approx 2.22, -0.22$

GG-85. b. $x = \pm\sqrt{24-6y}$; $y = \frac{24-x^2}{6}$
d. $x = \pm\sqrt{25 + y^2}$; $y = \pm\sqrt{x^2 - 25}$

GG-86. $\sqrt{164} = 2\sqrt{41} \approx 12.81$ cm

GG-87. One possibility is (6, −5). Find another one!

GG-88. The point (2, 6) could be A. If P is (x, y), then $\frac{x}{2} = \frac{3}{2\sqrt{10}}$, so $x = \frac{3}{\sqrt{10}} \approx 0.95$; $P \approx (0.95, 2.85)$.

GG-89. b. $x > -2.25$

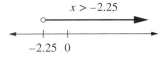
$x > -2.25$
$-2.25 \quad 0$

GG-98. c. Results will vary; one possibility is $y = 2.5x - 9$.
d. The mass will be 53 to 55 grams.

GG-99. Francie drives at 30 miles per hour, and Julia drives at 70 mph.

GG-100. One equation is $2L + 3S = 250$; $S = 44.4$ cm, $L = 58.4$ cm

GG-101. One equation is $2W + 2L = 100$, so $W + L = 50$. The pen is 13 ft by 37 ft.

GG-102. Define a variable and use equal ratios to write an equation; about 148.8 lb.

GG-103. a. Undo the square-root process, and then solve the resulting equation.
b. $x = -222.75$

GG-104. There are three solutions.

GG-105. **a.** To start, factor the numerator:
$$\frac{x^2(x^2 - 3x + 2)}{x(x - 2)} = \frac{x^2(x - 1)(x - 2)}{x(x - 2)};$$
$x \neq 0$ or 2

b. First factor a 3 out of the numerator, and then factor again! $x \neq -1$ or -2

GG-106. **a.** $4x^2 + 12x + 9 = (2x + 3)^2$

GG-107. **a.** ≈ 76.5 units **b.** 28 units

GG-108. Substitute for y and then use the quadratic formula to solve the resulting equation for x. The points are about $(7.057, 7.114)$ and $(-4.724, -16.447)$.

GG-109. **a.** all points on or below the line $y = \frac{-2}{3}x + 2$
b. all points on or below the line $y = \frac{10}{3}x + 5$

GG-110. The points lie on or above the parabola.

GG-111. The points lie below the parabola.

GG-130. **a.** approximately $y = 3x$
b. For diameter d, $C = \pi d$.

GG-131. **f.** The times would be the same around the year 2030.
g. about 43 sec

GG-132. One equation could be $W^2 + L^2 = 40^2$. The garden is about 11.77 m by about 38.23 m.

GG-133. **a.** $440 + 14\pi \approx 484$ yd
b. $14\pi \approx 44$ yd

GG-134. **a.** $x = \frac{2 \pm \sqrt{20}}{2} \approx -1.24, 3.24$
b. $x = \frac{-6 \pm \sqrt{20}}{2} \approx -5.24, -0.76$
c. $x = \frac{-2 \pm \sqrt{88}}{6} \approx -1.90, 1.23$

GG-136. **a.** $2x^2 - 7x + 3 = (2x - 1)(x - 3)$; keep going!
b. $x = \frac{1}{2}, 3$

GG-137. **a.** $x = 2, \frac{-5}{6}$;
so $6x^2 - 7x - 10 = (x - 2)(6x + 5)$

b. $x = -1, \frac{2}{9}$;
so $9x^2 + 11x + 2 = (x + 1)(9x + 2)$

GG-138. at about $(-9.48, 172.64)$ and $(1.48, -2.64)$

GG-139. Undo the square-root process to get $4x - 3 = x^2$, and then solve the resulting quadratic equation to get $x = 1, 3$.

GG-140. **a.** $\frac{2x^7(x - 2)^2}{(x - 4)^5}$ for $x \neq 0, 4$
b. Start with $\frac{(m - 9)(m - 3)}{m(m - 9)}$ for $m \neq 0, 9$
c. $8n^2$ for $n \neq 0$
d. Factor $x - 5$ out of the numerator and the denominator.

GG-141. $x = 3$

GG-142. If s represents the length of the short side, then $s^2 + (s + 5)^2 = (s + 10)^2$; 15 cm by 20 cm.

GG-143. Draw a diagram. What is the slope of each line?

GG-144. **a.** $\sqrt{41}$ units **b.** $\frac{-4}{5}$
c. $(-2.5, 0)$ **d.** $y = \frac{-4}{5}x - 2$

GG-145. $(-3, 4)$

GG-146. A parabola with vertex $(0, -6)$ and x-intercepts $(-4, 2)$ and $(4, 2)$

GG-147. **a.** $x^2 + 4xy + 4y^2 = (x + 2y)^2$
b. One factor is $x + 2y$.
c. One factor is x.

GG-148. **a.** all points to the right of the line $x = -2$
b. all points on or below the x–axis
c. all points on or above the line $y = \frac{1}{2}x - 3$
d. all points above the line $y = \frac{1}{4}x$

GG-149. **a.** all points on or below the line $y = \frac{-3}{2}x - 4$
b. all points above the line $y = 2x + 7$

GG-150. yes; $\left(\frac{-4}{9}, \frac{19}{9}\right)$

GG-151. Graph both inequalities on the same coordinate plane. Some possible points are $(0, 2)$, $(4, 1)$, and $(3, 5)$.

GG-152. Use the graph from GG-151. Some possible points are: $(-2, 0)$, $(-3, 1)$, and $(-8, 3)$.

GLOSSARY

A

Algebraic expression

an expression that uses mathematical operations and at least one variable; for example $3x^2y$, or $4m + 10$, or $5 - \frac{x}{2}$ **OP-41**

Algebraic fraction

a fraction that has an algebraic expression in its numerator, or denominator, or both; for example $\frac{12(x-2)^2}{x+5}$ and $\frac{37}{4x}$ are both algebraic fractions, but $\frac{37}{4}$ is not.

Average (or mean) of data values

An average is a way of describing the middle of a set of data. One way to find an average of a set of numbers is to add them and then divide the result by the number of data values (the result is called an "arithmetic mean"). Since $\frac{3+4+9+2}{4} = 4.5$, we say that 4.5 is an average of the four values 3, 4, 9, and 2.

OP-63, GG-96

B

Billion

in the U.S., the number 1,000,000,000 **OP-17**

Binomial

see LT-21

C

Coefficient of a term

a number that multiplies a monomial expression; the coefficient of $3x^2$ is 3, and the coefficient of $-x$ is -1. **WR-61**

Composite rectangle

a rectangle made with two or three types of algebra tiles: large squares (x^2 tiles), rectangles (x tiles) and small squares (1 tiles) **SP-76, LT-20**

Congruent

Two figures are congruent if they are the same size and the same shape. **OP-29**

Conjecture

a statement about what seems to be true based on patterns and reasoning **SP-61, SP-127**

Consecutive

Two numbers in a pattern are consecutive if one follows right after the other; for example, "1, 3, 5" are consecutive odd integers. **OP-47**

Consistent scales on coordinate axes

see SP-4

Constant term of a polynomial

a number alone, not multiplied by a variable; the constant term in the polynomial $3x^2 + 4x - 6$ is -6. **WR-61**

Coordinate axis (plural: axes)

a number line used to graph points **SP-4**

CRV

a sport-utility vehicle **WR-91**

D

Digit

Each of the symbols 0, 1, 2, 3, 4, 5, 6, 7, 8, and 9 is a digit in our numeration system. **OP-15**

E

Equation

a mathematical statement, such as $x + 5 = 7$, that includes an "equals" sign ($=$) **OP-15**

F

Fair standard die (plural: dice)

an evenly balanced cube that has dots on each of its 6 faces (sides). There is 1 dot on one face, 2 dots on another face, 3 dots on another face, and so on, so the sixth face has 6 dots. **OP-95**

G

Generic rectangle

a sketch of a composite rectangle that shows groups of the same kinds of tiles rather than individual tiles **LT-84**

Generic slope triangle

a slope triangle that is not drawn to scale **FF-2**

Gorp
Good Ol' Raisins and Peanuts—a high-energy mixture of nuts, dried fruit, and chocolate bits; it's also called "trail mix." **WR-92**

H

Hexagon
a flat figure (polygon) with six straight sides and six vertices **OP-30**

I

Input values
numbers used in a rule or pattern; they may be chosen independently and can be substituted for variables. **OP-37**

Integers
the numbers …, $-3, -2, -1, 0, 1, 2, 3, …$ **OP-15**

M

Mental math
You are doing mental math when you use your knowledge of arithmetic to perform calculations without the aid of a calculator or paper and pencil. **OP-71**

Million
in the U.S., the number 1,000,000 **OP-25**

Mixed number
a number made up of an integer and a fraction of an integer, such as $1\frac{2}{3}$, $16\frac{4}{7}$, and $-2\frac{1}{5}$ **EF-57**

Monomial
see LT-21

N

Number line
an endless line, often marked like a ruler with equally spaced marks that are often labeled with consecutive integers; helpful for visualizing relationships between numbers **SP-4**

O

Ones digit, ones place
in our numeration system, the digit (place) just to the left of the decimal point; the ones digit in 437.2 is 7, and 9 is in the ones place in 209. **SP-52**

Ordered pair of numbers
written in the form (x, y) in which the position of the numbers in the pair matters; the coordinates of a point are always given in order with the input (x-value) on the left and the output (y-value) on the right: (x, y) **SP-4**

Origin in the coordinate plane
The origin is the point (0, 0). **SP-4**

Outcome
a result of an experiment or trial **OP-95**

Output values
numbers that result when chosen values are used in a pattern or rule; an output value cannot be chosen freely, but depends on the choice of the input value. **OP-37**

P

Parabola
a U-shaped curve that is the graph of a quadratic equation **SP-92**

Perfect square
The first six perfect square numbers are 0, 1, 4, 9, 16, and 25.

Perpendicular lines
lines that meet at right (90 degree) angles **SP-4**

Polygon
a flat figure with straight sides, formed with line segments that meet end-to-end so that each segment meets one other segment at each of its endpoints. Some examples of polygons are squares, triangles, and rectangles; circles and ovals *are not* polygons. **SP-17**

Polynomial
see LT-21

Prime number
2, 3, 5, 7, 11, 13, 17, … are prime numbers. None can be written as a product of two positive integers both of which are smaller than itself.

Problem solving strategies
useful ways to go about solving mathematical problems, such as drawing a picture, trying a few numbers (guessing and checking), looking for a pattern, and solving a smaller, related problem

Product of numbers
When numbers are multiplied together—such as 6 times 3, or 2 times 5 times 48—they are a product of numbers. **OP-26, OP-32**

Q

Quadratic equation, in standard form
When written in standard form a quadratic equation reads $ax^2 + bx + c = 0$ where a, b, and c are numbers and $a \neq 0$. **GG-4**

Quadrilateral
a flat figure (polygon) with four straight sides and four vertices

R

Ratio of two numbers
The number $\frac{m}{n}$ is the ratio of the two numbers m and n (n cannot be zero). **LT-78**

Rule for calculating an output value
directions that specify what is done to obtain an output value from an input value **SP-1**

S

Scale (of coordinate axes)
numbering on the axes, as on a ruler, with the numbers increasing in size going to the right along the horizontal axis and going up along the vertical axis **SP-4**

Slope triangle
see WR-5

Square of a number
a number times itself, such as $7 \cdot 7$ or $(-2)(-2)$ **SP-1**

Standard deck of playing cards
There are 52 playing cards in a standard deck. They are divided into 4 types ("suits"): clubs, spades, hearts, and diamonds. The clubs and spades are black; the hearts and diamonds are red. Each suit has 13 different cards: 2 through 9, and A (ace), K (king), Q (queen), and J (jack). **OP-96**

Sum of numbers
When numbers are added together—such as 6 plus 3, or 2 plus 5 plus 48—they are a sum of numbers. **OP-26**

T

Term of a polynomial
see LT-21

Trinomial
see LT-21

V

Vertex (plural: vertices) of a polygon
an endpoint common to two adjacent sides of a polygon (a "corner point" of a polygon)

Vertex of a parabola
the highest point, or lowest point, of a parabola **SP-92**

MATHEMATICAL SYMBOLS

(3)(5), 3(5), (3)5, or $3 \cdot 5$ for "3 times 5"	**OP-6**
2^9 for "2 raised to the power 9"	**OP-32**
-5 for "negative 5"	**A-86**
x to represent a variable	**OP-38**
(x, y) for the coordinates of a point	**SP-4**
$\frac{1}{x}$ for "the reciprocal of a non-zero number x"	**SP-10**
$x \neq 0$ for "x is not equal to 0" or "x does not equal 0"	**SP-44**
… an ellipsis	**SP-83**
$H > 6$ for "H is greater than 6"	**SP-88**
$S < 6$ for "S is less than 6"	**SP-88**
$x \geq 6$ for "x is greater than or equal to 6"	**SP-88**
$x \leq -2$ for "x is less than or equal to -2"	**SP-88**
$-3 \leq x \leq 7$ for "x is between -3 and 7"	**SP-91**
$\sqrt{9}$ for "the square root of 9"	**SP-109**
$|AB|$ for "length of segment AB"	**EF-34**
π the Greek letter "pi," which represents the ratio of the circumference of a circle to its diameter	**EF-52**
$\approx$ "for is approximately equal to"	**EF-52**
$\triangle ABC$ for "triangle ABC"	**EF-63**
marks to indicate angles of the same size	**EF-63**
$\angle G$ for "angle G"	**EF-68**
$m \angle G$ for "the measure of angle G"	**EF-68**
a small square to indicate a right angle	**EF-86**
m to represent the slope of a line	**WR-8**
$\pm$ for "plus or minus"	**GG-6, GG-7**

INDEX

ABOUT THE INDEX

The entries in the index are listed in alphabetical order, and each entry is referenced by the problem in which it is first discussed. In some cases, a term or idea may be discussed in several of the immediately following problems as well. Each problem is listed by its two-letter code that references the chapter followed by the problem number. Whenever possible, terms are cross-referenced to make the search process quicker and easier. For example, *factoring* also appears under *polynomials*.

CHAPTER LABELS

Chapter 1	OP	Openings: Data Organization
Chapter 2	SP	Professor Speedi's Commute: Patterns and Graphs
Chapter 3	LT	Lions, Tigers, and Emus: Writing and Solving Equations
Chapter 4	EF	Estimating Fish Populations: Numeric, Geometric, and Algebraic Ratios
Chapter 5	WR	Managing Water Resources: Graphs, Slope Ratios, and Linear Equations
Chapter 6	FF	Fitness Finances: Solving Systems of Equations
Chapter 7	RT	The Buckled Railroad Track: From Words to Diagrams to Equations
Chapter 8	GG	The Grazing Goat: Dealing with Complicated Situations
Appendix A	A	Debits and Credits: An Introduction to Integer Tiles
Appendix B	B	Using Suproblems to Derive the Quadratic Formula
Appendix C	C	Exploring Quadratic Equations and Their Graphs

A

addition method
   for solving a system of equations   FF-92
   revisited   FF-94
algebra tiles   SP-72, SP-75, LT-19
Algebra Tool Kit   OP-9
The Algebra Walk problem   SP-1
algebraic fractions (see fractions)

angles
   measuring with a protractor   EF-63
   right   EF-86
Ant Thermometers problem   FF-128
area
   meaning of   OP-30
   of a circle   EF-52
   of a rectangle
     to model a product of two expressions   LT-128
     as a product and area as a sum   SP-20, SP-76, LT-20
   of a triangle   EF-26
arithmetic mean   GG-96
average
   arithmetic mean   GG-96
   average speed   GG-27
axes, coordinate   SP-4

B

Barbie Trivia problem   EF-38
binomials   LT-105
   factoring
     using generic rectangles   EF-50
   difference of squares   WR-116
Buckled Railroad Track problem   RT-125
Building Corrals problem   OP-1
Burning Candle Investigation   SP-123
   revisited   FF-122

C

calculator skills
   calculating the reciprocal of a number   SP-44

changing the sign of a
    number   OP-22
exponentiation key   OP-52
    for complicated
        expressions   GG-14
    entering negative integer
        exponents   LT-100
    scientific notation
        entering numbers in
            OP-77
        representation of   OP-77
    using the square root key
        RT-29
Calculator Tool Kit   OP-22
Changing Populations:
    Oakwood and Elm
    Creek   FF-22
Chapter Summaries   OP-104,
    SP-122, LT-123, EF-
    142, WR-141, FF-112,
    RT-123, GG-112
circumference of a circle
    EF-52
Cleopatra problem   RT-78
coefficient of linear thermal
    expansion   RT-18
collaborative work (see
    working in groups)
common factor   RT-72
composite rectangles and area
    SP-76, LT-20
conjectures
    making and testing   SP-73,
        EF-21
continuous graphs   SP-82
Cookie Cutter Investigation
    GG-114
coordinate axes   SP-4
coordinate graphs (see graphs)
coordinates of a point   SP-4
Country Road problem
    FF-119
Course Summary   WR-164
    Update   GG-154
Cricket Thermometers
    problem   WR-83

D

Diamond Problems   OP-5
    relationship to factoring and
        generic rectangles
        EF-148
difference of squares
    WR-116
    geometric approach
        FF-48
discrete graphs   SP-82
distance between two points
    RT-54, RT-55

distance, rate and time
    OP-93, SP-37
    calculating average speed
        GG-27
    word problems   RT-76
distributive property   SP-78
dollar bill problems
    Big Bucks   OP-82
    Big Spender   OP-25
    Dollar Bill Facts   OP-17
    More Dollar Bill Facts
        OP-81
    weight of one billion
        OP-99
Dollar Bill-Folds problem
    SP-108

E

Egg Toss problem   SP-92
    revisited   FF-98
Election Poster problem
    GG-54
ellipsis   SP-83
enlarging geometric figures
    EF-1
equations
    checking a solution   LT-36
    containing parentheses
        LT-65
        "Distribute First" method
            LT-66, LT-69
        "Divide First" method
            LT-67, LT-69
        "doing" and "un-doing"
            LT-57, LT-58
    fractions, dealing with
        EF-90
    linear
        complicating   LT-59
        modeling situations
            FF-81
        slope-intercept form
            WR-61
        solving   LT-36
    literal (with more than one
        variable)   LT-80
    quadratic equations   FF-74
        standard form   GG-4
        that can't be factored
            GG-3
    quadratic formula   GG-6,
        GG-7
    square root   GG-81
    systems of linear equations
        FF-1, FF-22
        solving by graphing
            FF-22, FF-61
        solving by setting
            y-forms equal
            FF-24

    solving by the addition
        method   FF-92,
        FF-94
    solving by the
        substitution
        method   FF-39
    to solve a word problem
        FF-46
    systems of nonlinear
        equations   FF-54
    to solve a quadratic   FF-74
Estimating Fish Populations
    Investigation   EF-130
Estimating Rocket Height
    problem   FF-55
evaluating algebraic
    expressions   OP-41
Expanding Metals problem
    RT-18
exponential form   OP-32
exponents
    negative integer   LT-100,
        EF-77, GG-50
    on a calculator   OP-52,
        LT-100
    positive integer   OP-32
    powers of 2   SP-50
    summary of rules   WR-144
    using meaning to rewrite an
        expression   LT-131
    zero   WR-99

F

F.O.I.L. acronym   WR-81
factoring
    common factor   RT-72
    "completely"   RT-72
    difference of squares
        WR-116
    geometric approach
        FF-48
    with algebra tiles   EF-43
    with generic rectangles
        EF-46
Fitness Finances problem
    FF-38
Five-Digit Problem   OP-15
four ways to make sense of a
    mathematical concept
    OP-6
Fraction Busters   EF-90
fractional forms of 1   EF-16
fractions
    algebraic fractions
        restricted values
            GG-34, GG-37
        simplest form   RT-43
        simplifying   EF-16
Fruit Punch   EF-112
    revisited   WR-4

G

generic rectangles   LT-84
    to find products   EF-51
    relationship to Diamond
        Problems   EF-148
geometric factoring   EF-43
Gorp problem   WR-92
graphs
    continuous graphs   SP-82
    coordinate axes   SP-4
    cubic graph $y = x^3$ SP-130
    discrete graphs   SP-82
    exponential graphs
        Bill-Folds   SP-108
        $y = 2^x$ EF-162
    human graphing (the
        Algebra Walk)   SP-1
    intersection of two graphs
        SP-16, FF-23
    linear graphs
        distinguishing from
            other graphs
            SP-97
        horizontal lines   EF-109
        line of best fit   GG-95
        point-slope method for
            graphing lines
            WR-65
        rates of change   SP-55
        slope-intercept form
            WR-61
        to model situations
            WR-83
        two-point slide method
            for graphing lines
            WR-34
        vertical lines   EF-109
        writing an equation of
            a line without
            graphing   FF-2
        writing an equation for
            a linear graph
            WR-123, WR-151
        $y = mx + b$ form
            WR-61
    linear vs. nonlinear graphs
        SP-19
    nonlinear graphs
        cubic graph $y = x^3$
            SP-130
        exponential graphs
            Bill-Folds   SP-108
            $y = 2^x$ EF-162
        quadratic graphs
            Egg Toss   SP-92
            parabola $y = x^2$
                SP-56,
                Appendix C
            $y = -x^2$ vs.
                $y = (-x)^2$
                SP-93

reciprocal graph $y = \frac{1}{x}$
   SP-45
square root graph
   $y = \sqrt{x}$ SP-109,
     RT-28, RT-
     107
making a graph from data
   OP-61
parabola $y = x^2$ SP-56,
   Appendix C
plotting points   SP-4
point of intersection SP-16,
   FF-23
quadratic graphs (see
   nonlinear)
reading and interpreting
   graphs   OP-83,
   OP-92
reciprocal graph $y = \frac{1}{x}$
   SP-45
setting up a coordinate
   graph   SP-4, WR-83
   scaling the axes   SP-4
sketching a reasonable
   graph to represent a
   situation   SP-94
square root graph
   $y = \sqrt{x}$ SP-109,
     RT-28, RT-107
x-intercept   SP-62
$y = -x^2$ and $y = (-x)^2$
   SP-93
y-form   WR-35
y-intercept   SP-102
Grazing Goat problem   GG-73
groups (see working in groups)
guess-and-check tables
   as related to a system of
     equations   FF-46
   Guide to Setting Up
     OP-44, LT-2
   to solve a problem   OP-43
   to solve an equation   LT-54
   to write an equation   LT-1
   with ratios   EF-29

H

Hawaiian Punch Mix problem
   SP-53
horizontal line, graphing
   EF-109
hypotenuse of a right triangle
   RT-2

I

inequalities
   graphing linear inequalities
     GG-51

regions in a plane GG-26
representing on a number
   line   RT-50
solving linear   RT-73
symbols for   SP-88
to express betweenness
   SP-91
using a two-pan scale to
   represent   RT-24
input   OP-37, SP-15
integer arithmetic   OP-72,
   Appendix A
intersection, point of SP-16,
   FF-23
investigations
   Burning Candle   SP-123
   Cookie Cutter   GG-114
   Estimating Fish   EF-130
   Lunch Bunch   GG-117
   Using Algebra to Model and
     Analyze Linear Data
     FF-113—FF-116

J

Jelly Bean Riddle problem
   LT-36

K

Kayak Rental problem
   LT-103
King's Reward problem
   OP-31

L

Leasing a Car problem
   WR-143
legs of a right triangle   RT-2
line of best fit   GG-95
linear equations (see
   equations)
linear situations   WR-83
   modeling with an equation
     LT-102, LT-103
linear thermal expansion
   RT-18
Lions, Tigers, and Emus
   problem   LT-102
Logo Contest problem
   GG-16
Lou's Shoes problem
   GG-125 (see also
   Predicting Shoe Size)
Lunch Bunch Investigation
   GG-117

M

mathematical model FF-82,
   FF-114
mean data point   GG-97
monomials   LT-21

N

National Debt problem
   OP-76
non-linear equations
   graphing (see graphs)
   solving systems of   FF-54
notebook
   how to organize   OP-11

O

Olympic Records problem
   FF-1
order of operations   OP-14
ordered pair   SP-4
organizing data
   by graphing   OP-62
   by making an organized
     list—tossing coins
     OP-95
   in a guess-and-check table
     OP-43
output   OP-37, SP-15

P

percent
   writing ratios as percents
     EF-9, EF-32
perimeter
   of a circle   EF-52
   of a figure   OP-8
point of intersection of two
   graphs   SP-16, FF-23
point-slope method for
   graphing lines   WR-65
Polygon Perimeters problem
   SP-17
polynomials
   factoring
     difference of squares
       WR-116
     using algebra tiles
       EF-43
     using generic rectangles
       EF-46
     using solutions from the
       quadratic formula
       GG-62
   finding products of more
     than 2 factors   RT-19
   meaning of LT-21
   perfect square trinomials
     GG-106
   representing with algebra
     tiles   LT-47
   using tiles to multiply
     LT-61, LT-64, LT-84,
     LT-105, LT-128
powers of 2—patterns with
   exponents   SP-50
Predicting Shoe Size problem
   OP-61

probability   OP-95
problem solving strategies
   drawing a diagram   RT-9,
     RT-76
   guess and check   OP-43,
     SP-112
   identifying subproblems
     EF-3
   looking for patterns   SP-17
   making a model   OP-1,
     RT-51
   making an organized list
     OP-95
Professor Speedi's Commute
   problem   SP-38
Proxima Centauri problem
   OP-108
Pythagorean relationship   RT-6
   and area   RT-53
   to find distance between
     two points   RT-54,
     RT-55

Q

quadratic equations (see
   equations)
quadratic formula   GG-6
   derivation of Appendix B
   using solutions to factor
     GG-62
quadratic graphs (see graphs)
quadratic polynomials,
   factoring   EF-47 (see
   also polynomials and
   factoring)

R

ratios
   and graphing Fruit Punch
     problem   EF-112
   definition   LT-78
   enlargement and reduction
     ratios summary
     EF-21
   enlargement ratios and
     graphing of side
     lengths in similar
     triangles   EF-64
   reduction ratios   EF-6
   slope ratio (see slope ratio)
   to describe steepness of a
     line   WR-4
   writing ratios as percents
     EF-9, EF-32
reciprocals
   calculating   SP-44
   definition of   SP-10
   graph of $y = \frac{1}{x}$ SP-45
reducing geometric figures
   EF-6

regions in a coordinate plane GG-26
right angle (see angle)
right triangle (see triangle)
rolling two fair standard dice SP-48

S

scientific notation
    definition of OP-76
    on your calculator OP-77
Sierra River Reservoir problem WR-123
similar triangles
    definition of EF-63
    ratios of side lengths EF-64, EF-69
Slicing a Pizza problem SP-73
slides
    A Way to Represent Change LT-119, LT-120
    to find other points on a line WR-2
slope ratio
    interpreting the slope WR-9
    of a line WR-4, WR-8, WR-28
    related to length? RT-62
slope triangle WR-5
slope-intercept form of a linear equation WR-61
Spiral of Right Triangles RT-25
square roots
    approximating RT-28, RT-29
    arithmetic RT-61, RT-100, RT-103
    picturing lengths RT-25, RT-26
    simplified square root form RT-105
    simplifying RT-104
    solving equations GG-81
    summary of rules RT-131
    undoing RT-65
standard decimal form OP-76
subproblems
    to find the area of a figure
        subdivide EF-3
        surround and subtract EF-3
    to solve complicated problems EF-41, EF-46, EF-90
substitution method for solving pairs of linear equations FF-39
Sums and Products Tables EF-96, EF-124, EF-134
symbols (see also Mathematical Symbols list)
    ellipsis SP-83
    for betweenness SP-91
    for inequality SP-88
    for the length of a line segment EF-34
    to represent an angle EF-68
    to represent multiplication OP-6, OP-39
systems of equations (see equations)

T

Temperature Conversion problem WR-124
    revisited FF-30
Teraflops, world's most powerful computer OP-80
Tool Kit Checks OP-109, SP-126, LT-132, EF-149, WR-145, FF-118, RT-132, GG-129
Tool Kits
    Algebra OP-9
    Calculator OP-22
trend line WR-83
triangles
    area of EF-26, GG-57
    right
        definition of EF-86
        relationship among sides RT-6
    similar EF-63
two-point slide method for graphing lines WR-34

U

Undoing, to solve an equation LT-57, LT-58

V

variables OP-38
vertex of a parabola SP-92
vertical lines, graphing EF-109

W

working in groups
    Building Corrals problem OP-1
    Guidelines for xiii
    reflections on group work OP-90
    roles of group members OP-58

X

x-intercept(s) of a graph SP-63
relations to solution(s) to equations FF-75

Y

$y = mx + b$ WR-61
y-form WR-35
y-intercept
    of a graph SP-102
    of a linear graph WR-28

Z

zero product method FF-76
zero product property LT-54, LT-72, LT-96, LT-97, LT-126, LT-127, FF-75